Physiology in Health and Disease

Published on behalf of the American Physiological
Society by Springer

Physiology in Health and Disease

This book series is published on behalf of the American Physiological Society (APS) by Springer. Access to APS books published with Springer is free to APS members.
APS publishes three book series in partnership with Springer: *Physiology in Health and Disease* (formerly *Clinical Physiology*), *Methods in Physiology*, and *Perspectives in Physiology* (formerly *People and Ideas*), as well as general titles.

More information about this series at http://www.springer.com/series/11780

Kirk L. Hamilton • Daniel C. Devor
Editors

Ion Transport Across Epithelial Tissues and Disease

Ion Channels and Transporters of Epithelia in Health and Disease - Vol. 2

Second Edition

Springer

american physiological society

Editors
Kirk L. Hamilton
Department of Physiology, School of Biomedical Sciences
University of Otago
Dunedin, Otago, New Zealand

Daniel C. Devor
Department of Cell Biology
University of Pittsburgh
Pittsburgh, PA, USA

ISSN 2625-252X ISSN 2625-2538 (electronic)
Physiology in Health and Disease
ISBN 978-3-030-55309-8 ISBN 978-3-030-55310-4 (eBook)
https://doi.org/10.1007/978-3-030-55310-4

This Springer imprint is published by the registered company Springer Nature Switzerland AG.
The registered company address is: Gewerbestrasse 11, 6330 Cham, Switzerland

We dedicate this second edition to our families . . . Judy, Nathan, and Emma for KLH, and Cathy, Caitlin, Emily, and Daniel for DCD.

Preface to Second Edition—Volume 2

Our ultimate goal for the first edition of *Ion Channels and Transporters of Epithelia in Health and Disease* was to provide a comprehensive and authoritative volume that encapsulated the most recent research findings in basic molecular physiology of epithelial ion channels and transporters of molecular diseases from the laboratory bench top to the bedside. Additionally, we envisioned that the book would be very exciting and useful to a range of readers from undergraduate and postgraduate students, to postdoctoral fellows, to research and clinical scientists providing a wealth of up-to-date research information in the field of epithelial ion channels and transporters in health and disease. We firmly believe that the first edition fulfilled a niche that was crucially required. We have been informed that the first edition of the book has proven to be the best performing APS/Springer book based on downloaded chapters, to date. This is a direct testament to the world-class scientists and clinicians who contributed excellent chapters to that edition. Of course, there were many epithelial ion channels and transporters which were not included in the first edition, but certainly warranted inclusion.

With our second edition, we have superseded our original expectations by increasing the number of chapters from 29 in the first edition to a three-volume second edition including 54 chapters, resulting in 25 new chapters. All of the original chapters have been expanded. Again, we were very fortunate to recruit "key" outstanding scientists and clinicians who contributed excellent chapters, some of whom were unable to commit to the first edition. In the end, the second edition has a total of 128 authors from 13 countries across four continents and both hemispheres. We truly believe that this book series represents a worldwide collaboration of outstanding international scientists and clinicians.

Volume 2: Ion Transport Across Epithelial Tissues and Disease

This is the second of three volumes highlighting the importance of epithelial ion channels and transporters in the basic physiology and pathophysiology of human diseases. For this volume, experts in their respective fields have contributed chapters on epithelial ion transport across a wide array of epithelial tissues and the use of organoids to study epithelial function. Volume 2 consists of 11 chapters (including 9 new chapters), which begins with a historical perspective of organoids as a model for intestinal ion transport physiology followed by two returning chapters focused on the roles of epithelia in the intestine and disease. These chapters are followed by new chapters on ion transport of the sweat gland, transporters of lactating mammary epithelia, lipid transport of the mammary gland, ion transport of the inner ear, retinal pigment epithelium, choroid plexus, and finally, transport function of ectoderm epithelial cells forming dental enamel. The chapters in this volume examine the overall epithelial physiology of various tissues of the body. In doing so, the authors inform the reader about the dynamic function of epithelial ion channels and transporters, in concert, to maintain normal cell physiology and changes in pathological states. This volume provides a holistic background for Volume 3 where chapters are focused on the background, structure of specific ion channels and transporters, and normal physiology and pathophysiology of disease.

It is our intent that the second edition continues to be the comprehensive and authoritative work that captures the recent research on the basic molecular physiology of epithelial ion channels and transporters of molecular diseases. We hope this new edition will be the "go-to" compendium that provides significant detailed research results about specific epithelial ion channels and transporters, and how these proteins play roles in molecular disease in epithelial tissues.

As stated in the preface of the first edition, the massive undertaking of a book of this enormity would certainly be an "Everest" of work. We want to sincerely thank all of our authors, and their families, who have spared time from their very busy work and non-work schedules to provide exciting and dynamic chapters, which provide depth of knowledge, informative description, and coverage of the basic physiology and pathophysiology of the topic of their individual chapters.

We want to, again, thank Dr. Dee Silverthorn who planted the "initial seed" that developed into the first edition, which stemmed from a Featured Topic session entitled "Ion Channels in Health and Disease" held during the Experimental Biology meetings in Boston in April 2013 (chaired by KLH). Then, based on the performance of that edition, Dee "twisted" our arms, with love, to attempt a second edition in 2017. We, once again, want to extend our huge thanks, gratitude, and appreciation to the members of the American Physiology Society Book Committee for their continued faith in us to pursue such a monumental second edition.

As with the first edition, this three-volume second edition would not have been possible without the excellent partnership between the American Physiological Society and Springer Nature and the publishing team in Heidelberg, Germany.

Many thanks to Markus Spaeth, Associate Editor (Life Science and Books), and Dr. Andrea Schlitzberger, Project Coordinator (Book Production Germany and Asia), who guided us on our second book publication journey never dreaming that this edition would be a three-volume book bonanza.

We extend special thanks to Anand Venkatachalam (Project Coordinator, Books, Chennai, India) at SPi Global who answered unending questions during the production process. We thank his production team who assisted us through the many stages of the publication of the second edition. We also thank Nancey Biswas (Project Management, SPi Content Solution, Puducherry, India), Nedounsejiane Narmadha (Production General, SPi Technologies, Puducherry, India), and Mahalakshmi Rajendran (Project Manager, SPi Technologies, Chennai, India) at SPi Global for their assistance for overseeing the production of the chapters during the final print and online file stages of the second edition.

We want to thank our mentors Douglas C. Eaton and the late Dale J. Benos for KLH; Michael E. Duffey and Raymond A. Frizzell for DCD; and our colleagues who guided us over the years to be able to undertake this book project.

Finally, and most importantly, we want to thank our families: Judy, Nathan, and Emma for KLH, and Cathy, Caitlin, Emily, and Daniel for DCD for all your love and support during this 8-year journey.

We dedicate this second edition to our families.

Dunedin, New Zealand
Pittsburgh, PA
July 2020

Kirk L. Hamilton
Daniel C. Devor

Preface

Ion channels and transporters play critical roles both in the homeostasis of normal function of the human body and during the disease process. Indeed, as of 2005, 16% of all Food and Drug Administration-approved drugs targeted ion channel and transporters, highlighting their importance in the disease process. Further, the Human Genome Project provided a wealth of genetic information that has since been utilized, and will again in the future, to describe the molecular pathophysiology of many human diseases. Over the recent years, our understanding of the pathophysiology of many diseases has been realized. The next great "step" is a combined scientific effort in basic, clinical, and pharmaceutical sciences to advance treatments of molecular diseases.

A number of unique ion channels and transporters are located within epithelial tissues of various organs including the kidney, intestine, pancreas, and respiratory tract, and all play crucial roles in various transport processes responsible for maintaining homeostasis. Ultimately, understanding the fundamentals of ion channels and transporters, in terms of function, modeling, regulation, molecular biology, trafficking, structure, and pharmacology, will shed light on the importance of ion channels and transporters in the basic physiology and pathophysiology of human diseases.

This book contains chapters written by notable world-leading scientists and clinicians in their respective research fields. The book consists of four sections. The first section of the book is entitled **Basic Epithelial Ion Transport Principles and Function** (Chapters 1–8) and spans the broad fundamentals of chloride, sodium, potassium, and bicarbonate transepithelial ion transport, the most recent developments in cell volume regulation, the mathematical modeling of these processes, the mechanisms by which these membrane proteins are correctly sorted to the apical and basolateral membranes, and protein folding of ion channels and transporters. The chapters in Section 1 provide the foundation of the molecular "participants" and epithelial cell models that play key roles in transepithelial ion transport function of epithelia detailed throughout the rest of this volume.

The second section is entitled **Epithelial Ion Channels and Transporters** and contains seventeen chapters (9–25) in which authors have concentrated their discussion on a particular ion channel or transporter ranging from chloride channels to the Na^+/K^+-ATPase, for example. Generally, the authors have initially provided a broad perspective of the physiology/biology of a particular ion channel or transporter in epithelial tissues, followed by a focused in-depth discussion of the latest physiology, cell biology, and molecular biology of the ion channel/transporter and then finish their discussion on aspects of pathophysiology and disease.

It will be appreciated following the discussion of the various ion channels and transporters that many of these transport proteins are potential pharmacological targets for possible treatment of disease. Therefore, the third section is entitled **Pharmacology of Potassium Channels** that consists of two chapters (26 and 27) that provide the latest developments on the pharmacology of calcium-activated potassium channels and small-molecule pharmacology of inward rectified potassium channels. It should be noted, however, that pharmacological information about various ion channels and transporters is also provided in some of the chapters found within Section 2 of this volume.

Finally, the last section in the book is entitled **Diseases in Epithelia** and consists of two chapters (28 and 29). These chapters are designed to bridge the basic cellular models and epithelial transport functions discussed throughout this volume with a compelling clinical perspective: from bench to bedside. In these chapters, Dr. Whitcomb discusses the role of ion channels and transporters in pancreatic disease, while Dr. Ameen and her colleagues similarly provide insights into the secretory diarrheas.

Our utmost goal, with this book, was to provide a comprehensive and authoritative volume that encapsulates the most recent research findings in the basic physiology of ion channels and transporters of molecular diseases from the laboratory bench top to the bedside. Additionally, we hope that the book will be very exciting and useful to a range of readers from students to research scientists providing a wealth of up-to-date research information in the field of epithelial ion channels and transporters in health and disease.

The undertaking of a book of this scale would always be a "mountain" of work. We want to give our heartfelt thanks to all of our authors who have taken time from their very busy work and non-work schedules to provide excellent chapters, which provided depth of knowledge, informative description, and coverage of the basic physiology and pathophysiology of the topic of their particular chapters.

We want to thank Dr. Dee Silverthorn who planted the "seed" that developed into this volume, which stemmed from a Featured Topic session entitled "Ion Channels in Health and Disease" held during the Experimental Biology meetings in Boston in April 2013 (chaired by KLH). We thank the members of the American Physiology Society (APS) Book Committee who had faith in us to pursue such an exciting book.

As with any book, this volume would not have been possible without the excellent partnership between the APS and Springer-Verlag and the publishing team at Heidelberg, Germany (Britta Mueller, Springer Editor, and Jutta Lindenborn, Project Coordinator). We wish to thank Portia Wong, our Developmental Editor at

Springer+Business Media (San Mateo, CA), and her team who assisted with the early stages of the publishing process that greatly added to this contribution. Finally, special thanks to Shanthi Ramamoorthy (Production Editor, Books) and Ramya Prakash (Project Manager) of Publishing—Springer, SPi Content Solutions—SPi Global and their production team who assisted us through the final stages of the publication of our book.

Finally, we want to thank our mentors Douglas C. Eaton and the late Dale J. Benos for KLH; Michael E. Duffey and Raymond A. Frizzell for DCD; and our colleagues who guided us over the years to be able to undertake this volume.

Dunedin, New Zealand — Kirk L. Hamilton
Pittsburgh, PA — Daniel C. Devor
June 2015

Contents

About the Editors

Kirk L. Hamilton was born in Baltimore, Maryland in 1953. He gained his undergraduate (biology/chemistry) and M.Sc. (ecology) degrees from the University of Texas at Arlington. He obtained his Ph.D. at Utah State University under the tutelage of Dr. James A. Gessaman, where he studied incubation physiology of barn owls. His first postdoctoral position was at the University of Texas Medical Branch in Galveston, Texas under the mentorship of Dr. Douglas C. Eaton where he studied epithelial ion transport, specifically, the epithelial sodium channel (ENaC). He then moved to the Department of Physiology at the University of Alabama, Birmingham for additional post-doctoral training under the supervision of the late Dr. Dale J. Benos where he further studied ENaC and nonspecific cation channels. He took his first academic post in the Department of Biology at Xavier University of Louisiana in New Orleans (1990–1994). He then joined the Department of Physiology at the University of Otago in 1994 and he is currently an Associate Professor. He has focused his research on the molecular physiology and trafficking of potassium channels (specifically KCa3.1). He has published more than 60 papers and book chapters. His research work has been funded by the NIH, American Heart Association, Cystic Fibrosis Foundation, and Lottery Health Board New Zealand. Dr. Devor and he have been collaborators since 1999. When he is not working, he enjoys playing guitar (blues and jazz) and volleyball. Kirk is married to Judith Rodda, a recent Ph.D. graduate in spatial ecology. They have two children, Nathan (b. 1995) and Emma (b. 1998).

Daniel C. Devor was born in Vandercook Lake, Michigan in 1961. His education took him through Southampton College of Long Island University, where he studied Marine Biology, before entering SUNY Buffalo for his Ph.D., under the guidance of Dr. Michael E. Duffey. During this time, he studied the role of basolateral potassium channels in regulating transepithelial ion transport. He subsequently did his postdoctoral work at the University of Alabama, Birmingham, under the mentorship of Dr. Raymond A. Frizzell, where he studied both apical CFTR and basolateral KCa3.1

in intestinal and airway epithelia. He joined the University of Pittsburgh faculty in 1995 where he is currently a professor of cell biology. During this time, he has continued to study the regulation, gating, and trafficking of KCa3.1 as well as the related family member, KCa2.3, publishing more than 50 papers on these topics. These studies have been funded by the NIH, Cystic Fibrosis Foundation, American Heart Association and pharmaceutical industry. When not in the lab, he enjoys photography and growing exotic plants. Dan is married to Catherine Seluga, an elementary school teacher. They have three children, Caitlin (b. 1990), Emily (b. 1993) and Daniel (b. 1997).

Chapter 1
Organoids as a Model for Intestinal Ion Transport Physiology

Hugo R. de Jonge, Marcel J. C. Bijvelds, Ashlee M. Strubberg, Jinghua Liu, and Lane L. Clarke

Abstract The advent of intestinal organoid culture in 2009 was a fortuitous development in the search for a valid marker of intestinal stem cells, and provided proof of murine intestinal stem cell regenerative potential. Intestinal organoid culture was preceded by key discoveries of the Wnt/β-catenin signaling pathway and the development of 3D culture matrices. The latter, involving a laminin-rich gel to provide an artificial basement membrane, was instrumental to primary intestinal epithelial culture by preventing anoikis, an immediate apoptotic event when intestinal epithelial cells detach from the basement membrane. One of the first physiological studies using 3D murine "mini-gut" structures showed cystic fibrosis transmembrane conductance regulator (CFTR) expression and anion channel activity in the crypt-like structures projecting from the epithelial-lined central cavity. Detailed investigations of ion transport physiology using human intestinal organoids, both primary and iPSC-derived, found close similarities to existing knowledge of ion transport physiology and included the development of the forskolin-induced swelling assay (FIS). The FIS assay using organoids cultured from rectal biopsies of cystic fibrosis patients provided an avenue for personalized medicine to test small-molecule modulators on different CFTR mutations. More recent research has led to the development of 2D primary intestinal epithelial monolayers, which provide easy access to the apical, lumen-facing membrane and the opportunity for traditional ion transport studies with Ussing chambers. Human 2D primary intestinal monolayers also demonstrate the dominance of CFTR in anion secretion and provide a quantitative

H. R. de Jonge · M. J. C. Bijvelds
Department of Gastroenterology and Hepatology, Erasmus MC University Medical Center, Rotterdam, The Netherlands

A. M. Strubberg · L. L. Clarke (✉)
Dalton Cardiovascular Research Center, Columbia, MO, USA

Department of Biomedical Sciences, University of Missouri, Columbia, MO, USA
e-mail: ClarkeL@missouri.edu

J. Liu
Dalton Cardiovascular Research Center, Columbia, MO, USA

K. L. Hamilton, D. C. Devor (eds.), *Ion Transport Across Epithelial Tissues and Disease*, Physiology in Health and Disease,
https://doi.org/10.1007/978-3-030-55310-4_1

evaluation of its chloride and bicarbonate secretory conductances. These aspects of ion transport physiology using 2D and 3D intestinal cultures are discussed along with the relative advantages and disadvantages of each culture method with respect to technical aspects and recapitulation of native intestinal epithelium.

Keywords Organoid · Enteroid · Colonoid · Intestine · Colon · CFTR · Cystic fibrosis · Precision medicine · Personalized medicine · Forskolin-induced swelling · Human · Mouse

1.1 Introduction

The culture of self-renewing primary intestinal organoids is one of the most significant experimental techniques developed for investigation of intestinal ion transport physiology. Before 2009, most research of ion transport physiology in intestinal epithelia involved studies of the whole animal (in vivo), short-term tissue cultures or intestinal cell lines, the latter enabling cell-based experiments and genetic manipulations. Studies of the native crypt, i.e., the stem and progenitor cell compartment, were particularly difficult because of the morphological structure and limited access in the intestine in vivo and ex vivo. Attempts at a regenerating primary culture were largely unsuccessful, with a few heroic exceptions (reviewed in Evans et al. 1994). In retrospect, it may be surmised that the technical difficulties were largely a consequence of the propensity of intestinal epithelia to undergo anoikis—an apoptotic event triggered when the epithelial cell detaches from the basement membrane. Short-term studies of isolated colonic crypts were possible (Robert et al. 2001; Singh et al. 1995; Greger et al. 1997; Reynolds et al. 2007; Mignen et al. 2000), apparently due to a higher resistance to anoikis, and provided important insight into the ion transport physiology of the crypt epithelium. These investigations were followed by the pioneering studies of Williams et al. in 2007 that extended the utility of isolated colonic crypts to several days by the provision of appropriate culture substrates (Reynolds et al. 2007). Around the turn of the century, a growing recognition of the importance of the interplay between the epithelium and the extracellular matrix leading to the development of 3D gel cultures in other epithelial cells set the stage for successful primary organoid culture of murine intestinal epithelium (Hofmann et al. 2007; Ootani et al. 2009). A breakthrough came with the recognition of R-spondin1 as an important mitogenic intestinal growth factor that was eventually found to be a secreted coagonist of Wnt/β-catenin signaling in intestinal stem cells (ISCs) (Kim et al. 2005). Using R-spondin1-supplemented medium, a robust, regenerating intestinal culture was developed using minced mouse intestine that gave rise to cysts in a 3D collagen gel (Ootani et al. 2009). The cysts were composed of a polarized, quasidifferentiated intestinal epithelium with crypt- and occasional villus-like structures in a surrounding layer of mesenchymal cells.

The development of a self-renewing, pure epithelial culture came from the search for unique markers of intestinal stem cells. Hans Clevers' laboratory reported the discovery of a Wnt target gene, leucine-rich repeat-containing G-protein coupled receptor 5 (Lgr5) that was expressed in cycling crypt-base columnar cells, i.e., cells originally described with potential for stem cell activity (Cheng and Leblond 1974; Barker et al. 2007). Lineage tracing experiments using mice expressing the knock-in alleles Lgr5-IRES-LacZ or the fusion Lgr5-EGFP-IRES-Cre-ERT2 showed that Lgr5+ cells generated all intestinal lineages, actively cycled and were located at both the crypt base (90%) and at the +4 cell position (10%), i.e., the locale of long-term label-retaining cells that are now referred to as quiescent stem cells (Barker et al. 2007). In 2009, Clevers' group provided positive verification of stem cell status for Lgr5+ cells and introduced the culture of well-differentiated, intestinal epithelial organoids to the field of gastroenterology research. In this study (Sato et al. 2009), a single Lgr5-GFP expressing cell from the Lgr5–EGFP–ires–CreERT2 small intestine was isolated by cell sorting and plated in laminin-rich Matrigel® in medium containing the Wnt signaling cofactor Rspondin1, epidermal growth factor (EGF), noggin (a bone morphogenic protein inhibitor), and 20% FBS. Over a 2-week period, the cells multiplied and gave rise to an intestinal epithelial organoid with a central cavity composed of differentiated cells (i.e., the villus-like domain) from which were outgrowths of organotypic crypts with the actively recycling Lgr5-positive stem cells localized at the crypt base. Fortunately, the isolation of individual Lgr5-positive stem cells was not necessary to culture mouse small intestinal organoids (enteroids, Stelzner et al. 2012) in that freshly isolated small intestinal crypts could be cultured in the same manner to form multiple intestinal epithelial organoids.

One of the favorable features of the enteroid model that was immediately apparent was the visual access provided by the 3D gel culture. Overcoming the difficulties of visualizing crypt epithelium in vivo and the onset of anoikis in epithelial cells of isolated crypts, the enteroid model enabled observations of individual epithelial cell types within the context of a model of native intestine. Evaluation of the crypt epithelium of the enteroid indicated polarization as denoted by an apical microvillus brush border that increased in length along cells in the upper crypt. Further, all the major cell lineages of the small intestine were represented, including absorptive enterocytes, goblet cells, enteroendocrine cells and Paneth cells in approximately the same percentages as found in the intestine in vivo (Sato et al. 2009). The organoids were devoid of nonepithelial cell types, could be passaged at weekly intervals for extended times (months), maintained euploidy and demonstrated a gene expression profile similar to freshly isolated crypts. Shortly after these studies, investigation of the fluorescent signature of crypt-base cells from mice expressing eGFP-labeled Sox9, a transcription factor enriched in Lgr5+ ISCs, recapitulated evidence that a single stem cell could generate differentiated organoids (Gracz et al. 2010). Subsequent analysis of the requirements for enteroid production led to the discovery that single, isolated Lgr5-positive stem cells have a low capacity to develop into enteroids (5%), whereas cell doublets of Lgr5-positive stem cells and lysozyme-positive Paneth cells have a high capacity (80%) for the

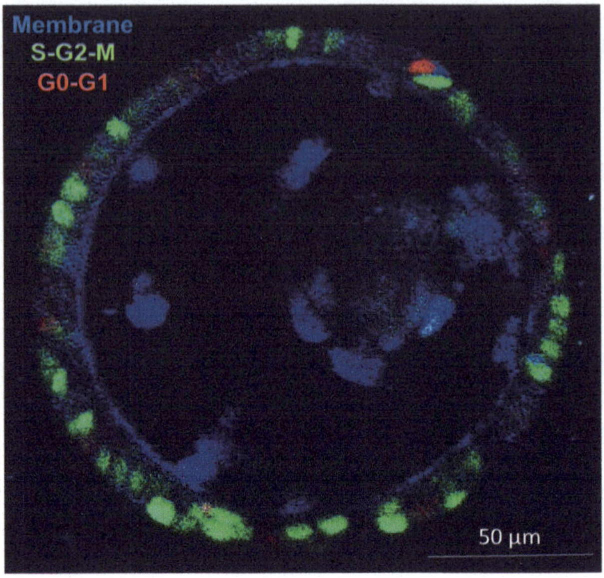

Fig. 1.1 FUCCI2 enterosphere. Optical cross-section of an intestinal stem cell-enriched enterosphere from a Fucci2 mouse intestine (Matsu-Ura et al. 2016), provided by RIKEN Center for Life Science Technologies (Kobe). Cell nuclei indicate quiescent cells (G0–G1 phases, red nuclei) and actively cycling cells (S-G2-M phases, green nuclei). Enterosphere cultured in Matrigel using a modified Sato method (Liu et al. 2012) and supplemented with Wnt3a 100 ng/mL for 48 h

development of enteroid structures (Sato et al. 2011b). Indeed, isolated Lgr5 + ISCs and Paneth cells mixed together in 3D gel culture avidly seek cell-to-cell contact between the two cell types by an, as yet, undescribed process of cell recognition and homing. A key feature of the enteroid model is the elaboration of Wnt3a by Paneth cells, which together with the addition of the Lrg5 ligand Rspondin-1, provides the continual self-renewal of primary enteroids and enables experiments of longer duration, along with the opportunity for genetic manipulations (Schwank et al. 2013).

The enteroid model system of both mouse and human intestine recapitulates many features of small intestinal specialization along the cephalocaudal axis, which is controlled by several transcription factors, in particular GATA4 that is expressed in the proximal intestine and suppresses the expression of distal-specific genes, e.g., the bile-acid transporter ASBT (Slc10a2) (VanDussen et al. 2015; Middendorp et al. 2014). Intestinal organoids also differ regionally with regard to the production of epithelial-autonomous Wnt in that enteroids from the terminal ileum proliferate and develop at slower rates than their more proximal counterparts. Organoids from large intestinal epithelium (aka colonoids) require exogenous Wnt supplementation for growth and development (Yui et al. 2012). Provision of supplemental Wnt ligand or Wnt agonists (e.g., CHIR 99021, an inhibitor of GSK 3β) to small intestinal enteroids also enhances proliferation and suppresses differentiation leading to the expansion of the ISC population (also termed undifferentiated cells) to form “enterospheres” that are composed of a single cell layer in a spheroid structure. As shown in Fig. 1.1, ISC-enriched enterospheres have a majority of cells in active stages of the cell cycle as shown by enterospheres from Fucci2 reporter mouse intestine. Together, these advances have enabled the expansion of intestinal epithelium from patients with intestinal disease that are proving to be valuable for translational studies and individualized medicine in gastroenterology (VanDussen et al. 2015; Sato et al. 2011a).

1.2 Ion Transport in 3D Mouse Intestinal Organoids

The earliest studies of ion transport physiology in the enteroid model were driven by an evaluation of the cystic fibrosis transmembrane conductance regulator anion channel CFTR (Liu et al. 2012). CFTR is the protein product of the gene that is mutated in the monogenic disease cystic fibrosis (CF), the most common lethal genetic mutation of people from a northern European background (Collins 1992). CFTR channel activity is the principal pathway for the secretion of Cl and HCO_3^- across the intestinal epithelium as well as the respiratory and pancreatic duct epithelia (Quinton 1999). Loss of CFTR function results in the dehydration of mucus secreted onto the epithelial surface, causing the accretion of abnormally viscous mucus, a condition known as mucoviscidosis. This pathogenic process underlies most CF disease manifestations, including the failure of mucociliary clearance from airways, intestinal impaction/constipation and plugging of pancreatic ducts with the sequela of pancreatic insufficiency. In contrast to CF, protracted hyperactivation of CFTR by microbial toxins strongly enhances fecal loss of salt and water to produce systemic dehydration and acidosis. Such "secretory" diarrheas typically ensue from colonization of the gut by enterotoxigenic bacteria, including Vibrio cholera (causing cholera) and specific *Escherichia coli* strains (causing colibacillosis, e.g., Traveler's diarrhea) (Barrett and Keely 2000).

CFTR activity in intestinal organoids was first assessed by comparing expression and functional responses in enteroids from wild-type (WT) and Cftr knockout (Cftr KO) mice (Liu et al. 2012). Using an adaptation of the culture method of Sato et al. (2011a), Liu and colleagues showed Cftr protein expression in passaged WT enteroids that was comparable in magnitude to expression in freshly isolated WT crypts and absent in enteroids from sex-matched littermate Cftr KO mice. Microelectrode analysis of Cftr function by impalements of crypt base epithelial cells was possible by gentle aspiration of the encasing Matrigel® via a micropipette to expose the basolateral side of the epithelium in enteroid crypts (Fig. 1.2a). As shown in Fig. 1.2b, c, a basolateral membrane potential of −40 mV measured in WT enteroid crypt epithelial cells abruptly depolarizes upon exposure to forskolin, a cyclic AMP agonist used to stimulate Cftr activity. The inward current produced by stimulation of the anion conductance was Cftr-dependent as shown by the failure of Cftr KO crypt epithelial cells to depolarize. Further, WT crypt epithelia exhibited partial repolarization of forskolin-stimulated enteroids upon acute treatment with the CFTR inhibitor $CFTR_{inh}172$ (Fig. 1.2d).

A second feature of Cftr function investigated in the mouse enteroid model was regulation of intracellular pH (pH_i) in the crypt epithelium. CFTR is conductive to both Cl^- and HCO_3^- anions with a relative permeability of ~4:1, respectively

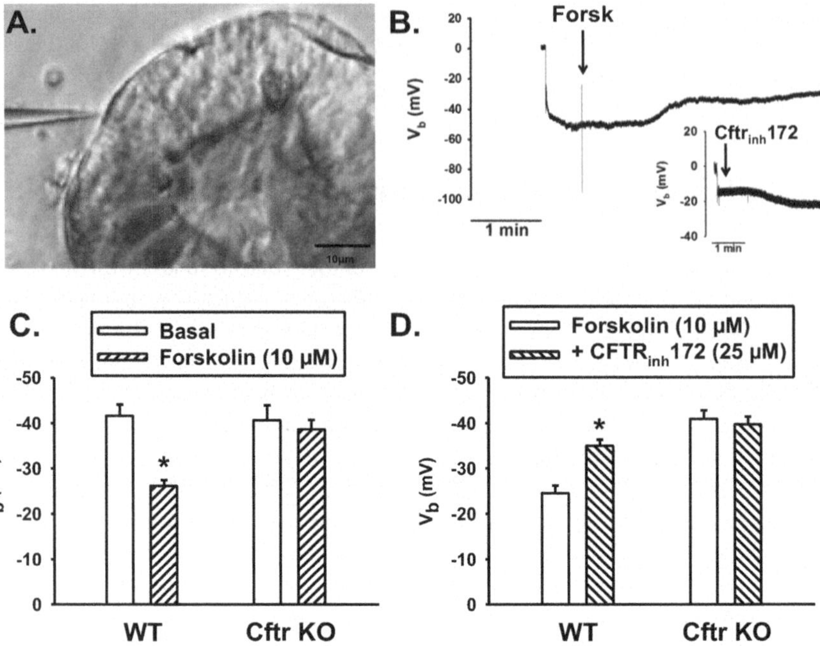

Fig. 1.2 Microelectrode analysis of Cftr-dependent changes in membrane potential in enteroid crypt epithelium. (**a**) Micrograph of enteroid crypt epithelial cell impaled with a conventional microelectrode (magnification: ×200). (**b**) Representative recording of the basolateral membrane potential (V_b) of an impaled WT cell before and after exposure to 10 μM forskolin. *Inset*: representative recording of V_b in an impaled cell from the same forskolin-treated WT enteroid crypt before and after exposure to 25 μM $Cftr_{inh}172$. Note abrupt voltage change from 0 mV upon microelectrode impalement and return toward 0 mV upon microelectrode retraction in both recordings. However, there was a ~4-mV electrode drift during the prolonged impalement showing forskolin-induced depolarization. (**c**) Cumulative data of mean V_b measured in WT and Cftr KO enteroid crypt epithelial cells before (basal) and after exposure to 10 μM forskolin. Impalements during forskolin were performed 10–30 min after treatment. Enteroids were from WT and Cftr KO sex-matched littermate mice. *$P < 0.05$, significantly different from basal within genotype; $n = 33–34$ impalements, 4–6 p0–p1 enteroids from 3 mice pairs. Mean V_b for Cftr KO, both basal and forskolin-treated, is significantly greater than WT treated with forskolin, $P < 0.05$. (**d**) Cumulative data of mean V_b measured in WT and Cftr KO forskolin-treated enteroid crypt epithelial cells before (forskolin: 10 μM) and after exposure to 25 μM $Cftr_{inh}172$. Enteroids were exposed to forskolin for 10 min before microelectrode impalements. Impalements during $Cftr_{inh}172$ were performed 10–30 min after treatment. Enteroids were from WT and Cftr KO sex-matched littermate mice. *$P < 0.05$, significantly different from forskolin within genotype; $n = 16–26$ impalements, 3–4 p0–p1 enteroids from 3 mice pairs (Liu et al. 2012)

(Poulsen et al. 1994). CFTR also facilitates HCO_3^- secretion by directly enhancing apical membrane Cl^-/HCO_3^- exchanger activity, notably Slc26a3 and Slc26a6 anion exchangers (Dorwart et al. 2008), by providing Cl^- recycling that prevents the development of an unfavorable inside to outside Cl^- concentration gradient for the exchange process (Simpson et al. 2005). Moreover, Slc26a9, a complex anion transporter whose function is CFTR-dependent in airway epithelium (Bertrand et al. 2017), has been shown to provide HCO_3^- secretion from a crypt-predominant location in the proximal duodenum (Liu et al. 2014). Thus, loss of CFTR, as a major pathway for HCO_3^- efflux from intestinal epithelial cells, sets the stage for pH_i dysregulation. Previous studies of cell lines and mouse intestinal villi have shown that CFTR expression/activity physiologically acidifies pH_i and, in its absence, cells exhibit an alkaline pH_i (Barriere et al. 2001; Elgavish 1991; Gottlieb and Dosanjh 1996; Simpson et al. 2005; Hirokawa et al. 2004). In accordance with those studies, Liu et al. (2012) found that the crypt epithelium of Cftr KO enteroids exhibited an alkaline pH (~7.5) relative to crypts from WT littermate enteroids (~7.2). WT crypts also acutely alkalized upon treatment with $CFTR_{inh}172$ (Fig. 1.3a, c). The observation that the alkaline pH_i in the Cftr KO crypt epithelium was not normalized by the activity of other acid-base transport processes led to further investigations using mouse enteroids. Walker et al. showed that the expression of anion exchanger 2 (AE2) is upregulated in Cftr KO enteroids, whereas the expression of Slc26a3, the Na^+/H^+ exchanger Nhe2, and the transmembrane carbonic anhydrase CA9 is decreased—all appropriate changes to compensate for an alkaline pH_i (Walker et al. 2016). However, the Cl^-/HCO_3^- exchange activity of Ae2 was reduced in the Cftr KO crypts due to coincident increases of intracellular Cl^-. Chloride retention in the Cftr KO mouse intestine is a finding that is consistent with previous X-ray microprobe analysis of human CF intestine (O'Loughlin et al. 1996). Chloride retention establishes an unfavorable $[Cl^-]_{in}$ to $[Cl^-]_{out}$ gradient, which retards the exchange process by Ae2 in Cftr KO crypt epithelium. Of note, a technical problem encountered in these studies was persistent retention of the Cl^- sensitive fluorescent dye MQAE within the Matrigel®, despite numerous washings. To overcome this obstacle Walker et al. removed the enteroids from Matrigel®, stabilizing the enteroids with a holding micropipette and included LY2763, a cell-permeant anoikis inhibitor, in the superfusate during the experiment.

The third piece of evidence for functional Cftr activity in the enteroid model is the role that Cftr plays in cell volume regulation. Earlier studies established that activation of CFTR reduces intestinal crypt epithelial cell volume (Valverde et al. 1995; MacLeod et al. 1994), which extends to villi of the duodenum that also express significant levels of CFTR (Gawenis et al. 2003; Strong et al. 1994). The volume reduction in villi also reduces NaCl absorption across villi by downregulating the activity of Na^+/H^+ exchanger isoform 3 (NHE3) (Gawenis et al. 2003; Szászi et al. 2001; Kapus et al. 1994), thereby contributing to the cAMP-induced inhibition of NHE3 mediated by the NHE regulatory protein NHERF (Avula et al. 2018; Seidler et al. 2009). Stimulation of CFTR reduces epithelial cell volume primarily by decreasing the intracellular Cl^- concentration as demonstrated in the elegant studies performed by Foskett and colleagues on airway serous gland epithelial cells (Lee and Foskett 2010; Foskett 1990). In the enteroid study, Liu et al. show that forskolin stimulation in WT enteroids causes a sustained decrease in the enteroid cell volume (~−25%), as indexed by the change in epithelial cell height, a response that is absent in the Cftr KO enteroid crypts (Fig. 1.4a, b) (Liu et al. 2012). Subsequent treatment of WT enteroids with $CFTR_{inh}172$ significantly reduced the cell shrinkage. Although

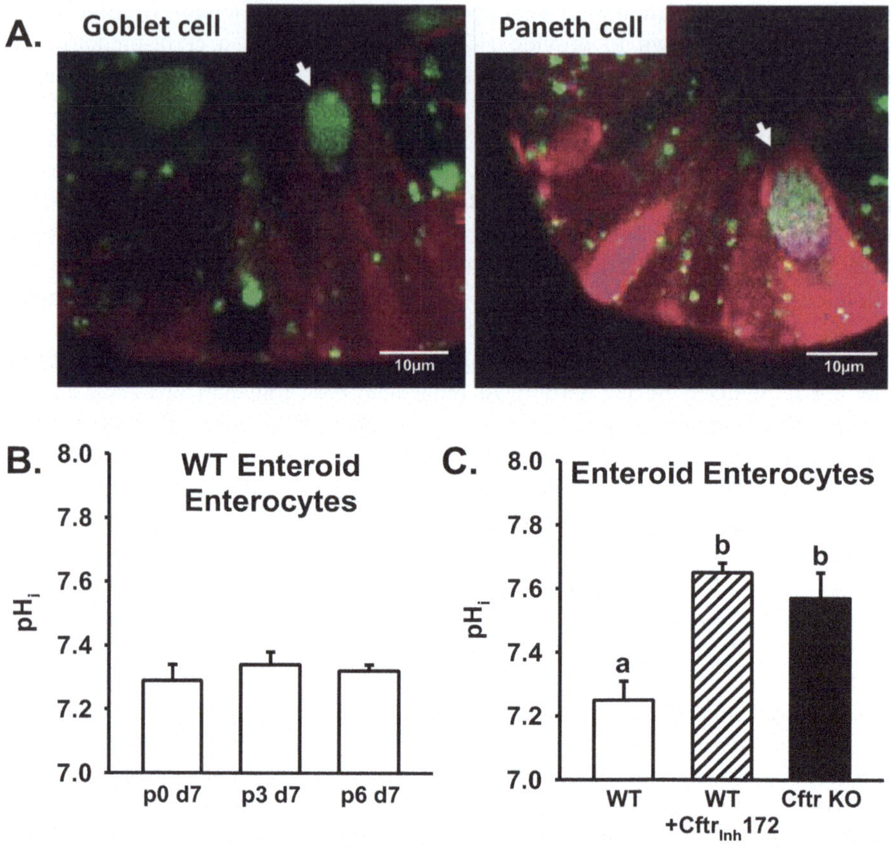

Fig. 1.3 Cftr-dependent effect on basal pH_i in enteroid crypt epithelium. (**a**) Merged confocal images of enteroid crypt epithelium stained with quinicrine (green) to identify granulated secretory cells (goblet and Paneth) and the pH-sensitive dye SNARF 5F (red) for measurement of pH_i (magnification: ×630, n.a. 1.2). (**b**) Mean enterocyte pH_i measured in p0, p3, and p6 enteroids after *day 7* (d7) in culture. Each group of enteroids were from the same WT mice ($n = 3$). (**c**) Mean enterocyte pH_i measured in WT, WT pretreated for 1 h with 25 μM $Cftr_{inh}172$ (WT + Cftrinh172) and Cftr KO enteroid crypts. Enteroids (p0–p1) were from WT and Cftr KO sex-matched littermate mice. $^{a,b}P < 0.05$, means with the same letter are not significantly different; $n = 6$ mouse pairs (Liu et al. 2012)

briefly mentioned, a technical difficulty encountered in studies of cell volume regulation was that WT enteroids would rapidly swell upon forskolin stimulation as a consequence of the luminally directed Cftr-mediated fluid secretion and reduced paracellular fluid leakage through well-developed tight junctions. This combination in the WT enteroid generated sufficient backpressure to flatten the epithelium uniformly, thereby obviating the measurements of cell volume after forskolin. Although enteroid swelling would later become the basis for screening CFTR modulator drugs (see Sect. 1.4, below), the enteroids in the studies by Liu et al.

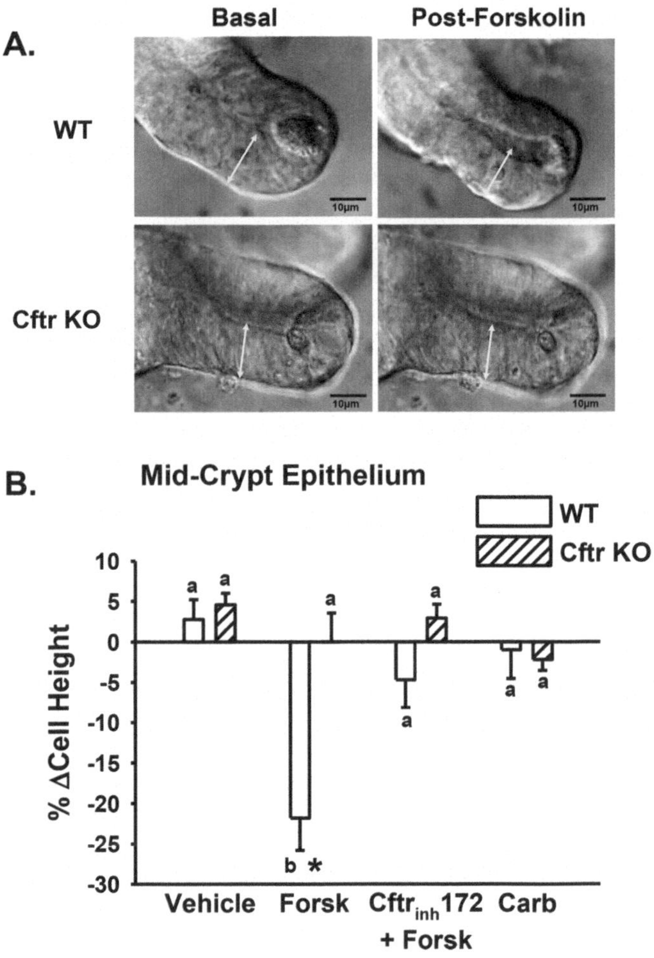

Fig. 1.4 Cftr-dependent cell shrinkage in enteroid crypts. (**a**) Photomicrographs of WT and Cftr KO enteroid crypts before (basal) and after 10-min exposure to 10 μM forskolin (post-forskolin). Arrowed bar indicates measurement of epithelial cell height as an index of cell shrinkage before forskolin treatment. Dotted white and solid white lines indicate diameters (*2r*) of crypt and crypt lumen, respectively, and dashed white line indicates height (*h*) for calculation of epithelial volume before forskolin treatment (magnification: ×400). Enteroids were from WT and Cftr KO sex-matched littermate mice. Mean cell height before treatments were WT = 21.1 ± 0.9 μm and Cftr KO = 21.3 ± 0.9 μm; $n = 18$. (**b**) Cumulative data of % change in epithelial volume after vehicle (Vehicle), forskolin (Forsk, 10 μM), pretreatment with 25 μM $Cftr_{inh}172$ + forskolin ($Cftr_{inh}172$ + Forsk), or carbachol (Carb, 100 μM) in midcrypt epithelium (between position +8 to +15) of paired WT and Cftr KO enteroids (p0–p1). Epithelial volume was calculated by subtracting the crypt luminal volume from the total crypt volume between cell positions +8 to +15, assuming each with a cylindrical shape, using the formula $\pi \cdot h \cdot r^2$ and averaged measurements of the height (*h*) and radius (*r*) of each in optical cross sections. %ΔEpithelial volume was calculated from the formula: change in epithelial volume (in μm^3)/basal epithelial volume (in μm^3) * 100. [a,b]Means with the same letter are not significantly different within genotype. $*P < 0.05$, significantly different from Cftr KO; $n = 6$ WT and Cftr paired enteroids (Liu et al. 2012)

were bisected manually before forskolin exposure to prevent enteroid swelling and, thereby, ensure the accuracy of the epithelial cell volume measurements during forskolin stimulation.

Soon after these studies, Engevik and others used the mouse enteroid model to evaluate the effects of electroneutral NaCl absorption on the microbiota of the distal intestine (Engevik et al. 2013). These investigators showed that Nhe3 knockout mice (Nhe3−/−) have increased intestinal lumen Na^+ concentration and pH, a consequence of lacking the Na^+/H^+ exchange activity at the luminal membrane. Nhe3−/− mice also had alterations in the composition of the microbiota in the distal intestine including increases in Bacteroidetes spp. and an increase in *fut2* expression causing surface fucosylation. To show that the abnormal increase in surface fucosylation was due to the altered microbiota and not a consequence of epithelial Nhe3 ablation per se, these investigators used Nhe3−/− ileal enteroids and found they did not develop surface fucosylation spontaneously, but did so after intraluminal injection of Bacteroidetes spp. into the enteroids. Although ion transport studies of the enteroids were not performed, this research showed two important aspects of the utility of intestinal organoids. First, it demonstrated the potential to investigate host-microbe interactions in organoids with regard to ion transport. Secondly, it demonstrated the opportunity to separate epithelial-autonomous functions such as ion transport from the influences of the intestinal environment, i.e., microbiota and aspects of humoral, neural, submucosal, and immunological regulation.

As mentioned in reference to Fig. 1.1, stem cell proliferation in enteroids can be greatly enhanced through treatment with Wnt3a supplementation resulting in the generation of ISC-enriched "enterospheres or enterospheroids" (Miyoshi and Stappenbeck 2013; Miyoshi et al. 2012). Using the enterosphere model, Strubberg et al. showed that the hyperproliferative state previously demonstrated in the intestinal epithelium of Cftr KO mice in vivo extended to the Cftr KO ISC population (Strubberg et al. 2018; Gallagher and Gottlieb 2001). Further evidence that Wnt3a supplementation yields a model for ISC investigation can be demonstrated in developed enteroids (4 days old) using a single dose of Wnt3a (100 ng/mL). As shown in Fig. 1.5, Wnt3a supplementation causes enteroids to assume an enlarged spheroid shape that, in time, gives rise to an extraordinary increase in crypt structures, an index of ISC proliferation (Fuller et al. 2012). The enterosphere model also demonstrates an important role for Cftr in regulating proliferation of ISCs. Fresh enterospheres from WT and Cftr KO mice generated in the presence of Wnt3a supplementation exhibit a marked difference in luminal volume (as indexed by spheroid diameter) and cell height (Fig. 1.6a), similar to that demonstrated previously in mouse colonoids treated with Wnt3a (Dekkers et al. 2013). Forskolin stimulation causes a rapid increase of sphere diameter in 100% of WT mouse enterospheres and 0% in Cftr KO enterospheres, indicating the presence of functional Cftr in these ISC-enriched structures (Fig. 1.6b, c). The demonstration of functional Cftr activity at an early stage of enterosphere development (2 days) lends support to previous evidence that CFTR is a target of Wnt signaling through an intestine-specific enhancer element located within the first intron (Paul et al. 2007). However, as shown in Fig. 1.6d, e, acute treatment of Wnt3a-treated WT

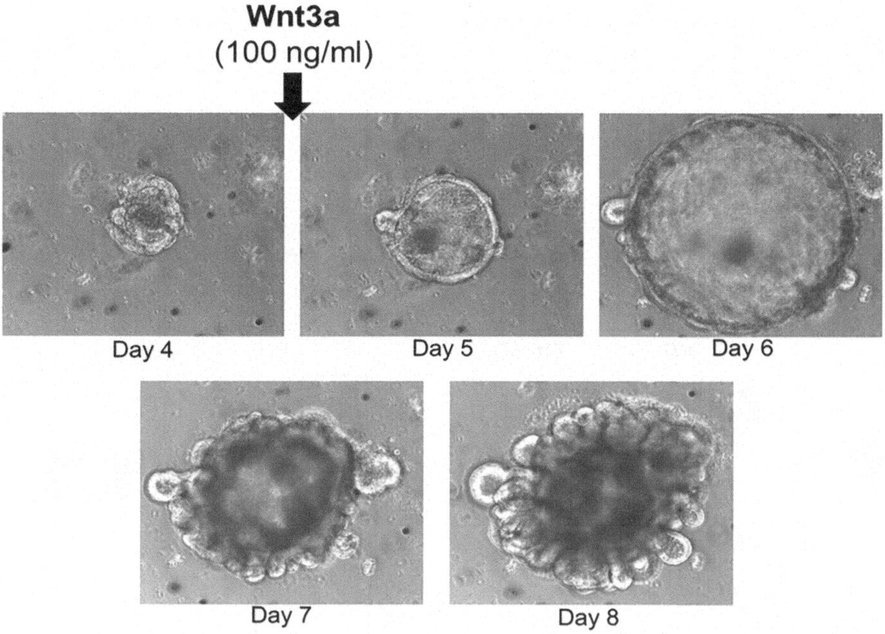

Fig. 1.5 Time course of WT enteroid treated with a single dose of Wnt3a (100 ng/mL on Day 4 in culture). Note expansion of organoid after 2 days (Day 6) and subsequent formation of multiple crypts (Day 8)

enterospheres to inhibit Cftr-mediated anion secretion either by blocking Cl^- uptake by $Na^+/K^+/2Cl^-$ cotransport with bumetanide or by treatment with $CFTR_{inh}$-172, caused a reduction in enterosphere diameter in only a fraction of the structures (30–35%). This observation requires further investigation in that it may reflect limitations in drug delivery through the Matrigel® for certain compounds or, may indicate changes to the Cftr-dependent anion secretory process when enterospheres attain a specific level of turgor or membrane stretch (see Sect. 1.4).

1.3 Ion Transport in 3D Human Intestinal Organoids

The next steps in the investigation of ion transport physiology using intestinal organoids came with the developments of intestinal-differentiated induced pluripotent stem cells (gut-iPSCs) and primary organoid culture of human intestinal epithelium (Sato et al. 2011a; Spence et al. 2011). Spence et al. used a temporal series of growth factor exposure to direct iPSCs from definitive endoderm to hindgut specification/morphogenesis and generate human intestinal organoids (HIOs) in a 3D Matrigel® prointestinal culture system (Spence et al. 2011). The HIOs showed proliferative crypt-like structures projecting from a central cavity with functional

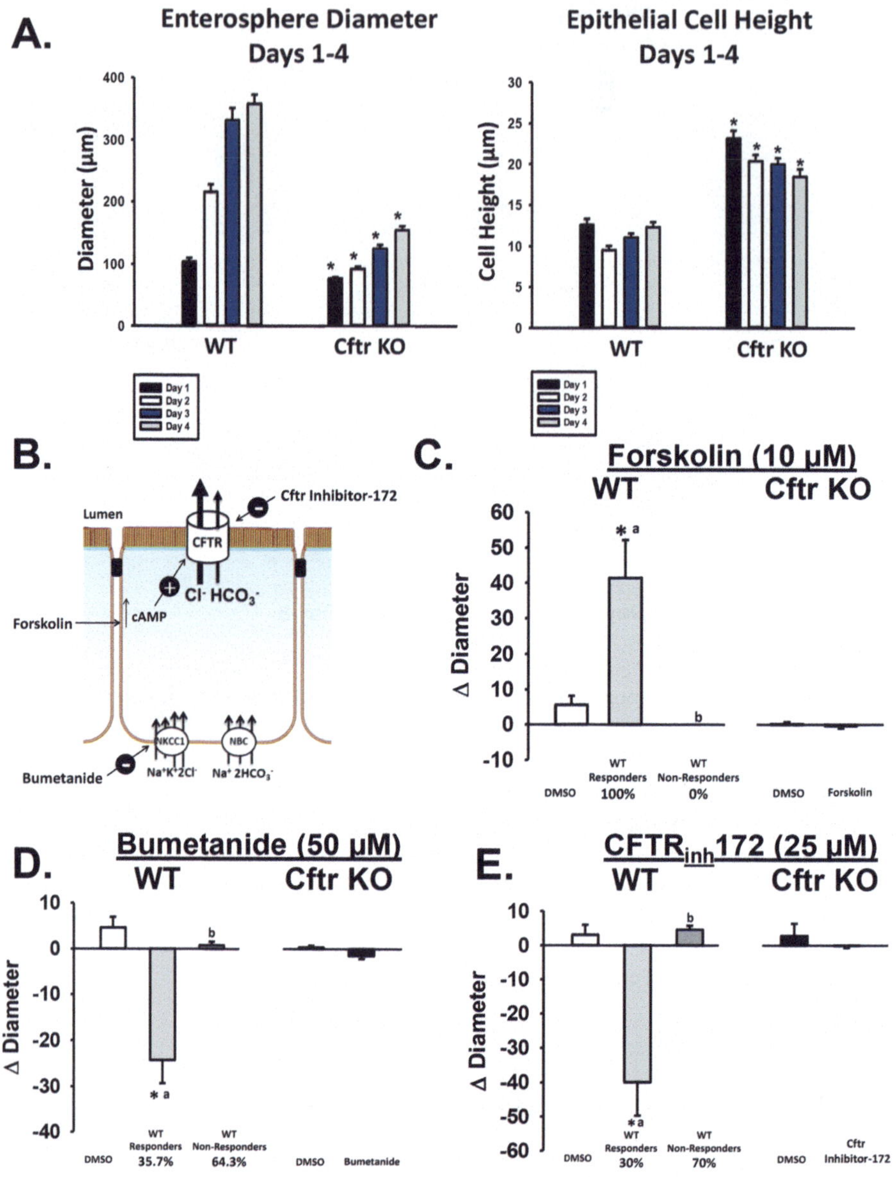

Fig. 1.6 Anion transport in murine enterospheres. (**a**) Time course of WT and Cftr KO enterospheres in culture showing enterosphere diameter and epithelial cell height. $^*P < 0.001$ vs. WT, $n = 35$–42 enterospheres from 3 WT/Cftr KO sex-matched littermates. (**b**) Schematic representation of the experiments on Day 2 enterospheres shown in **c**–**e**. (**c**) Effect of forskolin treatment (10 μM, 15 min) on the change in diameter of basal WT and Cftr KO enterospheres. $^*P < 0.05$ vs. forskolin-treated Cftr KO. [a,b]Means with different letters are significantly different vs. DMSO control, $P < 0.05$, $n = 8$–11 enterospheres from 3 WT/Cftr KO sex-matched littermates. (**d**) Effect of bumetanide (50 μM, 1 h) on change in diameter of basal WT and Cftr KO enterospheres. Only 35.7% of WT enterospheres responded to bumetanide treatment. $^*P < 0.05$ vs. bumetanide-treated Cftr KO. [a,b]Means with different letters are

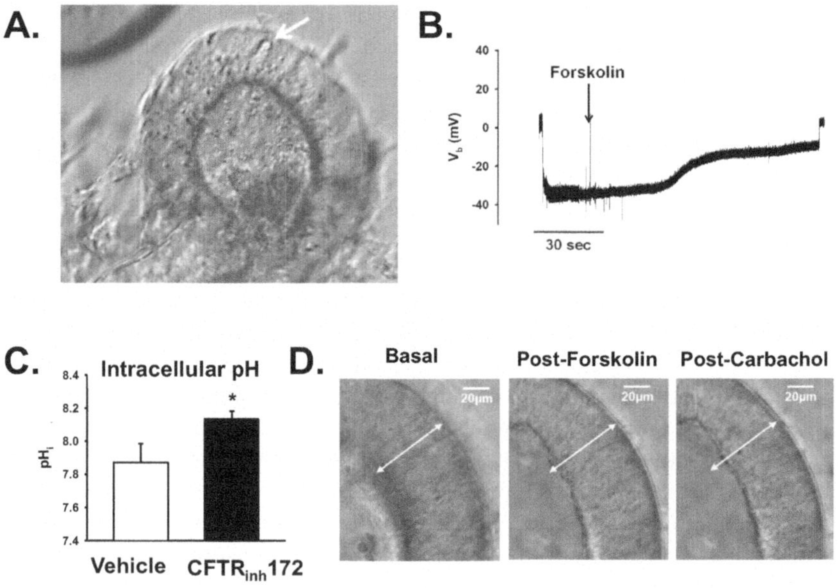

Fig. 1.7 Evidence of CFTR function in intestinal-differentiated iPSCs (HIOs). (**a**) Human intestinal organoid (HIO) from a non-CF, gut-differentiated iPSC (arrow, Paneth cell at crypt base). HIOs were a gift from Dr. James Wells, University of Cincinnati, Cincinnati Children's Medical Center. (**b**) Recording of a microelectrode impalement across the basolateral membrane of a non-CF HIO crypt epithelial cell showing depolarization of the basal membrane potential (V_b) after forskolin (10 μM) addition to the bath, representative of three experiments. (**c**) Effect of CFTR inhibition with CFTR$_{inh}$172 (10 μM, 24 h) on intracellular pH of non-CF HIO crypt epithelial cells. *Significantly different vs. vehicle control (Veh), $n = 4$ different passages. (**d**) Changes in HIO crypt epithelial volume as indexed by changes in cell height following sequential treatments with forskolin (10 μM, 15 min and carbachol (100 μM, 10 min). Bidirectional arrow indicates cell height under basal conditions in all images. Representative of three separate experiments

enterocytes, goblet cells, Paneth, and enteroendocrine cells (Fig. 1.7a). Enterocyte uptake of a fluorescently labelled dipeptide indicated an intact peptide transport system. To investigate CFTR function in the gut-iPSCs, a similar series of experiments were employed as in the murine enteroid study by (Liu et al. 2012). As shown in Fig. 1.7b, microelectrode impalements of non-CF HIO crypt epithelial cells exhibit membrane depolarization after activation of CFTR with forskolin. In

Fig. 1.6 (continued) significantly different vs. DMSO control, $P < 0.05$, $n = 6$–18 enterospheres from 3 WT/Cftr KO sex-matched littermates. (**e**) Effect of CFTR$_{inh}$172 (25 μM, 1 h) on change in diameter of basal WT and Cftr KO enterospheres. Only 30.0% of WT enterospheres responded to CFTR$_{inh}$172 treatment. *$P < 0.05$ vs. inhibitor-treated Cftr KO. [a,b]Means with different letters are significantly different vs. DMSO control, $P < 0.05$, $n = 6$–14 enterospheres from 4 WT/Cftr KO sex-matched littermates

Fig. 1.7c, the regulation of pH_i by CFTR function is assessed by treating the non-CF HIOs with $CFTR_{inh}$-172 for 24 h, which induced cellular alkalization. Figure 1.7d shows the effects of stimulating cAMP and Ca^{2+} mobilization on cell volume regulation in non-CF HIOs, as indexed by lumen expansion and epithelial cell shrinkage. Thus, like murine enteroids, a functional CFTR is present in iPSC-derived HIOs from healthy subjects.

Paralleling the development of HIOs were successful efforts to culture primary human enteroids. Sato et al. found that the method developed to culture mouse enteroids required additional medium supplementation for propagation and prolongation of culture. These supplements included gastrin, nicotinamide, and inhibitors of TGFβ-activin receptors ALK4/5/7and p38 (Sato et al. 2011a). Using similar culture methods, Dekkers et al. explored the ion transport properties of primary human organoids generated from intestinal biopsies of healthy and CF patients. Healthy human enteroids and colonoids responded to forskolin stimulation with organoid swelling, but organoids from CF patients did not (Dekkers et al. 2013). A high-throughput format was developed for CFTR modulator drug screening using intracellular uptake of green fluorescent calcein to accentuate the cellular borders and 2-dimensional imaging (*x–y*) to measure changes in organoid area, all within a single culture well. The method was termed the forskolin-induced swelling assay (FIS). This study set the stage for personalized medicine for CF patients by showing that the response of primary human intestinal organoids to FIS in vitro correlated well with ex vivo intestinal short-circuit current measurements in Ussing chambers. Both assays used intestinal biopsies from the same human subjects. In the same year (2013), Donowitz et al. demonstrated that the intestinal organoids from human patient biopsies also exhibited Na^+/H^+ exchanger 3 (NHE3) activity by measuring changes in intracellular pH_i. Treatment of enteroids with the specific NHE3 inhibitor S3226 prevented increased pH_i after an NH_4Cl prepulse used to acidify the cytosol. Because it is known that intracellular cAMP stimulation inhibits NHE3 activity (Donowitz and Welsh 1987), this outcome provided evidence that FIS assay of human enteroids may include inhibition of coupled NaCl absorption (Na^+/H^+, Cl^-/HCO_3^- exchange) in addition to CFTR-mediated anion secretion.

A comprehensive evaluation of ion transport physiology in 3D human intestinal organoids came from studies of upper small intestinal enteroids by Donowitz and others (Foulke-Abel et al. 2016). Studies focused on ion transport and markers of differentiation using Wnt3a stimulation to yield undifferentiated enteroids as a model for crypt-like epithelium and Wnt3a removal to yield differentiated enteroids as a model for villus-like epithelium. In the villus-like epithelial organoids, markers of stem cells (LGR5/achaete-scute complex homolog 2/olfactomedin 4) as well as proliferation indices were decreased relative to the undifferentiated crypt-like enteroids. As expected, markers of lineage commitment such as sucrose-isomaltase and trefoil factor 3 were increased in the villus-like enteroids. In this study, an evaluation of the Na^+/H^+ exchanger NHE3 by measurement of pH_i under conditions that isolate its activity demonstrated that this transporter plays a dominant role in Na^+/H^+ exchange of the enteroid epithelium. Interestingly, NHE3 activity was equivalent between the undifferentiated and differentiated organoids.

Foulke-Abel et al. also used the FIS assay to compare apical fluid secretion between the differentiated and undifferentiated human enteroids. Counterintuitively, the differentiated enteroids showed a large average FIS as compared to the undifferentiated enteroids. As pointed out by the authors, this comparison is confounded by the possibility that paracellular pore pathway in undifferentiated, crypt-like enteroids is leakier than in the differentiated, villus-like enteroids. This conjecture is supported by previous in vivo studies of rat intestine where it was shown that villous pores (<6 Å) of the paracellular pathway are an order of magnitude smaller than crypt epithelial pores (50–60 Å) (Fihn et al. 2000). In a comparison within groups, pretreatment with a combination of pharmacological inhibitors of CFTR ($CFTR_{inh}$-172, GlyH-101) indicated that CFTR is responsible for greater than 90% of FIS in both undifferentiated and differentiated enteroids. Interestingly, pretreatment before forskolin with pharmacological inhibitors of NHE3 (EIPA or S3226) caused a ~30% reduction of the FIS in both enteroid groups. CFTR activation and simultaneous NHE3 inhibition by increases in intracellular cAMP in intestinal epithelium is well documented (Gawenis et al. 2003; Kurashima et al. 1997; Zhao et al. 1999; Yun et al. 1997; Zizak et al. 1999). However, cAMP inhibition of ongoing NaCl absorption as a contribution to FIS is difficult to reconcile with almost complete prevention of FIS by CFTR inhibition. Beyond the possibility of off-target effects of CFTR or NHE3 inhibitors, previous studies have shown that NHE3 inhibition induces the partial activation of anion secretion (likely mediated by CFTR) in rat intestine (Furukawa et al. 2004). It is also known that an acidic pH_i can increase CFTR open probability by increasing the channel's affinity for MgATP (Chen et al. 2009). In the Foulke-Abel et al. study, the authors indicated that acute NHE3 inhibition did not affect enteroid diameter before forskolin, suggesting that inhibition of NaCl absorption may be offset by a suboptimal activation of CFTR. If so, the FIS magnitude during NHE3 inhibition may be reduced by pre-existing CFTR activation, a hypothesis that will require further investigation.

In the same study, bumetanide inhibition of Cl^- uptake by NKCC1 in the FIS assay reduced fluid secretion by 50% in the undifferentiated enteroids and 76% in the differentiated enteroids, indicating a dominant role for CFTR-mediated Cl^- secretion. The fact that inhibition by bumetanide of Cl^- secretion in the FIS assay was incomplete, led Foulke-Abel et al. to speculate that there may be additional Cl^- or HCO_3^- uptake mechanisms involved in FIS. To investigate this aspect, studies of pH_i in differentiated human enteroids were employed to show that cAMP activation of CFTR in the absence of HCO_3^- caused intracellular acidification of pH_i, which was not dependent on NHE3 activity. Forskolin-induced acidification of pH_i was abolished in the presence of HCO_3^-, but was recovered by inhibition of HCO_3^- uptake by the $NaHCO_3$ cotransporter NBCe1. These findings suggest that net HCO_3^- secretion via CFTR activity contributes to FIS of human enteroids, leading to the speculation that net HCO_3^- secretion in cases of secretory diarrhea (e.g., cholera, colibacillosis) may exacerbate systemic dehydration and acidosis.

In a more recent study (Yin et al. 2018), the Donowitz laboratory found that altering the culture conditions affected the ion transport physiology of the 3D human intestinal organoids in a way more consistent with intact intestine. Although the focus of this report was ion transport in 2D human intestinal enteroid monolayers, studies of transporter expression and FIS in 3D human enteroids were included. The principal change in culture conditions was removal of nicotinamide from the enteroid expansion medium. Although nicotinamide, the amide of the vitamin niacin, is often added to organoid culture to enhance stem cell proliferation, it also has additional activity as an inhibitor of Rho-associated protein kinase (ROCK), by a pathway different than standard ROCK inhibitory Y27632, and as an inhibitor of casein kinase 1 (CK1). The Yin et al. study found, in contrast to the previous Foulke-Abel et al. investigation, that the undifferentiated (crypt-like) 3D human enteroids have an ~two-fold greater FIS as compared to the differentiated (villus-like) enteroids. Moreover, expression profiling of ion transporters indicated that the undifferentiated enteroids had increased expression of CFTR, NKCC1, the cAMP-activated K^+ channel KCNE3 and decreased expression of SLC26A3, NBCe1, the electroneutral NaHCO3 cotransporter NBCn1, carbonic anhydrase 2, and carbonic anhydrase 4, as compared to the differentiated enteroids. With the notable exception of downregulation of NKCC1 in differentiated human enteroids, expression changes for CFTR, SLC26A3, and NBCe1 were not found between undifferentiated and differentiated human enteroids in the Foulke-Abel et al. (2016) study. Although these inconsistencies accentuate the shortcomings of enteroid culture, both studies showed that CFTR-mediated anion secretion and NaCl absorption were present to varying degrees in undifferentiated and differentiated human enteroids. This provides a new perspective on the longstanding controversy regarding "absorptive and secretory compartmentalization" in the small intestine. Some studies contend that an anatomical separation exists between net anion secretion by crypt epithelium and net NaCl absorption by villous epithelium (Hallbäck et al. 1982; Eklund et al. 1987). Other evidence supports the notion that these ion transport processes exist in both (Stewart and Turnberg 1989; De Jonge 1975; Kockerling and Fromm 1993; Singh et al. 1995; Ainsworth et al. 1996; Gawenis et al. 2003). The studies by the Donowitz group suggest the latter is true in isolated, regenerating intestinal epithelium. Although there are differences along the crypt-villus axis in the expression of transporters involved in these processes, it is apparent intestinal epithelial cells have the capacity for both anion secretion and NaCl absorption. The concept of "compartmentalization" of these two important transport processes may be a consequence of differences in activity along a crypt-villus continuum that is accentuated by differences in the epithelial cell number in crypts versus villi. Furthermore, the recent recognition that the small intestine has a subset of "CFTR high expressor" cells, including the villous epithelium, suggests that the concept of compartmentalization may be redefined at a cellular level (Jakab et al. 2013; see Sect. 1.12).

1.4 Precision Medicine of CF Using 3D Human Intestinal Organoid Transport Physiology

The earlier studies by Dekkers et al. (2013) led to the application of organoid ion transport physiology to screen compounds that modulate the activity and processing of mutant CFTR in human rectal organoids from CF patients (Dekkers et al. 2016). The ability to expand the organoids with multiple passages from a single rectal biopsy enabled the testing of different combinations and dosages of recently developed pharmaceutical modulators of CFTR, thus ushering in personalized medicine for CF patients. Over 2000 *CFTR* mutations have been identified worldwide (www.genet.sickkids.on.ca) with variances in phenotype, including single-organ CFTR-deficient diseases such as some cases of idiopathic chronic pancreatitis (Noone et al. 2001). Driven by the difficulties of testing pharmacologic therapy in CF patients with so many different mutations, especially those with rare mutations, these investigators used rectal organoids to test two CFTR modulators initially developed for a limited number of CFTR mutations. The first modulator registered for treating CF patients carrying *CFTR* mutations with defective channel gating (G551D, S1251N, others) was the CFTR potentiator ivacaftor (Kalydeco®, Vertex Pharmaceuticals) (Ramsey et al. 2011; Accurso et al. 2010; De Boeck et al. 2014). This was shortly followed by the introduction of the protein processing corrector lumacaftor, which in combination with ivacaftor, is known as ORKAMBI ® (Vertex Pharmaceuticals). The latter was shown in clinical trials to have beneficial effects on lung function in CF patients with the most common CF disease causing mutation, ΔF508 (Boyle et al. 2014; Wainwright et al. 2015).

The study employed the FIS assay as described here but found that the organoid phenotype (enlarged cysts) from biopsies of healthy patients or CF patients with mutations showing high residual function negatively correlated with the FIS response. To address this difficulty, Dekkers et al. (2016) introduced the steady-state lumen area assay (SLA), which basically involves confocal microscopic measurement of the total organoid area and the luminal area in the x–y plane in organoids under basal, steady-state conditions (without forskolin treatment). The SLA is expressed as the luminal organoid area as a percentage of the total organoid area. SLA could readily discriminate organoids from healthy subjects with at least one WT CFTR allele (SLA $\geq$ 50%) and organoids from CF patients with class I–V CFTR mutations (SLA 0–10%). Among CF organoids, the SLA assay could also discriminate between class I–III CFTR mutations (no CFTR synthesis, defective trafficking, and defective channel gating, respectively) with SLA = ~0–0.1% and class IV–V CFTR mutations (altered channel conductance or reduced amounts of CFTR protein) with SLA = ~5%. It was shown that large SLA organoids (>40%) from healthy subjects with at least one WT CFTR allele was negatively correlated with the increase in organoid area in the FIS. In contrast, the FIS assay, which extends over 60 min, was well correlated in organoids with lower SLA (<40%). This indicated that the FIS likely underestimates WT CFTR response to forskolin, thereby obviating a direct comparison to CF organoids. However, the SLA of all CF organoids was below 10%, therefore supporting the utility of the SLA assay for testing CFTR modulator therapy.

The poor response to FIS in Wnt-treated, non-CF rectal organoids (SLA >40%) in the Dekkers et al. (2016) studies has some parallels with observations of Wnt3a-treated WT murine enterospheres (see Fig. 1.6). These studies suggest that CFTR is active at some point during Wnt-stimulated growth and perhaps inhibited upon spheroid maturation. As shown in Fig. 1.6a, the rapid increase in enterosphere diameter during Days 2–3 in culture is nearing a plateau by Days 3–4. Indeed, the FIS assay of murine enterospheres was performed on culture Day 2 because older, enlarged WT enterospheres had diminished FIS responses similar to the findings with the human rectal organoids (data not shown). CFTR activity during stimulated growth is largely responsible for organoid swelling, but its activation negatively regulates and slows proliferation (Strubberg et al. 2018; Than et al. 2016). As mentioned previously, the nonresponders to Cftr inhibition (70%, Fig. 1.6e) in Day 2 WT enterospheres suggest that proliferation regulation may change with time in culture, likely in response to several factors during Wnt-stimulated organoid growth. For example, rapid growth of dividing cells (ISCs, transit-amplifying cells) may activate CFTR function to slow proliferation, but the consequent organoid swelling may exert negative influences on CFTR activity. Given the squamous morphology of cells constituting a mature, enlarged spheroid, the lack of response to forskolin may include alterations in CFTR or other transporters as a consequence of backpressure and cell stretching (Vitzthum et al. 2015). Further research in this area is needed because it is important to understand the underlying ion transport physiology of Wnt-stimulated organoids because of their utility for drug screening and, in the case of enterospheres, as an index of ISC function. However, it should be borne in mind that these studies involved *acute* pharmacological manipulation of organoids. Longer-term studies using SLA or FIS may be more difficult. Videomicroscopic recordings of enterospheres over a 24-h. period found that distended WT enterospheres undergo several episodes in which the epithelium ruptures, the organoid reduces in volume, the epithelium reseals, and the organoid swells again. Thus, the quantitative aspect of swelling assays may be confounded in longitudinal drug studies.

1.5 Advantages of Studies of Ion Transport in 3D Intestinal Organoids: A Summary

1. Organotypic morphology: One of the most exciting features of enteroid culture is the organotypic features of a central cavity from which crypts project, i.e., “mini-gut.” At the cell level is the formation of a polarized, well-differentiated intestinal epithelium with approximately the same percentage of cells of different lineages as found in vivo (Sato et al. 2009). Unlike cell lines, 3D organoids provide the opportunity to study the electrophysiology and ion transport capabilities of

intestinal epithelium in situ, in particular, the crypts. The verisimilar nature of the 3D organoids is highlighted by the capability for functional, long-term engraftment of intestinal organoids to sites of damaged colonic epithelium in a mouse model (Yui et al. 2012). This raises the possibility for autologous intestinal engraftment by in vivo differentiated HIOs produced from nondiseased tissues of a patient.

2. Visual access: A remarkable aspect of 3D intestinal cultures is the visual access to crypt function. The opportunity to view, by either light or fluorescence microscopy, the dynamics of a regenerating crypt epithelium in axial or coronal planes advances the ability to investigate ion transport properties in real time. As mentioned previously, it enables studies of ISCs for intracellular pH and ion concentrations using fluorescent dyes, provides electrophysiological access for microelectrode impalements or patch clamping of the basolateral membrane, and allows direct visualization of crypt and ISC cell volume regulation. Studies of ion transport physiology of other differentiated cell types are possible as shown by measurements of goblet cell pH_i, mucin granule pH, and the dynamics of degranulation in situ (Liu et al. 2015).
3. Isolation from the intestinal milieu: Intestinal organoids are growing epithelial structures removed for weeks or even months (in passage) from the intestinal environment. The advantage is that removal from neural, immune, humoral, and microbial influences enables the investigation of ion transport processes that are intrinsic to the epithelium and not secondary to other effects of disease, such as dysbiosis. We have found that some indicators of inflammation in the CF mouse intestine, e.g., Toll-Like receptors, and goblet cell hyperplasia are normalized by the first passage in enteroid culture (unpublished observations). This aspect alone raises the opportunity to assess epigenetic changes in ion transport physiology in disease states by following changes with time in culture.

1.6 Disadvantages for Studies of Ion Transport in 3D Intestinal Organoids

1. Access to the brush-border membrane: Given the vectorial nature of ion transport processes in a polarized epithelium, access to the brush-border membrane in intact organoids presents limitations to the efficiency and types of experiments. The containment of the brush-border membrane within the organoid prevents direct contact with micropipettes without disrupting the epithelial structure for access. Compounding this limitation is the presence of cell debris and even mucus (depending on the treatment protocol) within the organoid's lumen. Intestinal organoids are elastic and distensible, so it is possible to inject (Engevik et al. 2013) and even superfuse (Liu et al. 2015) the organoid lumen with compounds, isotopes, or different solutions to enable investigation of ion transport processes. However, these studies are time consuming and technically demanding. Bisection of organoids provides a means to allow access to the apical membrane for ion

transport probes or solutions, but with the limitation that both apical and basolateral membranes of the epithelium are exposed to the same milieu (Liu et al. 2015).

2. Limitations of the gel: The virtues of 3D culture of intestinal organoids within a gel (e.g., Matrigel®) include a laminin basement membrane that avoids anoikis and a gel structure enabling cell migration to self-organize into a 3D organoid. However, the gel provides limitations with regard to diffusibility of compounds or reagents, especially those of a hydrophobic nature. This limitation leads to heuristic efforts for finding optimal responses usually of an empirical nature such as dose-response or time-course studies for comparisons with published values from native tissue or cell line experiments. The same is also true about the washout of compounds/reagents from the gel in these studies. Though practical solutions are possible, the additional time required to develop an experiment presents an inefficiency. With regard to some classic approaches to ion transport studies, a qualitative rather than a quantitative outcome might also be a consequence of an experiment with 3D organoids. For example, a study of the basolateral Cl^-/HCO_3^- exchanger Ae2 in mouse intestinal organoids found an exchange rate that was lower than that established from studies of cell lines. The inconsistency was ascribed to the slow turnover of solutions within the gel during switches from Cl^--containing to Cl^--free solutions employed to drive the exchange process (Walker et al. 2016). In addition to the problem of drug/compound access to the organoid and their washout, the structure of the gel itself can be problematic in some treatments. For example, in the aforementioned study, the duration of the experiment was limited because the Matrigel® tended to dissolve and loosen in the culture dish during superfusion with the Cl^--free solution. Likewise, our laboratory has found that solutions with high K^+ concentrations to have a similar destabilizing effect on the gel.

Together, these limitations to ion transport studies in the 3D gel cultures of intestinal organoids have driven efforts to establish 2D monolayers of primary intestinal epithelium, i.e., 2D organoids.

1.7 Ion Transport in 2D Intestinal Organoids: Converting 3D into 2D Intestinal Organoids

Following their amplification as spheroids by multiple passaging, stem cell–derived 3D enteroids and colonoids from animal (mouse, pig) or human origin embedded in Matrigel droplets can be readily converted into confluent monolayers of polarized epithelial cells. Removal of the Matrigel is followed by trypsinization, mechanical disruption, straining through a 40-μm filter, and plating of the near-single cell fragments in expansion medium on a semipermeable substrate (Transwell or Snapwell filter inserts, Corning). Precoating of the inserts has been performed with a fibronectin-based peptide (Foulke-Abel et al. 2014), human collagen IV (In et al.

2016; Yin et al. 2018; Fernando et al. 2017; van der Hee et al. 2018), 1–2.5 vol% Matrigel (Wang et al. 2019; VanDussen et al. 2015), PureCol (Advanced Biomatrix) (Zomer-van Ommen et al. 2018), or Matrigel covered by a feeder layer of irradiated 3T3 fibroblasts (Wilke et al. 2019). In most protocols, expansion medium is supplemented with the ROCK inhibitor Y-27632 (to prevent anoikis) and the GSK3 inhibitor CHIR99021 (to promote Wnt/β-catenin signaling) during the first 2 days after seeding. Once monolayers become confluent, as judged by their morphology and transepithelial electrical resistance (TEER) requiring 7–14 days, they can be maintained in expansion medium for another 4–6 days, resulting in stem cell-enriched, undifferentiated 2D organoid monolayers. Of note, the cell composition of these undifferentiated monolayers differs from the native intestinal crypts or 3D enteroids in that secretory cell types, including mucus-secreting goblet cells and enteroendocrine cells (EEC), are underrepresented or completely missing.

However, 2D intestinal organoids can be enriched for absorptive and secretory cell types by growing them in differentiation medium made by withdrawing niche factors, in particular Wnt3a, the p38 MAPK inhibitor SB202190, and, occasionally, nicotinamide or R-spondin1 from the expansion medium (Yin et al. 2018; Zomer-van Ommen et al. 2018), or by replacing Wnt3a by the Wnt pathway inhibitor IWP-2 (Yin et al. 2014). During differentiation, the TEER value increases from ~200 to ~1400 Ohm.cm^2, i.e., appr. seven-fold (In et al. 2016). A further enrichment for goblet cells can be achieved by adding the Notch inhibitor DAPT (Yin et al. 2014) or by growing the monolayer under air-liquid interface (ALI) conditions, creating a thick mucus hydrogel layer. As shown in several recent studies, this condition offered a suitable platform for the efficient growth of *E. coli* microbes and *Cryptosporidium* parasites (Wang et al. 2019; Wilke et al. 2019; Heo et al. 2018). Alternatively, differentiation of stem cells into enteroendocrine cells (EEC), including GLP1-producing L-cells, can be reached by blockage of both the Notch and the EGFR/MAPK signaling pathway (Basak et al. 2017) or by inhibiting RhoA/ROCK/YAP signaling using the ROCK inhibitor Y-27632 (Petersen et al. 2018). It should be noted, however, that the latter studies have been carried out with murine enteroids only and that confirmation of these results for human organoids needs further experimentation.

1.8 Advantages and Limitations of 2D Organoids

Two-dimensional monolayers of enteroids and colonoids, in contrast to 3D organoids [oriented either basal-out or apical-out; cf. Co et al. (2019)], are freely accessible at both the basolateral (serosal) side and the apical (luminal) side, therefore bypassing the need for microinjection of test compounds or microbes. They greatly expand the applicability of intestinal organoids as ex vivo models in studies of epithelial permeability, mucus production and secretion, electrolyte and fluid transport, and interaction of epithelial cells with microbes, parasites, and microbial enterotoxins. Organoid monolayers enriched for specific cell types (e.g., goblet cells; EEC; tuft cells) can be used for studying apical vs. basolateral secretion of their content in response to secretory stimuli, including those of pathogens at an early stage of infection. Moreover, unlike transformed human intestinal cell lines (T84, HT29, CaCo2), they allow the study of segmental differences and spatial differences along the crypt-villus axis at both a macroscopic and microscopic level (Middendorp et al. 2014; Dutton et al. 2019).

Another advantage in comparison with most cell lines is the ability of goblet cell–enriched organoid monolayers to build up a mucus layer at their luminal side, which mimics the barrier function of native epithelium (Wang et al. 2019). By comparing CFTR-mediated anion currents across monolayers of undifferentiated, mucus-free organoids with mucus-enriched organoids, the impact of the mucus barrier on the efficacy of luminally added compounds, including CFTR inhibitors (relevant in diarrheal disease) and CFTR activators (relevant for CF), can be assessed. Moreover, approaches that facilitate transport of drugs across the mucus layer, such as alginates or nanoparticles, can be evaluated (Ermund et al. 2017; Davoudi et al. 2018).

Obvious shortcomings of the 2D minigut are the absence of a 3D crypt-villus architecture and consequently its inability to recapitulate in vivo cell migration, cell extrusion [a favorable infection site; cf. Co et al. (2019)], and hormone or growth factor gradients along the crypt-villus axis. Other limitations and pitfalls are the potential entrapment of compounds by the filter; the low throughput of transport assays in comparison with the FIS assay performed in 96-or 384-well plates; the lack of a mesenchyme, myofibroblasts, immune cells, endothelial cells, smooth muscle cells and enteric nerves; and the lack of peristalsis, luminal flow, and blood circulation (In et al. 2016). One approach to compensate for these limitations is to coculture 2D epithelial monolayers with such key cell types and to mimic peristalsis and flow in a microengineered gut-on-a-chip device (Kim and Ingber 2013; Kim et al. 2016, 2017; Bein et al. 2018).

1.9 Ion Transport Studies in Undifferentiated 2D Human Intestinal Organoids

As is apparent from their transcriptome profile and histological analysis, human rectal organoids grown in the presence of Wnt3a and other growth factors, similar to 2D human duodenal enteroids (Yin et al. 2018), are nearly devoid of mucin 2 (MUC2)-secreting goblet cells and express very low levels of the apical Na^+/H^+ exchanger NHE3 and the Cl^-/HCO_3^- exchanger DRA; in contrast, they are highly enriched in the stem cell marker Lgr5 and in ion transporters and channels involved in transepithelial secretion of chloride, i.e., NKCC1 and CFTR (Table 1.1; cf. Fig. 1.6b). Monolayers of colonoids are therefore excellently suited to study electrogenic anion secretion through short-circuit current (I_{sc}) measurements in conventional Ussing chambers or equivalent short-circuit current (I_{eq}) measurements in a 24-well MTECC system (EP-devices) (Zomer-van Ommen et al. 2018). They

Table 1.1 RNA-Seq of ion channels, transporters, and transport regulators expressed in undifferentiated, differentiated, and goblet cell–enriched human colonoids

	Transcript levels in human colonoids (RPKM)		
Gene	Undifferentiated	Differentiated	Goblet cell–enriched
LGR5	10.6	0.3	0.2
MUC2	4.4	44.2	136.1
CFTR	15.7	1.2	3.1
NKCC1/SLC12A2	125.2	8.8	14.6
NBC1/SLC4A4	1.6	3.9	3.4
NHE3/SLC9A3	2.1	8.2	10.1
DRA/SLC26A3	2.1	8.2	10.1
ANO1	0.2	ND	0.1
BEST2	ND	ND	ND
WNK1	16.2	5.5	6.7

Colonoids generated from rectal suction biopsies of healthy volunteers and grown in Matrigel droplets were cultured for 7–9 days in WENR medium (containing Wnt3a, EGF, Noggin, and R-spondin1, promoting the undifferentiated state), ENR medium (containing the WNT inhibitor IWP-2, resulting in differentiation), or ENR + IWP-2 + DAPT medium [promoting goblet cell-enrichment (Yin et al. 2014)]. RNA extraction and transcriptome sequencing was performed as described in detail elsewhere (Ikpa et al. 2020). Data depict reads per 1000 base pair transcript per million reads mapped (RPKM). *ND* not detected

can also be used for the screening of novel antidiarrheals, including CFTR inhibitors and guanylyl cyclase C (GUCY2C) inhibitors (Bijvelds et al. 2015). Moreover, 2D colonoids generated from rectal biopsies of CF individuals can be used to investigate the residual function of CFTR mutants and their response to CFTR modulators with more quantitative electrophysiological measurements (Zomer-van Ommen et al. 2018).

In comparison with intestinal short-circuit current measurements (I_{sc}) in freshly excised, near-native human rectal biopsies, 2D colonoids offer several major advantages: (1) their availability, through biobanking and expansion of 3D organoids, in almost unlimited amounts; (2) CFTR current responses to cAMP-linked agonists (e.g., forskolin, IBMX) are two- to threefold larger than in the biopsies of the same individual despite the flat structure of the monolayer, which comprised a much smaller number of cells per cm^2 (Fig. 1.8a vs. b). This difference is due, at least in part, to the enrichment of stem cells in the colonoids, which express high levels of CFTR (Strubberg et al. 2018), and to the finding that CFTR currents in biopsies, in contrast to colonoids, are no longer proportional to the amount of CFTR above ~20–30% of WT levels, as based on Isc and CFTR band C quantification in biopsies of CFTR splice mutants (De Boeck et al. 2014); (3) the near absence of goblet cells and mucus in the undifferentiated colonoids. The absence of a fully formed mucus layer may facilitate uptake of hydrophobic compounds, e.g., CFTR inhibitors and modulators, in cells from the luminal side. This, together with the lack of convectional washout of CFTR-targeted drugs from 3D crypts, may explain why CFTR inhibitors are highly effective in colonoids but poorly effective in rectal biopsies and

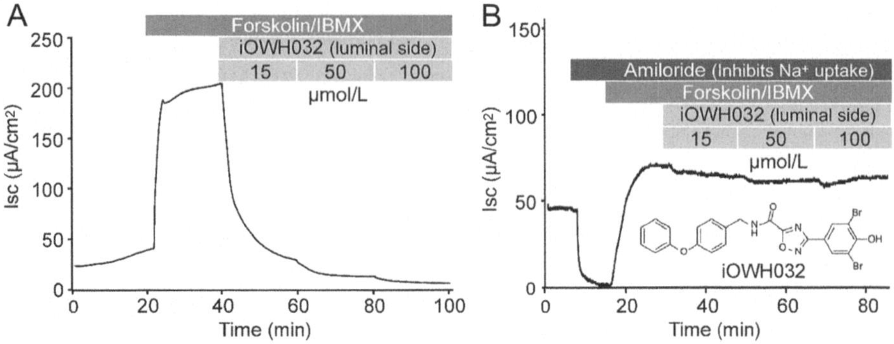

Fig. 1.8 Comparison of CFTR-mediated anion secretory currents in 2D human rectal colonoids (**a**) vs. human rectal biopsies (**b**). (**a**) Undifferentiated colonoid monolayers derived from human rectal biopsies were grown on Transwell filters and subsequently mounted in Ussing chambers. Short-circuit currents (I_{sc}), representing electrogenic ion transport, were assessed as described in detail elsewhere (Bijvelds et al. 2009). (**b**) Rectal biopsy specimens were obtained from healthy volunteers. Biopsies were mounted in Ussing chambers and the I_{sc} was measured as described for panel (**a**). Amiloride (10 µM) and the CFTR inhibitor iOWH032 were added only to the luminal bath; forskolin (10 µM) and IBMX (100 µM) were added to both the luminal and basolateral bath

in intact epithelium [Fig. 1.8a vs. b; cf. Thiagarajah and Verkman (2013)]. Furthermore, the absence of goblet cells in colonoids readily explains why electrogenic K^+ secretion in response to Ca^{2+}- or cAMP-linked secretagogues, a property attributed exclusively to goblet cells in distal colonic crypts (Linley et al. 2014), is very prominent in rectal biopsies of CF individuals but not manifest in 2D colonoids generated from these biopsies (Fig. 1.9a, b). The absence of active K^+-driven fluid secretion also explains why 3D colonoids expressing CFTR-null mutants do not swell in response to cAMP- or Ca^{2+}-linked secretagogues (Dekkers et al. 2016); (4) they can be used for prolonged ex vivo experiments, viz. for testing slow-acting small-molecule compounds that improve the folding of de novo synthesized mutant CFTR. Because optimal restoration of mutant CFTR activity by folding correctors typically requires >8 h incubation, such assays cannot be performed in intestinal biopsies, as these remain fully viable only for a limited period, typically <6 h.

1.10 CFTR Is the Major If Not Sole Apical Anion Channel Accounting for cAMP- and Ca^{2+}-Activated Electrogenic Anion Secretion in 2D Colonoids

In 2D, human colonoids both Cl^- and HCO_3^- secretory currents induced by the cAMP agonist forskolin are almost fully blocked (>90%) by the CFTR inhibitors iOWH-32 and $CFTR_{inh}$-172 (Fig. 1.10a–c). In accordance, these secretory currents are virtually absent from CF colonoids derived from patients with CFTR mutations

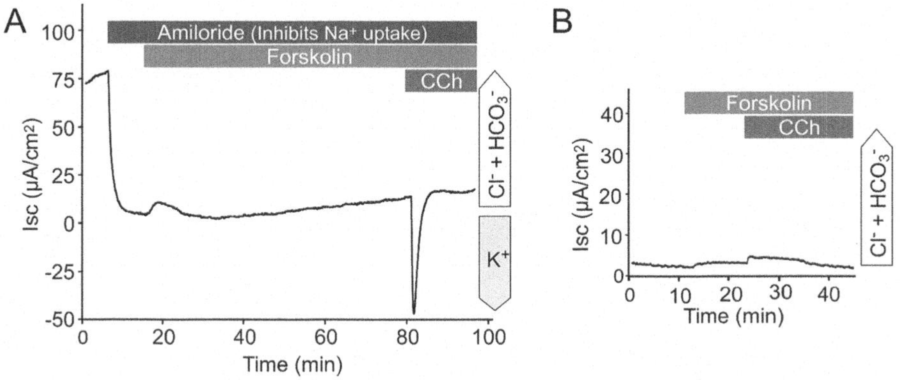

Fig. 1.9 Carbachol-activated K^+ secretion is prominent in CF rectal biopsies but absent in 2D CF colonoids. (**a**) Rectal biopsy specimens, obtained from CF patients (homozygous F508del), were mounted in Ussing chambers. Short-circuit currents (I_{sc}), representing electrogenic ion transport, were assessed as described in detail elsewhere (Bijvelds et al. 2009). (**b**) Undifferentiated colonoid monolayers derived from CF rectal biopsies (homozygous F508del) were grown on Transwell filters and subsequently mounted in Ussing chambers. The Isc was measured as described for panel (**a**). Amiloride (10 µM) was added only to the luminal bath; Carbachol (CCh; 100 µM) was added only to the basolateral bath; forskolin (10 µM) was added to both the luminal and basolateral bath

that severely impair ion transport function and/or protein maturation and membrane trafficking (e.g., homozygous F508del; Fig. 1.9b).

Addition of a Ca^{2+}-mobilizing secretagogue (carbachol; UTP) after CFTR channel blockade did not provoke anion secretion in undifferentiated colonoids (Fig. 1.11a) or enteroids (not shown), and triggered only a very small anion current in differentiated and goblet cell–enriched colonoids, arguing against a substantial contribution of apical Ca^{2+}-activated anion channels (CaCCs) to this process (Fig. 1.11a). In contrast, the application of UTP to 2D human biliary duct organoids clearly demonstrated the contribution of a Ca^{2+}-activated apical anion channel to Cl^- and HCO_3^- secretory currents in this model (Fig. 1.11b). Of note, the high remnant Ca^{2+}-dependent anion secretion previously observed in the colon of Cftr KO mice, thought to reflect bestrophin-2 (Best2) channel-mediated bicarbonate secretion from goblet cells (Yu et al. 2010), is not found in goblet cell–enriched human colonoids, in line with the virtual absence of *BEST2* transcripts (Table 1.1). Similar studies in mouse colonoids may reveal whether this finding reflects an interspecies difference or a limitation of the organoid model.

The outcome of these experiments on 2D human enteroids and colonoids is in line with a recent study showing that, whereas the CaCC TMEM16A/ANO1 is expressed in the mouse intestine, it is not directly involved in transepithelial anion secretion or required for mucus homeostasis (Vega et al. 2019). This study also suggests that the large contribution of CaCCs to Ca^{2+}-induced anion secretion reported previously for human colorectal tumor cells (T84; HT-29cl.19A) results from an upregulation of CaCCs during intestinal tumorigenesis (Bajnath et al. 1992; Namkung et al. 2011;

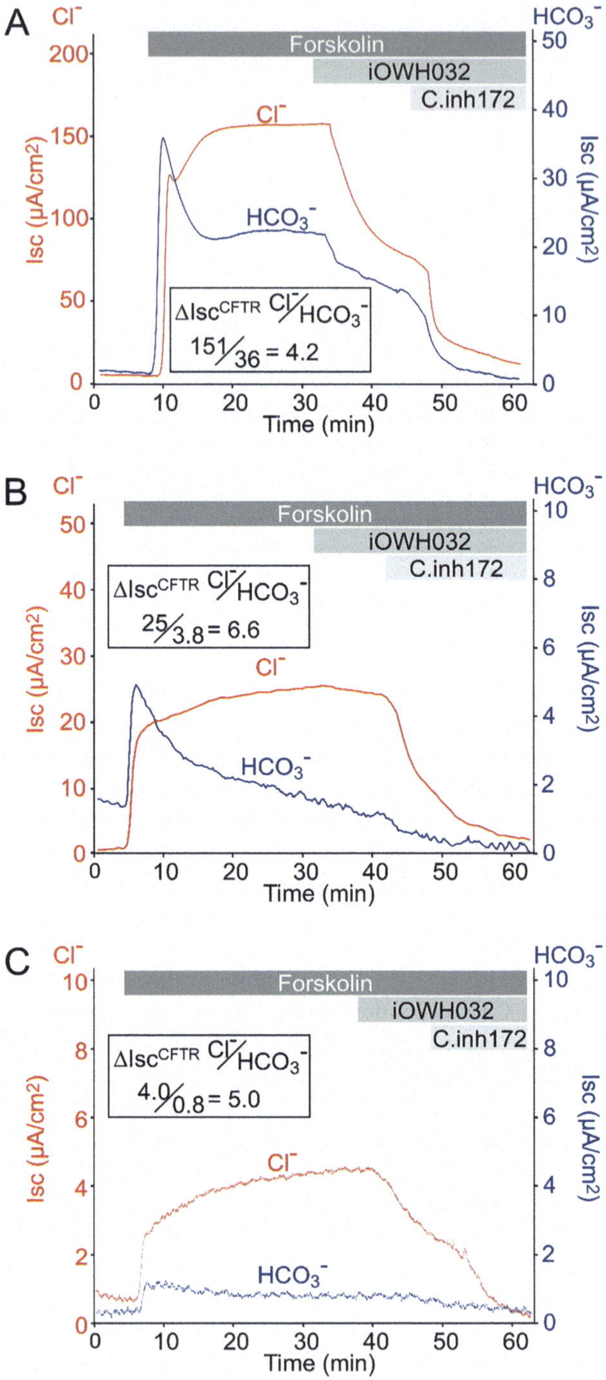

Fig. 1.10 CFTR-mediated Cl^- and HCO_3^- current responses to forskolin/cAMP in 2D human rectal organoids at various differentiation stages. (**a**) Undifferentiated; (**b**) Differentiated; (**c**) Goblet cell enriched, all derived from a healthy donor. Electrogenic Cl^- secretion was measured in a HEPES-buffered, HCO_3^--free (replaced by isethionate) isotonic solution. HCO_3^- secretion was measured in a Cl^--free (replaced by isethionate) isotonic solution. $CFTR_{inh}$-172 (C.inh172; 20 μM) and iOWH032 (30 μM) were added only to the luminal bath; forskolin (10 μM) was added to both the luminal and basolateral bath

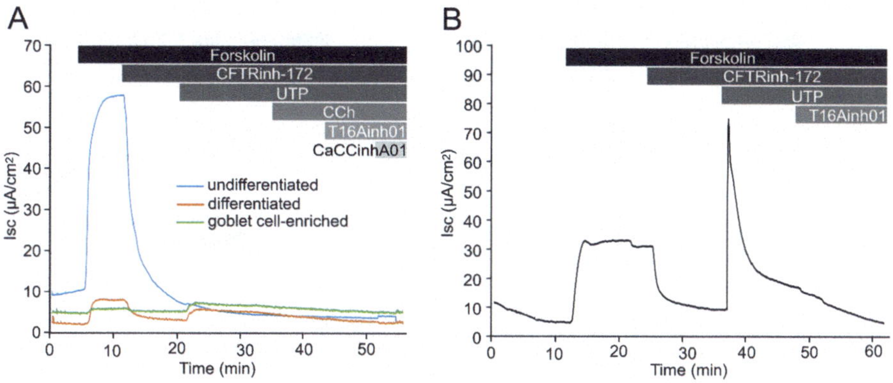

Fig. 1.11 Regulation of anion secretion in 2D cultured organoids. (**a**) Anion secretory currents mediated by cAMP- (forskolin) and Ca^{2+}- (carbachol, UTP) agonists in 2D human undifferentiated, differentiated, or goblet cell–enriched colonoids. (**b**) Anion secretory currents in human bile duct organoids. UTP (50 µM), $CFTR_{inh}$-172 (20 µM), TMEM16A inhibiter A01 ($T16A_{inh}01$; 25 µM), and CaCC inhibitor A01($CaCC_{inh}A01$, 25 µM) were added to the luminal bath only; forskolin (10 µM) was added to both the luminal and basolateral bath; carbachol (CCh; 100 µM) was added to the basolateral bath only

Sui et al. 2014). This view conflicts with other recent publications, arguing that, although ANO1 does not function as an intestinal apical chloride channel, it nevertheless is required for cAMP-dependent colonic chloride and fluid secretion, and that ANO1 blockers or activators can potentially be used to treat enterotoxin-related diarrhea or CF-related intestinal disease, respectively (Benedetto et al. 2017; Lee et al. 2019; Kunzelmann et al. 2019). Functional experiments in human intestinal organoids, using CRISPR-Cas9-based silencing of the ANO1 gene, may help to resolve this controversy. So far, the results of the organoid studies support the earlier notion that Ca^{2+} agonists act mainly, if not exclusively by activating the basolateral K^+ channel SK4 (Matos et al. 2007; Flores et al. 2007). The resulting membrane hyperpolarization enhances the electrical driving force for anion exit through apical CFTR channels (cf. Fig. 1.12; Greger 2000). Therefore, in contrast to bile ducts, human intestine cannot compensate functionally for defective CFTR channels by exploiting alternative channels, in line with the early onset and severity of CF intestinal disease (e.g., meconium ileus).

1.11 2D Organoids Allow the Separate Measurements of CFTR-Mediated Cl^- and HCO_3^- Secretion in Colonocytes at Different Stages of Maturation

To mimic physiological conditions, in short-circuit current measurements of intestinal biopsies or 2D organoids, the mucosal and serosal baths contain Cl^- and HCO_3^-, which both contribute to the anion secretory current. A simple approach

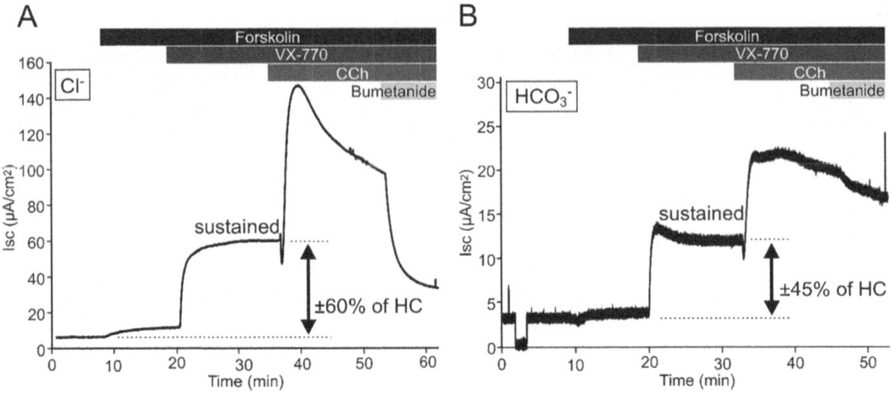

Fig. 1.12 VX-770 rescue of anion secretion in 2D human CF colonoids. Rescue of CFTR-mediated, forskolin (10 µM)-dependent Cl^- (panel A) and HCO_3^- (panel B) secretion in 2D human CF colonoids carrying the S1251N gating mutation by VX-770 (3 µM). The Ca^{2+} agonist carbachol (CCh; 100 µM) further stimulated the CFTR-mediated current. Bumetanide (50 µM) inhibited the Cl^--, but not the HCO_3^--current. The regain of the forskolin-activated Cl^- and HCO_3^- current in the S1251N organoids reached a value of 60% and 45%, respectively, of the current measured in non-CF organoids (HC) in the absence of CFTR modulators. See legend Fig. 1.10 for composition of bath solutions

to separate both currents is to perform I_{sc} measurement using a Cl^-- or HCO_3^- -free bath composition (Yin et al. 2018). Furthermore, the depletion of Cl^- in the luminal bath prevents the operation of the apical Cl^-/HCO_3^- exchanger SLC26A3 and thereby facilitates HCO_3^- exit through the CFTR channel. To perform this approach in 3D organoids is more cumbersome, because in this model, the ion composition of the luminal fluid is difficult to manipulate experimentally.

As shown in Fig. 1.10a–c, both the forskolin-activated, CFTR-mediated Cl^- current and HCO_3^- current decreased dramatically during differentiation and even further following DAPT treatment, in parallel with the drop in CFTR gene expression (Table 1.1). Remarkably, differentiation did not notably change the HCO_3^-/Cl^- current ratio (~0.2). This finding supports the notion that, in 2D colonoids, CFTR rather than basolateral Cl^- and HCO_3^- import through NKCC1 and NBCe1, respectively, is rate limiting for the current (cf. Table 1.1). Previous studies on ductal cells from the pancreas suggest that the permselectivity of CFTR for Cl^- and HCO_3^- ions is not invariable but dynamically regulated by WNK1, a protein kinase capable of binding to the N-terminus of CFTR at low intracellular Cl^- and shifting the HCO_3^-/Cl^- permeability ratio of the channel ~fivefold, i.e., from ~0.2 to 1.0 (Kim et al. 2020). Assuming that this mode of regulation is also operating in cultured intestinal colonoids (in which WNK1 is highly expressed; see Table 1.1), it appears that WNK1 activation does not notably enhance HCO_3^- secretion in this model. We speculate that the HCO_3^-/Cl^- current ratio measured under these conditions principally reflects a difference in the electrochemical driving force for extrusion of these ions across the apical membrane (i.e., $Cl^- >> HCO_3^-$). However, after selective permeabilization of the basolateral membrane by pore-forming peptides (e.g. nystatin), apical ion gradients can be experimentally manipulated. Under such conditions, the application of Cl^- or HCO_3^- concentration gradients across the monolayer allows for direct assessment of the HCO_3^-/Cl^- permeability ratio (Bijvelds et al. 2009). Such macroscopic current measurements may supplant measurements of the HCO_3^-/Cl^- permeability ratio of CFTR at a single cell level by the whole-cell patch-clamp technique and may reveal whether CFTR mutations or CFTR modulators, by affecting the conformation or accessibility of the channel pore, not only change the activity of CFTR but also modulate its permselectivity for anions (Ferrera et al. 2019).

1.12 Contribution of CFTR-High Expressor Cells to Ion Transport in Intestinal Organoids

A small subpopulation (~2.5%) of enterocyte-like epithelial cells that express extremely high levels of CFTR in their apical membrane named CFTR High Expressor (CHE) cells have been identified in duodenum and jejunum of human and rat, but not mouse intestine (Jakab et al. 2013). These cells resemble so-called pulmonary ionocytes identified more recently by single cell transcriptomics in human and mouse airways that have been claimed to function as a major source of CFTR activity in airway epithelium, despite representing only 1–2% of the epithelial cell population (Plasschaert et al. 2018). In both cases, the function of CFTR has not yet been finally resolved, awaiting new approaches to enrich cell preparations or organoids for these rare cell types. So far however the occurrence of CHE cells in human duodenal or jejunal enteroids has not been reported, hampering the use of organoids as a new and permanent source for studies of this intriguing cell type. It should be pointed out however that the level of CFTR-mediated anion secretory current measured in enteroids generated from the proximal (CHE-containing) and distal (CHE-less) small intestine is very similar, suggesting that CHE cells do not contribute much to transepithelial anion transport in the human intestine but are likely to serve other functions.

1.13 Testing of CFTR Correctors and Potentiators in 2D Colonoids and Enteroids Generated from Intestinal Biopsies of CF Patients

Two-dimensional intestinal organoids are not only suitable as an in vitro model for measuring the impact of CFTR mutations on intestinal Cl^- and HCO_3^- secretion, but, similar to the 3D organoids discussed earlier, can be exploited too, albeit at a much lower throughput, for theratyping CFTR mutants, in particular compound

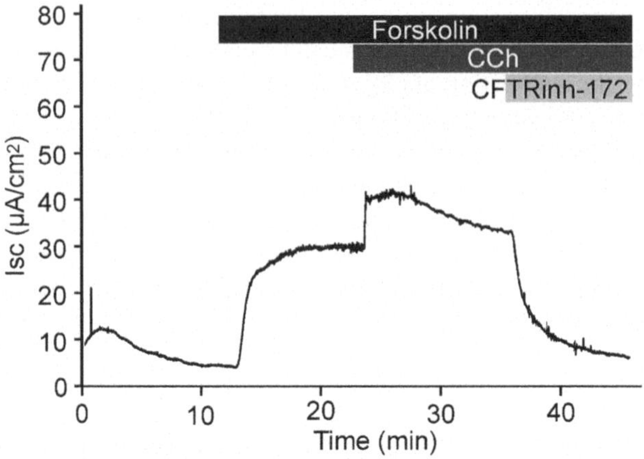

Fig. 1.13 TRIKAFTA rescue of anion secretion in 2D human CF colonoids. Rescue of CFTR-mediated anion secretion in 2D human colonoids (homozygous F508del; undifferentiated) by the triple combination of 2 Vertex correctors (VX-661 + VX-445) *plus* the Vertex potentiator (VX-770), named TRIKAFTA. The colonoid monolayers were preincubated for 20 h in the presence of TRIKAFTA or vehicle (0.1% DMSO). $CFTR_{inh}$-172 (25 µM) was added to the luminal bath only; forskolin (10 µM) was added to both the luminal and basolateral bath; carbachol (CCh; 100 µM) was added to the basolateral bath only

heterozygotes, using existing and candidate CFTR modulator drugs. A typical example is illustrated in Fig. 1.12a, b. It shows that the CFTR gating mutant S1251N can regain up to 60% of the Cl^- and 45% of the HCO_3^- secretory activity of healthy control-derived colonoids following addition of the potentiator VX-770/Ivacaftor (Vertex Pharmaceuticals). As shown in Fig. 1.13, a similar efficacy of CFTR rescue of the most common CFTR mutation F508del, suffering from multiple defects, including impaired protein folding, trafficking, and gating, required a combination of VX-770 with two folding correctors, VX-445/Elexacaftor and VX-661/Tezacaftor, named TRIKAFTA® (Vertex Pharmaceuticals). A high efficacy of CFTR rescue was also observed in 2D ileal organoids from homozygous F508del CF patients (not shown). Both CFTR modulator treatments are FDA approved and have shown to be highly effective in the clinic (Middleton et al. 2019; Heijerman et al. 2019), underscoring the predictive value of the in vitro functional tests in 2D organoids.

1.14 Concluding Remarks

Up until the advent of the organoid technology developed after 2000, intestinal epithelial cells were notoriously difficult to grow in vitro, and immortalized or cancer cell lines were the only models available. As discussed in this chapter, this new technique has revolutionized the field of intestinal transport physiology, pathology, and pharmacology and has removed many of the limitations inherent to the traditional models. Human intestinal organoids are also unique in their capacity to allow personalized theratyping of drugs that could be applied in ion transport diseases such as CF and diarrheal disease. Because human viruses (e.g., rotavirus; norovirus) and microbes very often replicate poorly in transformed cell lines and animal models as a host, the availability of biobanked human organoids allows a thorough exploration of the infection mechanism and repetitive preclinical testing of novel antiviral or antimicrobial drugs (Saxena et al. 2016; Yin et al. 2015, 2016; Estes et al. 2019). Furthermore, because intestinal organoids have now been generated successfully from many different animal species, they facilitate a cross-species comparison of ion transport properties in well-defined intestinal segments and may provide a better insight into the evolution of intestinal ion transporters and their regulation (Schwarz et al. 2015).

Although the enteroids and colonoids discussed in this chapter remain reductionist models, i.e., are missing the complexity of the native tissue, this limitation can be overcome by coculturing the organoids with other cell types such as microbes, immune cells, and mesenchymal cells (Kim and Ingber 2013; In et al. 2016). Finally, peristalsis and fluid flow can be mimicked in a microfluidic gut-on-a-chip device (Kim et al. 2016; Bein et al. 2018; Shin et al. 2019). By integrating multiple electrode pairs in the chip, impedance measurements can be carried out to measure barrier function and differentiation status of the 2D organoids (van der Helm et al. 2019). Such transepithelial impedance measurements could also be exploited in future studies to analyze anion secretion, as shown previously for monolayers of Calu-3 airway cells (Tamada et al. 2001). This and other new developments illustrate that we stand only at the beginning of an era in which organoid technology will further expand and miniaturize, resulting in an unprecedented enrichment of knowledge of intestinal ion transport based on the analysis of near-physiological model systems.

Acknowledgments The authors would like to thank Drs. Rowena Woode and Sarah Young, University of Missouri, for thoughtful review and comments on the manuscript. Supported by grants NIH R01DK048816 (LLC); Cystic Fibrosis Foundation grants CLARKE16P0, CLARKE17G0, CLARKE19XX0, DEJONG19GO, CF Foundation Therapeutics (HRdJ) and SRC011, UK-CF Trust (HRdJ).

References

Accurso FJ, Rowe SM, Clancy JP, Boyle MP, Dunitz JM, Durie PR, Sagel SD, Hornick DB, Konstan MW, Donaldson SH, Moss RB, Pilewski JM, Rubenstein RC, Uluer AZ, Aitken ML, Freedman SD, Rose LM, Mayer-Hambelett N, Dong Q, Zha J, Stone AJ, Olson ER, Ordonez CL, Campbell PW, Ashlock MA, Ramsey BW (2010) Effect of VX-770 in persons with cystic fibrosis and the G551D-CFTR mutation. N Engl J Med 343:1991–2003

Ainsworth MA, Amelsberg M, Hogan DL, Isenberg JI (1996) Acid-base transport in isolated rabbit duodenal villus and crypt cells. Scand J Gastroenterol 31:1069–1077

Avula LR, Chen T, Kovbasnjuk O, Donowitz M (2018) Both NHERF3 and NHERF2 are necessary for multiple aspects of acute regulation of NHE3 by elevated Ca2+, cGMP, and lysophosphatidic acid. Am J Physiol 314:G81–G90

Bajnath RB, Dekker K, Vaandrager AB, de Jonge HR, Groot JA (1992) Biphasic increase of apical Cl- conductance by muscarinic stimulation of HT-29cl.19A human colon carcinoma cell line: evidence for activation of different Cl- conductances by carbachol and forskolin. J Membr Biol 127:81–94

Barker N, Van Es JH, Kuipers J, Kujala P, Van der Born M, Cozijnsen M, Haegebarth A, Korving J, Begthel H, Peters PJ, Clevers H (2007) Identification of stem cells in small intestine and colon by marker gene Lgr5. Nature 449:1003–1008

Barrett KE, Keely SJ (2000) Chloride secretion by the intestinal epithelium: molecular basis and regulatory aspects. Annu Rev Physiol 62:535–572

Barriere H, Poujeol C, Tauc M, Blasi JM, Counillon L, Poujeol P (2001) CFTR modulates programmed cell death by decreasing intracellular pH in Chinese hamster lung fibroblasts. Am J Physiol 281:C810–C824

Basak O, Beumer J, Wiebrands K, Seno H, van Oudenaarden A, Clevers H (2017) Induced quiescence of Lgr5+ stem cells in intestinal organoids enables differentiation of hormone-producing enteroendocrine cells. Cell Stem Cell 20:177–190

Bein A, Shin W, Jalili-Firoozinezhad S, Park MH, Sontheimer-Phelps A, Tovaglieri A, Chalkiadaki A, Kim HJ, Ingber DE (2018) Microfluidic organ-on-a-chip models of human intestine. Cell Mol Gastroenterol Hepatol 5:659–668

Benedetto R, Ousingsawat J, Wanitchakool P, Zhang Y, Holtzman MJ, Amaral M, Rock JR, Schreiber R, Kunzelmann K (2017) Epithelial chloride transport by CFTR requires TMEM16A. Sci Rep 7:12397

Bertrand CA, Mitra S, Mishra SK, Wang X, Zhao Y, Pilewski JM, Madden DR, Frizzell RA (2017) The CFTR trafficking mutation F508del inhibits the constitutive activity of SLC26A9. Am J Physiol 312:L912–L925

Bijvelds MJC, Bot AG, Escher JC, de Jonge HR (2009) Activation of intestinal Cl- secretion by lubiprostone requires the cystic fibrosis transmembrane conductance regulator. Gastroenterology 137:976–985

Bijvelds MJC, Loos M, Bronsveld I, Hellemans A, Bongartz JP, Ver Donck L, Cox E, de Jonge HR, Schuurkes JA, De Maeyer JH (2015) Inhibition of heat-stable toxin-induced intestinal salt and water secretion by a novel class of guanylyl cyclase C inhibitors. J Infect Dis 212:1806–1815

Boyle MP, Bell SC, Konstan MW, McColley SA, Rowe SM, Rietschel E, Huang X, Waltz D, Patel NR, Rodman D (2014) A CFTR corrector (lumacaftor) and a CFTR potentiator (ivacaftor) for treatment of patients with cystic fibrosis who have a phe508del CFTR mutation: a phase 2 randomised controlled trial. Lancet Respir Med 2:527–538

Chen JH, Cai Z, Sheppard DN (2009) Direct sensing of intracellular pH by the cystic fibrosis transmembrane conductance regulator (CFTR) Cl - channel. J Biol Chem 284:35495–35506

Cheng H, Leblond CP (1974) Origin, differentiation and renewal of the four main epithelial cell types in the mouse small intestine. V. Unitarian theory of the origin of the four epithelial cell types. Am J Anat 141:537–561

Co JY, Margalef-Catala M, Li X, Mah AT, Kuo CJ, Monack DM, Amieva MR (2019) Controlling epithelial polarity: a human enteroid model for host-pathogen interactions. Cell Rep 26:2509–2520 e2504

Collins FS (1992) Cystic fibrosis: molecular biology and therapeutic implications. Science 256:774–779

Davoudi Z, Peroutka-Bigus N, Bellaire B, Wannemuehler M, Barrett TA, Narasimhan B, Wang Q (2018) Intestinal organoids containing poly(lactic-co-glycolic acid) nanoparticles for the treatment of inflammatory bowel diseases. J Biomed Mater Res A 106:876–886

De Boeck K, Munck A, Walker S, Faro A, Hiatt P, Gilmartin G, Higgins M (2014) Efficacy and safety of ivacaftor in patients with cystic fibrosis and a non-G551D gating mutation. J Cyst Fibros 13:674–680

De Jonge HR (1975) The response of small intestinal villous and crypt epithelium to choleratoxin in rat and guinea pig. Evidence against a specific role of the crypt cells in choleragen-induced secretion. Biochim Biophys Acta 381:128–143

Dekkers JF, Wiegerinck CL, de Jonge HR, Bronsveld I, Janssens HM, de Winter-de Groot KM, Brandsma AM, de Jong NWM, Bijvelds MJC, Scholte BJ, Nieuwenhuis EES, van den Brink S, Clevers H, van der Ent CK, Middendorp S, Beekman JM (2013) A functional CFTR assay using primary cystic fibrosis intestinal organoids. Nat Med 19:939–945

Dekkers JF, Berkers G, Kruisselbrink E, Vonk A, de Jonge HR, Janssens HM, Bronsveld I, van de Graaf EA, Nieuwenhuis EES, Houwen RHJ, Vleggaar FP, Escher JC, de Rijke YB, Majoor CJ, Heijerman HGM, de Winter-de Groot KM, Clevers H, van der Ent CK, Beekman JM (2016) Characterizing responses to CFTR-modulating drugs using rectal organoids derived from subjects with cystic fibrosis. Sci Transl Med 8:344ra384

Donowitz M, Welsh MJ (1987) Regulation of mammalian small intestinal electrolyte secretion. In: Johnson LR (ed) Physiology of the gastrointestinal tract. Raven Press, New York, pp 1351–1388

Dorwart MR, Shcheynikov N, Yang D, Muallem S (2008) The solute carrier 26 family of proteins in epithelial ion transport. Physiology 23:104–114

Dutton JS, Hinman SS, Kim R, Wang Y, Allbritton NL (2019) Primary cell-derived intestinal models: recapitulating physiology. Trends Biotechnol 37:744–760

Eklund S, Brunsson I, Jodal M, Lundgren O (1987) Changes in cyclic 3′5′-adenosine monophosphate tissue concentration and net fluid transport in the cat's small intestine elicited by cholera toxin, arachidonic acid, vasoactive intestinal polypeptide and 5-hydroxytryptamine. Acta Physiol Scand 129:115–125

Elgavish A (1991) High intracellular pH in CFPAC: a pancreas cell line from a patient with cystic fibrosis is lowered by retrovirus-mediated CFTR gene transfer. Biochem Biophys Res Commun 180:342–348

Engevik MA, Aihara E, Montrose MH, Shull GE, Hassett DJ, Worrell RT (2013) Loss of NHE3 alters gut microbiota composition and influences bacteroides thetaiotaomicron growth. Am J Physiol 305:G697–G711

Ermund A, Recktenwald CV, Skjak-Braek G, Meiss LN, Onsoyen E, Rye PD, Dessen A, Myrset AH, Hansson GC (2017) OligoG CF-5/20 normalizes cystic fibrosis mucus by chelating calcium. Clin Exp Pharmacol Physiol 44:639–647

Estes MK, Ettayebi K, Tenge VR, Murakami K, Karandikar U, Lin SC, Ayyar BV, Cortes-Penfield NW, Haga K, Neill FH, Opekun AR, Broughman JR, Zeng XL, Blutt SE, Crawford SE, Ramani S, Graham DY, Atmar RL (2019) Human norovirus cultivation in nontransformed stem cell-derived human intestinal enteroid cultures: Success and challenges. Viruses 11:1–12

Evans GS, Flint N, Potten CS (1994) Primary cultures for studies of cell regulation and physiology in intestinal epithelium. Annu Rev Physiol 56:399–417

Fernando EH, Dicay M, Stahl M, Gordon MH, Vegso A, Baggio C, Alston L, Lopes F, Baker K, Hirota S, McKay DM, Vallance B, MacNaughton WK (2017) A simple, cost-effective method for generating murine colonic 3D enteroids and 2D monolayers for studies of primary epithelial cell function. Am J Physiol 313:G467–G475

Ferrera L, Baroni D, Moran O (2019) Lumacaftor-rescued F508del-CFTR has a modified bicarbonate permeability. J Cyst Fibros 18:602–605

Fihn BM, Sjoqvist A, Jodal M (2000) Permeability of the rat small intestinal epithelium along the villus-crypt axis: effects of glucose transport. Gastroenterology 119:1029–1036

Flores CA, Melvin JE, Figueroa CD, Sepúlveda FV (2007) Abolition of Ca2+−mediated intestinal anion secretion and increased stool dehydration in mice lacking the intermediate conductance Ca2+−dependent K+ channel Kcnn4. J Physiol Lond 583:705–717

Foskett JK (1990) [Ca2+]i modulation of Cl- content controls cell volume in single salivary acinar cells during fluid secretion. Am J Physiol 259:C998–C1004

Foulke-Abel J, In J, Kovbasnjuk O, Zachos NC, Ettayebi K, Blutt SE, Hyser JM, Zeng XL, Crawford SE, Broughman JR, Estes MK, Donowitz M (2014) Human enteroids as an ex-vivo model of host-pathogen interactions in the gastrointestinal tract. Exp Biol Med (Maywood) 239:1124–1134

Foulke-Abel J, In J, Yin J, Zachos NC, Kovbasnjuk O, Estes MK, de Jonge HR, Donowitz M (2016) Human enteroids as a model of upper small intestinal ion transport physiology and pathophysiology. Gastroenterology 150:638–649.e638

Fuller MK, Faulk DM, Sundaram N, Shroyer NF, Henning SJ, Helmrath MA (2012) Intestinal crypts reproducibly expand in culture. J Surg Res 178:48–54

Furukawa O, Bi LC, Guth PH, Engel E, Hirokawa M, Kaunitz JD (2004) NHE3 inhibition activates duodenal bicarbonate secretion in the rat. Am J Physiol 286:G102–G109

Gallagher AM, Gottlieb RA (2001) Proliferation, not apoptosis, alters epithelial cell migration in small intestine of CFTR null mice. Am J Physiol 281:G681–G687

Gawenis LR, Franklin CL, Simpson JE, Palmer BA, Walker NM, Wiggins TM, Clarke LL (2003) cAMP inhibition of murine intestinal Na^+/H^+ exchange requires CFTR-mediated cell shrinkage of villus epithelium. Gastroenterology 125:1148–1163

Gottlieb RA, Dosanjh A (1996) Mutant cystic fibrosis transmembrane conductance regulator inhibits acidification and apoptosis in C127 cells: possible relevance to cystic fibrosis. Proc Natl Acad Sci U S A 93:3587–3591

Gracz AD, Ramalingam S, Magness ST (2010) Sox9 expression marks a subset of CD24-expressing small intestine epithelial stem cells that form organoids in vitro. Am J Physiol 298:G590–G600

Greger R (2000) Role of CFTR in the colon. Annu Rev Physiol 62:467–491

Greger R, Bleich M, Leipziger J, Ecke D, Mall M, Kunzelmann K (1997) Regulation of ion transport in colonic crypts. News Physiol Sci 12:62–66

Hallbäck D-A, Jodal M, Sjöqvist A, Lundgren O (1982) Evidence for cholera secretion emanating from the crypts: a study of villus tissue osmolality and fluid and electrolyte transport in the small intestine of the cat. Gastroenterology 83:1051–1056

Heijerman HGM, McKone EF, Downey DG, Van Braeckel E, Rowe SM, Tullis E, Mall MA, Welter JJ, Ramsey BW, McKee CM, Marigowda G, Moskowitz SM, Waltz D, Sosnay PR, Simard C, Ahluwalia N, Xuan F, Zhang Y, Taylor-Cousar JL, KS MC, Group VXT (2019) Efficacy and safety of the elexacaftor plus tezacaftor plus ivacaftor combination regimen in people with cystic fibrosis homozygous for the F508del mutation: a double-blind, randomised, phase 3 trial. Lancet 394:1940–1948

Heo I, Dutta D, Schaefer DA, Iakobachvili N, Artegiani B, Sachs N, Boonekamp KE, Bowden G, Hendrickx APA, Willems RJL, Peters PJ, Riggs MW, O'Connor R, Clevers H (2018) Modelling Cryptosporidium infection in human small intestinal and lung organoids. Nat Microbiol 3:814–823

Hirokawa M, Takeuchi T, Chu S, Akiba Y, Wu V, Guth PH, Engel E, Montrose MH, Kaunitz JD (2004) Cystic fibrosis gene mutation reduces epithelial cell acidification and injury in acid-perfused mouse duodenum. Gastroenterology 127:1162–1173

Hofmann C, Obermeier F, Artinger M, Hausmann M, Falk W, Schoelmerich J, Rogler G, Grossmann J (2007) Cell-cell contacts prevent anoikis in primary human colonic epithelial cells. Gastroenterology 132:587–600

Ikpa PT, Meijsen KF, Nieuwenhuijze NDA, Dulla K, De Jonge HR, Bijvelds MJC (2020) Transcriptome analysis of the distal small intestine of Cftr null mice. Genomics 112:1139–1150

In JG, Foulke-Abel J, Estes MK, Zachos NC, Kovbasnjuk O, Donowitz M (2016) Human mini-guts: new insights into intestinal physiology and host-pathogen interactions. Nat Rev Gastroenterol Hepatol 13:633–642

Jakab RL, Collaco AM, Ameen NA (2013) Characterization of CFTR high expresser cells in the intestine. Am J Physiol 305:G453–G465

Kapus A, Grinstein S, Wasan S, Kandasamy R, Orlowski J (1994) Functional characterization of three isoforms of the Na +/H + exchanger stably expressed in Chinese hamster ovary cells. ATP dependence, osmotic sensitivity and role in cell proliferation. J Biol Chem 269:23544–23552

Kim HJ, Ingber DE (2013) Gut-on-a-chip microenvironment induces human intestinal cells to undergo villus differentiation. Integr Biol (Camb) 5:1130–1140

Kim K-A, Kakitani M, Zhao J, Oshima T, Tang T, Binnerts M, Liu Y, Boyle B, Park E, Emtage P, Funk WD, Tomizuka K (2005) Mitogenic influence of human R-Spondin1 on the intestinal epithelium. Science 309:1256–1259

Kim HJ, Li H, Collins JJ, Ingber DE (2016) Contributions of microbiome and mechanical deformation to intestinal bacterial overgrowth and inflammation in a human gut-on-a-chip. Proc Natl Acad Sci U S A 113:E7–E15

Kim GA, Spence JR, Takayama S (2017) Bioengineering for intestinal organoid cultures. Curr Opin Biotechnol 47:51–58

Kim Y, Jun I, Shin DH, Yoon JG, Piao H, Jung J, Park HW, Cheng MH, Bahar I, Whitcomb DC, Lee MG (2020) Regulation of CFTR bicarbonate channel activity by WNK1: implications for pancreatitis and CFTR-related disorders. Cell Mol Gastroenterol Hepatol 9:79–103

Kockerling A, Fromm M (1993) Origin of cAMP-dependent Cl secretion from both crypts and surface epithelia of rat intestine. Am J Physiol 264:C1294–C1301

Kunzelmann K, Centeio R, Wanitchakool P, Cabrita I, Benedetto R, Saha T, Hoque KM, Schreiber R (2019) Control of ion transport by Tmem16a expressed in murine intestine. Front Physiol 10:1262

Kurashima K, Yu FH, Cabado AG, Szabo EZ, Grinstein S, Orlowski J (1997) Identification of sites required for down-regulation of Na +/H + exchanger NHE3 activity by cAMP-dependent protein kinase. J Biol Chem 272:28672–28679

Lee RJ, Foskett JK (2010) cAMP-activated Ca2+ signaling is required for CFTR-mediated serous cell fluid secretion in porcine and human airways. J Clin Invest 120:3137–3148

Lee B, Hong GS, Lee SH, Kim H, Kim A, Hwang EM, Kim J, Lee MG, Yang JY, Kweon MN, Tse CM, Mark D, Oh U (2019) Anoctamin 1/TMEM16A controls intestinal Cl(−) secretion induced by carbachol and cholera toxin. Exp Mol Med 51:1–14

Linley J, Loganathan A, Kopanati S, Sandle GI, Hunter M (2014) Evidence that two distinct crypt cell types secrete chloride and potassium in human colon. Gut 63:472–479

Liu J, Walker NM, Cook MT, Ootani A, Clarke LL (2012) Functional Cftr in crypt epithelium of organotypic enteroid cultures from murine small intestine. Am J Physiol 302:C1492–C1503

Liu X, Li T, Riederer B, Lenzen H, Ludolph L, Yeruva S, Tuo B, Soleimani M, Seidler U (2014) Loss of Slc26a9 anion transporter alters intestinal electrolyte and HCO3(−) transport and reduces survival in CFTR-deficient mice. Pflugers Arch 467:1261–1275

Liu J, Walker NM, Ootani A, Strubberg AM, Clarke LL (2015) Defective goblet cell exocytosis contributes to murine cystic fibrosis–associated intestinal disease. J Clin Invest 125:1056–1068

MacLeod RJ, Lembessis P, Hamilton JR (1994) Isotonic volume reduction associated with cAMP stimulation of ^{36}Cl efflux from jejunal crypt epithelial cells. Am J Physiol 267:G387–G392

Matos JE, Sausbier M, Beranek G, Sausbier U, Ruth P, Leipziger J (2007) Role of cholinergic-activated KCa1.1 (BK), KCa3.1 (SK4) and KV7.1 (KCNQ1) channels in mouse colonic Cl-secretion. Acta Physiol (Oxf) 189:251–258

Matsu-Ura T, Dovzhenok A, Aihara E, Rood J, Le H, Ren Y, Rosselot AE, Zhang T, Lee C, Obrietan K, Montrose MH, Lim S, Moore SR, Hong CI (2016) Intercellular coupling of the cell cycle and circadian clock in adult stem cell culture. Mol Cell 64:900–912

Middendorp S, Schneeberger K, Wiegerinck CL, Mokry M, Akkerman RDL, van Wijngaarden S, Clevers H, Nieuwenhuis EES (2014) Adult stem cells in the small intestine are intrinsically programmed with their location-specific function. Stem Cells 32:1083–1091

Middleton PG, Mall MA, Drevinek P, Lands LC, McKone EF, Polineni D, Ramsey BW, Taylor-Cousar JL, Tullis E, Vermeulen F, Marigowda G, CM MK, Moskowitz SM, Nair N, Savage J, Simard C, Tian S, Waltz D, Xuan F, Rowe SM, Jain R, Group VXS (2019) Elexacaftor-Tezacaftor-Ivacaftor for cystic fibrosis with a single Phe508del allele. N Engl J Med 381:1809–1819

Mignen O, Egee P, Liberge M, Harvey BJ (2000) Basolateral outward rectifier chloride channel in isolated crypts of mouse colon. Am J Physiol 279:G277–G287

Miyoshi H, Stappenbeck TS (2013) In vitro expansion and genetic modification of gastrointestinal stem cells in spheroid culture. Nat Protocols 8:2471–2482

Miyoshi H, Ajima R, Luo CT, Yamaguchi TP, Stappenbeck TS (2012) Wnt5a potentiates TGF-signaling to promote colonic crypt regeneration after tissue injury. Science 338:108–113

Namkung W, Yao Z, Finkbeiner WE, Verkman AS (2011) Small-molecule activators of TMEM16A, a calcium-activated chloride channel, stimulate epithelial chloride secretion and intestinal contraction. FASEB J 25:4048–4062

Noone PG, Zhou ZQ, Silverman LM, Jowell PS, Knowles MR, Cohn JA (2001) Cystic fibrosis gene mutations and pancreatitis risk: relation to epithelial ion transport and trypsin inhibitor gene mutations. Gastroenterology 121:1310–1319

O'Loughlin EV, Hunt DM, Bostrom TE, Hunter D, Gaskin KJ, Gyory A, Cockayne DJ (1996) X-ray microanalysis of cell elements in normal and cystic fibrosis jejunum: evidence for chloride secretion in villi. Gastroenterology 110:411–418

Ootani A, Li X, Sangiorgi E, Ho QT, Ueno H, Toda S, Sugihara H, Fujimoto K, Weissman IL, Capecchi MR, Kuo CJ (2009) Sustained in vitro intestinal epithelial culture within a Wnt-dependent stem cell niche. Nat Med 15:701–706

Paul T, Li S, Khurana S, Leleiko NS, Walsh MJ (2007) The epigenetic signature of CFTR expression is co-ordinated via chromatin acetylation through a complex intronic element. Biochem J 408:317–326

Petersen N, Frimurer TM, Terndrup Pedersen M, Egerod KL, Wewer Albrechtsen NJ, Holst JJ, Grapin-Botton A, Jensen KB, Schwartz TW (2018) Inhibiting RHOA signaling in mice increases glucose tolerance and numbers of enteroendocrine and other secretory cells in the intestine. Gastroenterology 155:1164–1176. e1162

Plasschaert LW, Zilionis R, Choo-Wing R, Savova V, Knehr J, Roma G, Klein AM, Jaffe AB (2018) A single-cell atlas of the airway epithelium reveals the CFTR-rich pulmonary ionocyte. Nature 560:377–381

Poulsen JH, Fischer H, Illek B, Machen TE (1994) Bicarbonate conductance and pH regulatory capability of cystic fibrosis transmembrane conductance regulator. Proc Natl Acad Sci U S A 91:5340–5344

Quinton PM (1999) Physiological basis of cystic fibrosis: a historical perspective. Physiol Rev 79 (Suppl. 1):S3–S22

Ramsey BD, Davies J, McElvaney NG, Tullis E, Bell SC, Drevinek P, Griese M, McKone EF, Wainwright CE, Konstan MW, Moss R, Ratjen F, Sermet-Gaudelus I, Rowe SM, Dong Q, Rodriguez S, Yen K, Ordonez C, Elborn JS (2011) A CFTR potentiator in patients with cystic fibrosis and the G551D mutation. N Engl J Med 365:1663–1672

Reynolds A, Parris A, Evans LA, Lindqvist S, Sharp P, Lewis M, Tighe R, Williams MR (2007) Dynamic and differential regulation of NKCC1 by calcium and cAMP in the native human colonic epithelium. J Physiol Lond 582:507–524

Robert ME, Singh SK, Ikuma M, Jain D, Ardito T, Binder HJ (2001) Morphology of isolated colonic crypts. Cells Tissues Organs 168:246–251

Sato T, Vries RG, Snippert HJ, van de Wetering M, Barker N, Stange DE, van Es JH, Abo A, Kujala P, Peters PJ, Clevers H (2009) Single Lgr5 stem cells build crypt-villus structures in vitro without a mesenchymal niche. Nature 459:262–265

Sato T, Stange DE, Ferrante M, Vries RGJ, van Es JH, van den Brink S, van Houdt WJ, Pronk A, van Gorp J, Siersema PD, Clevers H (2011a) Long-term expansion of epithelial organoids from human colon, adenoma, adenocarcinoma, and Barrett's epithelium. Gastroenterology 141:1762–1772

Sato T, Van Es JH, Snippert HJ, Stange DE, Vries RG, Van Den Born M, Barker N, Shroyer NF, van de Wetering M, Clevers H (2011b) Paneth cells constitute the niche for Lgr5 stem cells in intestinal crypts. Nature 469:415–419

Saxena K, Blutt SE, Ettayebi K, Zeng XL, Broughman JR, Crawford SE, Karandikar UC, Sastri NP, Conner ME, Opekun AR, Graham DY, Qureshi W, Sherman V, Foulke-Abel J, In J, Kovbasnjuk O, Zachos NC, Donowitz M, Estes MK (2016) Human intestinal enteroids: a new model to study human rotavirus infection, host restriction, and pathophysiology. J Virol 90:43–56

Schwank G, Koo B-K, Sasselli V, Dekkers Johanna F, Heo I, Demircan T, Sasaki N, Boymans S, Cuppen E, van der Ent Cornelis K, Nieuwenhuis Edward ES, Beekman Jeffrey M, Clevers H (2013) Functional repair of CFTR by CRISPR/Cas9 in intestinal stem cell organoids of cystic fibrosis patients. Cell Stem Cell 13:653–658

Schwarz JS, de Jonge HR, Forrest JN Jr (2015) Value of organoids from comparative epithelia models. Yale J Biol Med 88:367–374

Seidler U, Singh AK, Cinar A, Chen M, Hillesheim J, Hogema B, Riederer B (2009) The role of the NHERF family of PDZ scaffolding proteins in the regulation of salt and water transport. Ann N Y Acad Sci 1165:249–260

Shin W, Hinojosa CD, Ingber DE, Kim HJ (2019) Human intestinal morphogenesis controlled by transepithelial morphogen gradient and flow-dependent physical cues in a microengineered gut-on-a-chip. iScience 15:391–406

Simpson JE, Gawenis LR, Walker NM, Boyle KT, Clarke LL (2005) Chloride conductance of CFTR facilitates basal Cl^-/HCO_3^- exchange in the villous epithelium of intact murine duodenum. Am J Physiol 288:G1241–G1251

Singh SK, Binder HJ, Boron WF, Geibel JP (1995) Fluid absorption in isolated perfused colonic crypts. J Clin Invest 96:2373–2379

Spence JR, Mayhew CN, Rankin SA, Kuhar MF, Vallance JE, Tolle K, Hoskins EE, Kalinichenko VV, Wells SI, Zorn AM, Shroyer NF, Wells JM (2011) Directed differentiation of human pluripotent stem cells into intestinal tissue in vitro. Nature 470:105–109

Stelzner M, Helmrath M, Dunn JCY, Henning SJ, Houchen CW, Kuo C, Lynch J, Li L, Magness ST, Martin MG, Wong MH, Yu J (2012) A nomenclature for intestinal in vitro cultures. Am J Physiol 302:G1359–G1363

Stewart CP, Turnberg LA (1989) A microelectrode study of responses to secretagogues by epithelial cells on villus and crypt of rat small intestine. Am J Physiol 257:G334–G343

Strong TV, Boehm K, Collins FS (1994) Localization of cystic fibrosis transmembrane conductance regulator mRNA in the human gastrointestinal tract by in situ hybridization. J Clin Invest 93:347–354

Strubberg AM, Liu J, Walker NM, Stefanski CD, MacLeod RJ, Magness ST, Clarke LL (2018) Cftr modulates Wnt/beta-catenin signaling and stem cell proliferation in murine intestine. Cell Mol Gastroenterol Hepatol 5:253–271

Sui Y, Sun M, Wu F, Yang L, Di W, Zhang G, Zhong L, Ma Z, Zheng J, Fang X, Ma T (2014) Inhibition of TMEM16A expression suppresses growth and invasion in human colorectal cancer cells. PLoS One 9:e115443

Szászi K, Kurashima K, Kaibuchi K, Grinstein S, Orlowski J (2001) Role of the cytoskeleton in mediating cAMP-dependent protein kinase inhibition of the epithelial Na+/H+ Exchanger NHE3. J Biol Chem 276:40761–40768

Tamada T, Hug MJ, Frizzell RA, Bridges RJ (2001) Microelectrode and impedance analysis of anion secretion in Calu-3 cells. JOP 2:219–228

Than BLN, Linnekamp JF, Starr TK, Largaespada DA, Rod A, Zhang Y, Bruner V, Abrahante J, Schumann A, Luczak T, Niemczyk A, O'Sullivan MG, Medema JP, Fijneman RJA, Meijer GA, Van den Broek E, Hodges CA, Scott PM, Vermeulen L, Cormier RT (2016) CFTR is a tumor suppressor gene in murine and human intestinal cancer. Oncogene 35:4179–4187

Thiagarajah JR, Verkman AS (2013) Chloride channel-targeted therapy for secretory diarrheas. Curr Opin Pharmacol 13:888–894

Valverde MA, O'Brien JA, Sepulveda FV, Ratcliff RA, Evans MJ, Colledge WH (1995) Impaired cell volume regulation in intestinal crypt epithelia of cystic fibrosis mice. Proc Natl Acad Sci U S A 92:9038–9041

van der Hee B, Loonen LMP, Taverne N, Taverne-Thiele JJ, Smidt H, Wells JM (2018) Optimized procedures for generating an enhanced, near physiological 2D culture system from porcine intestinal organoids. Stem Cell Res 28:165–171

van der Helm MW, Henry OYF, Bein A, Hamkins-Indik T, Cronce MJ, Leineweber WD, Odijk M, van der Meer AD, Eijkel JCT, Ingber DE, van den Berg A, Segerink LI (2019) Non-invasive sensing of transepithelial barrier function and tissue differentiation in organs-on-chips using impedance spectroscopy. Lab Chip 19:452–463

VanDussen KL, Marinshaw JM, Shaikh N, Miyoshi H, Moon C, Tarr PI, Ciorba MA, Stappenbeck TS (2015) Development of an enhanced human gastrointestinal epithelial culture system to facilitate patient-based assays. Gut 64:911–920

Vega G, Guequen A, Johansson MEV, Arike L, Martinez-Abad B, Nystrom EEL, Scudieri P, Pedemonte N, Millar-Buchner P, Philp AR, Galietta LJ, Hansson GC, Flores CA (2019) Normal calcium-activated anion secretion in a mouse selectively lacking TMEM16A in intestinal epithelium. Front Physiol 10:694

Vitzthum C, Clauss WG, Fronius M (2015) Mechanosensitive activation of CFTR by increased cell volume and hydrostatic pressure but not shear stress. Biochim Biophys Acta 1848:2942–2951

Wainwright CE, Elborn JS, Ramsey BW, Marigowda G, Huang X, Cipolli M, Colombo C, Davies JC, De Boeck K, Flume PA, Konstan MW, McColley SA, McCoy K, McKone EF, Munck A, Ratjen F, Rowe SM, Waltz D, Boyle MP, Group TS, Group TS (2015) Lumacaftor-Ivacaftor in patients with cystic fibrosis homozygous for Phe508del CFTR. N Engl J Med 373:220–231

Walker NM, Liu J, Stein SR, Stefanski CD, Strubberg AM, Clarke LL (2016) Cellular chloride and bicarbonate retention alters intracellular pH regulation in Cftr KO crypt epithelium. Am J Physiol 310:G70–G80

Wang Y, Kim R, Sims CE, Allbritton NL (2019) Building a thick mucus hydrogel layer to improve the physiological relevance of in vitro primary colonic epithelial models. Cell Mol Gastroenterol Hepatol 8:653–655

Wilke G, Funkhouser-Jones LJ, Wang Y, Ravindran S, Wang Q, Beatty WL, Baldridge MT, VanDussen KL, Shen B, Kuhlenschmidt MS, Kuhlenschmidt TB, Witola WH, Stappenbeck TS, Sibley LD (2019) A stem-cell-derived platform enables complete cryptosporidium development in vitro and genetic tractability. Cell Host Microbe 26:123–134

Yin X, Farin HF, van Es JH, Clevers H, Langer R, Karp JM (2014) Niche-independent high-purity cultures of Lgr5+ intestinal stem cells and their progeny. Nat Methods 11:106–112

Yin Y, Bijvelds M, Dang W, Xu L, van der Eijk AA, Knipping K, Tuysuz N, Dekkers JF, Wang Y, de Jonge J, Sprengers D, van der Laan LJ, Beekman JM, Ten Berge D, Metselaar HJ, de Jonge H, Koopmans MP, Peppelenbosch MP, Pan Q (2015) Modeling rotavirus infection and antiviral therapy using primary intestinal organoids. Antivir Res 123:120–131

Yin Y, Wang Y, Dang W, Xu L, Su J, Zhou X, Wang W, Felczak K, van der Laan LJ, Pankiewicz KW, van der Eijk AA, Bijvelds MJC, Sprengers D, de Jonge HR, Koopmans MP, Metselaar HJ, Peppelenbosch MP, Pan Q (2016) Mycophenolic acid potently inhibits rotavirus infection with a high barrier to resistance development. Antivir Res 133:41–49

Yin J, Tse CM, Avula LR, Singh V, Foulke-Abel J, de Jonge HR, Donowitz M (2018) Molecular basis and differentiation-associated alterations of anion secretion in human duodenal enteroid monolayers. Cell Mol Gastroenterol Hepatol 5:591–609

Yu K, Lujan R, Marmorstein A, Gabriel S, Hartzell HC (2010) Bestrophin-2 mediates bicarbonate transport by goblet cells in mouse colon. J Clin Invest 120:1722–1735

Yui S, Nakamura T, Sato T, Nemoto Y, Mizutani T, Zheng X, Ichinose S, Nagaishi T, Okamoto R, Tsuchiya K, Clevers H, Watanabe M (2012) Functional engraftment of colon epithelium expanded in vitro from a single adult Lgr5+ stem cell. Nat Med 18:618–623

Yun CH, Oh S, Zizak M, Steplock D, Tsao S, Tse CM, Weinman EJ, Donowitz M (1997) cAMP-mediated inhibition of the epithelial brush border Na +/H + exhcanger, NHE3, requires an associated regulatory protein. Proc Natl Acad Sci U S A 94:3010–3015

Zhao H, Wiederkehr MR, Fan LZ, Collazo R, Crowder LA, Moe OW (1999) Acute inhibition of Na/H exchanger NHE-3 by cAMP. J Biol Chem 274:3978–3987

Zizak M, Lamprecht G, Steplock D, Tariq N, Shenolikar S, Donowitz M, Yun CHC, Weinman EJ (1999) cAMP-induced phosphorylation and inhibition of Na+/H+ exchanger 3 (NHE3) are dependent on the presence but not the phosphorylation of NHE regulatory factor. J Biol Chem 274:24753–24758

Zomer-van Ommen DD, de Poel E, Kruisselbrink E, Oppelaar H, Vonk AM, Janssens HM, van der Ent CK, Hagemeijer MC, Beekman JM (2018) Comparison of ex vivo and in vitro intestinal cystic fibrosis models to measure CFTR-dependent ion channel activity. J Cyst Fibros 17:316–324

Chapter 2
Secretory Diarrhea

Nadia Ameen, Sascha Kopic, Kaimul Ahsan, and Leandra K. Figueroa-Hall

Abstract Secretory diarrhea (SD) remains a major cause of morbidity and mortality globally particularly among infants and children. Enterotoxin-mediated diarrhea due to cholera, *Escherichia coli* (*E. coli*), and rotavirus remains the major cause of SD in developing countries, while genetic diseases that produce defects in intestinal ion transporters and motor proteins are increasingly implicated in diarrheal diseases in developed countries. In recent years, significant advances have been made in understanding various factors that contribute to the underlying pathogenesis of SD. These advances are the result of research progress in the fields of genetics, cell biology, and physiology and animal models that have culminated in development of new therapeutic agents for treating SD. The major advances in pathophysiology and treatment of enterotoxin and genetic causes of SD in childhood will be reviewed.

Keywords SD, secretory diarrhea · MVID, microvillus inclusion disease · CFTR, cystic fibrosis transmembrane conductance regulator · CCD, congenital chloride diarrhea · DRA, downregulated in adenoma · CSD, congenital sodium diarrhea · CHE, CFTR high expresser cell

N. Ameen (✉)
Department of Pediatrics/Cellular and Molecular Physiology, Yale University School of Medicine, New Haven, CT, USA
e-mail: nadia.ameen@yale.edu

S. Kopic
Department of Surgery and Cellular and Molecular Physiology, Yale University School of Medicine, New Haven, CT, USA

K. Ahsan
Department of Pediatrics, Yale University School Medicine, New Haven, CT, USA
e-mail: mdkaimul.ahsan@yale.edu

L. K. Figueroa-Hall
Department of Pediatrics, Yale University School Medicine, New Haven, CT, USA

Laureate Institute for Brain Research, Tulsa, OK, USA
e-mail: lfigueroa-hall@laureateinstitute.org

K. L. Hamilton, D. C. Devor (eds.), *Ion Transport Across Epithelial Tissues and Disease*, Physiology in Health and Disease,
https://doi.org/10.1007/978-3-030-55310-4_2

2.1 Introduction

Since this chapter was first published, the classification of diarrhea as osmotic vs secretory was recognized as misleading and a new classification has been proposed (Thiagarajah et al. 2018). Indeed, the new classification is based on the recognition that diarrheal diseases that fall under the umbrella of SD result from both defects in absorption and increased secretion of ions and thus the term is misleading. The old terminology of SD has been preserved in the revised chapter for purposes of continuity. The molecular physiology of ion transporters is discussed in detail in other chapters. This chapter will focus on advances in the underlying pathophysiology of enterotoxin-mediated and childhood secretory diarrhea. The roles of specific intestinal ion transporters relevant to SD pathogenesis will be reviewed and efforts to develop new therapies based on research advances in intestinal transporter physiology will be discussed.

The proximal small intestine is a major site for fluid absorption and secretion and the target site for enterotoxin-mediated secretory diarrhea (Cohen et al. 1989; Field and Semrad 1993). Under healthy conditions, more than 90% of the seven liters of fluid that is secreted into the intestine each day is absorbed. Fluid secretion is driven largely by active secretion of chloride into the lumen and is necessary to maintain fluidity of intestinal contents during various stages of digestion, allowing for diffusion of enzymes and nutrients to preserve digestive functions. However in disease states, dysregulation of intestinal transport mechanisms alters the balance between absorptive and secretory processes such that secretion predominates and results in diarrhea (Barrett 2000).

In the small intestine, mature columnar enterocytes line villi or fern-like structures that project into the lumen and increase the absorptive surface area. Immature enterocytes line the crypts that are positioned between villi and extend in the opposite direction toward the submucosa. Enterocytes possess polarized distribution of plasma membrane that is divided into two defined regions, the apical brush border membrane (BBM) and basolateral membrane (BLM). This division is functionally significant because segregation of specific membrane transport proteins to either the apical or basolateral membrane facilitates transepithelial transport processes (absorption and secretion) of electrolytes. Villus enterocytes are tall columnar epithelial cells endowed with a well-developed brush border and express major absorptive transport proteins such as Na^+-H^+-exchanger, (NHE3), and sodium glucose transporter 1 (SGLT1) on the apical BBM. Crypt cells are shorter and more immature compared to villus cells, possess a less developed brush border, and express abundant apical secretory (e.g., cystic fibrosis transmembrane conductance regulator, CFTR) and basolateral transporters (sodium potassium chloride cotransporter, NKCC1) to regulate anion secretion (Binder 2012; Binder and Reuben 2012) (Fig. 2.1).

In the healthy intestine, absorption and secretion occur simultaneously, but net absorption of electrolytes dominates. The geographic separation of villi and crypt supported separation of absorptive and secretory functions that were previously

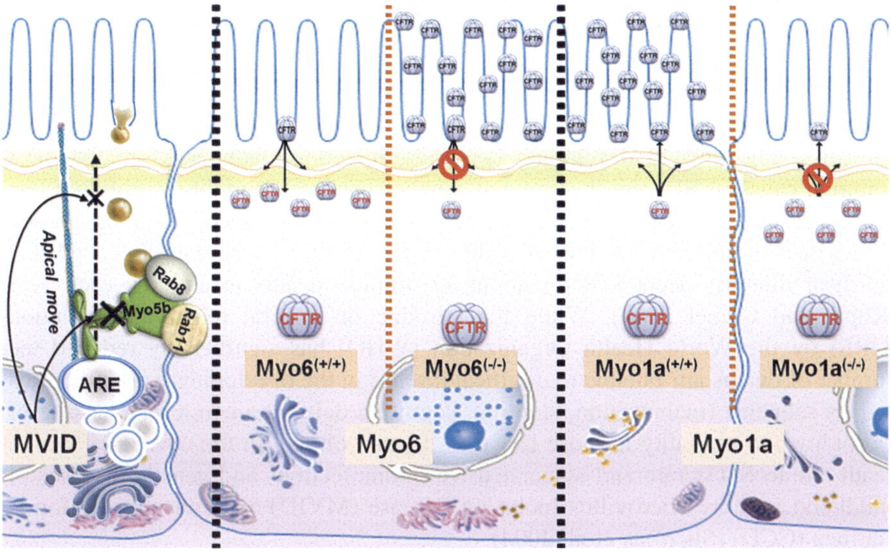

Fig. 2.1 Schematic diagram of myosin motor transport system–mediated recycling. Myo5b moves exocytic vesicles along actin filaments in the intestinal epithelial cells (*left panel*). Myo5b is mutated in MVID, leading to defective ARE-mediated recycling and exocytosis of apical proteins. Myo6 facilitates apical endocytosis of CFTR in the intestine. Deficiency of Myo6 leads to apical accumulation of CFTR (*middle panel*). Myo1a is important for the exocytosis and tethering of BB membrane-inserted CFTR in the villus enterocyte. Deficiency of Myo1a results in the accumulation of CFTR in the subapical compartment (*right panel*)

ascribed to villus and crypt, respectively. Recent data, however, suggest that both processes occur simultaneously indicating significant functional plasticity along enterocytes of the crypt villus axis (Jakab et al. 2011). Over the last decade, new intestinal ion transporters have been identified, and significant advances made in understanding their modes of regulation both in cultured cells and the native epithelium. This new knowledge has had significant impact on new understanding of the normal ion transport processes, the resulting pathophysiological changes that lead to diarrhea, and design of targeted therapies (Binder 2012).

Secretory diarrhea (SD) results when bacterial or viral enterotoxins signal enterocytes to increase intracellular second messengers (cAMP, cGMP, calcium). This process leads to downstream activation of membrane transporters that result in increased anion secretion (chloride, Cl^- and bicarbonate, HCO_3^-) that occurs with or without inhibition of sodium (Na^+) absorption. The increase in solute transport across the gut epithelium is associated with water transport into the lumen (Barrett and Keely 2000; Barrett 2000).

2.2 Clinical Relevance and Burden of Disease

SD is characterized by a large volume of watery stools that is present despite food withdrawal and most commonly follows infections by a host of bacterial, viral, and parasitic pathogens. In developing countries, infectious pathogens are the most common cause of SD, and this remains a major cause of mortality in children less than 5 years of age (Kopic and Geibel 2010). The first estimates of the global burden of diarrhea in children became available in the 1980s. Those data indicated that diarrheal illnesses accounted for about 4.6 million deaths in children each year (Kopic and Geibel 2010). While the introduction of oral rehydration solution (ORS) by the World Health Organization (WHO) has significantly reduced the number of deaths attributable to diarrheal disease in the developing world, diarrhea and its sequelae (malnutrition, stunting, cognitive delay) remain a major cause of morbidity and mortality in those less than 5 years of age. In the developed world, death due to SD is rare and associated with noninfectious and genetic diseases of childhood such as Microvillus Inclusion Disease (MVID) and congenital chloride diarrhea (CCD) (Sherman et al. 2004).

2.3 Infectious SD

Vibrio cholerae (*V. cholerae*) and *Escherichia coli* (*E. coli*) are the two most common bacterial pathogens responsible for SD. Of all enteric pathogens responsible for SD in humans, *V. cholerae* induces the most severe SD affecting humans and results in a rapidly fatal condition (Kopic and Geibel 2010). Significant advances have been made in understanding the mechanisms by which intestinal ion transporters contribute to the underlying pathogenesis of *V. cholerae* and *E. coli*-elicited SD. Translation of the research advances is currently being tapped for design of new therapies, some of which are in clinical trials for treatment of SD (Donowitz et al. 2012). Rotavirus is another common cause of SD and an important contributor to mortality in children. There has been recent advances in understanding rotavirus pathogenesis (Buccigrossi et al. 2014; Crawford et al. 2017; Lauciricia et al. 2017); however, less is understood of the mechanisms underlying rotavirus-induced diarrhea compared to bacterial secretory enterotoxins (Lorrot and Vasseur 2007).

2.4 Channels and Transporters Relevant to SD

2.4.1 CFTR

CFTR chloride channels represent the primary apical exit pathway for chloride secretion from the enterocyte BBM. Detailed discussion of CFTR is provided in

Chap. 16 of Volume 3. CFTR is the gene product responsible for the genetic disease Cystic Fibrosis (CF). It is also central to the underlying pathogenesis of *V. cholerae* and *E. coli*–induced diarrhea. In CF, mutations in *CFTR* lead to trafficking defects that result in reduced or absent functional CFTR on the enterocyte BBM. The net result is failure to secrete chloride and bicarbonate, constipation, and intestinal obstruction. In SD, intracellular second messenger (cyclic nucleotides) signaling leads to increased CFTR abundance and anion secretion on the BBM (Golin-Bisello et al. 2005). Long before the CFTR gene was discovered, CF-affected populations were noted to be protected against cholera epidemics and SD, but the underlying explanation for these observations came only after the CFTR gene was cloned and functional characterization conducted in vivo (Riordan et al. 1989; Field and Semrad 1993). Indeed, studies in genetically engineered mice confirmed that levels of expression of functional CFTR on the enterocyte brush border correlated with fluid secretory responses and diarrhea following exposure to cholera toxin (CTX). $CFTR^{-/-}$ mice fail to secrete fluid similar to CF-affected humans, while heterozygote mice expressing 50% of wild-type (WT) CFTR secreted 50% of the normal chloride and fluid following CTX exposure. These studies confirm the selective advantage of the CF heterozygote to cholera and SD (Gabriel et al. 1994).

2.4.1.1 CFTR Localization in Intestine: Relevance to SD

Chloride is taken up from the bloodstream across the basolateral membrane via NKCC1 that is driven secondarily by low intracellular sodium concentrations established by the active sodium/potassium ATPase pump (Na^+/K^+-ATPase) on the basolateral membrane. This arrangement allows chloride to accumulate in the cell above its electrochemical equilibrium. When apical chloride channels are opened, chloride flows out of the cell down this electrochemical gradient resulting in chloride secretion (Kopic and Geibel 2010). Early electrophysiological studies indicated that CFTR was confined to the apical membrane, consistent with a model of channel opening as the only mechanism regulating fluid secretion (Field and Semrad 1993). This model presumed that CFTR was confined to crypt epithelial cells, consistent with the paradigm of functional segregation of secretion and absorptive functions along the crypt-villus axis. However, generation of superior specific anti-CFTR antibodies enabled detailed visualization of the subcellular distribution and mapping of CFTR distribution and its expression gradient along the proximal-distal and crypt-villus axis of the intestine (Ameen et al. 1995; Jakab et al. 2011). CFTR is present in the apical domain of epithelial cells of the stomach (Weis et al. 2013), Brunner glands of the proximal duodenum (Jakab et al. 2011), crypt and villus enterocytes (Ameen et al. 2000) and very high levels are present in a subpopulation of crypt and villus enterocytes, called CFTR High Expresser (CHE) in the duodenum and jejunum (Ameen et al. 1995; Jakab et al. 2013). CFTR is absent from goblet cells. Highest levels of CFTR are observed in the proximal small intestine and lowest levels in the colon (Jakab et al. 2011).

Surprisingly, localization studies revealed that while CFTR was indeed present on the apical membranes of enterocytes as predicted, it was not confined to the apical membrane. CFTR is present in subapical vesicles and on the apical membrane in all CFTR-expressing epithelial cells in the intestine (Ameen et al. 2000). Ultrastructural localization studies in native intestine revealed that about one-third of the total CFTR in the enterocyte was localized to the apical membrane, while the majority of channels resided on the membranes of endosomal vesicles within the subapical cytoplasm of both crypt and villus enterocytes (Ameen et al. 2000). This subcellular distribution in the native intestine failed to reconcile with the existing paradigm of a singular regulation of CFTR by cyclic nucleotide-dependent phosphorylation on the apical membrane as suggested by electrophysiological studies. Indeed, further studies in native intestine confirmed that this subcellular distribution in native enterocytes was critical for acute regulation of CFTR and fluid secretory response to enterotoxins under conditions of SD (Jakab et al. 2012; Golin-Bisello et al. 2005). CFTR distribution in the intestine resembled that of other transporters such as the insulin stimulated glucose transporter 4 (Glut4). Localization and ultrastructural studies demonstrated that in adipose tissues endogenously expressing this transporter, Glut4 acutely and tightly controls glucose transport by shuttling the transporter from subapical vesicles into and out of the plasma membrane, thereby regulating the number and function of transporters on the plasma membrane (Rowland et al. 2011). Since then, immunolocalization approaches have been employed and confirmed that similar mechanisms of regulation are at work for other ion transporters in vivo (Jakab et al. 2012). Enterotoxin-dependent regulation of CFTR-mediated fluid secretion by traffic is discussed in the sections dedicated to *V. cholera* and *E. coli* pathogenesis.

2.4.1.2 CFTR and Bicarbonate

Clinicians have long noted defects in intestinal alkalization in the human CF intestine (Pratha et al. 2000; Gelfond et al. 2013). In CF, increased acidity due to absent CFTR on the enterocyte BBM contributes to bile acid-induced diarrhea, malabsorption, maldigestion, bacterial overgrowth, poor motility, and accumulation of thick inspissated mucus (Baxter et al. 1988). In addition to chloride, CTX and *E. coli* Heat-Stable enterotoxin (STa) elicit significant bicarbonate secretion and contribute to metabolic acidosis in affected individuals (De Jonge 1975b). Until recently, chloride secretion was central to CFTR function on the apical membrane of epithelia. Of note, the importance of CFTR to bicarbonate secretion is increasingly recognized and more attention given to this aspect of its function following the identification of chloride bicarbonate exchangers on the BBM in the intestine. In the intestine, CFTR-mediated bicarbonate secretion is functionally linked to members of the SLC26 family of apical chloride bicarbonate exchangers that are predominantly found on the villus epithelium (SLC26A3: DRA, downregulated in adenoma; SLC26A6: PAT1, putative anion transporter1; SLC4A2: anion exchanger AE2) (Singh et al. 2010). Apical bicarbonate secretion is dependent on basolateral

HCO_3^- entry transporters such as NBC, sodium bicarbonate cotransporter. Mutations in the apical SLC26A3 (DRA) are directly linked to CCD (congenital chloride diarrhea), an important cause of childhood diarrhea that is discussed later in this section (Hoglund et al. 1996b; Makela et al. 2002). Immunolocalization studies were critical in mapping specific cellular localization sites of bicarbonate/chloride entry and exit pathways in the native intestine. These studies contributed to unraveling an important role for villus enterocytes of the proximal small intestine in CFTR-mediated bicarbonate secretion, and provide a plausible explanation for the observed increase in duodenal luminal acidity in the CF human intestine (Jakab et al. 2012).

2.4.2 Sodium Hydrogen Exchanger NHE3

The fundamentals of sodium absorption in epithelia are discussed in Chap. 9 of Volume 1. In the intestine, NHE3 plays a critical role in electroneutral sodium chloride absorption and contributes to the pathogenesis of SD (Singh et al. 2014; Binder 2012). NHE3 is present on the apical domain of villus enterocytes of the small intestine and surface cells of the colon (Jakab et al. 2011). NHE3 function on the apical membrane is linked to CFTR and the Cl^-/HCO_3^- exchanger (Singh et al. 2010). Like CFTR, NHE3 is distributed on the apical membrane and in subapical vesicles of enterocytes and regulated by traffic into and out of the plasma membrane (Jakab et al. 2012). Under conditions of enterotoxin-induced SD, cyclic nucleotides (cAMP, cGMP) and second messengers maximize fluid secretion by simultaneously regulating traffic of NHE3 and CFTR in opposing directions on the enterocyte brush border (Fig. 2.2). Cyclic nucleotide signaling leads to rapid traffic of endosomal-associated CFTR into the brush border membrane to increase CFTR abundance and function, while simultaneously inhibiting sodium absorption by blocking NHE3 internalization, leading to increased NHE3 abundance on the BBM (Jakab et al. 2012).

2.5 CFTR High Expresser (CHE) Cells

In 1995, our group identified a small (2%) subpopulation of enterocytes expressing very high levels of CFTR in the human and rat intestine (Ameen et al. 1995). These cells, called CFTR High Expresser (CHE) cells, are present in the crypt and villus of the duodenum and jejunum, but are more abundant in villus epithelium. CHE cells face the lumen between the long fern like villi where they are positioned to respond to luminal milieu (Jakab et al. 2013). Immunoelectron microscopic studies reveal that villus CHE cells possess prominent cytoskeleton, but express lower levels of the BBM myosin Myo1a, possess slightly shorter microvilli, endocytic and exocytic trafficking machinery, and a large pool of subapical endosomes. In distinct contrast to villus enterocytes of the small intestine, CHE cells possess unique anti-absorptive

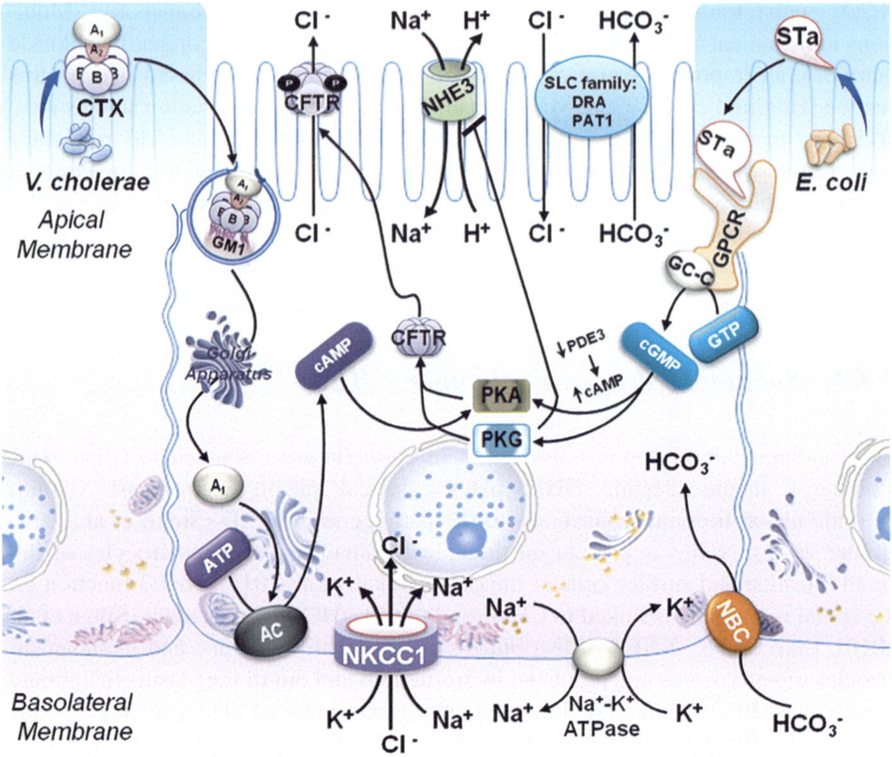

Fig. 2.2 Schematic illustration of ion transporters and cellular pathways relevant to secretory diarrhea pathogenesis. On the left, CTX binds to the surface ganglioside GM1 f.b. internalization, Golgi processing, and release of A subunit of the toxin. The released A subunit stimulates adenylate cyclase (AC), leading to increased intracellular levels of cAMP. On the right, heat-stable enterotoxin (STa) binds to G-protein coupling receptor (GPCR) and activates guanylyl cyclase C (GC-C), leading to increased intracellular levels of cGMP. Increased intracellular cAMP/cGMP increases intestinal fluid secretion by regulating apical trafficking of CFTR and/or NHE3 through activation of PKG or PKA signaling pathway, resulting in secretory diarrhea

features, in that they lack expression of absorptive apical proteins, including NHE3 and sucrose isomaltase (Ameen et al. 1995; Jakab et al. 2013;). CHE cells also lack apical chloride/bicarbonate exchangers that regulate bicarbonate secretion in conjunction with CFTR/NHE3. Interestingly, CHE cells possess higher levels of the basolateral NKCC1. Moreover, cAMP agonists, including vasoactive intestinal peptide (VIP) and CTX, trigger robust exocytosis of CFTR from the subapical endosomal pool into the BBM of CHE cells that can be reversed upon washout (Ameen et al. 1999), supporting a prosecretory behavior in CHE cells.

In the absence of direct evidence to demonstrate chloride secretion originating from CHE cells, morphological characterization supports a role for robust CFTR-mediated fluid secretion from villus CHE cells following cyclic nucleotide stimulation, suggesting that these cells contribute to enterotoxin-mediated diarrhea. There has been recent renewed interest and recognition of the importance of these cells as the major source of Cl- secretion, suggesting that they may be the target cells in CF lung and gut disease (Plasschaert et al. 2018; Montoro et al. 2018).

2.6 Motors/Binding Partners in Brush Border Membrane CFTR Traffic: Relevance to Secretory Diarrhea

CFTR has a long half-life (>72 h), and efficiently maintains its surface abundance in the intestine by apical endocytosis, recycling, and return to the plasma membrane through the exocytic pathway (Silvis et al. 2009). This constitutive recycling has been observed in a number of CFTR-expressing epithelia, including the intestine, and requires no second messenger stimulation (Ameen et al. 2007). In the intestine, CFTR also undergoes robust regulated trafficking in enterocytes following second messenger signaling (cAMP, cGMP) (Ameen et al. 1999; Golin-Bisello et al. 2005). This is a unique feature of its behavior that is prominent in the proximal small intestine and appears to be critical for acute responses to bacterial enterotoxins in eliciting the massive increase in fluid secretion that results in SD. Enterocytes possess a prominent cytoskeleton support network rich in actin and microtubule motors that facilitates movement or traffic of CFTR into and out of the BBM. The role of molecular motors in CFTR traffic is reviewed more comprehensively in (Kravtsov and Ameen 2013). Both cAMP- (CTX, Heat-Labile *E. coli* toxin) and cGMP- (Heat-Stable *E. coli* toxin) mediated enterotoxins utilize motors to redistribute CFTR into and out of the BBM in the native intestine. While much is left to understand about the roles of motors in CFTR traffic, the molecular motors do not appear to distinguish between second messengers, as available data indicate that cAMP and cGMP agonists employ similar motors to move CFTR. Cell-type specificity is a recognized characteristic of CFTR traffic that is linked to its function in native epithelial cells, and dictated largely by specific adaptor signaling (Silvis et al. 2009; Guggino and Stanton 2006). In addition to cell-specific adaptors, CFTR traffic into and out of the BBM in the intestine is differentially powered by specific actin motors in crypt vs. villus enterocytes (Kravtsov and Ameen 2013). The role of actin motors, and adaptors in SD (both cAMP and cGMP) pathogenesis will be discussed briefly.

The actin-binding motor, Myosin VI (Myo6), is present in many epithelial cells, including crypt and villus enterocytes (Ameen and Apodaca 2007). Myo6 is a unique minus end-directed motor that regulates clathrin-mediated endocytosis of a specific class of receptors (low-density lipoprotein receptor, LDLR) from the plasma membrane (Hasson 2003; Buss et al. 2004; Roberts et al. 2004). Studies in transgenic mice lacking Myo6 provided important insights into the physiological role that Myo6 plays in apical endocytosis of CFTR from the BBM and its contribution to STa-elicited diarrhea pathogenesis (Ameen and Apodaca 2007). Receptors for STa binding (guanylyl cyclase, GCC) are enriched on the apical membranes of villus enterocytes in the jejunum but also present in the crypt. STa-elicited fluid secretion is exclusively mediated by CFTR (De Jonge 1975a; Vaandrager et al. 1997). Absence of Myo6 in mice leads to defective apical endocytosis and accumulation of CFTR on the surface of crypt and villus enterocytes in the jejunum and increased STa-mediated fluid secretion, indicating the importance of this minus end-directed actin motor in regulating the enterocyte response to enterotoxin-elicited diarrhea (Ameen and Apodaca 2007) (Fig. 2.1; Myo6, *middle panel*).

LDLR receptor family members utilize Myo6 and the endocytic adaptor, Dab2, to facilitate clathrin-mediated endocytosis (Buss et al. 2001; Morris et al. 2002; Roberts et al. 2004). However, in contrast to airway epithelial cells, in the intestine, Dab2 does not play a direct role in CFTR endocytosis (Collaco et al. 2010; Cihil et al. 2012; Madden and Swiatecka-Urban 2012). Instead, the specific binding partner, AP2-alpha, directly binds to CFTR, while Dab2 is an indirect binding partner in directing Myo6-dependent endocytosis from the enterocyte BBM (Collaco et al. 2010).

A key step in enterotoxin-mediated increase in CFTR abundance and fluid secretion on the BBM is cAMP- or cGMP-regulated exocytosis of endosomal-associated CFTR into the BBM of crypt and villus enterocytes. CFTR exocytosis requires molecular motors. The specific exocytic motors and binding proteins responsible for CFTR exocytosis in the crypt has not been identified to date, but is likely to be a combination of microtubule- and actin-based motors. CFTR exocytosis into the villus enterocyte BBM of the proximal small intestine was recently shown to be dependent on the exocytic actin-binding motor Myosin Ia (Myo1a), a BBM myosin family member (Kravtsov et al. 2012). Myo1a is expressed on the apical domain of mature villus enterocytes of the small intestine, but is not in the crypt as it is critical for BBM development. Transgenic mice lacking Myo1a fail to traffic CFTR from subapical endosomes into the BBM under steady-state conditions but also following cAMP agonists. These findings are intriguing and may be crucial for the pathogenesis of CTX-mediated bicarbonate secretion in light of the new studies indicating the importance of CFTR in villus bicarbonate secretion in the proximal small intestine (Jakab et al. 2011; Kravtsov et al. 2012) (Fig. 2.1; Myo1a, *right panel*).

These observations suggest that targeting CFTR trafficking (either endocytosis or exocytosis) mechanisms into the enterocyte brush border membrane may be key to reducing fluid secretion in secretory diarrhea. The role of the actin motor Myosin VB in SD is discussed in the section under Microvillus Inclusion Disease (MVID).

2.7 The Pathophysiology of Cholera Toxin-Mediated Diarrhea

Cholera still remains a major healthcare concern. According to a WHO report, over 190,000 cases of cholera were reported in 2008. However, the actual number of cases is higher than reported because many countries lack surveillance systems to properly track cases and others fail to report. WHO estimates that official reports only capture 5–10% of actual cases. WHO estimates between 2008–2012 suggest that there were 2.9 million cases of cholera annually after adjusting for underreporting. As a result of the feco-oral transmission route, cholera outbreaks are typically associated with poor sanitary conditions occurring in the wake of natural or humanitarian disasters. The disease is characterized by its fulminant disease progression with water losses of up to 1 liter per hour resulting in severe and potentially fatal dehydration within hours (Kopic and Geibel 2010).

Cholera is caused by an enterotoxin secreted by the bacterium *V. cholera*. Cholera toxin (CTX) belongs to the family of AB_5 toxins. Of note, the heat-labile enterotoxin LT-I secreted by ETEC is a member of the same toxin family (see later). AB_5 toxins are large multimeric proteins consisting of an enzymatically active A subunit and five identical B subunits, which are assembled in a ring-like fashion. The A subunit is further divided into an A1 and A2 domain, the latter of which is responsible for linking the entire A subunit to the pentameric B subunit ring (Gill et al. 1981; Sixma et al. 1993).

Following secretion, the toxin attaches to GM1-type membrane gangliosides. Gangliosides are glycosphingolipids in the plasma membranes of cells. GM1 is particularly enriched in so-called caveloae of the BBM (Parton 1994). Caveolae are distinct plasma membrane invaginations characterized by the presence of the protein caveolin and can be internalized into the cell in a process that is distinct from clathrin-mediated endocytosis (Singh et al. 2003). Alternative clathrin-dependent and clathrin- and caveolin-independent internalization processes have been reported; however, in the intestine, caveolae-mediated uptake seems to be the primary mechanism of toxin internalization (Singh et al. 2003; Massol et al. 2004; Kirkham et al. 2005). The crystal structure of GM1-bound CTX is available and reveals that the B subunit of the holotoxin mediates the binding interaction (Merritt et al. 1994). Given the toxin's pentameric B subunit structure, up to five GM1 gangliosides can be bound by the holotoxin.

Once internalized, the toxin undergoes complex intracellular trafficking before exerting its hypersecretory activity. The holotoxin is first transported via endosomes to the Trans-Golgi network (Richards et al. 2002; Pelkmans et al. 2004) where it then enters the ER for further processing (Majoul et al. 1998; Richards et al. 2002). Here the protein is unfolded leading to dissociation of the A1 subunit from the toxin, a process that is catalyzed by the enzyme protein disulfide isomerase (PDI) (Tsai et al. 2001). Alternatively, it has been suggested that PDI is only responsible for the dissociation process of A1 and that the unfolding takes place spontaneously without enzymatic catalysis (Taylor et al. 2011). The A1 subunit then exits the ER (endoplasmic reticulum), most likely utilizing the ER-associated degradation system, which physically translocates misfolded proteins from the ER into the cytosol for subsequent degradation (Pande et al. 2007; Hazes and Read 1997; Bernardi et al. 2008; Massey et al. 2009; Taylor et al. 2010). Misfolded proteins bound for degradation are usually ubiquitinated; however, the A1 subunit escapes this fate given its lack of lysine residues, which are necessary for attachment of the ubiquitin tag (Hazes and Read 1997; Rodighiero et al. 2002). Once released into the cytosol, the A1 subunit exerts its toxic activity by transferring an ADP-ribose moiety from NAD+ to the alpha-subunit of a G-protein (Moss and Richardson 1978; Lee et al. 1991). The activated G-protein then stimulates adenylate cyclase, resulting in increased cAMP levels, PKA activation, and subsequent CFTR opening.

CTX-induced increase in cAMP levels drives robust traffic of endosomal-associated CFTR into the apical membrane to increase its abundance by at least twofold, that is accompanied by a fourfold increase in fluid secretion in native intestine. PKA phosphorylation is necessary for CFTR traffic from subapical vesicles into the enterocyte BBM, as well as channel opening on the plasma membrane (Golin-Bisello et al. 2005) (Fig. 2.2). However, the contribution of endosomal traffic versus phosphorylation of resident BBM CFTR channels to chloride secretion remains to be defined. The preferential abundance of CFTR in endosomes within enterocytes of the small intestine coupled with robust PKA-dependent traffic into the enterocyte BBM suggest that regulated traffic is critical to cholera-induced fluid secretion and secretory diarrhea pathogenesis.

2.8 Enterotoxigenic *E. coli* in Secretory Diarrhea

Enterotoxigenic *E. coli* (ETEC) is one of the major pathogens responsible for acute diarrheal illness in the developing world. Infants being particularly vulnerable to infection, it has been estimated that 280 million cases of symptomatic ETEC infections occur annually in children under the age of five (Kopic and Geibel 2010). Infection is associated with a significant mortality of 370,000 deaths per year, the majority of which are children (Kopic and Geibel 2010). Furthermore, ETEC is one of the main pathogens responsible for the development of traveler's diarrhea. Out of an estimated 40,000 daily cases of traveler's diarrhea, 50–60% are attributable to ETEC (Kopic and Geibel 2010).

ETEC can produce two molecularly distinct types of enterotoxins, which have historically been characterized by their heat stability. The STa's are short peptide chains, whereas the heat-labile enterotoxins (LT) are complex hetero-oligomeric proteins. Although LT exists in two classes: LT-I and LT-II, LT-II has low relevance for intestinal disease in humans. LT-I consists of one A subunit, conferring actual intracellular toxicity and five identical B subunits arranged in a ring, which are necessary for toxin uptake into the enterocyte. This distinct toxin structure is shared by other bacterial toxins, such as Shiga toxin (STX) and CTX with which LT-I shares 78% sequence homology (Kopic and Geibel 2010). Given the close structural relationship of LT-I to CTX, these toxins share an identical pathophysiological cellular mode of action (Kopic and Geibel 2010). However, despite the close structural and functional homology between both toxins, the disease progression of CTX is in general more severe, which has been attributed to a more stable linkage between the A and B subunits of CTX, thereby increasing toxin effectiveness (Rodighiero et al. 1999). The pathophysiological mode of action of LT-I is further described in the section on CTX (Sect. 2.7).

The heat-stable enterotoxin of ETEC is produced in two variations: STa and STb. Both toxins can be experimentally discerned by their methanol solubility. STb is mostly associated with diarrheal disease in piglets and is not relevant for human pathology (Alderete and Robertson 1978; Burgess et al. 1978; Takeda et al. 1979). STa consists of a short 18 or 19 amino acid (AA) chain and can be further classified into STIa and STIb depending on the respective AA sequence (So and McCarthy 1980; Staples et al. 1980; Aimoto et al. 1982; Moseley et al. 1983). The STa peptide contains three characteristic disulfide bonds, which confer the characteristic heat stability to the toxin. The six cysteine residues are part of a 13AA sequence called the toxic domain, which is essential for eliciting a hypersecretory response in the enterocyte. (Aimoto et al. 1983; Takao et al. 1985b; Gariepy and Schoolnik 1986; Gariepy et al. 1986) Analogous to LT-I and CTX, similar toxin architectures can be found in the heat-stable toxins from *V. cholerae* O1/non-O1 and *Yersinia enterocolitica* (Takao et al. 1985a, b; Yoshino et al. 1993).

The differences between LT-I and STa are not restricted to structural aspects but also predictably extend to divergent pathophysiological modes of action. Rather than relying on internalization and intracellular processing, STa exerts its toxic effects by binding to a receptor localized on the BBM of the enterocyte. First indications for the involvement of receptor binding emerged in the 1980s when radioisotope labeling of STa revealed characteristic binding kinetics, including competitive inhibition (Giannella et al. 1983). A few years earlier, it was also observed that STa exposure is associated with an increase in intracellular cGMP levels, which was attributed to increased guanylate cyclase (GC) activity (Field et al. 1978; Guerrant et al. 1980. Today, we know that that the long elusive STa receptor is in fact a membrane-bound GC, which was initially characterized as receptors of atrial natriuretic peptide (isoforms A and B) and were only later identified (isoform C) as STa receptors (Chinkers et al. 1989; Lowe et al. 1989; Schulz et al. 1990). The central role of GC-C in ETEC pathophysiology is effectively underlined by GC-C deficient animals, which are immune to the hypersecretory effects of STa (Mann et al. 1997; Schulz et al. 1997). Interestingly, receptor levels are reported to decrease with age, which may explain higher disease burden in children (Cohen et al. 1988). Attempts at characterizing the binding site of STa to GC-C have been made utilizing receptor mutagenesis and photoaffinity labeling (Wada et al. 1996; Hasegawa et al. 1999). In lack of a precise crystal structure of GC-C, homology modeling with GC-A has also been conducted (Hasegawa et al. 2005; Lauber et al. 2009).

GC-C also has physiological ligands. Uroguanylin and guanylin are produced by local intestinal cells and are endogenous activators of GC-C, thereby serving as physiological secretagogues (Joo et al. 1998; Currie et al. 1992; de Sauvage et al.

1992; Nakazato 2001; Lorenz et al. 2003). STa shares significant structural homology to both peptides, including the characteristic disulfide bonds (Skelton et al. 1994).

GC-C possesses an uneven distribution pattern along the intestine. Its density was shown to decrease from the proximal to distal small intestine (Krause et al. 1994). Expression also seems to be heterogeneous along the vertical crypt-villus axis; however, only inconclusive evidence about expression characteristics is available (Cohen et al. 1992; Krause et al. 1994).

STa binding to GC-C results in increased catalytic activity of the intracellular cyclase domain of the receptor, thereby elevating intracellular cGMP levels. cGMP in turn activates two intracellular signaling pathways: cGMP-dependent protein kinase isoform II (cGKII aka PKG) pathway and PKA pathway (Vaandrager et al. 2000). Preferential activation of these pathways varies along the intestine (the PKA pathway seems to be more specific to the colon); however, a functional interplay between both pathways is most likely (Vaandrager et al. 2000). cGKII activation causes increased chloride secretion through CFTR and concomitant inhibition of sodium absorption through NHE3, thereby increasing luminal osmolarity and water loss (Field et al. 1978; Pfeifer et al. 1996; Chao et al. 1994) (Fig. 2.2). In line with this hypothesis, cGKII-deficient animals demonstrate an impaired secretory response to STa exposure (Pfeifer et al. 1996; Vaandrager et al. 2000). The PKA-mediated pathway is less well characterized. It has been suggested that activation of cGMP leads to increased phosphodiesterase activity thereby leading to an accumulation of cAMP and PKA activation (Vaandrager et al. 2000).

2.9 Treatment of SD

The role of ORS and current treatment approaches to SD can be found in a number of excellent reviews (Farthing 1994, 2001; Thiagarajah et al. 2015; Das et al. 2018). ORS has remained the primary therapeutic modality for fluid replacement and rehydration for SD since its development in the 1970s. ORS formulation takes advantage of glucose-dependent sodium absorption and the stoichiometry of SGLT1 transport from the BBM of enterocytes. While ORS does not directly address the underlying pathophysiology of SD and significantly reduce stool volume and frequency, its implementation has saved millions of lives particularly in the developing world. A recent study examined the reasons for continued fluid secretion and diarrhea with ORS. Interestingly, Yin et al. reported that glucose increases intracellular calcium and stimulates active chloride secretion in the small intestine (Yin et al. 2014). These studies have implications for current ORS therapy and suggest that modification to the current formulation may be possible to help reduce diarrhea and fluid secretion.

Being the common molecular endpoint of various SD entities, CFTR has been identified as a potential pharmacological target. To date several specific, non-absorbable small-molecule CFTR inhibitors have been discovered with the aim of abrogating the hypersecretion of chloride occurring during SD (Ma et al. 2002; Muanprasat et al. 2004; Sonawane et al. 2005). Despite their excellent experimental efficacy of reducing chloride and water flux through CFTR, clinical applications have not been successful in the light of toxicity issues due to impairment of mitochondrial function (Kelly et al. 2010). Crofelemer is a new antisecretory agent that has had more success in progress toward clinical use (Crutchley et al. 2010). This compound is a purified proanthocyanidin oligomer extracted from the latex of a South American plant that has been used by locals for many years to treat diarrhea. Crofelemer has been shown to inhibit both CFTR and calcium-activated chloride channels (CaCCs) at the luminal membranes in enterocytes (Tradtrantip et al. 2010). This compound is currently in clinical trials for treatment of infectious secretory diarrhea, including cholera, and diarrhea predominant irritable bowel syndrome, and, was recently approved for HIV diarrhea (Tradtrantip et al. 2010; Cottreau et al. 2012; Crutchley et al. 2010; Yeo et al. 2013). Recent advances have identified other peptide inhibitors of chloride channels, including Plumbagin and the encephalinase inhibitor Racecdotril for treatment of SD (Yu et al. 2019; Liang et al. 2019).

2.10 CFTR/NHE3 in Secretory Diarrhea

CTX-triggered elevations of cAMP levels further inhibit Na^+ uptake on the villus enterocyte brush border membranes by NHE3, thereby increasing the osmotic driving force for fluid secretion (Donowitz and Welsh 1986). Net fluid secretion results from the simultaneous coordinated increase in CFTR activation and inhibition of NHE3-mediated Na^+ absorption through rapid internalization of NHE3 from the BBM that is facilitated by signaling through the cytoskeleton, Na^+/H^+ exchange regulatory cofactor, and adaptors (Singh et al. 2014). CFTR and NHE3 are expressed in the villus epithelium of the proximal small intestine, although NHE3 is more abundant than CFTR with the exception of the CHE subpopulation (Jakab et al. 2011).

While the evidence is undisputable that CFTR and NHE3 are major mediators of CTX-induced diarrhea, the role/contribution of specific cell types, crypt and villus enterocytes in CTX-induced secretion is not clear. CTX elicits secretion of mucus, chloride, and bicarbonate in the intestine. Mucus is secreted by goblet cells, but it is unclear whether CFTR is present in this cell type. How CTX activates chloride/bicarbonate secretion, whether specific cell types are associated with chloride versus bicarbonate secretion, and the details of how these are functionally linked to mucus secretion and NHE3 function are emerging but not fully understood. The crypt is a major site of CFTR expression and presumed target for CTX and chloride secretion. However, early studies in native epithelium revealed that CTX exerts its action first by elevating cAMP levels in villus enterocytes and it elicits fluid secretion from both villus and crypt cells (De Jonge 1975b). The observation that villus enterocytes of the proximal small intestine express CFTR and chloride/bicarbonate exchangers on the apical membrane and contribute significantly to cAMP activated bicarbonate secretion supports a model whereby CTX elicits chloride secretion predominantly from crypt cells, while villus cells secrete bicarbonate and inhibit sodium absorption via NHE3 (Jakab et al. 2012). Moreover, this model is consistent with recent data indicating that CFTR-mediated bicarbonate secretion from enterocytes is critical for mucus release from neighboring goblet cells (Yang et al. 2013). This scenario contrasts with conditions where CFTR is absent from the apical membranes of enterocytes, as in CF. Absence of CFTR in the enterocyte brush border is associated with thick acidic adherent mucus, lack of fluid secretion, and intestinal obstruction (Eggermont 1996). Thus, a picture is unfolding in support of cell-specific functions in CTX-induced diarrhea, but the details will require further studies aimed at mapping cellular sites along the crypt-villus epithelium using intact epithelium. In addition to its direct effects on enterocytes, *V. cholera* stimulates fluid secretion indirectly through its effects on neural pathways by stimulation of enterochromaffin cells to release serotonin, which in turn activates release of vasoactive intestinal ploypeptide (VIP) from local enteric neurons (Navaneethan and Giannella 2008; Camilleri et al. 2012). Through its release of cAMP, VIP in turn activates signaling and insertion of CFTR associated with endosomes into the enterocyte brush border to increase anion secretion (Ameen et al. 1999).

2.11 Rotavirus

Rotavirus is the most common cause of gastroenteritis and diarrhea in young children. In healthy children, the disease is self-limiting; however, rotavirus is associated with significant morbidity. The number of reported rotavirus-associated deaths in young children was 453,000 in 2008 (Srivastava et al. 1999; Bohles et al. 2014; Srivastava et al. 1999). Morbidity and mortality is increased under conditions of poor sanitation, absence of clean water supplies, overcrowding, and malnutrition. Rotavirus is a non-enveloped virus that is stable against environmental influences. Rotavirus targets enterocytes of the upper two-thirds of the villus epithelium in the small intestine, where it produces an enterotoxin called NSP4 (Lorrot and Vasseur 2007; Lorrot et al. 2006). Rotavirus-induced diarrhea was previously thought to be primarily anti-absorptive, as the virus is known to induce functional defects in villus BBM disaccharidase activities; glucose and sodium transport that contributes to massive water loss into the lumen without histologic damage. Moderate net chloride secretion coincident with diarrhea has been reported, but the increase in luminal chloride following rotavirus is much lower compared to that in pure SDs elicited by bacterial enterotoxins of *V. cholerae* or *E. Coli*. Other studies reported both chloride influx and efflux across the villus BBM following rotavirus infection (Lorrot et al. 2006).

The precise source of chloride secretion in rotavirus diarrhea remains unclear. Rotavirus is able to elicit diarrhea in mice lacking CFTR, suggesting that alternate chloride transport pathways are implicated. CaCCs and a chloride channel ClC-2 are suggested as alternate chloride efflux pathways to CFTR in the intestine. With regards to rotavirus pathogenesis, existing data indicate that the mechanisms are quite different compared to bacterial enterotoxins that cause a pure secretory diarrhea. Studies have focused on CaCCs. The rotavirus nonstructural protein NSP4 and the synthetic peptide $NSP4_{114-135}$ were shown to elicit diarrhea in young rodents (Ousingsawat et al. 2011). Infected intestinal cells release NSP4 into the lumen where it acts as a viral toxin, induces calcium-dependent chloride secretion and inhibition of SGLT1. New data using the synthetic peptide $NSP4_{114-135}$ suggest that rotavirus may elicit diarrhea by activating the calcium-activated chloride channel TMEM16A, while inhibiting the epithelial Na^+ channel ENaC and the SGLT1 (Ousingsawat et al. 2011). However, the role of CaCCs in the native intestine remains controversial and their contribution to secretory diarrhea pathogenesis is currently unclear. More detailed discussions of CaCCs can be found in excellent reviews (Barrett and Keely 2000; Pedemonte and Galietta 2014).

2.11.1 Treatment for Rotavirus

The development of effective rotavirus vaccines has already markedly reduced diarrhea, morbidity, and mortality from rotavirus gastroenteritis in affected children worldwide (Gray 2011). In addition to vaccines, a number of groups are investigating the role of CaCCs in rotavirus-induced diarrhea with a view to therapeutic intervention.

2.12 Congenital Diarrhea

This section reviews the secretory diarrheal diseases that are not induced by any infectious organisms, but rather have the following features in common: they all manifest early in life, often immediately after birth, they are a result of inborn genetic defects, and often, are notoriously difficult to treat. Such entities are collectively called Congenital Diarrheas.

2.12.1 Congenital Chloride Diarrhea (CCD)

Gamble et al. and Darrow (Gamble et al. 1945; Darrow 1945) first described infants with diarrhea from birth, metabolic alkalosis, stool chloride levels in excess of stool sodium values, and failure to thrive. They termed the disorder "congenital alkalosis with chloride diarrhea." It was later renamed CCD (McReynolds et al. 1974), an autosomal recessive form of severe chronic diarrhea characterized by secretion of large amounts of watery stool containing high levels of chloride, leading to dehydration, hypokalemia, and metabolic alkalosis. The electrolyte disorder resembles the renal disorder Bartter syndrome, except that CCD is not associated with calcium abnormalities. Hyponatremia, hypochloremia, and metabolic alkalosis are the notable biochemical abnormalities in affected children and infants (Choi et al. 2009). Voluminous watery stools are present from birth, and polyhydramnios is a typical feature, presumably due to intrauterine diarrhea (Holmberg et al. 1975). The disease is most common in Finnish (incidence 1:20,000), Polish (incidence 1:200,000), and Arabian populations (incidence up to 1:5000) (Kere et al. 1999). Defective Cl^-/HCO_3^- exchange was proposed as the pathophysiological mechanism as early as the 1970s (Holmberg et al. 1975; Holmberg 1978), but it was not until 1990 that a gene defect was implicated in CCD. Linkage disequilibrium analysis in Finnish families (Kere et al. 1993) positioned the candidate gene on chromosome 7, close to but distinct from the CFTR gene, and more detailed genetic and physical mapping studies suggested a candidate gene known as SLC26A3 (Hoglund et al. 1995, 1996a, b). Each of the three high-incidence populations has its own founder mutation: in Finland, the V317del mutation is present in all but one chromosome studied to date (98%); in Saudi Arabia and Kuwait, the G187X mutation is present in 17 of 18 chromosomes (94%); and in Poland the most common mutation I675–676 accounts for 16 of 34 disease chromosomes (47%) (Kere et al. 1999). DRA (SLC26A3) encodes a Na^+-independent Cl^-/HCO_3^- (or OH^-) exchanger. It is a member of the SLC26 sulfate permease/anion transporter family and it is expressed mainly in the apical brush border of the intestinal epithelium. The basic defect in CCD is loss of SLC26A3-mediated transport in the surface epithelium of ileum and colon. This results in defective intestinal absorption of Cl^- and secretion of HCO_3^-. Secondarily, the coupled Na^+/H^+ transport through the Na^+/H^+ exchangers (NHE2 or NHE3) is defective, leading to loss of both sodium and chloride and watery Cl^--rich diarrhea (Fig. 2.3; CCD, *left panel*). The only extraintestinal tissues that express SLC26A3 are eccrine sweat glands and seminal vesicles (Makela et al. 2002).

Interestingly, DRA is the exit pathway for SO_4^{2-} (Tyagi et al. 2001; Whittamore et al. 2013) as well as HCO_3^-; DRA was shown to be upregulated in the colon of mice lacking NHE3 exchanger (Melvin et al. 1999). Taken together with the reports that CFTR markedly activates Cl^- and OH^-/HCO_3^- by DRA, SLC26A6 and pendrin (Ko et al. 2002), it may be safe to postulate that the intestinal machinery responsible for Cl^- and HCO_3^- transport is regulated as a functional unit.

2.12.1.1 Treatment for CCD

Treatment of CCD consists of replacement with a high chloride intake to prevent volume depletion. Determining the optimal replacement dose is challenging because inadequate salt substitution paradoxically decreases diarrhea volume, and excessive salt administration also increases diarrhea by osmotic mechanisms (Wedenoja et al. 2010). In patients with CCD, oral intake of chloride, sodium, and potassium must exceed fecal output (i.e., there must be a positive gastrointestinal balance) so that

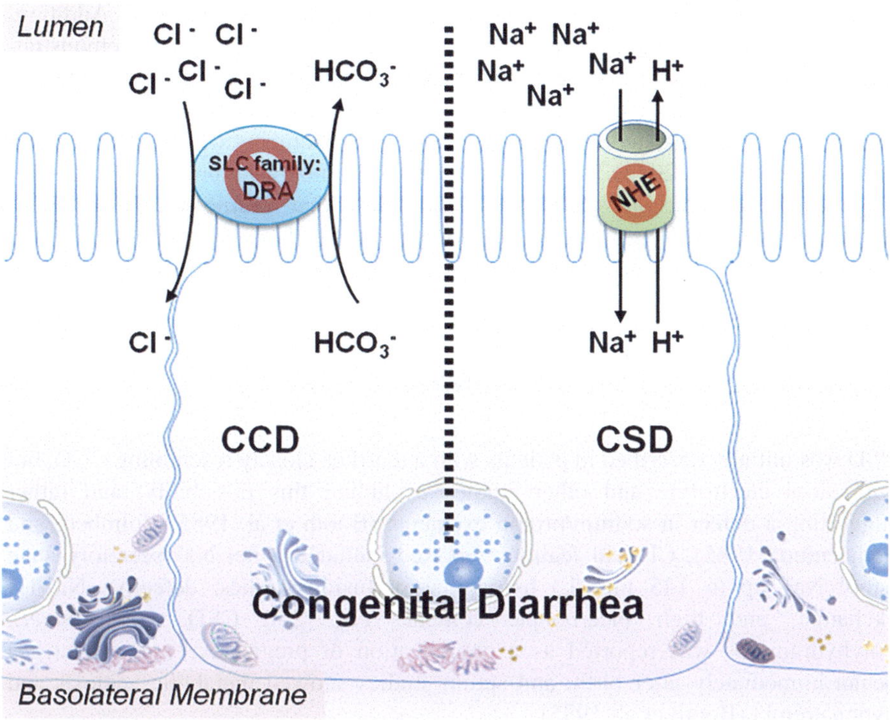

Fig. 2.3 Schematic diagram of pathogenesis of Congenital Chloride (CCD) and Sodium Diarrhea (CSD). DRA (SLC26A3) is a member of the SLC26 sulfate permease/anion transporter family that is primarily expressed on the apical BBM of the intestinal (ileum, colon) epithelium and regulates Na^+-independent Cl^-/HCO_3^- (or OH^-) exchange. Defect or dysfunction of DRA leads to accumulation of Cl^- in the lumen resulting in large volume of watery Cl^--rich diarrhea (*CCD*). Na^+/H^+ Exchangers (NHEs) regulate the Na^+/H^+ exchange across the plasma membrane. Mutations in NHEs are thought to result in malabsorption of Na^+ manifesting as Na^+-rich watery diarrhea (*CSD*)

obligatory losses in sweat can be replaced. A positive balance can best be ensured by a high intake of chloride, even though it exacerbates diarrhea.

Aichbichler and colleagues successfully used a proton-pump inhibitor, Omeprazole, for treatment of CCD in a 34-year-old male. Omeprazole reduced the number of stools from 6–12 per day down to 2–4 per day and prevented episodes of fecal incontinence. However, it did not reduce the need for careful monitoring of dietary intake, serum electrolyte concentrations, and urinary chloride excretion (Aichbichler et al. 1997).

Canani and coworkers examined butyrate as a possible treatment for CCD in 2004, concluding that butyrate could be effective in treating CCD (Canani et al. 2004). It is easily administered, useful in preventing severe dehydration episodes, and may be a promising therapeutic approach for a long-term treatment in this rare and severe condition (Canani et al. 2004). Wedenoja et al. reported on the controversies associated with butyrate treatment (Wedenoja et al. 2008). Additionally, Canani and colleagues revisited their study of butyrate efficiency, demonstrating its dependence on genotype in seven CCD children; the effect of butyrate was related in part to a different modulation of the expression of the two main apical membrane Cl^- exchangers of epithelial cells, both members of the SLC26 anion family (Canani et al. 2013). To improve the early diagnosis and prevent unnecessary surgical interventions for conditions like CCD that mimic meconium ileus (Krzemien et al. 2013), genetic and ultrasonographic methods are useful for prenatal diagnosis.

2.12.2 Congenital Sodium Diarrhea (CSD)

CSD was initially described in patients with a diarrhea closely resembling CCD, but with stool electrolyte and other studies excluding this possibility, and rather suggesting a defect in sodium/proton exchange (Booth et al. 1985; Holmberg and Perheentupa 1985). Clinical features reported included a net Na^+ secretory state (stool Na^+ up to 145 mEq/L) by jejunal perfusion studies, defective Na^+/H^+ exchange, and high bicarbonate content (Fig. 2.3; CSD, *right panel*). Polyhydramnios was reported as a complication of pregnancy, watery diarrhea began immediately after birth, and serum studies showed metabolic acidosis and hyponatremia (Booth et al. 1985).

The specific member of the NHE family that is directly responsible for the disease remains to be identified. Muller et al. conducted analysis of clinical and laboratory data from 5 affected patients and concluded that CSD is an autosomal recessive disorder, but is not related to mutations in the NHE1, NHE2, NHE3, and NHE5 genes encoding for currently known sodium/proton exchangers (Muller et al. 2000). NHE8, another member of the NHE family, was cloned and its expression was found to be higher in the neonatal intestine compared to adult, suggesting that it may be implicated in CSD. However, genetic studies of 5 patients failed to identify mutations in NHE8 (Baum et al. 2011).

Several variants of the disease have been reported (Keller et al. 1990; Fell et al. 1992), including the variant associated with morphological malformations, such as anal atresia and double kidney (Bird et al. 2007; Heinz-Erian et al. 2009). The latter group applied a genome-wide SNP scan and identified a homozygous c.593-1G→A splicing mutation in SPINT2, encoding a Kunitz-type serine-protease inhibitor, in one extended kindred with syndromic CSD. The same mutation and four distinct, homozygous or compound heterozygous mutations (p.Y163C, c.1A→T, c.337+2T→C, c.553+2T→A), were identified in all syndromic patients. No SPINT2 mutations were found in classic-CSD patients, leading the authors to delineate syndromic CSD as a distinct disease entity caused by SPINT2 loss-of-function mutations (Heinz-Erian et al. 2009).

Treatment of CSD remains supportive in the absence of more detailed understanding of the mechanisms leading to disease.

2.12.3 Microvillus Inclusion Disease (MVID)

MVID is a rare, life-threatening, autosomal recessive disorder of childhood that presents with severe intractable diarrhea. Without total parenteral nutritional support (TPN) or intestinal transplantation, death from diarrhea and dehydration is inevitable. Fecal sodium and chloride concentration approximate those of serum and the volume of stool loss often exceeds that seen in cholera. In contrast to CCD, oligohydramnios or prenatal diarrhea is not a feature of MVID and diarrhea is noted most often after feeding in the first hours and days of life (Cutz et al. 1989; Sherman et al. 2004). MVID has been identified in infants from multiple ethnic backgrounds, is associated with consanguinity, and clusters in certain populations, including infants of Navajo Indian descent (Pohl et al. 1999). Two variants of MVID are recognized: an early and late onset. Late-onset MVID can present within the first month of life and is usually less severe than the early variant (Croft et al. 2000).

In 1978, Davidson and coworkers first recognized MVID in a report of 5 infants with persistent severe diarrhea from birth and marked abnormalities of absorption with failure to thrive. The diarrhea eventually led to death in 4 out of 5 children due to failure to control and correct fluid and nutritional abnormalities. Ion flux studies demonstrated a net secretory state that accounts for the voluminous diarrhea. Marker perfusion studies of the proximal jejunum also revealed abnormal glucose absorption and a blunted response of Na^+ absorption. Net secretion of Na^+ and H_2O was also reported in MVID patients (Davidson et al. 1978).

Initial unifying morphological findings in the small intestine of patients suffering from MVID included villus atrophy, crypt hypoplasia without an increase in mitoses or inflammatory cell infiltrate in the lamina propria and absence of a brush border in villus enterocytes, increase in lysosome-like inclusions, and autophagocytosis (Davidson et al. 1978). Follow-up reports confirmed the ultrastructural presence of microvilli decorating intracytoplasmic inclusions (microvillus inclusions, MIs) in the enterocytes, and led to the designation MVID and the recognition of this feature as a pathognomonic hallmark of the disease.

In the following decade, several other cases were identified (Phillips et al. 1985; Cutz et al. 1989) and characterized. Cutz's group concluded that MVID might very well represent the most common cause of severe refractory diarrhea in the neonatal period. First notions were made of the possible pathogenesis of the disease, implicating defective intracellular transport leading to aberrant assembly or differentiation of the components of the enterocyte brush border. Rectal biopsy was proposed as a dependable and relatively easy method for early diagnosis (Cutz et al. 1989).

The clear definition of the clinical presentation and diagnostic features facilitated further analysis of the pathogenesis of MVID. Rhoads and coworkers conducted a study on the transport abnormalities of the MVID small intestine (Rhoads et al. 1991). They analyzed proximal and distal tract outputs via a proximal jejunostomy placement in one patient showing that less fluid was lost from mouth to jejunum (60 ml/kg^{1-}) than from jejunum to anus (100 ml/kg/day). They also performed Na^+, Cl^- , and conductance analysis in the piece of jejunum excised during jejunostomy and reported that compared with intestine of normal infants, this infant's epithelium showed transmural conductance and unidirectional ion fluxes that were only 30% of normal. With respect to both Na^+ and Cl^-, the excised jejunum was in a net secretory state.

Phillips and Schmitz surveyed 23 cases of MVID and described the pattern of the diarrhea, defining congenital and late-onset cases, and the volume and ion content of stools in affected patients (Phillips and Schmitz 1992). Late-onset cases appeared to have a better prognosis, which was still very grim overall. The authors also reported that the presence of inclusions is characteristic of the mature enterocytes, while in the less differentiated cells in the crypts a counterpart of the inclusions would be the abundant "secretory granules." Electron micrographs presented in their paper demonstrated that there were at least three different and distinctive appearances of the proximal small intestinal mucosa in MVID, and that not all enterocytes completely lacked microvilli, but enterocytes possessing mature brush borders were not abundant. These authors documented MIs attached to the apical plasma membrane and to the apical cell-cell contact areas (Phillips and Schmitz 1992).

The presence of intracellular vacuolar apical compartment with microvilli was noted previously in the cultured MDCK cell line following the disruption of cell-cell contacts (Vega-Salas et al. 1988). Chronic treatment of cultured cells with microtubular transport inhibitors such as Vincristine, Vinblastine, and Cytochalasin resulted in the formation of basolateral inclusions laden with microvilli and the collapse of apical polarity (Carruthers et al. 1986; Ellinger and Pavelka 1986; Achler et al. 1989). By the mid-90s, MVID was considered among the disorders resulting from alterations in epithelial polarity (Fish and Molitoris 1994). During this period, intestinal transplantation emerged as a recognized treatment for MVID (Oliva et al. 1994; Randak et al. 1998; Bunn et al. 2000).

Attempts at more detailed characterization of the molecular defect leading to diarrhea in MVID began to appear in the late 1990s—early 2000s. Michail et al. observed reduction in the expression levels of NHE2, NHE3, and SGLT1, but not NHE1 in the intestinal samples from MVID patients. They concluded that the patients with MVID had defects in apical but not basolateral membrane transport systems, and proposed that these defects were related to the pathogenesis of the disease (Michail et al. 1998). In 2000, Ameen and Salas executed a large-scale comparative analysis of protein distribution in MVID intestine, comparing apical (CFTR, actin, NHE3, sucrase-isomaltase, alkaline phosphatase, and cGKII) and basolateral (Na^+/K^+-ATPase) protein localization patterns. They demonstrated that all apical proteins examined were mislocalized in MVID enterocytes, and redistributed from the brush-border area into the cytosol and MIs. The effect was more pronounced in mature villus enterocytes. Distribution of the basolateral marker Na^+/K^+-ATPase was not affected (Ameen and Salas 2000), pointing to an apical specific trafficking defect; however, the latter discovery could be case-specific, as subsequent studies identified redistribution of basolateral transferrin receptor as well as Na^+/K^+-ATPase in MVID tissues (Dong et al. 2012; Thoeni et al. 2014b).

Increased paracellular macromolecular transport and reduced glucose uptake in duodenal biopsies of patients with MVID was demonstrated by Bijlsma et al. (2000).

Reinshagen and colleagues used immunogold electron microscopy to demonstrate that labeled ovalbumin appears in the inclusions after apical treatment, concluding that MIs are the result of autophagocytosis of the apical membrane in MVID (Reinshagen et al. 2002). Studies of small GTP-binding protein Rab8 in the knock-out mouse model by Sato et al. initially seemed very promising for deciphering the MVID pathogenesis, because the absence of Rab8 leads to the formation of BB-decorated inclusions of KO mice resembling those of MIs in human MVID intestine (Sato et al. 2007). However, Muller and coworkers identified the gene of Myosin Vb (Myo5b), a 214-kDa molecular motor, as the target gene responsible for MVID (Muller et al. 2008). This was consistent with previous observations of a 200-kDa band disappearance from intestinal immunoblots by Carruthers and coworkers (Carruthers et al. 1985). The group also demonstrated that the absence of Myo5b affects cell polarity (Muller et al. 2008). In 2010, Ruemmelle et al. confirmed that Myo5b was implicated in MVID pathogenesis and identified 15 novel Myo5b mutations in MVID patients. They also provided the first in vitro cellular model of MVID based on siRNA silencing of Myo5b expression (Ruemmele et al. 2010).

Current research focuses on the characterization of apical transport pathways connecting apical recycling endosomes (ARE) to the apical plasma membrane. MVID is assigned to the category of diseases associated with defects in "enterocyte differentiation and polarization" (Canani and Terrin 2011). Myo5b-binding partners, including Rab8 and Rab11, are uniformly accepted as small GTP-ases that are involved in the pathogenesis of MVID (Talmon et al. 2012; Dhekne et al. 2014; Dong et al. 2012) (Fig. 2.1; MVID, *left panel*). Recently, aberrations in the phosphorylation of ezrin and defects in apical PDK1 signaling were identified giving an insight into the mechanism of apical BB abnormalities and defects in apical sodium transport in human MVID intestine and an intestinal Caco2 cell model (Kravtsov et al. 2014; Dhekne et al. 2014). In 2014, Wiegerinck et al. provided an insight into the early- and late-onset variants of MVID by demonstrating that the latter is the result of mutations in the Syntaxin 3 gene (Wiegerinck et al. 2014). The involvement of a different gene may explain the difference in the patient presentation in the variant form of MVID.

The Rab11/Slp4/Myo5b/Munc18 complex carries apically bound vesicles to the surface and mediates their binding to STX3, as a first step for membrane fusion. This pathway mediates apical delivery of a subset of apical membrane proteins, including NHE3, the sodium-hydrogen exchanger that is necessary for sodium absorption in the small intestine, CFTR, and other apical membrane proteins (Vogel et al. 2015). However, apical surface expression of these proteins is severely decreased in MVID patients (Ameen et al. 2000; Kravtsov et al. 2016). Yet, defects in this exocytic pathway lead to selectivity in apical protein targeting to the membrane. While there is a consensus that NHE3 is not expressed on the apical surface, alkaline phosphatase and sucrase isomaltase appear both on the surface and in subapical compartments (Ameen et al. 2000). Furthermore, CFTR, the main Cl^- channel responsible for cAMP-dependent secretory diarrhea, is only partially decreased in human MVID intestine (Kravtsov et al. 2016). It has been proposed that MVID diarrhea results from an imbalance of only partially decreased CFTR (secretion) and totally absent NHE3 (absorption). However, absence of NHE3 on the MVID enterocyte apical membrane cannot explain the magnitude of the secretory diarrhea in these patients, because NHE3-deficient mice (Schultheis et al. 1998) and humans with congenital sodium diarrhea due to NHE3 mutations (Janecke et al. 2015) display milder diarrhea even in the presence of normal CFTR expression. In fact, it is counterintuitive that decreased CFTR as a consequence of the Myo5b loss of function contributes to secretory diarrhea. Provocative recent studies by Ameen and Salas provide insights that explain the onset of diarrhea after birth in MVID. These studies implicate perinatal stress, serum and gluococortoicoid-induced kinase (SGK) signaling, and downstream PKA activation of CFTR by phosphorylation leading to fluid secretion, resulting from loss of function of Myo5B and defective polarity (Forteza et al. 2019).

Glucocorticoids are known to regulate ion transport by nongenomic mechanisms (Lang et al. 2014). Our most recent work provides further evidence in support of an important role for SGK and downstream kinase signaling in diarrheal diseases. In vivo data from rat intestine studies revealed a profound role for short-term nongenomic glucocorticoid-induced regulation of CFTR protein, surface expression and function, that are linked to activation of the ubiquitin ligase neural precursor cell expressed developmentally downregulated 4-like (Nedd4–2) and 14–3-3ε, leading to stability and retention of CFTR on the apical membrane (Ahsan et al. 2020). Thus, in addition to its role in MVID diarrhea, kinase signaling is emerging as an important new druggable pathway for stress-related diarrheal diseases.

Significant progress over the past 5 years has been made in understanding the genetic, molecular, and signaling mechanisms involved in the pathogenesis of MVID; however, several important questions remain unanswered. Myo5b is expressed in all epithelial cells (Rodriguez and Cheney 2002), but the reason that the major clinical disease manifestation targets the small intestine is not understood. Recent studies provide insights into the role of Myo5b in the genesis of cholestatic liver disease (Qiu et al. 2017) and extraintestinal disease (Golachowska et al. 2012; Jajawardena et al. 2019). Ongoing and future studies should identify intestine-specific factors to explain selective targeting of disease in the intestine.

2.12.3.1 Treatment MVID

In 1994, the first successful intestinal transplantation was performed for MVID (Oliva et al. 1994). Since then, there have been no advances in therapeutic modalities largely because of lack of understanding of the underlying pathogenesis of the secretory diarrhea. Thus, TPN and intestinal transplantation are the only currently available treatment options. While supportive, these are not ideal therapeutic modalities. Long-term use of TPN is associated with many complications, including growth retardation, septicemia, cirrhosis, and cholelithiasis all of which contribute to impaired survival in MVID patients. Intestinal transplantation is expensive, available only in specialized tertiary case centers, and associated with high mortality.

Given the recent identification of glucocorticoids and the SGK1 kinase signaling pathway in MVID, it is conceivable that kinase inhibitors can be developed to reduce MVID diarrhea in the future.

2.12.4 Tufting Enteropathy

Congenital tufting enteropathy (CTE) or intestinal epithelial dysplasia is a rare autosomal recessive disease affecting multiple epithelial tissues. CTE presents with intractable diarrhea of infancy characterized by villous atrophy, absence of inflammation, focal epithelial tufts involving the small intestine and colon (Goulet et al. 1995). CTE presents early in life with chronic watery diarrhea and failure to thrive, and most affected individuals require parenteral nutrition for normal growth and development (Sivagnanam et al. 2008).

The disease was first described by Reifen et al. (1994) in a group of 3 children, each of whom presented early in life with protracted watery diarrhea with volumes in excess of 1500 ml/day. The diarrhea was refractory to a variety of diets, necessitating the use of TPN. Characteristic histopathologic findings included partial villous atrophy accompanied by crypt hyperplasia and an increased number of mitotic figures in the crypts. The most striking findings in the small intestine of the three children were observed in the surface epithelium, which showed focal epithelial "tufts." Tufts were composed of closely packed enterocytes with apical rounding of the plasma membrane, resulting in a teardrop configuration to the cells. Approximately 80–90% of epithelial surface contained tufts, versus <10% in the normal jejunum. Colonic biopsies showed no histopathologic abnormalities: specifically, no tufting lesions were observed (Reifen et al. 1994).

Although CTE was initially considered a defect of a2b1 integrin (Goulet et al. 1995; Patey et al. 1997), genotyping studies by Sivagnanam et al. supported by direct sequencing of genes on chromosome 2p21, discovered a homozygous mutation in epithelial cell adhesion molecule (EpCAM) of affected patients (Sivagnanam et al. 2008). Immunohistochemistry and immunoblotting of patient intestinal tissue revealed decreased expression of EpCAM. The authors concluded that EpCAM mutations are responsible for CTE, and presented further evidence for EpCAM as the gene for congenital tufting enteropathy by identifying a novel homozygous C > T substitution at c.412 of exon 3 of EpCAM resulting in amino acid change from arginine (CGA) to a stop codon (UGA) (Sivagnanam et al. 2010b).

As more cases of CTE were identified, a variant form of the disease associated with various malformations (most consistently, keratitis and choanal atresia) (Abely et al. 1998; El-Matary et al. 2007; Bird et al. 2007; Al-Mayouf et al. 2009) was described and found to be caused by mutations of SPINT2, a gene associated with a variant form of CSD (Sivagnanam et al. 2010a). In the review of 57 cases of CTE, an EpCAM mutation was involved in 41 patients (73%) who mainly displayed isolated digestive symptoms. 12 other patients (21%) carried mutations in SPINT2, and were phenotypically characterized by morphological malformations. Finally, 4 patients

(7%) with isolated digestive symptoms had no detectable EpCAM or SPINT2 mutation (Salomon et al. 2014). The precise mechanism of diarrhea following the loss of EpCAM is still unknown. Dysfunctional epithelial barrier is one of the possible explanations (Thoeni et al. 2014a), and EpCAM can possibly interact with the tight junction protein Claudin 7 in the human intestine (Mueller et al. 2014). Treatment of CTE thus relies on TPN and bowel transplantation as major modes of life support.

2.13 Summary

SD remains a major global health problem that contributes to significant morbidity and mortality in children. Until recently, no new therapies were developed for treating SD, as pharmaceutical industry resources have been focused on developing therapies for diseases affecting the developed world. Fortunately, over the last decade, there have been significant advances in elucidating the underlying pathophysiology of infectious and genetic causes of secretory diarrhea. Genetic advances contributed to identification of several new intestinal ion transporters and molecules such as molecular motors relevant for infectious and inherited SD pathogenesis. Cell biological approaches combined with conventional electrophysiological ion transport methods have elucidated novel cell specific mechanisms (endocytosis, exocytosis, adaptors, molecular motors) that regulate intestinal ion transporters and contribute to SD. These advances have led to development of new therapies such as Crofelemer that is currently approved and in use for HIV-related diarrhea. The identification of CFTR as the major apical exit pathway for chloride in the intestine led to development of inhibitors that are in the pipeline for cholera and *E. coli*-based SD. The rapid pace of recent research advances in the pathogenesis of congenital diarrheas, including CCD, CSD, and MVID, will undoubtedly lead to development of new and specific therapies in the coming years. This will prove a relief for the many affected families of inherited childhood diarrheas since intestinal transplant remains the only hope of survival, but is far from ideal.

Acknowledgments The authors wish to acknowledge Anne Collaco for assistance and technical support and the NIH-R01 DK 077065 to NA.

References

Abely M, Hankard GF, Hugot JP et al (1998) Intractable infant diarrhea with epithelial dysplasia associated with polymalformation. J Pediatr Gastroenterol Nutr 27(3):348–352

Achler C, Filmer D, Merte C et al (1989) Role of microtubules in polarized delivery of apical membrane proteins to the brush border of the intestinal epithelium. J Cell Biol 109(1):179–189

Aichbichler BW, Zerr CH, Santa Ana CA et al (1997) Proton-pump inhibition of gastric chloride secretion in congenital chloridorrhea. N Engl J Med 336(2):106–109

Aimoto S, Takao T, Shimonishi Y et al (1982) Amino-acid sequence of a heat-stable enterotoxin produced by human enterotoxigenic *Escherichia coli*. Eur J Biochem 129(2):257–263
Aimoto S, Watanabe H, Ikemura H et al (1983) Chemical synthesis of a highly potent and heat-stable analog of an enterotoxin produced by a human strain of enterotoxigenic *Escherichia coli*. Biochem Biophys Res Commun 112(1):320–326
Alderete JF, Robertson DC (1978) Purification and chemical characterization of the heat-stable enterotoxin produced by porcine strains of enterotoxigenic *Escherichia coli*. Infect Immun 19 (3):1021–1030
Al-Mayouf SM, Alswaied N, Alkuraya FS et al (2009) Tufting enteropathy and chronic arthritis: a newly recognized association with a novel EpCAM gene mutation. J Pediatr Gastroenterol Nutr 49(5):642–644
Ameen N, Apodaca G (2007) Defective CFTR apical endocytosis and enterocyte brush border in myosin VI-deficient mice. Traffic 8(8):998–1006
Ameen NA, Salas PJ (2000) Microvillus inclusion disease: a genetic defect affecting apical membrane protein traffic in intestinal epithelium. Traffic 1(1):76–83
Ameen NA, Ardito T, Kashgarian M et al (1995) A unique subset of rat and human intestinal villus cells express the cystic fibrosis transmembrane conductance regulator. Gastroenterology 108 (4):1016–1023
Ameen NA, Martensson B, Bourguinon L et al (1999) CFTR channel insertion to the apical surface in rat duodenal villus epithelial cells is upregulated by VIP in vivo. J Cell Sci 112(Pt 6):887–894
Ameen NA, van Donselaar E, Posthuma G et al (2000) Subcellular distribution of CFTR in rat intestine supports a physiologic role for CFTR regulation by vesicle traffic. Histochem Cell Biol 114(3):219–228
Ameen N, Silvis M, Bradbury NA (2007) Endocytic trafficking of CFTR in health and disease. J Cyst Fibros 6(1):1–14
Ahsan, Md K, Figueroa-Hall L, Baratta V et al (2020) Glucocorticoids and Serum and Glucocorticoid-inducible kinase 1 (SGK1) are potent regulators of CFTR in the native intestine: implications for stress-induced diarrhea. Am J Physiol 319(2):G121–G132
Barrett KE (2000) New insights into the pathogenesis of intestinal dysfunction: secretory diarrhea and cystic fibrosis. World J Gastroenterol 6(4):470–474
Barrett KE, Keely SJ (2000) Chloride secretion by the intestinal epithelium: molecular basis and regulatory aspects. Annu Rev Physiol 62:535–572
Baum M, Martin MG, Booth IW et al (2011) Nucleotide sequence of the Na^+/H^+ exchanger-8 in patients with congenital sodium diarrhea. J Pediatr Gastroenterol Nutr 53(5):474–477
Baxter PS, Dickson JA, Variend S et al (1988) Intestinal disease in cystic fibrosis. Arch Dis Child 63(12):1496–1497
Bernardi KM, Forster ML, Lencer WI et al (2008) Derlin-1 facilitates the retro-translocation of cholera toxin. Mol Biol Cell 19(3):877–884
Bijlsma PB, Kiliaan AJ, Scholten G et al (2000) Increased paracellular macromolecular transport and subnormal glucose uptake in duodenal biopsies of patients with microvillus inclusion disease. Comparisons to other chronic diarrhea patients and to nondiarrhea patients. Ann N Y Acad Sci 915:267–269
Binder HJ (2012) Intestinal fluid and electrolyte movement In: Boron WF, Boulpaep EL (eds) Medical physiology a cellular and molecular approach, vol 1 (Updated 2nd edn). Saunders Elsevier, Philadelphia, p 933
Binder HJ, Reuben A (2012) Nutrient digestion and absorption. In: Boron WF, Boulpaep EL (eds) Medical physiology a cellular and molecular approach, vol 1 (Updated 2nd edn). Saunders Elsevier, Philadelphia, p 949
Bird LM, Sivagnanam M, Taylor S et al (2007) A new syndrome of tufting enteropathy and choanal atresia, with ophthalmologic, hematologic and hair abnormalities. Clin Dysmorphol 16 (4):211–221
Bohles N, Busch K, Hensel M (2014) Vaccines against human diarrheal pathogens: current status and perspectives. Hum Vaccin Immunother 10(6):1522–1535

Booth IW, Stange G, Murer H et al (1985) Defective jejunal brush-border Na^{+}/H^{+} exchange: a cause of congenital secretory diarrhoea. Lancet 1(8437):1066–1069

Buccigrossi V, Laudiero G, Russo C et al (2014) Chloride secretion induced by rotavirus is oxidative stress-dependent and inhibited by *Saccharomyces boulardii* in human enterocytes. PLoS One 9(6):e99830

Bunn SK, Beath SV, McKeirnan PJ et al (2000) Treatment of microvillus inclusion disease by intestinal transplantation. J Pediatr Gastroenterol Nutr 31(2):176–180

Burgess MN, Bywater RJ, Cowley CM et al (1978) Biological evaluation of a methanol-soluble, heat-stable *Escherichia coli* enterotoxin in infant mice, pigs, rabbits, and calves. Infect Immun 21(2):526–531

Buss F, Arden SD, Lindsay M et al (2001) Myosin VI isoform localized to clathrin-coated vesicles with a role in clathrin-mediated endocytosis. EMBO J 20(14):3676–3684

Buss F, Spudich G, Kendrick-Jones J (2004) Myosin VI: cellular functions and motor properties. Annu Rev Cell Dev Biol 20:649–676

Camilleri M, Nullens S, Nelsen T (2012) Enteroendocrine and neuronal mechanisms in pathophysiology of acute infectious diarrhea. Dig Dis Sci 57(1):19–27

Canani RB, Terrin G (2011) Recent progress in congenital diarrheal disorders. Curr Gastroenterol Rep 13(3):257–264

Canani RB, Terrin G, Cirillo P et al (2004) Butyrate as an effective treatment of congenital chloride diarrhea. Gastroenterology 127(2):630–634

Canani RB, Terrin G, Elce A et al (2013) Genotype-dependency of butyrate efficacy in children with congenital chloride diarrhea. Orphanet J Rare Dis 8:194

Carruthers L, Phillips AD, Dourmashkin R et al (1985) Biochemical abnormality in brush border membrane protein of a patient with congenital microvillus atrophy. J Pediatr Gastroenterol Nutr 4(6):902–907

Carruthers L, Dourmashkin R, Phillips A (1986) Disorders of the cytoskeleton of the enterocyte. Clin Gastroenterol 15(1):105–120

Chao AC, de Sauvage FJ, Dong YJ et al (1994) Activation of intestinal CFTR Cl^{-} channel by heat-stable enterotoxin and guanylin via cAMP-dependent protein kinase. EMBO J 13(5):1065–1072

Chinkers M, Garbers DL, Chang MS et al (1989) A membrane form of guanylate cyclase is an atrial natriuretic peptide receptor. Nature 338(6210):78–83

Choi M, Scholl UI, Ji W et al (2009) Genetic diagnosis by whole exome capture and massively parallel DNA sequencing. Proc Natl Acad Sci U S A 106(45):19096–19101

Cihil KM, Ellinger P, Fellows A et al (2012) Disabled-2 protein facilitates assembly polypeptide-2-independent recruitment of cystic fibrosis transmembrane conductance regulator to endocytic vesicles in polarized human airway epithelial cells. J Biol Chem 287(18):15087–15099

Cohen MB, Guarino A, Shukla R et al (1988) Age-related differences in receptors for *Escherichia coli* heat-stable enterotoxin in the small and large intestine of children. Gastroenterology 94 (2):367–373

Cohen MB, Thompson MR, Giannella RA (1989) Differences in jejunal and ileal response to *E. coli* enterotoxin: possible mechanisms. Am J Physiol Gastrointest Liver Physiol 257(1):G118–G123

Cohen MB, Mann EA, Lau C et al (1992) A gradient in expression of the *Escherichia coli* heat-stable enterotoxin receptor exists along the villus-to-crypt axis of rat small intestine. Biochem Biophys Res Commun 186(1):483–490

Collaco A, Jakab R, Hegan P et al (2010) Alpha-AP-2 directs myosin VI-dependent endocytosis of cystic fibrosis transmembrane conductance regulator chloride channels in the intestine. J Biol Chem 285(22):17177–17187

Cottreau J, Tucker A, Crutchley R et al (2012) Crofelemer for the treatment of secretory diarrhea. Expert Rev Gastroenterol Hepatol 6(1):17–23

Crawford SE, Ramani S, Tate JE et al (2017) Rotavirus infection. Nat Rev Dis Primers 3:17083

Croft NM, Howatson AG, Ling SC et al (2000) Microvillous inclusion disease: an evolving condition. J Pediatr Gastroenterol Nutr 31(2):185–189

Crutchley RD, Miller J, Garey KW (2010) Crofelemer, a novel agent for treatment of secretory diarrhea. Ann Pharmacother 44(5):878–884

Currie MG, Fok KF, Kato J et al (1992) Guanylin: an endogenous activator of intestinal guanylate cyclase. Proc Natl Acad Sci U S A 89(3):947–951

Cutz E, Rhoads JM, Drumm B et al (1989) Microvillus inclusion disease: an inherited defect of brush-border assembly and differentiation. N Engl J Med 320(10):646–651

Darrow DC (1945) Congenital alkalosis with diarrhea. J Pediatr 26:519–532

Das S, Jayaratne R, Barrett KE (2018) The role of ion transporters in the pathophysiology of infectious diarrhea. Cell Mol Gastroenterol Hepatol 6:33–45

Davidson GP, Cutz E, Hamilton JR et al (1978) Familial enteropathy: a syndrome of protracted diarrhea from birth, failure to thrive, and hypoplastic villus atrophy. Gastroenterology 75 (5):783–790

De Jonge HR (1975a) The localization of guanylate cyclase in rat small intestinal epithelium. FEBS Lett 53(2):237–242

De Jonge HR (1975b) The response of small intestinal villous and crypt epithelium to choleratoxin in rat and guinea pig. Evidence against a specific role of the crypt cells in choleragen-induced secretion. Biochem Biophys Acta 381(1):128–143

de Sauvage FJ, Keshav S, Kuang WJ et al (1992) Precursor structure, expression, and tissue distribution of human guanylin. Proc Natl Acad Sci U S A 89(19):9089–9093

Dhekne HS, Hsiao NH, Roelofs P et al (2014) Myosin Vb and Rab11a regulate phosphorylation of ezrin in enterocytes. J Cell Sci 127(Pt 5):1007–1017

Dong W, Chen X, Chen P et al (2012) Inactivation of MYO5B promotes invasion and motility in gastric cancer cells. Dig Dis Sci 57(5):1247–1252

Donowitz M, Welsh MJ (1986) Ca^{2+} and cyclic AMP in regulation of intestinal Na, K and Cl transport. Annu Rev Physiol 48:135–150

Donowitz M, Alpers DH, Binder HJ et al (2012) Translational approaches for pharmacotherapy development for acute diarrhea. Gastroenterology 142(3):e1–e9

Eggermont E (1996) Gastrointestinal manifestations in cystic fibrosis. Eur J Gastroenterol Hepatol 8 (8):731–738

Ellinger A, Pavelka M (1986) Colchicine-induced tubular, vesicular and cisternal organelle aggregates in absorptive cells of the small intestine of the rat. II. Endocytosis studies. Biol Cell 58 (1):31–41

El-Matary W, Dalzell AM, Kokai G et al (2007) Tufting enteropathy and skeletal dysplasia: is there a link? Eur J Pediatr 166(3):265–268

Farthing MJ (1994) Oral rehydration therapy. Pharmacol Ther 64(3):477–492

Farthing MJ (2001) Treatment of gastrointestinal viruses. Novartis Found Symp 238:289–300. discussion 300–285

Fell JM, Miller MP, Finkel Y et al (1992) Congenital sodium diarrhea with a partial defect in jejunal brush border membrane sodium transport, normal rectal transport, and resolving diarrhea. J Pediatr Gastroenterol Nutr 15(2):112–116

Field M, Semrad CE (1993) Toxigenic diarrheas, congenital diarrheas, and cystic fibrosis: disorders of intestinal ion transport. Annu Rev Physiol 55:631–655

Field M, Graf LH Jr, Laird WJ et al (1978) Heat-stable enterotoxin of *Escherichia coli*: in vitro effects on guanylate cyclase activity, cyclic GMP concentration, and ion transport in small intestine. Proc Natl Acad Sci U S A 75(6):2800–2804

Fish EM, Molitoris BA (1994) Alterations in epithelial polarity and the pathogenesis of disease states. N Engl J Med 330(22):1580–1588

Forteza R, Ahsan K, Garcia FC, Arango D et al (2019) Glucocorticoids and myosin 5b loss of function induce heightened PKA signaling in addition to membrane trafic defects. Mol Biol Cell 30(26):3076–3089. https://doi.org/10.1091/mbc.E18-07-0415

Gabriel SE, Brigman KN, Koller BH et al (1994) Cystic fibrosis heterozygote resistance to cholera toxin in the cystic fibrosis mouse model. Science 266(5182):107–109

Gamble JL, Fahey KR, Appleton J et al (1945) Congenital alkalosis with diarrhea. J Pediatr 26 (6):509–518
Gariepy J, Schoolnik GK (1986) Design of a photoreactive analogue of the *Escherichia coli* heat-stable enterotoxin STIb: use in identifying its receptor on rat brush border membranes. Proc Natl Acad Sci U S A 83(2):483–487
Gariepy J, Lane A, Frayman F et al (1986) Structure of the toxic domain of the *Escherichia coli* heat-stable enterotoxin ST I. Biochemistry 25(24):7854–7866
Gelfond D, Ma C, Semler J et al (2013) Intestinal pH and gastrointestinal transit profiles in cystic fibrosis patients measured by wireless motility capsule. Dig Dis Sci 58(8):2275–2281
Giannella RA, Luttrell M, Thompson M (1983) Binding of *Escherichia coli* heat-stable enterotoxin to receptors on rat intestinal cells. Am J Physiol Gastrointest Liver Physiol 8(4):G492–G498
Gill DM, Clements JD, Robertson DC et al (1981) Subunit number and arrangement in *Escherichia coli* heat-labile enterotoxin. Infect Immun 33(3):677–682
Golachowska MR, van Dael CM, Keuning H et al (2012) MYO5B mutations in patients with microvillus inclusion disease presenting with transient renal Fanconi syndrome. J Pediatr Gastroenterol Nutr 54(4):491–498
Golin-Bisello F, Bradbury N, Ameen N (2005) STa and cGMP stimulate CFTR translocation to the surface of villus enterocytes in rat jejunum and is regulated by protein kinase G. Am J Physiol Cell Physiol 289(3):C708–C716
Goulet O, Kedinger M, Brousse N et al (1995) Intractable diarrhea of infancy with epithelial and basement membrane abnormalities. J Pediatr 127(2):212–219
Gray J (2011) Rotavirus vaccines: safety, efficacy and public health impact. J Intern Med 270 (3):206–214
Guerrant RL, Hughes JM, Chang B et al (1980) Activation of intestinal guanylate cyclase by heat-stable enterotoxin of *Escherichia coli*: studies of tissue specificity, potential receptors, and intermediates. J Infect Dis 142(2):220–228
Guggino WB, Stanton BA (2006) New insights into cystic fibrosis: molecular switches that regulate CFTR. Nat Rev Mol Cell Biol 7(6):426–436
Hasegawa M, Hidaka Y, Matsumoto Y et al (1999) Determination of the binding site on the extracellular domain of guanylyl cyclase C to heat-stable enterotoxin. J Biol Chem 274 (44):31713–31718
Hasegawa M, Matsumoto-Ishikawa Y, Hijikata A et al (2005) Disulfide linkages and a three-dimensional structure model of the extracellular ligand-binding domain of guanylyl cyclase C. Protein J 24(5):315–325
Hasson T (2003) Myosin VI: two distinct roles in endocytosis. J Cell Sci 116(Pt 17):3453–3461
Hazes B, Read RJ (1997) Accumulating evidence suggests that several AB-toxins subvert the endoplasmic reticulum-associated protein degradation pathway to enter target cells. Biochemistry 36(37):11051–11054
Heinz-Erian P, Muller T, Krabichler B et al (2009) Mutations in SPINT2 cause a syndromic form of congenital sodium diarrhea. Am J Hum Genet 84(2):188–196
Hoglund P, Sistonen P, Norio R et al (1995) Fine mapping of the congenital chloride diarrhea gene by linkage disequilibrium. Am J Hum Genet 57(1):95–102
Hoglund P, Haila S, Scherer SW et al (1996a) Positional candidate genes for congenital chloride diarrhea suggested by high-resolution physical mapping in chromosome region 7q31. Genome Res 6(3):202–210
Hoglund P, Haila S, Socha J et al (1996b) Mutations of the Down-regulated in adenoma (DRA) gene cause congenital chloride diarrhoea. Nat Genet 14(3):316–319
Holmberg C (1978) Electrolyte economy and its hormonal regulation in congenital chloride diarrhea. Pediatr Res 12(2):82–86
Holmberg C, Perheentupa J (1985) Congenital Na^+ diarrhea: a new type of secretory diarrhea. J Pediatr 106(1):56–61
Holmberg C, Perheentupa J, Launiala K (1975) Colonic electrolyte transport in health and in congenital chloride diarrhea. J Clin Invest 56(2):302–310

Jajawardena D, Alrefai WA, Dudeja D, Gill RK (2019) Recent advances in understanding and managing malabsorption: focus on microvillus inclusion disease. F1000 Res 8:2061

Jakab RL, Collaco AM, Ameen NA (2011) Physiological relevance of cell-specific distribution patterns of CFTR, NKCC1, NBCe1, and NHE3 along the crypt-villus axis in the intestine. Am J Physiol Gastrointest Liver Physiol 300(1):G82–G98

Jakab RL, Collaco AM, Ameen NA (2012) Cell-specific effects of luminal acid, bicarbonate, cAMP, and carbachol on transporter trafficking in the intestine. Am J Physiol Gastrointest Liver Physiol 303(8):G937–G950

Jakab RL, Collaco AM, Ameen NA (2013) Characterization of CFTR high expresser cells in the intestine. Am J Physiol Gastrointest Liver Physiol 305(6):G453–G465

Janecke A, Heinz-Erian P, Yin J et al (2015) Reduced sodium/proton exchanger NHE3 activity causes congenital sodium diarrhea. Hum Mol Gen 24(23):6614–6623

Joo NS, London RM, Kim HD et al (1998) Regulation of intestinal Cl^- and $HCO3^-$secretion by uroguanylin. Am J Phys 274(4 Pt 1):G633–G644

Keller KM, Wirth S, Baumann W et al (1990) Defective jejunal brush border membrane sodium/proton exchange in association with lethal familial protracted diarrhoea. Gut 31(10):1156–1158

Kelly M, Trudel S, Brouillard F et al (2010) Cystic fibrosis transmembrane regulator inhibitors CFTR(inh)-172 and GlyH-101 target mitochondrial functions, independently of chloride channel inhibition. J Pharmacol Exp Ther 333(1):60–69

Kere J, Sistonen P, Holmberg C et al (1993) The gene for congenital chloride diarrhea maps close to but is distinct from the gene for cystic fibrosis transmembrane conductance regulator. Proc Natl Acad Sci U S A 90(22):10686–10689

Kere J, Lohi H, Hoglund P (1999) Genetic disorders of membrane transport III. Congenital chloride diarrhea. Am J Physiol Gastrointest Liver Physiol 276(1):G7–G13

Kirkham M, Fujita A, Chadda R et al (2005) Ultrastructural identification of uncoated caveolin-independent early endocytic vehicles. J Cell Biol 168(3):465–476

Ko SB, Shcheynikov N, Choi JY et al (2002) A molecular mechanism for aberrant $CFTR^-$dependent $HCO3^-$ transport in cystic fibrosis. EMBO J 21(21):5662–5672

Kopic S, Geibel JP (2010) Toxin mediated diarrhea in the 21 century: the pathophysiology of intestinal ion transport in the course of ETEC, V. cholerae and rotavirus infection. Toxins 2 (8):2132–2157

Krause WJ, Cullingford GL, Freeman RH et al (1994) Distribution of heat-stable enterotoxin/guanylin receptors in the intestinal tract of man and other mammals. J Anat 184(Pt 2):407–417

Kravtsov DV, Ameen NA (2013) Molecular motors and apical CFTR traffic in epithelia. Int J Mol Sci 14(5):9628–9642

Kravtsov DV, Caputo C, Collaco A et al (2012) Myosin Ia is required for CFTR brush border membrane trafficking and ion transport in the mouse small intestine. Traffic 13(8):1072–1082

Kravtsov D, Mashukova A, Forteza R et al (2014) Myosin 5b loss of function leads to defects in polarized signalling: implication for microvillus inclusion disease pathogenesis and treatment. Am J Physiol Gastrointest Liver Physiol 307:G992–G1001

Kravtsov D, Ahsan K, Kumari V et al (2016) Identification of intestinal ion transport defects in microvillus inclusion disease. Am J Physiol Gastrointest Liver Physiol 311:G142–G155

Krzemien G, Szmigielska A, Jankowska K et al (2013) Congenital chloride diarrhea mimicking meconium ileus in newborn. Med Wieku Rozwoj 17(4):320–323

Lang F, Stournaras C, Alesutan I et al (2014) Regulation of transport across cell membranes by the serum and glucocorticoid-indicible kinase SGK1. Mol Membr Biol 31(1):29–36

Lauber T, Tidten N, Matecko I et al (2009) Design and characterization of a soluble fragment of the extracellular ligand-binding domain of the peptide hormone receptor guanylyl cyclase-C. Protein Eng Des Sel 22(1):1–7

Lauciricia DR, Triantis V, Schoemaker R et al (2017) Milk oligosaccharides inhibit human rotavirus infectivity in MA 104 cells. J Nutr 147(9):1709–1714

Lee CM, Chang PP, Tsai SC et al (1991) Activation of *Escherichia coli* heat-labile enterotoxins by native and recombinant adenosine diphosphate-ribosylation factors, 20-kD guanine nucleotide-binding proteins. J Clin Invest 87(5):1780–1786

Liang Y, Zhang L, Zeng L et al (2019) Racecadotril for acute diarrhoea in children. Cochrane Database of Syst Rev 2019(12):CD009359

Lorenz JN, Nieman M, Sabo J et al (2003) Uroguanylin knockout mice have increased blood pressure and impaired natriuretic response to enteral NaCl load. J Clin Invest 112(8):1244–1254

Lorrot M, Vasseur M (2007) How do the rotavirus NSP4 and bacterial enterotoxins lead differently to diarrhea? Virol J 4:31

Lorrot M, Benhamadouche-Casari H, Vasseur M (2006) Mechanisms of net chloride secretion during rotavirus diarrhea in young rabbits: do intestinal villi secrete chloride? Cell Physiol Biochem 18(1–3):103–112

Lowe DG, Chang MS, Hellmiss R et al (1989) Human atrial natriuretic peptide receptor defines a new paradigm for second messenger signal transduction. EMBO J 8(5):1377–1384

Ma T, Thiagarajah JR, Yang H et al (2002) Thiazolidinone CFTR inhibitor identified by high-throughput screening blocks cholera toxin-induced intestinal fluid secretion. J Clin Invest 110 (11):1651–1658

Madden DR, Swiatecka-Urban A (2012) Tissue-specific control of CFTR endocytosis by Dab2: Cargo recruitment as a therapeutic target. Commun Integr Biol 5(5):473–476

Majoul I, Sohn K, Wieland FT et al (1998) KDEL receptor (Erd2p)-mediated retrograde transport of the cholera toxin A subunit from the Golgi involves COPI, p23, and the COOH terminus of Erd2p. J Cell Biol 143(3):601–612

Makela S, Kere J, Holmberg C et al (2002) SLC26A3 mutations in congenital chloride diarrhea. Hum Mutat 20(6):425–438

Mann EA, Jump ML, Wu J et al (1997) Mice lacking the guanylyl cyclase C receptor are resistant to STa-induced intestinal secretion. Biochem Biophys Res Commun 239(2):463–466

Massey S, Banerjee T, Pande AH et al (2009) Stabilization of the tertiary structure of the cholera toxin A1 subunit inhibits toxin dislocation and cellular intoxication. J Mol Biol 393 (5):1083–1096

Massol RH, Larsen JE, Fujinaga Y et al (2004) Cholera toxin toxicity does not require functional Arf6- and dynamin-dependent endocytic pathways. Mol Biol Cell 15(8):3631–3641

McReynolds EW, Roy S 3rd, Etteldorf JN (1974) Congenital chloride diarrhea. Am J Dis Child 127 (4):566–570

Melvin JE, Park K, Richardson L et al (1999) Mouse down-regulated in adenoma (DRA) is an intestinal Cl^-/HCO_3^- exchanger and is up-regulated in colon of mice lacking the NHE3 Na^+/H^+ exchanger. J Biol Chem 274(32):22855–22861

Merritt EA, Sarfaty S, Van den Akker F et al (1994) Crystal structure of cholera toxin B-pentamer bound to receptor G(M1) pentasaccharide. Protein Sci 3(2):166–175

Michail S, Collins JF, Xu H et al (1998) Abnormal expression of brush-border membrane transporters in the duodenal mucosa of two patients with microvillus inclusion disease. J Pediatr Gastroenterol Nutr 27(5):536–542

Montoro DT, Haber AL, Biton M et al (2018) A revised airway epithelial hierachy includes CFTR-expressing ionocytes. Nature 560(7718):319–324

Morris SM, Arden SD, Roberts RC et al (2002) Myosin VI binds to and localises with Dab2, potentially linking receptor-mediated endocytosis and the actin cytoskeleton. Traffic 3 (5):331–341

Moseley SL, Hardy JW, Hug MI et al (1983) Isolation and nucleotide sequence determination of a gene encoding a heat-stable enterotoxin of *Escherichia coli*. Infect Immun 39(3):1167–1174

Moss J, Richardson SH (1978) Activation of adenylate cyclase by heat-labile *Escherichia coli* enterotoxin. Evidence for ADP-ribosyltransferase activity similar to that of choleragen. J Clin Invest 62(2):281–285

Muanprasat C, Sonawane ND, Salinas D et al (2004) Discovery of glycine hydrazide pore-occluding CFTR inhibitors: mechanism, structure-activity analysis, and in vivo efficacy. J Gen Physiol 124(2):125–137

Mueller JL, McGeough MD, Pena CA et al (2014) Functional consequences of EpCam mutation in mice and men. Am J Physiol Gastrointest Liver Physiol 306(4):G278–G288

Muller T, Wijmenga C, Phillips AD et al (2000) Congenital sodium diarrhea is an autosomal recessive disorder of sodium/proton exchange but unrelated to known candidate genes. Gastroenterology 119(6):1506–1513

Muller T, Hess MW, Schiefermeier N et al (2008) MYO5B mutations cause microvillus inclusion disease and disrupt epithelial cell polarity. Nat Genet 40(10):1163–1165

Nakazato M (2001) Guanylin family: new intestinal peptides regulating electrolyte and water homeostasis. J Gastroenterol 36(4):219–225

Navaneethan U, Giannella RA (2008) Mechanisms of infectious diarrhea. Nat Clin Pract Gastroenterol Hepatol 5(11):637–647

Oliva MM, Perman JA, Saavedra JM et al (1994) Successful intestinal transplantation for microvillus inclusion disease. Gastroenterology 106(3):771–774

Ousingsawat J, Mirza M, Tian Y et al (2011) Rotavirus toxin NSP4 induces diarrhea by activation of TMEM16A and inhibition of Na^+ absorption. Pflugers Arch 461(5):579–589

Pande AH, Scaglione P, Taylor M et al (2007) Conformational instability of the cholera toxin A1 polypeptide. J Mol Biol 374(4):1114–1128

Parton RG (1994) Ultrastructural localization of gangliosides; GM1 is concentrated in caveolae. J Histochem Cytochem 42(2):155–166

Patey N, Scoazec JY, Cuenod-Jabri B et al (1997) Distribution of cell adhesion molecules in infants with intestinal epithelial dysplasia (tufting enteropathy). Gastroenterology 113(3):833–843

Pedemonte N, Galietta LJ (2014) Structure and function of TMEM16 proteins (anoctamins). Physiol Rev 94(2):419–459

Pelkmans L, Burli T, Zerial M et al (2004) Caveolin-stabilized membrane domains as multifunctional transport and sorting devices in endocytic membrane traffic. Cell 118 (6):767–780

Pfeifer A, Aszodi A, Seidler U et al (1996) Intestinal secretory defects and dwarfism in mice lacking cGMP-dependent protein kinase II. Science 274(5295):2082–2086

Phillips AD, Schmitz J (1992) Familial microvillous atrophy: a clinicopathological survey of 23 cases. J Pediatr Gastroenterol Nutr 14(4):380–396

Phillips AD, Jenkins P, Raafat F et al (1985) Congenital microvillous atrophy: specific diagnostic features. Arch Dis Child 60(2):135–140

Plasschaert LW, Zilionis R, Choo-Wing R et al (2018) A single-cell atlas of the airway epithelium reveals the CFTR-rich pulmonary ionocyte. Nature 56(7718):377–381

Pohl JF, Shub MD, Trevelline EE et al (1999) A cluster of microvillous inclusion disease in the Navajo population. J Pediatr 134(1):103–106

Pratha VS, Hogan DL, Martensson BA et al (2000) Identification of transport abnormalities in duodenal mucosa and duodenal enterocytes from patients with cystic fibrosis. Gastroenterology 118(6):1051–1060

Qui Y-L, Gong JY, Feng JY et al (2017) Defects in Myosin VB are associated with a spectrum of previously undiagnosed low γ-glutamyltransferase cholestasis. Hepatology 65(5):1655–1669

Randak C, Langnas AN, Kaufman SS et al (1998) Pretransplant management and small bowel-liver transplantation in an infant with microvillus inclusion disease. J Pediatr Gastroenterol Nutr 27 (3):333–337

Reifen RM, Cutz E, Griffiths AM et al (1994) Tufting enteropathy: a newly recognized clinicopathological entity associated with refractory diarrhea in infants. J Pediatr Gastroenterol Nutr 18 (3):379–385

Reinshagen K, Naim HY, Zimmer KP (2002) Autophagocytosis of the apical membrane in microvillus inclusion disease. Gut 51(4):514–521

Rhoads JM, Vogler RC, Lacey SR et al (1991) Microvillus inclusion disease. In vitro jejunal electrolyte transport. Gastroenterology 100(3):811–817

Richards AA, Stang E, Pepperkok R et al (2002) Inhibitors of COP-mediated transport and cholera toxin action inhibit simian virus 40 infection. Mol Biol Cell 13(5):1750–1764

Riordan JR, Rommens JM, Kerem B et al (1989) Identification of the cystic fibrosis gene: cloning and characterization of complementary DNA. Science 245(4922):1066–1073

Roberts R, Lister I, Schmitz S et al (2004) Myosin VI: cellular functions and motor properties. Philos Trans R Soc Lond Ser B Biol Sci 359(1452):1931–1944

Rodighiero C, Aman AT, Kenny MJ et al (1999) Structural basis for the differential toxicity of cholera toxin and *Escherichia coli* heat-labile enterotoxin: Construction of hybrid toxins identifies the A2-domain as the determinant of differential toxicity. J Biol Chem 274 (7):3962–3969. https://doi.org/10.1074/jbc.274.7.3962

Rodighiero C, Tsai B, Rapoport TA et al (2002) Role of ubiquitination in retro-translocation of cholera toxin and escape of cytosolic degradation. EMBO Rep 3(12):1222–1227

Rodriguez OC, Cheney RE (2002) Human myosin-Vc is a novel class V myosin expressed in epithelial cells. J Cell Sci 115(Pt 5):991–1004

Rowland AF, Fazakerley DJ, James DE (2011) Mapping insulin/GLUT4 circuitry. Traffic 12 (6):672–681

Ruemmele FM, Muller T, Schiefermeier N et al (2010) Loss-of-function of MYO5B is the main cause of microvillus inclusion disease: 15 novel mutations and a CaCo-2 RNAi cell model. Hum Mutat 31(5):544–551

Salomon J, Goulet O, Canioni D et al (2014) Genetic characterization of congenital tufting enteropathy: epcam associated phenotype and involvement of SPINT2 in the syndromic form. Hum Genet 133(3):299–310

Sato T, Mushiake S, Kato Y et al (2007) The Rab8 GTPase regulates apical protein localization in intestinal cells. Nature 448(7151):366–369

Schultheis P, Clarke LL, Meneton P et al (1998) Renal and intestinal absorptive defects in mice lacking the NHE3 Na^+/H^+ exchanger. Nat Genet 19:282–285

Schulz S, Green CK, Yuen PS et al (1990) Guanylyl cyclase is a heat-stable enterotoxin receptor. Cell 63(5):941–948

Schulz S, Lopez MJ, Kuhn M et al (1997) Disruption of the guanylyl cyclase-C gene leads to a paradoxical phenotype of viable but heat-stable enterotoxin-resistant mice. J Clin Invest 100 (6):1590–1595

Sherman PM, Mitchell DJ, Cutz E (2004) Neonatal enteropathies: defining the causes of protracted diarrhea of infancy. J Pediatr Gastroenterol Nutr 38(1):16–26

Silvis MR, Bertrand CA, Ameen N et al (2009) Rab11b regulates the apical recycling of the cystic fibrosis transmembrane conductance regulator in polarized intestinal epithelial cells. Mol Biol Cell 20(8):2337–2350

Singh RD, Puri V, Valiyaveettil JT et al (2003) Selective caveolin-1-dependent endocytosis of glycosphingolipids. Mol Biol Cell 14(8):3254–3265

Singh AK, Riederer B, Chen M et al (2010) The switch of intestinal Slc26 exchangers from anion absorptive to HCOFormula secretory mode is dependent on CFTR anion channel function. Am J Physiol Cell Physiol 298(5):C1057–C1065

Singh V, Yang J, Chen TE et al (2014) Translating molecular physiology of intestinal transport into pharmacologic treatment of diarrhea: stimulation of Na^+ absorption. Clin Gastroenterol Hepatol 12(1):27–31

Sivagnanam M, Mueller JL, Lee H et al (2008) Identification of EpCAM as the gene for congenital tufting enteropathy. Gastroenterology 135(2):429–437

Sivagnanam M, Janecke AR, Muller T et al (2010a) Case of syndromic tufting enteropathy harbors SPINT2 mutation seen in congenital sodium diarrhea. Clin Dysmorphol 19(1):48

Sivagnanam M, Schaible T, Szigeti R et al (2010b) Further evidence for EpCAM as the gene for congenital tufting enteropathy. Am J Med Genet A 152A(1):222–224

Sixma TK, Kalk KH, Van Zanten BAM et al (1993) Refined structure of *Escherichia coli* heat-labile enterotoxin, a close relative of cholera toxin. J Mol Biol 230(3):890–918
Skelton NJ, Garcia KC, Goeddel DV et al (1994) Determination of the solution structure of the peptide hormone guanylin: observation of a novel form of topological stereoisomerism. Biochemistry 33(46):13581–13592
So M, McCarthy BJ (1980) Nucleotide sequence of the bacterial transposon Tn1681 encoding a heat-stable (ST) toxin and its identification in enterotoxigenic *Escherichia coli* strains. Proc Natl Acad Sci U S A 77(7):4011–4015
Sonawane ND, Muanprasat C, Nagatani R Jr et al (2005) In vivo pharmacology and antidiarrheal efficacy of a thiazolidinone CFTR inhibitor in rodents. J Pharm Sci 94(1):134–143
Srivastava R, Zinman HM, McKenzie MD et al (1999) Rotavirus vaccine. Pediatrics 104 (6):1419–1420
Staples SJ, Asher SE, Giannella RA (1980) Purification and characterization of heat-stable enterotoxin produced by a strain of *E. coli* pathogenic for man. J Biol Chem 255(10):4716–4721
Takao T, Shimonishi Y, Kobayashi M et al (1985a) Amino acid sequence of heat-stable enterotoxin produced by Vibrio cholerae non-01. FEBS Lett 193(2):250–254
Takao T, Tominaga N, Yoshimura S et al (1985b) Isolation, primary structure and synthesis of heat-stable enterotoxin produced by Yersinia enterocolitica. Eur J Biochem 152(1):199–206
Takeda Y, Takeda T, Yano T et al (1979) Purification and partial characterization of heat-stable enterotoxin of enterotoxigenic *Escherichia coli*. Infect Immun 25(3):978–985
Talmon G, Holzapfel M, DiMaio DJ et al (2012) Rab11 is a useful tool for the diagnosis of microvillous inclusion disease. Int J Surg Pathol 20(3):252–256
Taylor M, Navarro-Garcia F, Huerta J et al (2010) Hsp90 is required for transfer of the cholera toxin A1 subunit from the endoplasmic reticulum to the cytosol. J Biol Chem 285(41):31261–31267
Taylor M, Banerjee T, Ray S et al (2011) Protein-disulfide isomerase displaces the cholera toxin A1 subunit from the holotoxin without unfolding the A1 subunit. J Biol Chem 286 (25):22090–22100
Thiagarajah JR, Donowitz M, Verkman AS (2015) Secretory diarrhoea: mechanisms and emerging therapies. Nat Rev Gastroenterol Hepatol 12:446–457
Thiagarajah JR, Kamin DS, Acra S et al (2018) Advances in evaluation of chronic diarrhea in infants. Gastroenterology 154:2045–2059
Thoeni C, Amir A, Guo C et al (2014a) A novel nonsense mutation in the EpCAM gene in a patient with congenital tufting enteropathy. J Pediatr Gastroenterol Nutr 58(1):18–21
Thoeni CE, Vogel GF, Tancevski I et al (2014b) Microvillus inclusion disease: loss of Myosin vb disrupts intracellular traffic and cell polarity. Traffic 15(1):22–42
Tradtrantip L, Namkung W, Verkman AS (2010) Crofelemer, an antisecretory antidiarrheal proanthocyanidin oligomer extracted from Croton lechleri, targets two distinct intestinal chloride channels. Mol Pharmacol 77(1):69–78
Tsai B, Rodighiero C, Lencer WI et al (2001) Protein disulfide isomerase acts as a redox-dependent chaperone to unfold cholera toxin. Cell 104(6):937–948
Tyagi S, Kavilaveettil RJ, Alrefai WA et al (2001) Evidence for the existence of a distinct SO_4^{--}-OH^- exchange mechanism in the human proximal colonic apical membrane vesicles and its possible role in chloride transport. Exp Biol Med 226(10):912–918
Vaandrager AB, Bot AG, De Jonge HR (1997) Guanosine 3′,5′-cyclic monophosphate-dependent protein kinase II mediates heat-stable enterotoxin-provoked chloride secretion in rat intestine. Gastroenterology 112(2):437–443
Vaandrager AB, Bot AG, Ruth P et al (2000) Differential role of cyclic GMP-dependent protein kinase II in ion transport in murine small intestine and colon. Gastroenterology 118(1):108–114
Vega-Salas DE, Salas PJ, Rodriguez-Boulan E (1988) Exocytosis of vacuolar apical compartment (VAC): a cell-cell contact controlled mechanism for the establishment of the apical plasma membrane domain in epithelial cells. J Cell Biol 107(5):1717–1728
Vogel GF, Klee KM, Janecke AR et al (2015) Cargo selective apical exocytosis in epithelial cells is conducted by Myo5B, Slp4a, Vamp7 and Syntaxin3. J Cell Biol 211:587–604

Wada A, Hirayama T, Kitaura H et al (1996) Identification of ligand recognition sites in heat-stable enterotoxin receptor, membrane-associated guanylyl cyclase C by site-directed mutational analysis. Infect Immun 64(12):5144–5150

Wedenoja S, Holmberg C, Hoglund P (2008) Oral butyrate in treatment of congenital chloride diarrhea. Am J Gastroenterol 103(1):252–254

Wedenoja S, Hoglund P, Holmberg C (2010) Review article: the clinical management of congenital chloride diarrhoea. Aliment Pharmacol Ther 31(4):477–485

Weis VG, Sousa JF, LaFleur BJ et al (2013) Heterogeneity in mouse spasmolytic polypeptide-expressing metaplasia lineages identifies markers of metaplastic progression. Gut 62 (9):1270–1279

Whittamore JM, Freel RW, Hatch M (2013) Sulfate secretion and chloride absorption are mediated by the anion exchanger DRA (Slc26a3) in the mouse cecum. Am J Physiol Gastrointest Liver Physiol 305(2):G172–G184

Wiegerinck CL, Janecke AR, Schneeberger K et al (2014) Loss of syntaxin 3 causes variant microvillus inclusion disease. Gastroenterology 147(1):65–68 e10

Yang N, Garcia MA, Quinton PM (2013) Normal mucus formation requires cAMP-dependent $HCO3^-$ secretion and Ca2+−mediated mucin exocytosis. J Physiol 591:4581–4593

Yeo QM, Crutchley R, Cottreau J et al (2013) Crofelemer, a novel antisecretory agent approved for the treatment of HIV-associated diarrhea. Drugs Today 49(4):239–252

Yin L, Vijaygopal P, MacGregor GG et al (2014) Glucose stimulates calcium-activated chloride secretion in small intestinal cells. Am J Physiol Cell Physiol 306(7):C687–C696

Yoshino K, Miyachi M, Takao T et al (1993) Purification and sequence determination of heat-stable enterotoxin elaborated by a cholera toxin-producing strain of Vibrio cholerae O1. FEBS Lett 326(1–3):83–86

Yu B, Zhu X, Yang X et al (2019) Plumbagin prevents secretory diarrhea by inhibiting CaCC and CFTR channel activities. Front Pharmacol 10:1181

Chapter 3
Role of the Epithelium in Diseases of the Intestine

Jörg D. Schulzke and Michael Fromm

Abstract The human intestinal mucosa is lined by a single layer of epithelial cells covering a complexly folded area of up to 200 m^2. Regarding transport, this epithelium fulfills two contrasting functions: (1) absorption and secretion of ions, enzymes, nutrients, and water, and (2) formation of a selective barrier between the circulation and the gut lumen which prevents, e.g., uptake of infectious or immunological relevant solutes. Almost all intestinal diseases affect, or are affected by, transepithelial transport or paracellular barrier functions. An impaired intestinal barrier is preferably caused by modified tight junctions or altered expression of their proteins. This has two functional effects: (1) ions and water diffuse into the intestinal lumen and cause a leak-flux diarrhea, and (2) uptake of luminal pathogens induces an immune response and maintains an inflammatory process. In the second part of this chapter, intestinal diseases are described in pathophysiological and clinical detail and are grouped in three categories. The first are the inflammatory bowel diseases Crohn's disease, ulcerative colitis, and microscopic colitis. Another category are the gluten-sensitive autoimmune disease celiac disease and the non-celiac gluten-sensitivity. A final group contains diseases caused by bacterial infection, giardia lamblia infection and Whipple's disease.

Keywords Transport of nutrients · Segmental heterogeneity · Trans- and paracellular pathway · Leaky and tight epithelia · Tight junction · Barrier function · Claudin-based channels · Crohn's disease · Ulcerative colitis · Microscopic colitis · Celiac disease · Non-celiac wheat-sensitivity · Giardia lamblia infection · Whipple's disease

---

J. D. Schulzke (✉) · M. Fromm
Institute of Clinical Physiology/Nutritional Medicine, Charité – Universitätsmedizin Berlin, Berlin, Germany
e-mail: joerg.schulzke@charite.de; michael.fromm@charite.de

K. L. Hamilton, D. C. Devor (eds.), *Ion Transport Across Epithelial Tissues and Disease*, Physiology in Health and Disease,
https://doi.org/10.1007/978-3-030-55310-4_3

3.1 Introduction: Basics of Intestinal Epithelial Transport

Dual Role of the Intestinal Epithelium The intestinal epithelium consists of one single cell layer covering the surface of the gut wall, which is about 200 m^2. Regarding transport, this single cell sheet is challenged by two main tasks, (1) absorption and secretion of ions, nutritives, digestive juices, and water, and (2) formation of a reliable barrier against unwanted uptake from the lumen of potentially infectious or immunologically relevant solutes and against fluid loss from blood to the lumen.

The amount of exchanged fluid per day is large. While oral intake adds up to 2 l, gastrointestinal secretion amounts 7 l per day and absorption averages nearly 9 l per day. Compared to these numbers, fluid excretion by feces is small, 0.1 to 0.2 l per day. Despite these large transport rates, the barrier enfaces a gigantic amount of bacteria (up to 10 times the number of body cells) and its products, which have to be kept out of the mucosa and circulation.

3.1.1 Transcellular Transport of Ions, Nutrients, and Water

Channels and carriers are distributed along the gut axis forming a typical segmental heterogeneity, which is principally also present in other tubular epithelia like that of the kidney. Regarding ions and water, in a simplified scheme proximal segments (e.g., small intestine, proximal nephron) absorb and secrete large amounts. This occurs for most solutes and for water against no major concentration gradients so that the luminal content remains roughly isoosmotic. In contrast, in distal segments (e.g., large intestine, distal nephron), smaller amounts of ions and water are transported, but if applicable against larger gradients and under hormonal control (aldosterone; vasopressin in kidney only).

Further details on ion transport are not discussed here, interested readers are directed to, Chap. 2 of Volume 1 by Silverthorn. There might be more cross-referencing to other chapters in the three volumes.

Segmental Heterogeneity of Nutrient Transport For the transport sites of nutrients, an axial segmental heterogeneity exists, as they are absorbed in the proximal but not the distal segments. This rule of the thumb is true for the kidney, but the intestine exhibits one significant exception, the absorption of short-chain fatty acids (SCFA) which takes place in the upper segments of the large intestine (Wang et al. 2013).

Although nutrient transport is not a main scope of this book, the transport mechanisms are introduced here, because they are related or even directly coupled to ion transport. This becomes evident in diseases of the intestine, where in most cases, ion as well as solute and water transport is affected.

Intestinal Transport of Sugars Sugars are absorbed by specific membrane transporters exclusively in form of monosaccharides and not—at least to a significant quantity—as disaccharides or even larger entities. Monosaccharide carriers exist in the apical membrane of epithelia like small intestine or proximal kidney tubule for glucose, galactose, and fructose. This means that if disaccharides like lactose (consisting of glucose + galactose) are not properly hydrolyzed in the intestine by lactase, no significant absorption takes place, a common condition called lactose intolerance (Bayless et al. 2017). The lactose molecules are axially transported into the colon and are digested thereby luminal bacteria.

SGLT1 (SLC5A1) and SGLT2 (SLC5A2) Two distinct monosaccharide symport carriers exist which both are driven by luminal uptake of Na^+, but in the intestine, only one of them, SGLT1, is found (Wright et al. 2011). It cotransports glucose or galactose against high concentration gradients, as it is coupled in a 1:2 stoichiometry to Na^+, which drives the transport cycle due to its own electrochemical gradient. This Na^+ gradient across the apical cell membrane is continuously provided by the Na^+/K^+-ATPase. While the transport of Na^+ is called "primary active," because the metabolic energy (ATP) is directly used for its translocation, the transport of glucose is called "secondary active," since its transport is indirectly driven by the primary active transporter.

Regarding the SGLT1 and SGLT2, the kidney proximal tubule exhibits a functionally very meaningful segmental heterogeneity. In the most proximal segments, S1 and S2, SGLT2 is present and works a high transport rates but limited maximum gradients due to its 1:1 Na^+ to monosaccharide coupling, while SGLT1, due to its 2:1 coupling, is able to take up the remaining monosaccharides from the lumen until to a very low concentration (Ghezzi et al. 2018). Full detail of SGLTs is given in Volume 3 in the Chap. 6.

Glucose/Galactose Malabsorption A genetic defect of SGLT1 causes the syndrome of glucose/galactose malabsorption in the small intestine and a mild glucosuria. The mildness of the effect in the kidney is caused by the compensatory action of SGLT2 (Turk et al. 1991). A genetic defect of SCLT2 causes the syndrome of **familial renal glucosuria**, which is restricted to the kidney as SGLT2 is not expressed in the intestine (Santer and Calado 2010).

SGLT1 as a Water Channel Water is known to be transported through cell membranes by various water channels, the aquaporins. However, besides being a carrier for Na^+ and monosaccharides, the SGLT1 protein of the small intestine has proven to act also as a water channel (Zeuthen et al. 2016). This has been under discussion for long time, since it was difficult to explain how a membrane transporter can perform at the same time as symport carrier and channel. The current model postulates spontaneous formation of transient water-conducting states which are permeable to water but occluded to other substrates, thereby not hindering the uphill transport of glucose. It is proposed that water-conducting states may represent an universal phenomenon in substrate carriers (Li et al. 2013).

GLUT5 (SLC2A5) and GLUT2 (SLC2A2) The third of the monosaccharides, fructose, is taken up from the small intestinal lumen (as well as from proximal tubular lumen) into the cell by a uniporter carrier, GLUT5 (Wright et al. 2011). As no coupling to Na^+ exists, fructose uptake is by definition "passive" and works as long as the luminal concentration is higher than the intracellular one. If GLUT5 is defective or not properly expressed, fructose is insufficiently absorbed, and symptoms of a fructose intolerance may occur, if fructose intake is above a critical level.

The exit of the monosaccharides across the basolateral membrane occurs passively via another uniporter, **GLUT2**. This carrier accepts all three monosaccharides (Ghezzi et al. 2018). A genetic defect of GLUT2 causes the **Fanconi-Bickel syndrome**, which is characterized by hepatomegaly, glucosuria, and impaired intestinal monosaccharide absorption (Santer et al. 2002).

Intestinal Transport of Proteins by PepT1 (SLC15A1) and PepT2 (SLC15A2) In contrast to sugars, protein components are transported as single amino acids, but also as dipeptides or even tripeptides. Details are described in Volume 3 in Chap. 3?. Briefly, for single amino acids, different carriers exist with different specificity, e.g., for anionic, neutral, and cationic amino acids. Most of the amino acid carriers are symporters with Na^+, which causes the amino acids to be taken up against their electrochemical gradient (secondary active transport).

The uptake of di- or tripeptides is even more sophisticated. Here, symport carriers with H^+ take the di-/tripeptides from the lumen, PepT1 in the small intestine, and the S1 segment of proximal tubule and PepT2 in S2–S3 segments of proximal tubule (Smith et al. 2013). The driving force for di-/tripeptide uptake is provided by a H^+ gradient across the apical membrane, which in turn is produced by a Na^+/H^+ antiport carrier, NHE3. While NHE3 mediates a secondary active transport, the transport by PepT1 and PepT2 is defined as "tertiary active transport," because it is driven by a secondary active transport (Ganapathy et al. 1987). Di- and tripeptides are intracellularly hydrolyzed by dipeptidases and then expedited passively across the basolateral membrane by various amino acid uniporter carriers. Hereditary defects of amino acid carriers are extremely seldom and do not necessarily produce intestinal symptoms, possibly because of redundancy of transport mechanisms.

3.1.2 Transcellular and Paracellular Pathways

It was clear for decades that transport in epithelia and endothelia does not solely occur through the apical and basolateral cell membranes or by transcytosis (transcellular pathway) but also through the clefts between the cells (paracellular pathway) (Fig. 3.1). The paracellular pathway leads through a nominally sealing structure, the tight junction. The concept of trans- and paracellular pathways through epithelia and endothelia was discovered long before the molecular basis of paracellular permeability was discovered (Clarkson 1967). In the seventies of the last century, it has been demonstrated using sophisticated electrophysiological

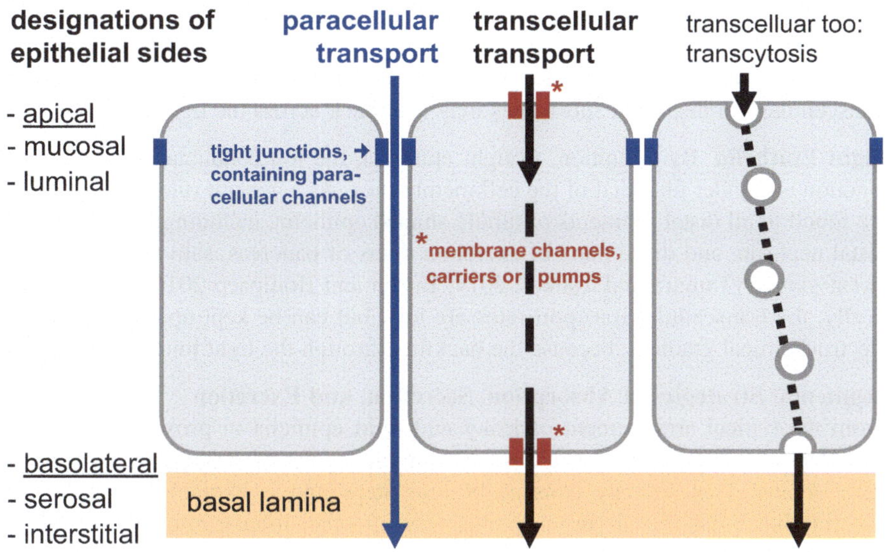

Fig. 3.1 Scheme of an epithelial tissue layer. Shown here is the absorptive direction; however, secretion follows the same pathways. Left: common terms for side declaration. Blue: paracellular pathway through the tight junction. Black: Transcellular pathway trough apical and basolateral membrane channels, carriers and pumps (symbolized in red). Dotted: Transcellular pathway by vesicular transport, called transcytosis. In "leaky epithelia" (e.g., small intestine, proximal nephron), the paracellular pathway is more permeable to ions than the transcellular ones, while in "tight epithelia" (e.g., large intestine, distal nephron) the opposite applies

techniques that in some epithelia, a high ion permeability exists at the site of the tight junctions (Frömter and Diamond 1972).

3.1.3 Leaky and Tight Epithelia

The concept of "leaky" and "tight epithelia" (Diamond 1974, 1978) is based on the ratio of paracellular over transcellular ion conductance, thus mainly representing Na^+, K^+, and Cl^-.

Leaky Epithelia By definition, in leaky epithelia, the ion conductance of the tight junction is higher than that of the cell membranes. This by no means implies that permeabilities for other solutes as, e.g., for nutrients or macromolecules behave in the same way. As a general rule, leaky epithelia are represented in all proximal segments of tubule-shaped epithelia, including small intestine, proximal nephron, and proximal parts of excretion ducts of pancreas, salivary glands, and sweat glands. The epithelium of the gallbladder is also an example of a leaky epithelium. Characteristically, the transcellular transport rates are high and are even enhanced by paracellular, osmotically driven, water diffusion which in turn carries more solutes by solute drag. This causes high transport rates without consuming much metabolic energy (Yu 2017). However, this only works as long as there is no significant electrochemical gradient against paracellular diffusion. If a gradient originates, transcellularly transported substances may leak back across the tight junction.

Tight Epithelia By definition, in tight epithelia, the ion conductance of the tight junction is smaller than that of the cell membranes. As a general rule, tight epithelia are found in all distal segments of tubule-shaped epithelia, including large intestine, distal nephron, and distal parts of excretion ducts of pancreas, salivary glands, and sweat glands (Tamura and Tsukita 2014; Boron and Boulpaep 2016). Characteristically, the transcellular transport rates are low, but can be kept up also against an electrochemical gradient, because the backflux through the tight junction is limited.

Segmental Strategies of Absorption, Secretion, and Excretion
From the typical arrangement of leaky and tight epithelia in proximal and distal segments of tubular epithelia follows a common segmental strategy of transport in such organs. This strategy consists of four steps and is explained here for the intestine but holds true in an analogous way for other tubular epithelia, including the kidney tubular system:

1. ***Formation of an isoosmotic primary content***. The luminal content becomes isoosmotic to the plasma already in the stomach or shortly after leaving it by water influx or efflux as appropriate.

2a. ***Isoosmotic mass transport***. The leaky epithelia of the small intestine absorb large amounts of solutes and water in a nearly isoosmotic way, without significant regulation by hormones.

2b. **Nutrient absorption against large gradients**. In the same segment, although leaky for small ions, monosaccharides and amino acids (including small peptides) are absorbed against high concentration gradients because the tight junction is not permeable for nutrients.

3. ***Fine adjustment of the excretion product***. The tight epithelia of the large intestine absorb smaller amounts of ions and water than the proximal segments, but if necessary against considerable gradients. Na^+ absorption is regulated by aldosterone, in order to help maintaining homeostasis of the body fluids. As an exception to the rule, short-chain fatty acids (SCFA) are absorbed in the proximal colon. As the colon accommodates an enormous amount of bacteria, any nutrients which were not absorbed by the small intestine are subject of bacterial breakdown and digestion.
4. ***Temporary storage of the excretion product***. The rectum is tighter than the colon and thus continues to extract ions and water. Depending on storage time, feces are converted from semi-fluid to formed stool before defecation.

3.1.4 The Tight Junction

Morphologically, one can discriminate between the **bicellular tight junction**, which connects two adjacent cells at the apex of their lateral cell membranes and the **tricellular tight junction**, which is the site where three (or even four) cells meet (Fig. 3.2a).

Currently, we know two families of tetraspan tight junction proteins which are direct effectors of barrier or channel function, the **claudin** family (claudin-1 to -27 in mammals; gene names CLDN1 to -27) and the **TAMP** family (tight junction-associated MARVEL proteins: **occludin**, **tricellulin,** and **marvelD3**) (Morita et al. 1999; Günzel and Fromm 2012). In addition, there are two sorts of single-span tight junction proteins without barrier or channel function, the JAM family (junctional adhesion proteins 1 to 3) and the angulin family (angulin-1 to -3, equivalent to LSR, ILDR1, and ILDR2) (Higashi et al. 2013).

The abundance of the tight junction proteins varies among and within epithelial organs, and therefore, it is obvious that the specific pattern of tight junction proteins determines the functional properties of the tight junction. The tetraspan tight junction proteins consist of two extracellular loops, a short intracellular loop, and intracellular C- and N-terminals. With their extracellular loops, they interact with partners from opposing lateral membranes (Fig. 3.2c).

Paracellular Ion and Water Channels Only in the beginning of this century, it became clear that some protein components of the tight junction, instead of fully sealing that structure, form paracellular channels (Furuse et al. 2001). These channels do not cross any cell membrane as transmembranal channels do, but pass extracellularly through the tight junction between the apical and the basolateral compartment (Fig. 3.2b). In the following, some aspects of these channels are described.

The bicellular tight junction is the site where paracellular channels are found. These channels can be divided regarding their selectivity into three main groups (Günzel and Yu 2013):

1. **Cation channels**. Channels selective for small cations are comprised of **claudin-2** (Amasheh et al. 2002), **claudin-10b** (van Itallie et al. 2001; Günzel et al. 2009), **claudin-15** (Van Itallie et al. 2003), **claudin-16** in conjunction with **claudin-19** (Gong et al. 2015), and probably **claudin-21** (Tanaka et al. 2016).
2. **Anion channels**. Channels selective for small anions are formed by **claudin-10a** (van Itallie et al. 2001; Günzel et al. 2009) and by **claudin-17** (Krug et al. 2012; Conrad et al. 2016).
3. **Water channels**. Two of the cation channels conduct also water, **claudin-2** (Rosenthal et al. 2010, 2017; Muto et al. 2010) and **claudin-15** (Rosenthal et al. 2020).

The majority of all other claudins contribute to sealing the tight junction. Overexpression of these barrier-forming claudins decreases the permeability to

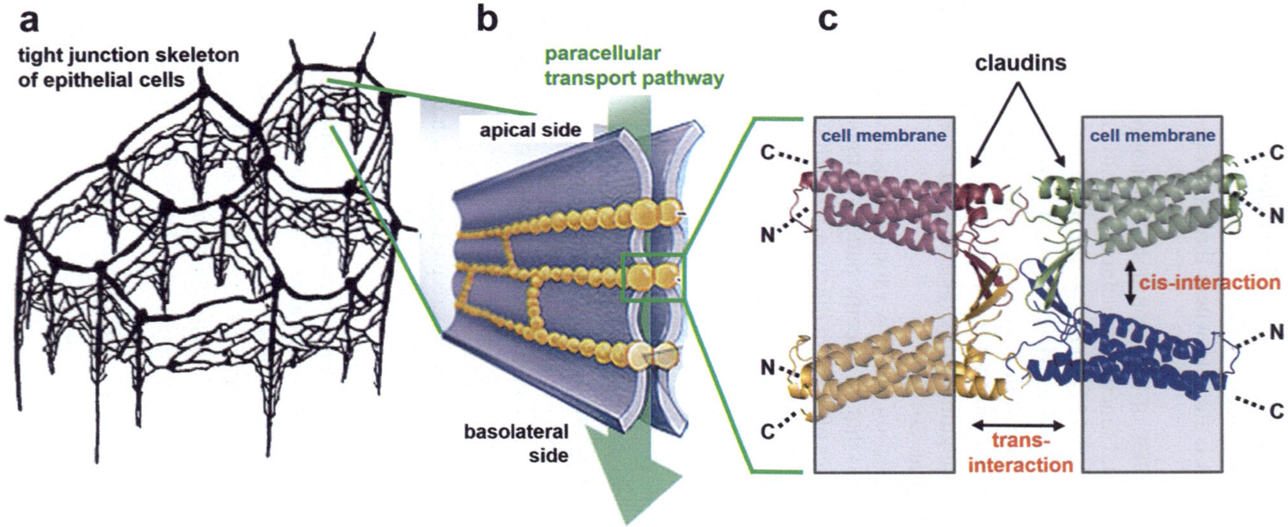

Fig. 3.2 Tight junction (TJ) skeleton, paracellular pathway, and arrangement of claudins. (**a**) Strand meshwork of the bicellular and tricellular TJ. Cells were omitted so that the skeleton of TJ strands remains. (**b**) Magnified detail showing strands of TJ proteins. In the presence of channel-forming TJ proteins, paracellular transport driven by electrochemical gradients takes place. (**c**) Magnified detail showing strand formation by claudins of two adjacent lateral cells (scheme is based on a claudin-15 polymer model (Suzuki et al. 2015)). Claudins consist of two extracellular loops, four transmembrane regions, and a short intracellular loop and N- and C-terminals. They form either pure barriers or paracellular channels for cations, anions, or water. TJ proteins interact with each other within the same cell membrane (cis-interaction) and across neighboring cell membranes (trans-interaction). Modified from Fromm et al. (2017; Pflügers Arch 469:877–887)

ions, either equally for cations and anions or, in some cases, for cations more than for anions (Krug et al. 2014).

Of note, many of the barrier-forming claudins are able to replace each other regarding barrier function in case of isolated downregulation. A most prominent example for this mechanism was the puzzling finding that occludin knockout mice did not alter ion permeability (Saitou et al. 2000). The molecular structure of a first tight junction protein, claudin-15, has been published (Suzuki et al. 2014), and a first model of claudin-15 channel structure has followed (Suzuki et al. 2015).

The tricellular tight junction is considered to be a potential weak point of the tight junctional seal. An example of a tight junction protein not being easily replaced in function if downregulated is **tricellulin** (Riazuddin et al. 2006). While all other tight junction proteins are predominantly localized within the border of two adjacent cells (bicellular tight junction), tricellulin forms the protein basis of the site where three cells meet (tricellular tight junction) (Ikenouchi et al. 2005). At low tricellulin concentration, the tricellular tight junction becomes permeable for small and, most importantly for qualitative reasons, for large molecules (Krug et al. 2009). The same mechanism comes into effect in ulcerative colitis, where tricellulin is downregulated via the interleukin-13-receptor α2 (Krug et al. 2018).

Along the same lines, short-lived partial opening of the tight junction is applied in an attempt to enhance drug absorption across physiological barriers of the intestinal epithelium (Cording et al. 2017; Krug et al. 2017) and the blood-brain barrier for another example (Zeniya et al. 2018).

Clinical Impact of Disturbed Tight Junctional Function In several intestinal disorders (as well as in some diseases of other organs), tight junction proteins are involved, either as a primary cause or secondary to the disease. Chronic inflammatory diseases or infections of the intestine may cause a dysregulation of tight junction proteins, which may aggravate the symptoms by impairing the epithelial barrier function. This has two functional effects (Krug et al. 2014):

1. The barrier defect causes ions and water to diffuse from the blood to the intestinal lumen and leading to the mechanism of a **leak-flux diarrhea** (Sandle 2005). A main player of such barrier defects is an upregulated claudin-2 cation and water channel.
2. The barrier defect concerns increased permeability to large molecules. As a consequence, an **uptake of luminal pathogens** like food antigens and bacterial lipopolysaccharides occurs and, by inducing an immune response, initiates or maintains the inflammatory process (Martini et al. 2017).

3.2 Diseases Related to Intestinal Epithelial Transport

In the second part of this chapter, intestinal diseases are described in more detail. For an overview, the changes in functional parameters are summarized in Table 3.1.

3.2.1 Crohn's Disease

Clinical Data Crohn's disease is an inflammatory bowel disease (IBD) (Crohn et al. 1932). The etiology is complex. The destructive mucosal immune response due to a pro-inflammatory genetic background is directed against the own microbiota (Duchmann et al. 1995). External stimuli like GI-infections and stress influences can trigger or start the inflammatory activity. It presents as segmental inflammation of small and/or large intestine but is often not affecting the rectum. The ileum is often involved, either exclusively which led to the name ileitis terminalis or in combination with the right-handed colon (ileocolitis). Another 20–30% of the patients present with inflammation restricted to the colon (colitis Crohn) (Triantafillidis et al. 2000). Different segments can be independently involved ("skip lesions"). However, Crohn's disease can also spread from the intestine to the stomach or esophagus. Furthermore, extraintestinal manifestations are frequently seen as, e.g., in skin (pyoderma gangrenosum and erythema nodosum), at the eye (iridocyclitis), or on joints (arthritis) (Triantafillidis et al. 2000). Many patients have abdominal complaints, particular in stenotic disease, and non-bloody diarrhea. However, the clinical picture is heterogeneous and depends on course and extent of the disease. It can include weight loss (small intestinal disease), stenosis, and fistula formation. In contrast to ulcerative colitis, Crohn's disease tends to start in younger people and even in childhood but can begin at any age including late-onset disease in elderly people (Triantafillidis et al. 2000). Therapy comprises immune suppressives like prednisolone, azathioprine, and methotrexate as well as biologics like antibodies against TNFα, interleukin-12, and $\alpha_4\beta_7$-integrin (Reinglas et al. 2018).

Barrier Function In the inflamed mucosa of Crohn's disease, the colonic epithelium exhibits pronounced barrier dysfunction (Zeissig et al. 2007a). This is evident from a decrease in epithelial resistance indicating increased permeability of tight junctions for ions, as a result of which leak-flux diarrhea is a prominent feature. In addition, strong upregulation of epithelial apoptosis contributes to this type of diarrhea (Gitter et al. 2001). However, barrier dysfunction in Crohn's disease facilitates also uptake of luminal antigens and bacteria. Antigen uptake may occur also paracellularly but in addition transcytotically after enforced epithelial endocytosis, which is evident from an accelerated ovalbumin and horseradish peroxidase uptake into IBD epithelium with subsequent vesicular transport to the basolateral membrane (Söderholm et al. 2004). This is of high functional relevance as indicated by animal models of inflammation, where the barrier defect and the luminal microbiota are a crucial component in the ongoing inflammation (Bücker et al. 2014).

Table 3.1 Changes of functional transport and barrier parameters in diseases of small and large intestine

Parameter	Function	Crohn's disease	Ulcerative colitis	Collagenous colitis	Lymphocytic colitis	Celiac disease	G. lamblia infection	Whipple's disease
Villi/ Surface	Absorption					↓	↓	↓
Crypts	Secretion	↑	↓	=	=	↑	↑	↑
Apoptoses	Leak	↑	↑	=	=	↑	↑	↑
Claudin-1	Barrier	=	↓	=	=	=	↓	↓
Claudin-2	Cation channel water channel	↑/=	↑	=		↑		↑
Claudin-3	Barrier	↓	↓/=	=	=	↓		↓
Claudin-4	Barrier anion channel	=	↓	↓	↓	=	=	=
Claudin-5	Barrier	↓		=	↓	↓		=
Claudin-7	Barrier anion channel	=	=				=	=
Claudin-8	Barrier	↓			↓			
Claudin-15	Cation channel					↑		=
Occludin	Barrier		↓				=	=
Tricellulin	Barrier	=	↓					↓
ENaC	Na^+ channel	↓	↓	↓	↓			
NHE3/ DRA	NaCl-uptake		↓	↓		↓		
SGLT-1	Glucose uptake					↓	↓	↓
CFTR	Cl^--secretion	↓	↓	↑	=	↓	↑	↑

There has been a long debate whether or not barrier dysfunction in Crohn's disease is only secondary or could be a primary phenomenon. However, there is no final evidence for a primary barrier dysfunction so far. Even if a permeability increase is present, e.g., for ^{51}Cr-EDTA in healthy monozygotic co-twins of twins with Crohn's disease when compared to control (Keita et al. 2018), this does not rule out that these apparently healthy co-twins have starting inflammation in a mid-small intestinal segment which escapes diagnosis in an early stage of the disease. Thus, the conclusion on a primary barrier defect requires molecular identification of a genetically altered barrier component in Crohn's disease. Up to now, IBD (inflammatory bowel disease) genes are immune-regulatory factors, which finally affect also barrier features.

Histology and Electron Microscopy Histologically, the colon mucosa in Crohn's disease is characterized by crypt rarefaction and crypt elongation, as a result of which the mucosal surface area is not reduced, but is slightly increased by 24% (Zeissig et al. 2007a).

With the use of freeze-fracture electron microscopy, Zeissig et al. (2004) reported that the epithelial tight junction between the enterocytes in Crohn's disease is characterized by a reduction in horizontally oriented TJ strands with a concomitant reduction in TJ meshwork depth. In addition, a higher number of TJ strand breaks of >25 nm have been observed (Zeissig et al. 2004).

Tight Junction (TJ) Proteins To identify the structural correlate for impaired barrier function in Crohn's disease, occludin and claudins 1–5, 7, 8, 11, 12, and 14–16 were studied as molecular components of the tight junction (Fig. 3.3). While claudins 11, 12, and 14–16 were not detectable in the colon, occludin and the "tightening" claudins 3, 5, and 8 were expressed to a lesser extent in Crohn's disease (Zeissig et al. 2004). The downregulation of "tightening" TJ proteins in combination with an upregulation of the pore-forming TJ protein claudin-2 (Prasad et al. 2005; Zeissig et al. 2007a) is one feature of the epithelial barrier defect. Finally, the "tightening" of claudins 5 and 8, in addition to being less abundant, undergoes subcellular redistribution off the tight junction into the cytoplasm and the basolateral membrane, further increasing the permeability of the tight junctions (Zeissig et al. 2007a). Occludin reduction does not increase ionic permeability but contributes to higher macromolecule passage (Al-Sadi et al. 2011).

Functionally, an increase in ionic permeability as result of a higher claudin-2 expression has been discussed to be beneficial during intestinal inflammation. This became apparent when modulation of claudin-2 expression in mice was studied with the Citrobacter rodentium infection model (Tsai et al. 2017). The inherent mechanism could be that antigens are rinsed off the mucosal surface (like "tears of the gut"). Thus, while barrier dysfunction restricted to ions, as in case of claudin-2 overexpression, can cause leak-flux diarrhea and may attenuate inflammation, an unrestricted barrier defect intensifies antigen uptake and will intensify the mucosal immune response.

In contrast to active Crohn's disease, Crohn's patients in remission have no significant changes in epithelial barrier function or tight junction protein expression

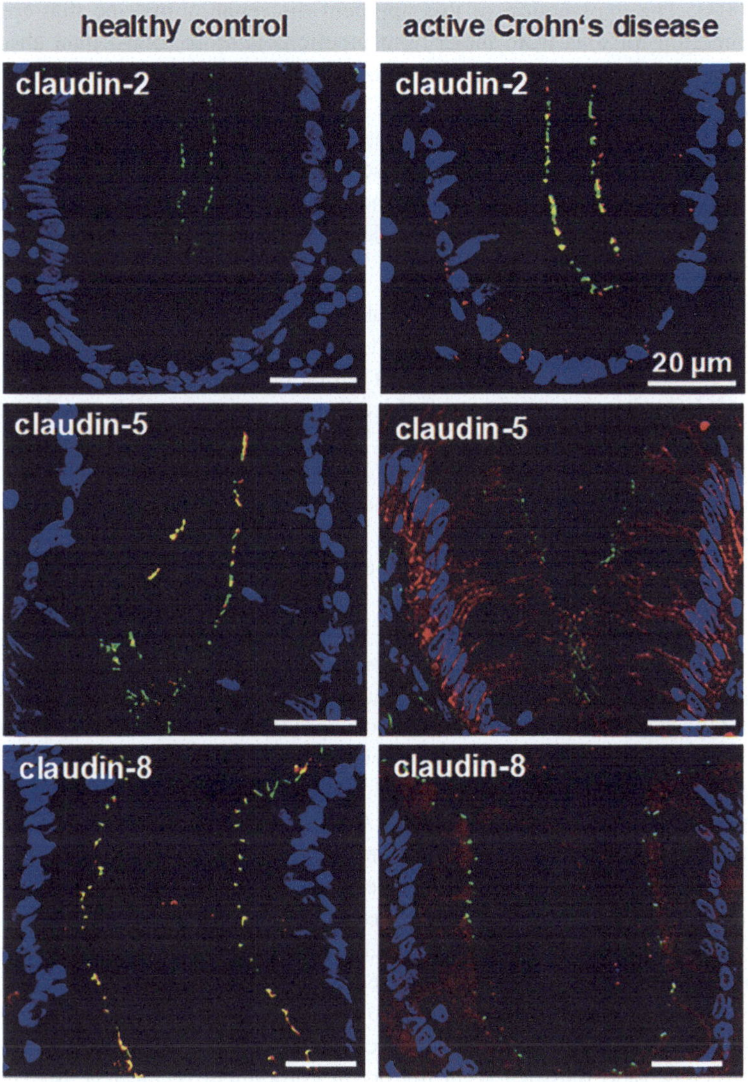

Fig. 3.3 Confocal laser scanning microscopy of tight junction proteins in Crohn's disease. Merged pictures of ZO-1 (green), the respective claudin (red), and nuclei (blue) from the upper part of a crypt of sigmoid colon biopsies. The tightening claudins 5 and 8 are downregulated and distributed off the tight junction domain of the cells, while the pore-forming claudin-2 is upregulated. Bar = 20 mm. Modified from Zeissig et al. (2007a; Gut 56:61–72)

underlining the importance of claudins for the normal function of the epithelial barrier (Zeissig et al. 2007a). This does also differ from transport dysfunction discussed below, which is hampered even in not macroscopically and not microscopically affected colon segments in patients with active ileitis terminalis (Zeissig et al. 2008). The explanation of this apparent discrepancy may be that epithelial barrier function for ions from intact tight junctions is more robust than absorptive transport processes and low-grade inflammatory signaling affects absorption much earlier than barrier function.

The pro-inflammatory effector cytokine profile toward the epithelium in active Crohn's disease is characterized by release of TNFα and IFNγ, both of which can reduce barrier function. Furthermore, the pattern of the tight junction protein changes seen in response to these cytokines resembles that observed in active Crohn's disease (Zeissig et al. 2007a).

Epithelial Ion Transport Electroneutral NaCl cotransport is among the most important resorptive transport mechanism in large intestine and is present also in the small intestine. Two antiporters act in parallel, the sodium proton exchanger-3 (NHE3) (SLC9A3) and the chloride bicarbonate antiporter (DRA) (SLC26A3). Transport energy is provided by the Na^+/K^+-ATPase, which reduces the intracellular sodium concentration below the electrochemical equilibrium. In the inflamed mucosa of Crohn's patients, electroneutral sodium absorption is reduced (U. Seidler, personal communication).

The epithelial sodium channel (ENaC) (SCNN1) is the second important sodium absorption system in the large intestine, and it is seriously impaired in Crohn's colitis, which contributes to the diarrhea in these patients, as well (Zeissig et al. 2008). Interestingly, this impairment is also seen outside macroscopically affected intestinal segments, which can explain that extensive watery diarrhea is observed also in Crohn's disease patients with macroscopic inflammation restricted to a small limited segment in the small intestine (Zeissig et al. 2008). Inactivation of ENaC-dependent sodium absorption is induced by the pro-inflammatory cytokines like TNFα and mediated by activation of the MAP-kinase ERK1/2, since ERK-inhibition can restore transport activity in intestinal biopsies in vitro (Zeissig et al. 2008). On the one hand, ERK1/2 downregulates the expression of the β- and γ-ENaC subunit. On the other hand, ERK1/2 decreases the expression of the serum glucose-activated kinase 1 (SGK1). Decreased SGK1 keeps the ubiquitinase NEDD4–2 activated, as a result of which ENaC subunits are removed from the apical enterocyte membrane.

The diarrhea in Crohn's disease can also be caused by other mechanisms. Besides the loss of absorptive intestinal surface area due to respective surgery (short bowel syndrome), bile acid diarrhea is an important component (Camilleri 2015). When reabsorption of conjugated bile acids in the distal ileum is hampered by inflammation, bile acids enter the colon and stimulate active electrogenic anion secretion (Dharmsathaphorn et al. 1989). Furthermore, anti-inflammatory drugs can contribute to the diarrhea in Crohn's patients. For example, olsalazine (Dipentum®) is a dimeric mesalazine derivative which is released from of its capsula in the proximal colon and after bacterial cleavage causes osmotically driven water entry into the lumen (Berglindh et al. 1988).

Anti-inflammatory Therapy Therapeutically, glucocorticoids improve intestinal barrier dysfunction and this is accompanied by lower claudin-2 and an increase in claudin-4 expression mediated through activation of MAPK phosphatase-1 (Fischer et al. 2014). Furthermore, glucocorticoids can induce ENaC-expression. This therapeutic effect of glucocorticoids is especially pronounced during Th1-cytokine activation, since TNFα causes an upregulation of the glucocorticoid receptor level via p38 (Bergann et al. 2009).

The TNFα antibody infliximab (Remicade®) has been shown to rescue barrier dysfunction in active Crohn's disease by normalization of barrier function including normalization of elevated epithelial apoptosis (Zeissig et al. 2004). However, this therapeutic effect is rather not the result of immune neutralization of circulating cytokine but is the consequence of antibody binding to TNFα exposed at the cell surface of mucosal immune cells, which as the result of this binding undergo cell death induction (Lügering et al. 2001). This interpretation is supported by the fact that TNFα binding proteins as etanercept (Enbrel®) are not effective in Crohn's disease.

A comparable influence was observed for the TNFα antibody adalimumab (Humira®) which during TNFα and IFNγ exposure inhibited the decrease in electrical resistance and the changes of claudin-1, -2, and -4 as well as of occludin (Fischer et al. 2013). Thus, both glucocorticoids and anti-TNFα antibodies have an anti-inflammatory therapeutic effect by protecting the mucosal barrier, an important prerequisite for "mucosal healing."

3.2.2 Ulcerative Colitis

Clinical Data The other important inflammatory bowel disease (IBD) besides Crohn's disease is ulcerative colitis. The mean age of onset is later than in Crohn's disease with a peak age in the third decade (da Silva et al. 2014). It affects the large intestine, starting in the rectum and spreading backward to the proximal colon. Thus, in contrast to Crohn's disease, the rectum is involved in the inflamed intestinal region (da Silva et al. 2014). The transition from inflamed to healthy is continuous and not sharp as in Crohn's disease. In contrast to Crohn's disease, only the mucosa and submucosa are inflammatorily affected and not deeper layers of the intestinal wall. Thus, fistulas and stenoses are missing (da Silva et al. 2014). Since the epithelial cover is missing in severely inflamed mucosae due to ulcer formation, bleeding and diarrhea are frequent clinical manifestations.

A life-threatening complication is the toxic megacolon (Autenrieth and Baumgart 2012). The extent of the inflammation can be small as in ulcerative proctitis or may include the entire large intestine (pancolitis) (da Silva et al. 2014). In some of the patients with pancolitis, back wash ileitis is spreading a few centimeters into the most distal part of the small intestine. Another important complication in a long-term course of the disease is development of colorectal cancer, which is assumed to be due to the accelerated epithelial cell turn over during chronic inflammation (Loddenkemper et al. 2006). As in Crohn's disease, anti-inflammatory drugs like prednisolone and azathioprine represent the standard therapy. In recent years, also biologicals like TNFα antibodies were successfully applied (Reinglas et al. 2018).

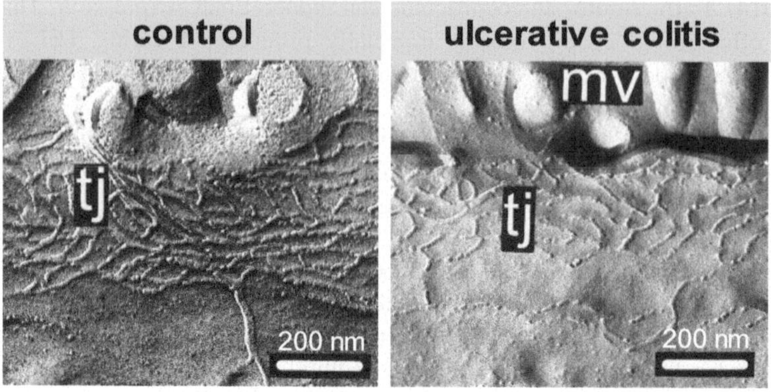

Fig. 3.4 Freeze-fracture electron microscopy of epithelial tight junctions in ulcerative colitis compared to healthy controls. Tissue specimens are frozen down to −100 °C, fractured and covered with platinum. Analysis is performed by transmission electron microscopy. In this manner, the tight junction strand network between adjacent enterocytes is visible. Tight junction structure in ulcerative colitis is impaired as indicated by a reduced number of tight junction strands in distal colon biopsies modified from Schmitz et al. (1999, Gastroenterology 116:301–309)

Histology Morphology of the inflamed mucosa in ulcerative colitis is altered with a rarefaction of crypts (Schmitz et al. 1999), as a consequence of which epithelial surface area is reduced. Epithelial apoptosis is more frequent, and gross lesions like erosions are an early finding. The thickness of the mucosa and the submucosa is increased due to mucosal edema, infiltration with inflammatory cells, and an increase in thickness of the muscularis mucosae (Schmitz et al. 1999).

Freeze-Fracture Electron Microscopy The epithelial cell layer is tightened against the intestinal lumen by tight junctions. Epithelial tight junction structure in active ulcerative colitis is impaired (Schmitz et al. 1999) with a decrease in the mean number of tight junction strands, as a result of which the depth of the main tight junctional strand meshwork is reduced (Fig. 3.4). Additionally, in a significant amount of tight junctional regions analyzed in ulcerative colitis, only 1 or 2 stands were left which differs from control. On the other hand, strand discontinuities were not more frequent in ulcerative colitis than in control. The structural basis for these EM changes is the altered molecular composition of tight junction proteins.

Tight Junction Proteins Claudin-2 expression is increased in colonic biopsies from patients with active ulcerative colitis. Western blot analysis showed a tenfold higher expression of claudin-2 when compared to control (Heller et al. 2005), an increase which is much more pronounced than in Crohn's disease. This is considered to represent extensive epithelial repair in ulcerative colitis, and its normalization has even been proposed as surrogate marker for "barrier healing" as an endpoint of biochemical normalization. As far as the other tight junction proteins are concerned, occludin, claudin-1, and claudin-4 expressions were lower than in controls (Heller et al. 2005). Furthermore, membrane protein expression of tricellulin was reduced in ulcerative colitis in comparison with control patients (Krug et al. 2018). Tricellulin is an important factor in tightening the epithelial contact points where 3 or more cells meet within an epithelium, a natural weak point of transepithelial permeation for ions, small molecules, and macromolecules <20 kDa. The tricellulin reduction in ulcerative colitis induces an important early weakness of epithelial barrier function, before gross lesions as ulcers cause a global barrier breakdown. Taken together, these changes cause a higher permeability for ions and water and contribute as leak-flux mechanism to the diarrhea in patients with ulcerative colitis.

Barrier Function Functional analysis of the epithelial tight junction is usually done by electrical resistance measurements, which indicate the ionic permeability for sodium and chloride. While methods, which measure resistance with direct current, are suitable for cell models without subepithelial tissue layers, in native epithelia especially with inflammatory alterations in the subepithelium alternating current impedance spectroscopy is necessary for characterization of epithelial barrier function, since only by this approach compensation for subepithelial resistance contributions is possible (Gitter et al. 1998). Biopsy specimens from patients with active ulcerative colitis were studied (Schmitz et al. 1999) and epithelial resistance turned out to be reduced, although total resistance of the tissues was similar to that of controls, since subepithelial resistance was increased due to the inflammatory changes in lamina propria, muscularis mucosae, and submucosa. Thus, increased ionic permeability is one of the functional correlates of the structural tight junction changes reported above. In addition, permeability for the paracellular flux markers fluorescein and 4 kDa-FITC-dextran (FD4) was increased indicating an impaired paracellular barrier also for macromolecules (Krug et al. 2018). This is important for the entry of antigens into the mucosa, which is important for the perpetuation of inflammatory conditions in the intestine.

Active Secretion Another diarrheal mechanism is active anion secretion. It is very abundant in all types of infectious diarrhea, e.g., activated by toxin of vibrio cholera via cAMP as second messenger (Fenton et al. 2019). In biopsy specimens from patients with ulcerative colitis, however, active electrogenic chloride secretion was not found to be activated, since baseline short circuit current was not different from control (Schmitz et al. 1999). When the maximum transport capacity of this transport system was determined as the chloride-dependent increase in short circuit current during simultaneous stimulation with carbachol, PGE_2, and theophylline, however, V_{max} was not significantly diminished in ulcerative colitis confirming that the viability of the epithelium was preserved. These findings together are somewhat surprising, since pro-inflammatory mediators as cytokines are highly upregulated and released in ulcerative colitis (Heller et al. 2005). A possible explanation may be the severe barrier breakdown in ulcerative colitis mentioned above, since the epithelial barrier is a prerequisite for any net ion transport including active anion secretion.

Sodium Absorption Ulcerative colitis always involves the rectum, more or less extending throughout the colon. In distal colon and rectum, electrogenic sodium absorption via the epithelial sodium channel (ENaC) is the dominant mechanism of electrolyte absorption (Epple et al. 1995), while electroneutral NaCl absorption via NHE3 and DRA is prominent more in the proximal and mid-colon (Talbot and Lytle 2010). In comparison with healthy controls, the inflamed mucosa in active ulcerative colitis is functionally severely compromised. ENaC-dependent electrogenic sodium absorption, which represents an important absorptive transport process during diarrheal diseases, is completely absent. The inherent mechanisms of regulation resemble those discussed above for Crohn's colitis, but in addition, the pro-inflammatory cytokine interleukin-13 is important in ulcerative colitis by inhibiting ENaC-function via the STAT6/p38 MAPK pathways (Dames et al. 2015).

In addition, electroneutral NaCl absorption can be disturbed, if the inflammation zone extends to the proximal or mid-colon. While expression of the transporter NHE3 itself is not affected, transport activity is rather compromised by regulatory inputs arising from the NHE3 scaffold protein PDZK1 (Yeruva et al. 2015).

In an attempt to restore sodium absorption transport toward normal, glucocorticoids were tested. Glucocorticoid therapy restored sodium transport activity within a few hours in patients with ulcerative colitis as confirmed in transport studies in vivo (Sandle et al. 1986). Furthermore, there is evidence from in vitro studies that this pro-absorptive effect is intensified by short-chain fatty acids like butyrate or propionate, which are produced by the colonic microbiota from dietary carbohydrates or soluble fibers. This is due to histone deacetylase inhibition, thereby intensifying the binding of the nuclear binding factor SP3 to the β- and γ-ENaC promoters, which activates ENaC subunit expression (Zeissig et al. 2007b).

3.2.3 Microscopic Colitis

Definition Microscopic colitis is the generic term about two chronic inflammatory intestinal diseases of the colon characterized by watery diarrhea (Fernandez-Banares et al. 1999). However, in contrast to Crohn's disease and ulcerative colitis, macroscopic signs of inflammation are lacking during endoscopy. Thus, diagnosis is based on histological analysis, which is characterized by an elevated intraepithelial lymphocyte count. Depending on the additional presence of a collagenous band below the epithelium, the disease is either subclassified as collagenous colitis or lymphocytic colitis (Fernandez-Banares et al. 1999).

Clinical Data Clinically, mainly mid-aged to older women are affecting (Fernandez-Banares et al. 1999). The etiology is often unclear and most likely rather heterogeneous, intake of NSAIDs and chronic pancreatic diseases have been reported (Fernandez-Banares et al. 1999). The natural course is variable. Milder courses with sometimes spontaneous recovery are observed in lymphocytic colitis, while collagenous colitis often shows more severe symptoms and longer courses and often requires long-term anti-inflammatory treatment to prevent relapse, e.g., with topic steroids like budesonide (Fernandez-Banares et al. 1999).

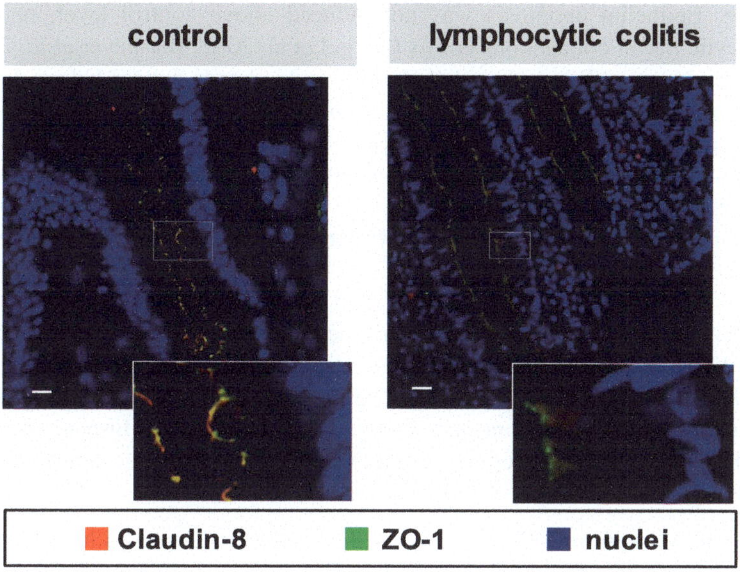

Fig. 3.5 Diminished claudin-8 in tight junctions in crypt epithelium of microscopic colitis. Claudin-8 was stained against the tight junction strand protein ZO-1. Nuclei were stained with DAPI. Specimens are from distal colon biopsies modified from Barmeyer et al. (2017a, J Gastroenterol 52:1090–1100)

Intestinal Ion Transport Epithelial pathophysiology has been intensively studied in microscopic colitis. Evidence is available in collagenous colitis for sodium and chloride malabsorption via disturbed electroneutral NaCl cotransport (Bürgel et al. 2002). Furthermore, electrogenic sodium absorption via the ENaC, which represents an absorptive reserve capacity in terms of ongoing diarrhea, is also seriously impaired and this has been detected in both collagenous and lymphocytic colitis (Barmeyer et al. 2016, 2017b). The mechanism behind this latter effect has been extensively studied, and the disturbance is the result of ERK1/2 activation in response to the pro-inflammatory cytokines. When Ingenuity Pathway Analysis (IPA, Qiagen Bioinformatics) was performed in lymphocytic colitis, TNFα, IFNγ, and IL-15 were found to be upregulated and a cytokine cocktail of these cytokines could mimic the effect observed in colon biopsies from patients with lymphocytic colitis (Barmeyer et al. 2016).

Epithelial Barrier Function Epithelial electrical resistance is diminished in both collagenous and lymphocytic colitis. The molecular basis is not explored in much detail in collagenous colitis, where only a reduction in occludin and claudin-4 expression has been identified (Bürgel et al. 2002). In lymphocytic colitis, however, epithelial tight junctions have been characterized in more detail recently (Barmeyer et al. 2017a). It turned out that claudin-4, -5, and -8 expression was reduced and that claudins 5 and 8 were distributed off the tight junction domain of the enterocytes toward an intracellular vesicular compartment (Fig. 3.5), an effect assumed to result from activation of endocytosis of tight junction proteins in response to inflammatory signaling. Epithelial apoptosis was not induced above control level, neither in lymphocytic nor in collagenous colitis (Bürgel et al. 2002; Barmeyer et al. 2017a).

In collagenous colitis, subepithelial collagenous bands appear below the basement membrane and represent a significant diffusion barrier for actively absorbed ions on their way to mucosal blood vessels, as a result of which epithelial net absorption is handicapped. This is due to a partial back diffusion of already absorbed ions to the intestinal lumen. A quantitative calculation of the barrier properties resulted in an estimated reduction of net absorption by a factor of 2 as a result of this phenomenon (Bürgel et al. 2002).

In collagenous colitis, there is a significant change in the expression of claudin-2 in comparison with healthy controls, which acts as sodium and water channel in the epithelial tight junction meshwork (Bürgel et al. 2002). The functional meaning of this finding, however, is circumstantial. Due to the lack of active sodium absorption as discussed above, there is no net water absorption initiated. Instead, water will passively cross the colonic epithelium and will get lost into the intestinal lumen contributing to diarrhea by this leak-flux mechanism.

Therapy of Microscopic Colitis One of the standard therapies is long-term treatment with the topic glucocorticoid budesonide which in addition to suppressing inflammation in the long run has an immediate direct influence on epithelial ion transport by improving ENaC-dependent electrogenic sodium absorption (Barmeyer et al. 2012).

3.2.4 Celiac Disease

Definition Celiac disease is a gluten-sensitive autoimmune disease. The autoimmune component is the contribution of the enzyme tissue transglutaminase 2 (TTG-2), which is released in active celiac disease (Dieterich et al. 1997). The symptoms often begin in early childhood, usually several months after the onset of wheat intake, in which the alcohol-soluble fraction of the gluten, the gliadins, is the toxic component. After activation by deamidation through TTG-2 and presentation by dendritic cells equipped with the HLA type 2 antigen DQ2, t-cells are the main player of the immune response in the mucosa (Stein and Schuppan 2014). This immune activation causes damage to the mucosa with typical hyper-regenerative transformation of the epithelium (see below).

Clinical Data Active celiac disease in adult patients is characterized by diarrhea either without or often with fat stool (steatorrhea) and a loss of body weight (Murray 1999). Depending on the extent of the change in mucosal architecture either none, only some or even most vitamins and trace elements are concerned (partial to global malabsorption syndrome) (Deora et al. 2017). This leads to a plethora of different clinical aspects including microcytic anemia (iron deficiency), acrodermatitis enteropathica (zinc deficiency), and many others (Murray 1999). Additionally, atypical symptoms can arise (Murray 1999) due to the systemic inflammation including aphthoid lesions in the mouth, elevated liver laboratory parameters, joint complaints, and/or skin alterations (Morbus herpetiformis Duhring) (Caproni et al. 2012). These changes disappear under gluten-free diet.

Mucosal Morphology The immune activation of the celiac mucosa causes damage to the epithelium. This is characterized by higher intraepithelial lymphocyte counts, elongated crypts, and a reduction of villus height (Murray 1999). Epithelial cell division is compensatory upregulated but cannot catch up with cell loss and thus cannot prevent the overall reduction of the size of the villus cell compartment, at the end of which a partial or even total villus atrophy is observed (flat mucosa) (Wright et al. 1973). Since migration rate of epithelial cells becomes fourfold faster in active celiac disease when compared to control small intestine, enterocytes cannot fully maturate during their way along the basal membrane to the extrusion zone at the surface (Wright et al. 1973). Therefore, absorptive transport mechanisms like the sodium-dependent glucose transporter 1 (SGLT1) or the sodium-coupled amino acid transporters are not present in their active form as usual.

Epithelial Transport and Barrier Function and Diarrheal Mechanisms The abovementioned reduction of the presence of active absorptive transport proteins for glucose and amino acids is the most important mechanism of the malabsorption in celiac disease. Whether or not diarrhea is a prominent symptom in an individual patient depends on the extent of mucosal transformation along the small intestine (Doyev et al. 2019). For example, when a significant part of the distal small intestine is unaffected and functionally intact, diarrhea can be lacking, since this situation offers enough reserve ion transport capacity to ensure an almost complete uptake of ions and water along the gastrointestinal tract (Murray 1999). Then, monosymptomatic celiac disease may be a typical result, e.g., hypochromic microcytic anemia due to iron deficiency, since iron absorption normally takes place in the most proximal small intestine (Freeman 2015). In addition to the malabsorptive mechanisms, maldigestion is another important pathomechanism due to the lack of brush border disaccharidases (Peters et al. 1975).

Furthermore, epithelial barrier function is seriously affected in celiac disease. In freeze-fracture electron microscopy (Fig. 3.6), epithelial tight junctions exhibit a reduced number of strands and an increase in strand discontinuities (Schulzke et al. 1998). This is the result of a downregulation of the barrier-forming tight junction proteins claudins 3, 5, and 7 and occludin, while the channel-forming tight junction proteins claudin-2 and claudin-15 are upregulated (Schumann et al. 2012). An increase in intracellular claudin-5 staining indicates a shift from tight junctions into the cytoplasm. Taken together, this points to disturbance of epithelial barrier for ions in active celiac disease, which represents an important mechanism of diarrhea (leak-flux diarrhea). The reduction in claudin-5 causes barrier dysfunction, while the redistribution of claudin-15 off the tight junction domain of the epithelial cells may rather be relevant for intestinal glucose malabsorption, which requires the continuous supply of sodium to the SGLT-1 in the apical membrane of surface enterocytes (Tamura et al. 2011).

Surprisingly, active electrogenic anion secretion is not activated in active celiac disease (Schulzke et al. 1995), although TNFα is an inductor of this transport and is

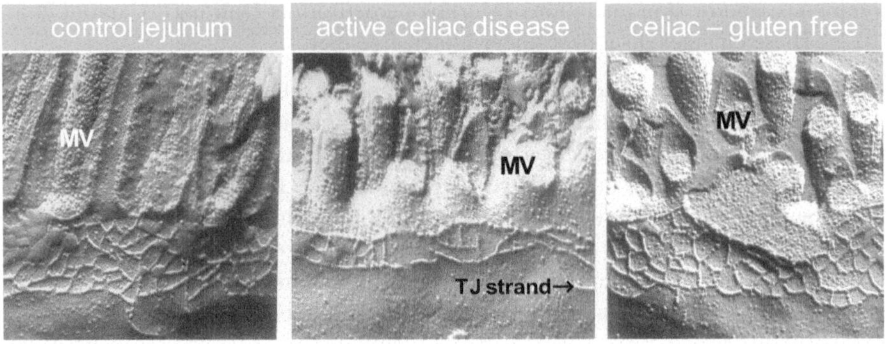

Fig. 3.6 Freeze-fracture electron microscopy of epithelial tight junctions in active celiac disease. Tight junction structure is impaired as indicated by a reduced number of tight junction strands and the appearance of tight junction strand discontinuities. *MV* microvilli. Modified from Schulzke et al. (1998, Pediatr Res 43:435–441)

released to a higher extent than in healthy controls. Thus, either the abovementioned cell maturation defect is also relevant for this (lacking) transport event, even if active chloride secretion is expressed already in the crypt cell compartment, or the severe loss of epithelial barrier function interferes with the transport efficiency of active chloride secretion. Hence, epithelial barrier function is a prerequisite for each vectorial transport process, e.g., for net absorption of ions and water. Thus, a loss of barrier function compromises active ion transport mechanisms. Therefore, the severe barrier disturbance in celiac disease has to be assumed to be a relevant co-component of malabsorption.

Epithelial Barrier and Disease Perpetuation by Elevated Antigen Uptake Tight junction alterations in active celiac disease as discussed above predominantly induce a passive loss of ions and water into the intestinal lumen and lead to leak-flux diarrhea. However, barrier impairment in active celiac disease concerns also molecules of higher molecular weight like gliadin and gliadin cleavage products and this increased macromolecule permeability is important for disease perpetuation (Schumann et al. 2008). The mechanisms behind this barrier disturbance for macromolecules in active celiac disease are complex. The first and earliest event is elevated transcellular uptake of gliadin by endocytosis with a lack of the ability to degrade gliadin during transcytosis (Ménard et al. 2012). In later stages, also paracellular leakage may contribute to the barrier defect as epithelial tight junctions show severe structural defects (Schulzke et al. 1998). In addition, a higher apoptotic rate in the epithelium may contribute.

Barrier repair in response to gluten-free diet is an essential aspect in healing celiac disease. Although transient antigen suppression is often sufficient to re-induce immune tolerance to gluten, patients should follow a lifelong gluten-free diet, in order to minimize the risk of intestinal T-cell lymphoma development (Catassi et al. 2005).

Non-celiac Wheat-Sensitivity From celiac disease, other wheat-sensitive disorders have to be discriminated. This has recently been taken into account by using the name "non-celiac wheat-sensitivity" or "non-celiac gluten-sensitivity" (Skodje et al. 2018). In addition to allergic reactions against wheat proteins including gliadin, which affect intestinal functions via IgE and mast cell activation, reactions to fermented oligo-, di-, and monosaccharides and polyols (FODMAP) and to amylase trypsin inhibitors (ATI) via the innate immune system have been reported (Junker et al. 2012). In contrast to classic celiac disease, much less is known about epithelial transport and barrier function in these conditions. In patients presenting with diarrhea-dominant irritable bowel syndrome (IBS-d), increased small intestinal myosin light chain phosphorylation as well as increased colonic claudin-15 expression accompanying elevated small intestinal mannitol uptake has been reported under a gluten-containing diet (Wu et al. 2017).

3.2.5 *Giardia Lamblia Infection*

Definition *Giardia lamblia* is a parasite, which is found in tropic and subtropical regions around the world, but infection can easily occur also in Europe or North America. Infection usually starts with uptake of parasite cysts, which develop in the human small intestine to the typical disk-shaped protozoon attaching to epithelial cells (Greenson 2015). Acute infection causes diarrhea, which is among the most frequent sources of traveler's diarrhea, e.g., seen in returnees from India. Chronic colonization of the small intestine with damage to the underlying epithelial cells is the reason for malabsorptive diarrhea and weight loss (Troeger et al. 2007). Since microscopic stool examination has turned out to be difficult and results were often false negative, only the introduction of immune detection techniques in stool samples could overcome this problem (Janoff et al. 1989).

Clinical Data Clinically, patients suffer from diarrhea, weight loss, anemia, and the lack of different vitamins, trace elements, and ions. Diagnostically also a flattened mucosa similar to celiac disease is found in small intestinal biopsies, although often much less severe than in celiac disease (usually partial villus atrophy) (Greenson 2015; Troeger et al. 2007). Therapeutically antibiotics as metronidazole are effective, although more and more nitro-imidazole-resistant infections have been reported (Carter et al. 2018).

Mucosal Alterations Partial villus atrophy with or without hyperplastic crypts (Fig. 3.7) and an increased number of intraepithelial lymphocytes (IEL) can be observed in giardia lamblia infection (Troeger et al. 2007). However, especially in acute disease, mucosal architecture is only slightly affected and parasites are seen in between the almost intact villi in histological sections ("falling leaf phenomenon") (Greenson 2015). Most important, however, these patients with malabsorptive diarrhea and mucosal atrophy, although morphologically resembling the mucosa in celiac disease, do not respond to a gluten-free diet but only become better on

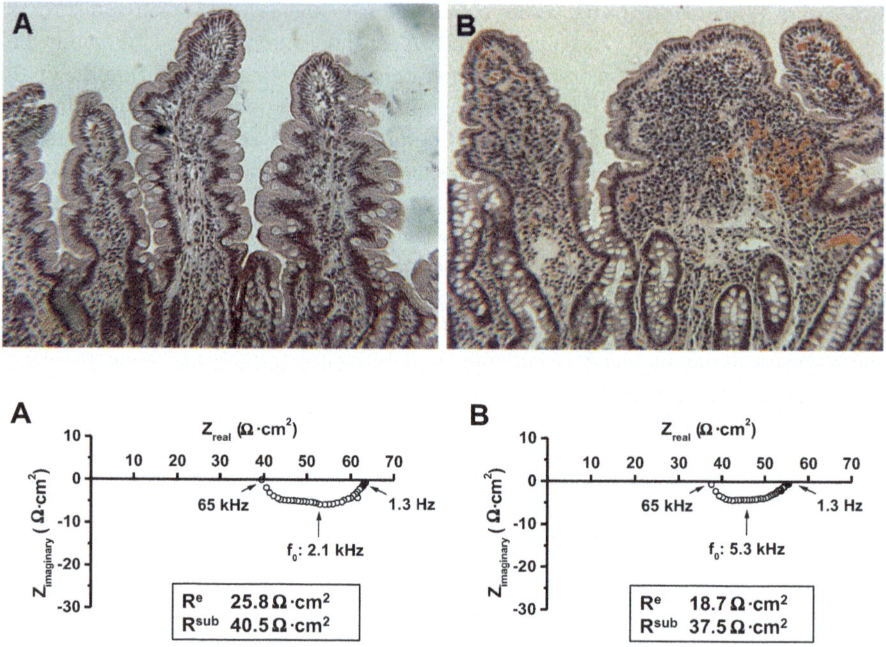

Fig. 3.7 Reduced surface area and barrier dysfunction in giardia lamblia infection. HE-stained histology and impedance spectroscopy from active lambliasis presenting with diarrhea (**a**) in comparison with a healthy control, (**b**) R^e and R^{sub} are the epithelial and subepithelial electrical resistance of a duodenal biopsy specimen modified from Troeger et al. (2007, Gut 56:328–335)

antibiotics. This picture, however, resembles the findings reported for tropical sprue. When having the difficulties with giardia lamblia diagnosis at a time without immune detection in mind, it seems reasonable to assume that many cases of tropical sprue may have been due to chronic giardia lamblia infection (Khokhar and Gill 2004).

Functional Analysis There are only a few functional data available for this disease. Duodenal specimens from patients with Giardia lamblia infection were obtained by forceps biopsies during upper gastrointestinal endoscopy and studied in miniaturized Using-chambers (Troeger et al. 2007). Reduced glucose-dependent sodium absorption was detected as a diminished glucose-dependent increase in short circuit current, blockable by phlorizin. This represents direct evidence for an impaired sodium glucose transporter 1 (SGLT-1) in the small intestine (Troeger et al. 2007). This is in good agreement with clinical data of impaired D-xylose blood tests in 50% of the patients with giardia lamblia infection, which indicates a malabsorptive diarrheal mechanism as well (Horvath et al. 1990). While this finding suggests a reduction of villus surface area or cellular damage after direct contact of parasites to cells in the villus compartment, there is also evidence for an upregulation of active electrogenic chloride secretion in this biopsy study (Troeger et al. 2007). This, however, is not compatible with a simple loss of cell viability after infection. Instead, maturation arrest of the cells in the villus compartment could lead to the SGLT1 defect, while secretory cells in the crypt compartment are functionally still intact and become activated.

In addition, epithelial barrier defects (Fig. 3.7) contribute to the diarrhea (by a leak-flux mechanism) (Troeger et al. 2007). Both, epithelial tight junction changes and abundant apoptoses of epithelial cells, are the structural correlate. A decrease in expression of the barrier-forming tight junction protein claudin-4 is the most prominent features in tight junction barrier biology (Troeger et al. 2007).

There are reports in the literature that infection of epithelial Caco-2 cells with Giardia lamblia is a suitable model for studying the induction of barrier disturbances in vitro (Humen et al. 2011). This at least indicates a direct pathogenic influence of the parasites on the enterocytes. However, an additional contribution of the mucosal immune system or of other factors to giardia lamblia pathology may be possible.

3.2.6 Whipple's Disease

Whipple's disease is a chronic infectious disease, which usually starts in the small intestine. It can remain restricted to the small intestine but is later on often also disseminated causing, e.g., arthritis, endocarditis, and central nervous manifestations (Marth et al. 2016). The causative agent of this disease is *Tropheryma whipplei*, a rod-shaped gram-negative bacterium that is very common all over the world. It is detected by PCR in the intestine in a significant percentage of the population, also under healthy conditions. That manifestation of Whipple's disease is a rare event points to a special pre-deposition of the patients, most likely some type of an immune deficiency (Marth et al. 2016). However, this immunological defect is rather specific for *T. whipplei*, since patients are not more frequently affected by other infections (Marth et al. 2016).

For cultivating *T. whipplei* in fibroblasts, interleukin-4 exposure turned out to be essential. Immune cells from patients with Whipple's disease produce less interferon-gamma, interleukin-2, and interleukin-12 than controls (Schoedon et al. 1997). Clinically, significant malabsorption and diarrhea occur (Marth et al. 2016). Antibiotic treatment is effective, e.g., with cotrimoxazole but needs long treatment periods in the range of a year or longer (Marth et al. 2016). In case of extraintestinal manifestations, the disease can easily become life threatening.

Mucosal Alterations Beyond the detection of the bacteria incorporated into macrophages by Periodic acid–Schiff stain (PAS) staining or electron microscopy, the main finding in the small intestinal mucosa is blunted and shortened villi causing a reduction in mucosal surface area and elongated crypts (hyper-regenerative transformation of the mucosa) (Epple et al. 2017). In some patients, these villus alterations are discontinuously distributed along the small intestine. If the proximal duodenum is not involved, either enteroscopy or double-/single-balloon endoscopy is necessary for diagnosing Whipple's disease (Nagasue et al. 2016), since stool tests with PCR detection of *T. whipplei* do not make sense due to the high bacterial prevalence also in healthy people.

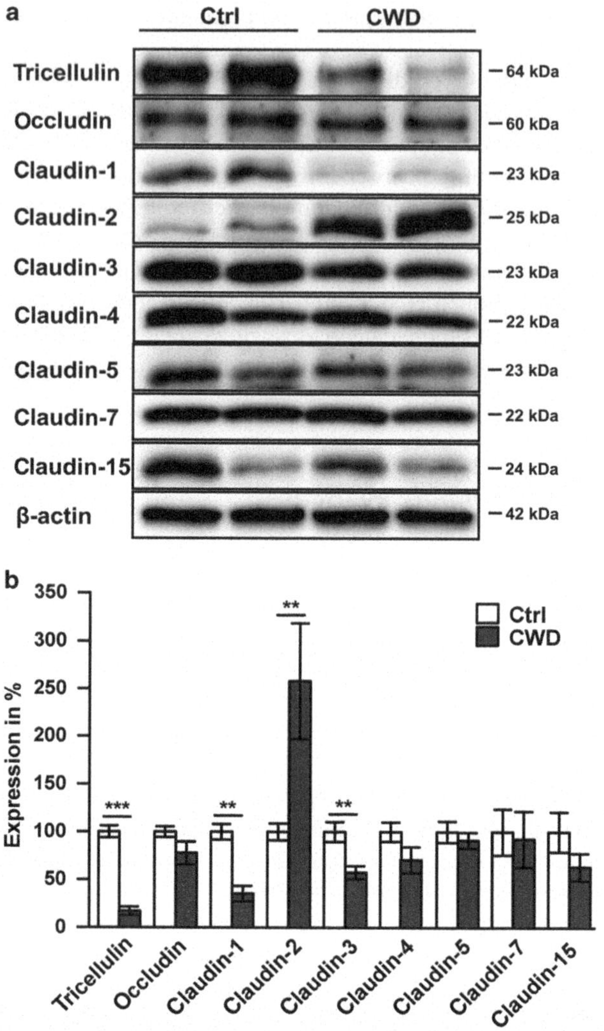

Fig. 3.8 Epithelial barrier dysfunction in Whipple's disease. Quantitative expression analysis of epithelial tight junction proteins obtained from Western blotting and densitometry. *Ctrl* control, *CWD* chronic Whipple's disease, modified from Epple et al. (2017, Mucosal Immunol 10:1542–1552)

Functional Data Functional studies of this disease are rare. Biopsy specimens from patients with acute intestinal Whipple's disease were studied in miniaturized Ussing-chambers (Epple et al. 2017). Reduced glucose-dependent sodium absorption via SGLT1 was detected. In addition, active electrogenic anion secretion was stimulated as indicated by a higher bumetanide-sensitive short circuit current. Finally, alternating current impedance spectroscopy revealed a decrease in epithelial resistance, as a result of which sodium and water get lost into the intestinal lumen contributing to the diarrhea by a leak-flux mechanism. As shown in Fig. 3.8, this is on the one hand due to a lower number of barrier-forming epithelial tight junction proteins claudin-1, claudin-3, and tricellulin, and on the other hand to an upregulation of the pore-forming tight junction protein claudin-2. Furthermore, epithelial apoptosis is more frequent, which contributes to the barrier defect (Epple et al. 2017).

The type of mediation of this cytopathic effect and the mechanisms responsible for the hyper-regenerative mucosal transformation are not fully understood. Most likely, pro-inflammatory cytokines as, e.g., TNFα and interleukin-13 released from mucosal immune cells may contribute (Epple et al. 2017). This view is especially supported by the fact that direct effects of bacteria on epithelial cell in cell models are lacking. However, infection experiments with coculture cell models or organoids are urgently needed.

3.3 Summary

In order to establish transport- and barrier-specific therapy concepts, the pathophysiology of immunologically driven intestinal diseases has to be known in detail. Especially, mechanisms that trigger disease onset and propagation or recurrence of disease activity are highly relevant, since they are often related to barrier dysfunction of the intestinal epithelium. While originally features and genes were considered which had been suspected to weaken this barrier function when altered, recently more and more effort was made to understand the role of environmental influences like the role of the own intestinal microbiota or infectious agents. Therefore, this chapter presents the pathophysiology of a group of immune-related intestinal diseases like IBD or celiac disease and of intestinal infections like Whipple's disease or the intestinal infection with giardia lamblia. Main focus was placed on epithelial transport and barrier function, since these pathophysiological details are important for the future development of diagnostic and therapeutic approaches.

Acknowledgments Own original work reported in this article was granted by DFG Research Unit 721, DFG Research Training Group 2318, and DFG Research Grants Schu 559/11 and Fr 652/12.

References

Al-Sadi R, Khatib K, Guo S, Ye D, Youssef M, Ma T (2011) Occludin regulates macromolecule flux across the intestinal epithelial tight junction barrier. Am J Physiol Gastrointest Liver Physiol 300:G1054–G1064

Amasheh S, Meiri N, Gitter AH, Schöneberg T, Mankertz J, Schulzke JD, Fromm M (2002) Claudin-2 expression induces cation-selective channels in tight junctions of epithelial cells. J Cell Sci 115:4969–4976

Autenrieth DM, Baumgart DC (2012) Toxic megacolon. Inflamm Bowel Dis 18:584–591

Barmeyer C, Erko I, Fromm A, Bojarski C, Allers K, Moos V, Zeitz M, Fromm M, Schulzke JD (2012) Ion transport and barrier function are disturbed in microscopic colitis. Ann N Y Acad Sci 1258:143–148

Barmeyer C, Erko I, Fromm A, Bojarski C, Loddenkemper C, Dames P, Kerick M, Siegmund B, Fromm M, Schweiger MR, Schulzke JD (2016) ENaC dysregulation through activation of

MEK1/2 contributes to impaired Na^+ absorption in lymphocytic colitis. Inflamm Bowel Dis 22:539–547
Barmeyer C, Erko I, Awad K, Fromm A, Bojarski C, Meissner S, Loddenkemper C, Kerick M, Siegmund B, Fromm M, Schweiger MR, Schulzke JD (2017a) Epithelial barrier dysfunction in lymphocytic colitis through cytokine-dependent internalization of claudin-5 and -8. J Gastroenterol 52:1090–1100
Barmeyer C, Fromm M, Schulzke JD (2017b) Active and passive involvement of claudins in the pathophysiology of intestinal inflammatory diseases. Pflügers Arch 469:15–26
Bayless TM, Brown E, Paige DM (2017) Lactase non-persistence and lactose intolerance. Curr Gastroenterol Rep 19(5):23
Bergann T, Zeissig S, Fromm A, Richter JF, Fromm M, Schulzke JD (2009) Glucocorticoids and tumor necrosis factor-alpha synergize to induce absorption by the epithelial sodium channel in the colon. Gastroenterology 136:933–942
Berglindh T, Landström E, Bergqvist E (1988) Induction of and adaptation to olsalazine-induced intestinal volume load. Scand J Gastroenterol Suppl 148:24–28
Boron WF, Boulpaep EL (2016) Medical physiology, 3rd edn. Elsevier, Philadelphia
Bücker R, Schulz E, Günzel D, Bojarski C, Lee IF, John LJ, Wiegand S, Janßen T, Wieler LH, Dobrindt U, Beutin L, Ewers C, Fromm M, Siegmund B, Troeger H, Schulzke JD (2014) α-Haemolysin of Escherichia coli in IBD: a potentiator of inflammatory activity in the colon. Gut 63:1893–1901
Bürgel N, Bojarski C, Mankertz J, Zeitz M, Fromm M, Schulzke JD (2002) Mechanisms of diarrhea in collagenous colitis. Gastroenterology 123:433–443
Camilleri M (2015) Bile acid diarrhea: prevalence, pathogenesis, and therapy. Gut Liver 9:332–339
Caproni M, Bonciolini V, D'Errico A, Antiga E, Fabbri P (2012) Celiac disease and dermatologic manifestations: many skin clue to unfold gluten-sensitive enteropathy. Gastroenterol Res Pract 2012:952753. https://doi.org/10.1155/2012/952753
Carter ER, Nabarro LE, Hedley L, Chiodini PL (2018) Nitroimidazole-refractory giardiasis: a growing problem requiring rational solutions. Clin Microbiol Infect 24:37–42
Catassi C, Bearzi I, Holmes GK (2005) Association of celiac disease and intestinal lymphomas and other cancers. Gastroenterology 128:S79–S86
Clarkson TW (1967) The transport of salt and water across isolated rat ileum. Evidence for at least two distinct pathways. J Gen Physiol 50:695–727
Conrad MP, Piontek J, Günzel D, Fromm M, Krug SM (2016) Molecular basis of claudin-17 anion selectivity. Cell Mol Life Sci 73:185–200
Cording J, Arslan B, Staat C, Dithmer S, Krug SM, Krüger A, Berndt P, Günther R, Winkler L, Blasig IE, Haseloff RF (2017) Trictide, a tricellulin-derived peptide to overcome cellular barriers. Ann N Y Acad Sci 1405:89–101
Crohn BB, Ginzburg L, Oppenheimer GD (1932) Regional ileitis: a pathologic and clinical entity. Mt Sinai J Med 67:263–268
da Silva BC, Lyra AC, Rocha R, Santana GO (2014) Epidemiology, demographic characteristics and prognostic predictors of ulcerative colitis. World J Gastroenterol 20:9458–9467
Dames P, Bergann T, Fromm A, Bücker R, Barmeyer C, Krug SM, Fromm M, Schulzke JD (2015) Interleukin-13 affects the epithelial sodium channel in the intestine by coordinated modulation of STAT6 and p38 MAPK activity. J Physiol 593:5269–5282
Deora V, Aylward N, Sokoro A, El-Matary W (2017) Serum vitamins and minerals at diagnosis and follow-up in children with celiac disease. J Pediatr Gastroenterol Nutr 65:185–189
Dharmsathaphorn K, Huott PA, Vongkovit P, Beuerlein G, Pandol SJ, Ammon HV (1989) Chloride secretion induced by bile salts. A study of the mechanism of action based on a cultured colonic epithelial cell line. J Clin Invest 84:945–953
Diamond JM (1974) Tight and leaky junctions of epithelia: a perspective on kisses in the dark. Fed Proc 33:2220–2224
Diamond JM (1978) Channels in epithelial cell membranes and junctions. Fed Proc 37:2639–2643

Dieterich W, Ehnis T, Bauer M, Donner P, Volta U, Riecken EO, Schuppan D (1997) Identification of tissue transglutaminase as the autoantigen of celiac disease. Nat Med 3:797–801

Doyev R, Cohen S, Ben-Tov A, Weintraub Y, Amir A, Galai T, Moran-Lev H, Yerushalmy-Feler A (2019) Ultra-short celiac disease is a distinct and milder phenotype of the disease in children. Dig Dis Sci 64:167–172

Duchmann R, Kaiser I, Hermann E, Mayet W, Ewe K, Meyer zum Büschenfelde KH (1995) Tolerance exists towards resident intestinal flora but is broken in active inflammatory bowel disease (IBD). Clin Exp Immunol 102:448–455

Epple HJ, Schulzke JD, Schmitz H, Fromm M (1995) Enzyme and mineralocorticoid receptor controlled electrogenic sodium absorption in the human rectum, in vitro. Am J Phys 269:G42–G48

Epple HJ, Friebel J, Moos V, Troeger H, Krug SM, Allers K, Schinnerling K, Fromm A, Siegmund B, Fromm M, Schulzke JD, Schneider T (2017) Architectural and functional alterations of the small intestinal mucosa in classical Whipple's disease. Mucosal Immunol 10:1542–1552

Fenton RA, Murali SK, Kaji I, Akiba Y, Kaunitz JD, Kristensen TB, Poulsen SB, Dominguez Rieg JA, Rieg T (2019) Adenylyl cyclase 6 expression is essential for cholera toxin-induced diarrhea. J Infect Dis 220(11):1719–1728. https://doi.org/10.1093/infdis/jiz013

Fernandez-Banares F, Salas A, Forne M, Esteve M, Espinos J, Viver JM (1999) Incidence of collagenous and lymphocytic colitis: a 5-year population-based study. Am J Gastroenterol 94:418–423

Fischer A, Gluth M, Pape UF, Wiedenmann B, Theuring F, Baumgart DC (2013) Adalimumab prevents barrier dysfunction and antagonizes distinct effects of TNFα on tight junction proteins and signaling pathways in intestinal epithelial cells. Am J Physiol Gastrointest Liver Physiol 304:G970–G979

Fischer A, Gluth M, Weege F, Pape UF, Wiedenmann B, Baumgart DC, Theuring F (2014) Glucocorticoids regulate barrier function and claudin expression in intestinal epithelial cells via MKP-1. Am J Physiol Gastrointest Liver Physiol 306:G218–G228

Freeman HJ (2015) Iron deficiency anemia in celiac disease. World J Gastroenterol 21:9233–9238

Fromm M, Piontek J, Rosenthal R, Günzel D, Krug SM (2017) Tight junctions of the proximal tubule and their channel proteins. Pflügers Arch 469:877–887

Frömter E, Diamond J (1972) Route of passive ion permeation in epithelia. Nat New Biol 235:9–13

Furuse M, Furuse K, Sasaki H, Tsukita S (2001) Conversion of zonulae occludentes from tight to leaky strand type by introducing claudin-2 into Madin-Darby canine kidney I cells. J Cell Biol 153:263–272

Ganapathy V, Miyamoto Y, Leibach FH (1987) Driving force for peptide transport in mammalian intestine and kidney. Beitr Infusionther Klin Ernahr 17:54–68

Ghezzi C, Loo DDF, Wright EM (2018) Physiology of renal glucose handling via SGLT1, SGLT2 and GLUT2. Diabetologica 61:2087–2097

Gitter AH, Fromm M, Schulzke JD (1998) Impedance analysis for the determination of epithelial and subepithelial resistance in intestinal tissues. J Biochem Biophys Methods 37:35–46

Gitter AH, Wullstein F, Fromm M, Schulzke JD (2001) Epithelial barrier defects in ulcerative colitis: characterization and quantification by electrophysiological imaging. Gastroenterology 121:1320–1328

Gong Y, Renigunta V, Zhou Y, Sunq A, Wang J, Yang J, Renigunta A, Baker LA, Hou J (2015) Biochemical and biophysical analyses of tight junction permeability made of claudin-16 and claudin-19 dimerization. Mol Biol Cell 26:4333–4346

Greenson JK (2015) Diagnostic pathology: gastrointestinal, 2nd edn. Elsevier, Oxford

Günzel D, Fromm M (2012) Claudins and other tight junction proteins. Compr Physiol 2:1819–1852

Günzel D, Yu AS (2013) Claudins and the modulation of tight junction permeability. Physiol Rev 93:525–569

Günzel D, Stuiver M, Kausalya PJ, Haisch L, Krug SM, Rosenthal R, Meij IC, Hunziker W, Fromm M, Müller D (2009) Claudin-10 exists in six alternatively spliced isoforms which exhibit distinct localization and function. J Cell Sci 122:1507–1517

Heller F, Florian P, Bojarski C, Richter JF, Christ M, Hillenbrand B, Mankertz J, Gitter AH, Bürgel N, Fromm M, Zeitz M, Fuss I, Strober W, Schulzke JD (2005) Interleukin-13 is the key effector Th2 cytokine in ulcerative colitis that affects epithelial tight junctions, apoptosis and cell restitution. Gastroenterology 129:550–564

Higashi T, Tokuda S, Kitajiri S, Masuda S, Nakamura H, Oda Y, Furuse M (2013) Analysis of the 'angulin' proteins LSR, ILDR1 and ILDR2 – tricellulin recruitment, epithelial barrier function and implication in deafness pathogenesis. J Cell Sci 126:966–977

Horvath K, Horn G, Bingadeem H, Nemes Nagy A, Bodanszky H (1990) The value of the d-xylose loading test in the diagnosis of malabsorption syndromes. Orv Hetil 131:1803–1806

Humen MA, Pérez PF, Liévin-Le Moal V (2011) Lipid raft-dependent adhesion of Giardia intestinalis trophozoites to a cultured human enterocyte-like Caco-2/TC7 cell monolayer leads to cytoskeleton-dependent functional injuries. Cell Microbiol 13:1683–1702

Ikenouchi J, Furuse M, Furuse K, Sasaki H, Tsukita S, Tsukita S (2005) Tricellulin constitutes a novel barrier at tricellular contacts of epithelial cells. J Cell Biol 171:939–945

Janoff EN, Craft JC, Pickering LK, Novotny T, Blaser MJ, Knisley CV, Reller LB (1989) Diagnosis of Giardia lamblia infections by detection of parasite-specific antigens. J Clin Microbiol 27:431–435

Junker Y, Zeissig S, Kim SJ, Barisani D, Wieser H, Leffler DA, Zevallos V, Libermann TA, Dillon S, Freitag TL, Kelly CP, Schuppan D (2012) Wheat amylase trypsin inhibitors drive intestinal inflammation via activation of toll-like receptor 4. J Exp Med 209:2395–2408

Keita ÅV, Lindqvist CM, Öst Å, Magana CDL, Schoultz I, Halfvarson J (2018) Gut barrier dysfunction – a primary defect in twins with Crohn's disease predominantly caused by genetic predisposition. J Crohns Colitis 12:1200–1209

Khokhar N, Gill ML (2004) Tropical sprue: revisited. J Pak Med Assoc 54:133–134

Krug SM, Amasheh S, Richter JF, Milatz S, Günzel D, Westphal JK, Huber O, Schulzke JD, Fromm M (2009) Tricellulin forms a barrier to macromolecules in tricellular tight junctions without affecting ion permeability. Mol Biol Cell 20:3713–3724

Krug SM, Günzel D, Conrad MP, Rosenthal R, Fromm A, Amasheh S, Schulzke JD, Fromm M (2012) Claudin-17 forms tight junction channels with distinct anion selectivity. Cell Mol Life Sci 69:2765–2778

Krug SM, Schulzke JD, Fromm M (2014) Tight junction, selective permeability, and related diseases. Semin Cell Dev Biol 36:166–176

Krug SM, Hayaishi T, Iguchi D, Watari A, Takahashi A, Fromm M, Nagahama M, Takeda H, Okada Y, Sawasaki T, Doi T, Yagi K, Kondoh M (2017) Angubindin-1, a novel paracellular absorption enhancer acting at the tricellular tight junction. J Control Release 260:1–11

Krug SM, Bojarski C, Fromm A, Lee IM, Dames P, Richter JF, Turner JR, Fromm M, Schulzke JD (2018) Tricellulin is regulated via interleukin-13-receptor α2, affects macromolecule uptake, and is decreased in ulcerative colitis. Mucosal Immunol 11:345–356

Li J, Shaikh SA, Enkavi G, Wen PC, Huang Z, Tajkhorshid E (2013) Transient formation of water-conducting states in membrane transporters. Proc Natl Acad Sci U S A 110:7696–7701

Loddenkemper C, Keller S, Hanski ML, Cao M, Jahreis Stein HG, Zeitz M, Hanski C (2006) Prevention of colitis-associated carcinogenesis in a mouse model by diet supplementation with ursodeoxycholic acid. Int J Cancer 118:2750–2757

Lügering A, Schmidt M, Lügering N, Pauels HG, Domschke W, Kucharzik T (2001) Infliximab induces apoptosis in monocytes from patients with chronic active Crohn's disease by using a caspase-dependent pathway. Gastroenterology 121:1145–1145

Marth T, Moos V, Müller C, Biagi F, Schneider T (2016) Tropheryma whipplei infection and Whipple's disease. Lancet Infect Dis 16:e13–e22

Martini E, Krug SM, Siegmund B, Neurath MF, Becker C (2017) Mend your fences: The epithelial barrier and its relationship with mucosal immunity in inflammatory bowel disease. Cell Mol Gastroenterol Hepatol 4:33–46

Ménard S, Lebreton C, Schumann M, Matysiak-Budnik T, Dugave C, Bouhnik Y, Malamut G, Cellier C, Allez M, Crenn P, Schulzke JD, Cerf-Bensussan N, Heyman M (2012) Paracellular versus transcellular intestinal permeability to gliadin peptides in active celiac disease. Am J Pathol 180:608–615

Morita K, Furuse M, Fujimoto K, Tsukita S (1999) Claudin multigene family encoding four-transmembrane domain protein components of tight junction strands. Proc Natl Acad Sci U S A 96:511–516

Murray JA (1999) The widening spectrum of celiac disease. Am J Clin Nutr 69:354–365

Muto S, Hata M, Taniguchi J, Tsuruoka S, Moriwaki K, Saitou M, Furuse K, Sasaki H, Fujimura A, Imai M, Kusano E, Tsukita S, Furuse M (2010) Claudin-2-deficient mice are defective in the leaky and cation-selective paracellular permeability properties of renal proximal tubules. Proc Natl Acad Sci U S A 107:8011–8016

Nagasue T, Kurahara K, Yaita H, Tanaka T, Oshiro Y, Kuno N, Harada A, Kameda M, Iwasaki K, Tsutsumi Y (2016) Whipple's disease diagnosed using electron microscopy and polymerase chain reaction. Nihon Shokakibyo Gakkai Zasshi 113:1894–1900

Peters TJ, Heath JR, Wansbrough-Jones MH, Foe WF (1975) Enzyme activities and properties of lysosomes and brush borders in jejunal biopsies from control subjects and patients with coeliac disease. Clin Sci Mol Med 48:259–267

Prasad S, Mingrino R, Kaukinen K, Hayes KL, Powell RM, MacDonald TT, Collins JE (2005) Inflammatory processes have differential effects on claudins 2, 3 and 4 in colonic epithelial cells. Lab Investig 85:1139–1162

Reinglas J, Gonczi L, Kurt Z, Bessissow T, Lakatos PL (2018) Positioning of old and new biologicals and small molecules in the treatment of inflammatory bowel diseases. World J Gastroenterol 24:3567–3582

Riazuddin S, Ahmed ZM, Fanning AS, Lagziel A, Kitajiri S, Ramzan K, Khan SN, Chattaraj P, Friedman PL, Anderson JM, Belyantseva IA, Forge A, Riazuddin S, Friedman TB (2006) Tricellulin: a novel tight junction protein necessary for hearing. Am J Hum Genet 79:1040–1051

Rosenthal R, Milatz S, Krug SM, Oelrich B, Schulzke JD, Amasheh S, Günzel D, Fromm M (2010) Claudin-2, a component of the tight junction, forms a paracellular water channel. J Cell Sci 123:1913–1921

Rosenthal R, Günzel D, Krug SM, Schulzke JD, Fromm M, Yu ASL (2017) Claudin-2-mediated cation and water transport share a common pore. Acta Physiol 219:521–536

Rosenthal R, Günzel D, Piontek J, Krug SM, Ayala-Torres C, Hempel C, Theune D, Fromm M (2020) Claudin-15 forms a water channel through the tight junction with distinct function compared to claudin-2. Acta Physiol 228(1):e13334

Saitou M, Furuse M, Sasaki H, Schulzke JD, Fromm M, Takano H, Noda T, Tsukita S (2000) Complex phenotype of mice lacking occludin, a component of tight junction strands. Mol Biol Cell 11:4131–4142

Sandle GI (2005) Pathogenesis of diarrhea in ulcerative colitis: new views on an old problem. J Clin Gastroenterol 39:S49–S52

Sandle GI, Hayslett JP, Binder HJ (1986) Effect of glucocorticoids on rectal transport in normal subjects and patients with ulcerative colitis. Gut 27:309–316

Santer R, Calado J (2010) Familial renal glucosuria and SGLT2: from a mendelian trait to a therapeutic target. Clin J Am Soc Nephrol 5:133–141

Santer R, Groth S, Kinner M, Dombrowski A, Berry GT, Brodehl J, Leonard JV, Moses S, Norgren S, Skovby F, Schneppenheim R, Steinmann B, Schaub J (2002) The mutation spectrum of the facilitative glucose transporter gene SLC2A2 (GLUT2) in patients with Fanconi-Bickel syndrome. Hum Genet 110:21–29

Schmitz H, Barmeyer C, Fromm M, Runkel N, Foss HD, Bentzel CJ, Riecken EO, Schulzke JD (1999) Altered tight junction structure contributes to the impaired epithelial barrier function in ulcerative colitis. Gastroenterology 116:301–309

Schoedon G, Goldenberger D, Forrer R, Gunz A, Dutly F, Höchli M, Altwegg M, Schaffner A (1997) Deactivation of macrophages with interleukin-4 is the key to the isolation of Tropheryma whippei. J Infect Dis 176:672–677

Schulzke JD, Schulzke I, Fromm M, Riecken EO (1995) Epithelial barrier and ion transport in celiac sprue: electrical measurements on intestinal aspiration biopsies. Gut 37:777–782

Schulzke JD, Bentzel CJ, Schulzke I, Riecken EO, Fromm M (1998) Epithelial tight junction structure in the jejunum of children with acute and treated celiac sprue. Pediatr Res 43:435–441

Schumann M, Richter JF, Wedell I, Moos V, Zimmermann-Kordmann M, Schneider T, Daum S, Zeitz M, Fromm M, Schulzke JD (2008) Mechanisms of epithelial translocation of the a2-gliadin-33mer in celiac sprue. Gut 57:747–754

Schumann M, Günzel D, Buergel N, Richter JF, Troeger H, May C, Fromm A, Sorgenfrei D, Daum S, Bojarski C, Heyman M, Zeitz M, Fromm M, Schulzke JD (2012) Cell polarity-determining proteins Par-3 and PP-1 are involved in epithelial tight junction defects in coeliac disease. Gut 61:220–228

Skodje GI, Sarna VK, Minelle IH, Rolfsen KL, Muir JG, Gibson PR, Veierod MB, Henriksen C, Lundin KEA (2018) Fructan, rather than gluten, induces symptoms in patients with self-reported non-celiac gluten sensitivity. Gastroenterology 154:529–539

Smith DE, Clémençon B, Hediger MA (2013) Proton-coupled oligopeptide transporter family SLC15: physiological, pharmacological and pathological implications. Mol Asp Med 34:323–336

Söderholm JD, Streutker C, Yang PC, Paterson C, Singh PK, McKay DM, Sherman PM, Croitoru K, Perdue MH (2004) Increased epithelial uptake of protein antigens in the ileum of Crohn's disease mediated by tumour necrosis factor alpha. Gut 53:1817–1824

Stein J, Schuppan D (2014) Coeliac disease – new pathophysiological findings and their implications for therapy. Viszeralmedizin 30:156–165

Suzuki H, Nishizawa T, Tani K, Yamazaki Y, Tamura A, Ishitani R, Dohmae N, Tsukita S, Nureki O, Fujiyoshi Y (2014) Crystal structure of a claudin provides insight into the architecture of tight junctions. Science 344:304–307

Suzuki H, Tani K, Tamura A, Tsukita S, Fujiyoshi Y (2015) Model for the architecture of claudin-based paracellular ion channels through tight junctions. J Mol Biol 427:291–297

Talbot C, Lytle C (2010) Segregation of Na/H exchanger-3 and Cl/HCO3 exchanger SLC26A3 (DRA) in rodent cecum and colon. Am J Physiol Gastrointest Liver Physiol 299:G358–G367

Tamura A, Tsukita S (2014) Paracellular barrier and channel functions of TJ claudins in organizing biological systems: advances in the field of barriology revealed in knockout mice. Semin Cell Dev Biol 36:177–185

Tamura A, Hayashi H, Imasato M, Yamazaki Y, Hagiwara A, Wada M, Noda T, Watanabe M, Suzuki Y, Tsukita S (2011) Loss of claudin-15, but not claudin-2, causes Na^+ deficiency and glucose malabsorption in mouse small intestine. Gastroenterology 140:913–923

Tanaka H, Yamamoto Y, Kashihara H, Yamazaki Y, Tani K, Fujiyoshi Y, Mineta K, Takeuchi K, Tamura A, Tsukita S (2016) Claudin-21 has a paracellular channel role at tight junctions. Mol Cell Biol 36(6):954–964

Triantafillidis JK, Emmanouilidis A, Manousos O, Nicolakis D, Kogevinas M (2000) Clinical patterns of Crohn's disease in Greece: a follow-up study of 155 cases. Digestion 61:121–128

Troeger H, Epple HJ, Schneider T, Wahnschaffe U, Ullrich R, Burchard GD, Jelinek T, Zeitz M, Fromm M, Schulzke JD (2007) Effect of chronic Giardia lamblia infection on epithelial transport and barrier function in human duodenum. Gut 56:328–335

Tsai PY, Zhang B, He WQ, Zha JM, Odenwald MA, Singh G, Tamura A, Shen L, Sailer A, Yeruva S, Kuo WT, Fu YX, Tsukita S, Turner JR (2017) IL-22 upregulates epithelial claudin-2 to drive diarrhea and enteric pathogen clearance. Cell Host Microbe 21:671–681

Turk E, Zabel B, Mundlos S, Dyer J, Wright EM (1991) Glucose/galactose malabsorption caused by a defect in the Na^+/glucose cotransporter. Nature 350(6316):354–356

Van Itallie C, Rahner C, Anderson JM (2001) Regulated expression of claudin-4 decreases paracellular conductance through a selective decrease in sodium permeability. J Clin Invest 107:1319–1327

Van Itallie CM, Fanning AS, Anderson JM (2003) Reversal of charge selectivity in cation or anion-selective epithelial lines by expression of different claudins. Am J Physiol Renal Physiol 285: F1078–F1084

Wang TY, Liu M, Portincasa P, Wang DQH (2013) New insights into the molecular mechanism of intestinal fatty acid absorption. Eur J Clin Investig 43:1203–1223

Wright N, Watson A, Morley A, Appleton D, Marks J (1973) Cell kinetics in flat (avillous) mucosa of the human small intestine. Gut 14:701–710

Wright EM, Loo DD, Hirayama BA (2011) Biology of human sodium glucose transporters. Physiol Rev 91:733–794

Wu RL, Vazquez-Roque MI, Carlson P, Burton D, Grover M, Camilleri M, Turner JR (2017) Gluten-induced symptoms in diarrhea-predominant irritable bowel syndrome are associated with increased myosin light chain kinase activity and claudin-15 expression. Lab Investig 97:14–23

Yeruva S, Chodisetti G, Luo M, Chen M, Cinar A, Ludolph L, Lünnemann M, Goldstein J, Singh AK, Riederer B, Bachmann O, Bleich A, Gereke M, Bruder D, Hagen S, He P, Yun C, Seidler U (2015) Evidence for a causal link between adaptor protein PDZK1 downregulation and Na^+/H^+ exchanger NHE3 dysfunction in human and murine colitis. Pflügers Arch 467:1795–1807

Yu ASL (2017) Paracellular transport as a strategy for energy conservation by multicellular organisms? Tissue Barriers 5(2):e1301852

Zeissig S, Bojarski C, Buergel N, Mankertz J, Zeitz M, Fromm M, Schulzke JD (2004) Downregulation of epithelial apoptosis and barrier repair in active Crohn's disease by tumour necrosis factor alpha antibody treatment. Gut 53:1295–1302

Zeissig S, Bürgel N, Günzel D, Richter J, Mankertz J, Wahnschaffe U, Kroesen AJ, Zeitz M, Fromm M, Schulzke JD (2007a) Changes in expression and distribution of claudin 2, 5 and 8 lead to discontinuous tight junctions and barrier dysfunction in active Crohn's disease. Gut 56:61–72

Zeissig S, Fromm A, Mankertz J, Weiske J, Zeitz M, Fromm M, Schulzke JD (2007b) Butyrate induces intestinal sodium absorption via Sp3-mediated transcriptional up-regulation of epithelial sodium channels. Gastroenterology 132:236–248

Zeissig S, Bergann T, Fromm A, Bojarski C, Heller F, Guenther U, Zeitz M, Fromm M, Schulzke JD (2008) Altered ENaC expression leads to impaired sodium absorption in the noninflamed intestine in Crohn's disease. Gastroenterology 134:1436–1444

Zeniya S, Kuwahara H, Daizo K, Watari A, Kondoh M, Yoshida-Tanaka K, Kaburagi H, Asada K, Nagata T, Nagahama M, Yagi K, Yokota T (2018) Angubindin-1 opens the blood-brain barrier in vivo for delivery of antisense oligonucleotide to the central nervous system. J Control Release 283:126–134

Zeuthen T, Gorraitz E, Her K, Wright EM, Loo DD (2016) Structural and functional significance of water permeation through cotransporters. Proc Natl Acad Sci U S A 113:E6887–E6894

Chapter 4
Diseases of the Pancreas Involving Channels/Transporters

Brandon M. Blobner and David C. Whitcomb

Abstract The pancreas is a digestive gland that is functionally divided into the exocrine pancreas producing digestive enzymes and the endocrine pancreas producing insulin and other metabolic hormones. Diseases of the exocrine pancreas are primarily inflammatory, such as acute pancreatitis and chronic pancreatitis. These diseases are symptom-defined clinical syndromes reflecting the final pathologic pathway of multiple complex gene x gene x environmental interactions. Diseases of the pancreas originating with channel–transporter dysfunction are dominated by dysfunction in the cystic fibrosis transmembrane conductance regulator (CFTR). Risk for chronic pancreatitis is also linked to claudin-2, a tight junction protein that forms a cation–water channel between cells. The dysfunction, however, is unrelated to channel function. New genetic discoveries of a unique class of CFTR variants that specifically alter CFTR bicarbonate permeability and conductance are discussed, along with their implications for human disease.

Keywords Cystic fibrosis transmembrane conductance regulator · CFTR · Claudin-2 · CLDN2 · Cationic trypsinogen · PRSS1 · Pancreatitis · Pancreatic duct · Cicarbonate · Cystic fibrosis · CFTR-related disorders

---

B. M. Blobner
Division of Gastroenterology, Hepatology and Nutrition, Department of Medicine, University of Pittsburgh, and UPMC, Pittsburgh, PA, USA
e-mail: bmb83@pitt.edu

D. C. Whitcomb (✉)
Division of Gastroenterology, Hepatology and Nutrition, Departments of Medicine, Cell Biology and Molecular Physiology, and Human Genetics, University of Pittsburgh, and UPMC, Pittsburgh, PA, USA
e-mail: whitcomb@pitt.edu

K. L. Hamilton, D. C. Devor (eds.), *Ion Transport Across Epithelial Tissues and Disease*, Physiology in Health and Disease,
https://doi.org/10.1007/978-3-030-55310-4_4

4.1 Introduction

The pancreas is a digestive gland located in the retroperitoneum of the abdomen. The pancreas has two functional components: the endocrine pancreas secreting hormones into the blood stream and the exocrine pancreas secreting juices into the proximal intestine just downstream of the stomach. The endocrine pancreas consists of clusters of islet cells that form roughly one million islets of Langerhans, suspended in a sea of exocrine pancreas cells. Four major and one minor type of islet cells each produce different hormones including alpha cells (glucagon), beta cells (insulin), delta cells (somatostatin), and PP/gamma cells (pancreatic polypeptide) as well as scarce epsilon cells (ghrelin). These cells release their respective hormones into the blood stream in a highly regulated way to regulate intermediate metabolism. The exocrine pancreas consists of the acinar cells that make pancreatic digestive pro-enzymes (zymogens) and duct cells that produce bicarbonate-rich fluid to alkalinize and flush the zymogens into the duodenum through the pancreatic duct system. The exocrine pancreas responds to neuro-hormonal stimuli through multi-level, integrated, and well-regulated control systems (Noble and Liddle 2005; Deng and Whitcomb 1998; Mussa and Verberne 2013).

Different types of medical and surgical specialists manage human diseases of the endocrine and exocrine pancreas, even though there is significant interaction between the endocrine and exocrine pancreas. This chapter focuses on the exocrine pancreas and diseases associated with the acinar and duct cell pathobiology linked to ion channels.

4.2 Human Pancreatic Diseases

The majority of the pancreatic mass is composed of acinar cells (~80%) with only a small percentage of duct cells (~5%). Compared to other organs, the exocrine pancreas is relatively simple. The acinar cells are polarized epithelial cells that serve one primary function—to manufacture precursors to pancreatic digestive enzymes (zymogens) and pack them into vesicles and secrete them into the pancreatic duct system on demand by exocytosis. The duct cells form an arborizing network of ducts linking the acinar cells to the intestine and secrete fluid as a result of ion transport through specific apical channels and transporters. The microanatomy of the exocrine pancreas is the acinus, which is a cluster of acinar cells joined at the center and connected by tight junction proteins that allow the apical surface to form a vestibule where secreted zymogens accumulate (Fig. 4.1). The vestibule is connected to the duct system forming a cul-de-sac. Within the pocket are duct-like cells called centroacinar cells. These cells connect with duct cells to form the pancreatic ductal system.

The primary diseases of the exocrine pancreas from a medical perspective are acquired inflammatory disorders (pancreatitis) and neoplastic disorders (e.g.,

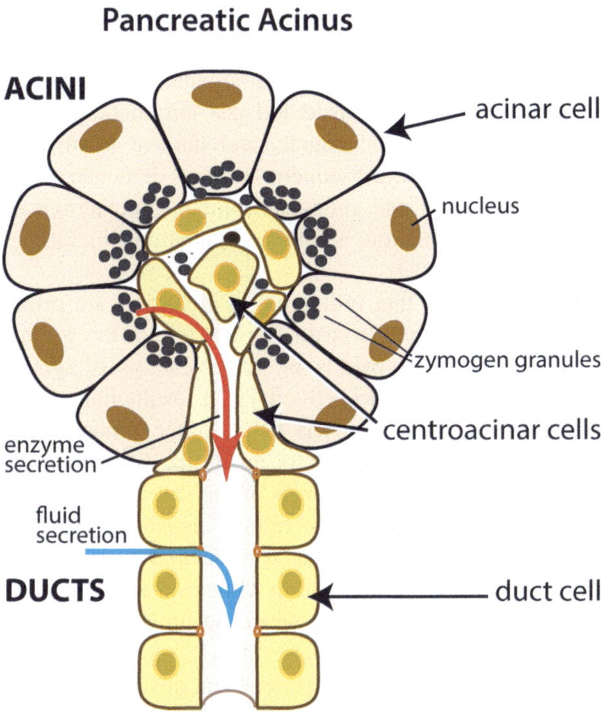

Fig. 4.1 Microanatomy of the exocrine pancreas. The functional unit of the exocrine pancreas is the pancreatic acinus, composed of acinar cells and duct cells organized to form a cul-du-sac linked to an upstream duct. The acinar cells are polarized epithelial cells with the basolateral membrane away from the duct, and the apical membrane facing the lumen of the cul-du-sac. The acinar cells are protein factories rich in endoplasmic reticulum and mitochondria to produce digestive enzyme zymogens. The zymogens are concentrated in zymogen granules at the apical membrane (black circles). With stimulation of the acinar cells on the basolateral side there is calcium-mediated excitation–secretion coupling with zymogen exocytosis into the lumen (red arrow). The duct cells begin as centroacinar cells and then form a duct that links with ducts from other acini. Duct cells transport ions (chloride and/or bicarbonate) from the basolateral membrane to the apical membrane where high ionic concentrations are generated. Water is drawn into the lumen by osmotic forces (blue arrow) resulting in hydrostatic pressure which forces the pancreatic juice (ionic solution and proteins) through the ducts and into the duodenum

pancreatic ductal adenocarcinoma). The inflammatory diseases are broadly divided into acute pancreatitis and chronic pancreatitis, which are complex syndromes defined by clinical criteria. Acute pancreatitis is defined as sudden inflammation of the pancreas with at least two of three criteria: (1) typical pain, (2) elevated concentrations of pancreatic digestive enzymes such as amylase or lipase in the serum (greater than three times normal values), and/or (3) evidence of pancreatic inflammation on abdominal imaging. Chronic pancreatitis is a complex syndrome of pancreatic injury and inflammation resulting in recurrent and persistent pathologic features which include pathologic fibrosis, atrophy, calcifications, pain, diabetes mellitus (from islet destruction), and increased risk of pancreatic cancer (Sarles et al. 1989; Sarner and Cotton 1984; Witt et al. 2007). Thus, these pancreatic diseases have been defined by symptoms rather than pathogenic mechanisms. Indeed, as late as 1995, chronic pancreatitis was considered an "enigmatic process of uncertain pathogenesis, unpredictable clinical course, and unclear treatment" (Steer et al. 1995). However, studies over the subsequent 20 years demonstrate that pancreatitis is a complex set of gene–gene and gene–environment interactions that predispose the pancreas to injury, inflammation, and its complications, including cancer (Chen and Ferec 2012; Whitcomb 2013).

Recent studies demonstrate that inflammatory disorders are primarily linked to risk factors that cause unregulated trypsin activity within the pancreas (Chen and Ferec 2012; Whitcomb 2013). Trypsin is the master catalyst of zymogen activation. Activation of inactive trypsinogen to active trypsin results in an activation cascade with activation of additional trypsinogen as well as other zymogens to become active digestive enzymes. When activation occurs inside the pancreas rather than being delayed until the zymogens reach the lumen of the small intestine the prematurely activated digestive enzymes cause parenchymal digestion with pancreatic injury signaling [e.g., release of damage-associated molecular patterns, DAMPs (Hoque et al. 2012)] and stimulate acute inflammation. Other mechanisms of immune system activation include acinar stress signaling from the unfolded protein response and likely other stressors and autoimmune pancreatitis (Rosendahl et al. 2008; Beer et al. 2012; Witt et al. 2013). A number of rare genetic and developmental disorders of the pancreas have also been identified.

Unraveling the complex genetics of human exocrine pancreatic diseases has progressed rapidly because of the singular purpose of the exocrine pancreas to manufacture, store, and secrete digestive enzymes. The mechanisms of zymogen synthesis, packaging, and secretion are known, and the mechanism of duct cell fluid secretion is also known. This allows candidate gene approaches to understand human genetics, in contrast to "unbiased" statistical approaches, such as genome-wide association studies, to identify disease-associated or clearly pathogenic genetic variants (Whitcomb et al. 2012; Rosendahl et al. 2017). To date, about 180 genetic variants have been linked to pancreatitis, but of the genes these variants are mapped to, four encode an ion channel or transporter—the cystic fibrosis transmembrane conductance transporter gene (*CFTR*) (Sharer et al. 1998; Cohn et al. 1998), claudin-2 gene (*CLDN2*) (Whitcomb et al. 2012; Derikx et al. 2015; Masamune et al. 2015), K^+ voltage-gated channel subfamily J member 3 (*KCNJ3*), and transient receptor potential cation channel subfamily V member 6 (*TRPV6*) (Masamune et al. 2020).

The common pathway linking genetic defects to human pancreatic inflammatory diseases is in genes linked to regulation of and protection from trypsin within the pancreas. The variants with the strongest pathogenic effects are called *pathogenic variants* and are disease-causing. The strongest pathogenic variants are *gain-of-function* mutations in the cationic trypsinogen gene (*PRSS1*) (Whitcomb et al. 1996; Gorry et al. 1997). Premature activation of mutant trypsinogen to trypsin results in a persistently active protease that cannot be degraded by the calcium-regulated trypsin autodegradation process (Fig. 4.2). This results in recurrent attacks

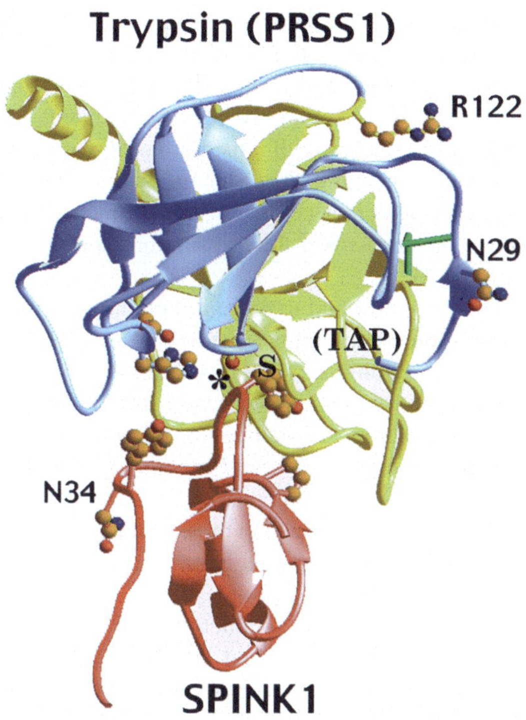

Fig. 4.2 Trypsin–SPINK1 complex. Ribbon diagram of trypsin (yellow and green) and SPINK1 (red) showing the interaction of the two molecules. Trypsin is painted yellow and blue to illustrate the two globular domains attached by the connecting chain (top). Arginine 122 (numbered R117 using a protein numbering system) is the target of autodegradation by another trypsin molecule when the side chain flexes away from the molecule. Mutations R122H and N29I are associated with altered activity and risk of pancreatitis. N34 is the site of an SPINK1 (pancreatic secretory trypsin inhibitor) variant (N34S) that is linked to the risk of pancreatitis, although the functional variant is likely in a regulatory element since the amino acid substitution does not affect protein function. TAP indicates the location of the trypsinogen activation peptide, an 8 amino acid sequence that is cleaved by trypsin or enterokinase to change inactive trypsinogen to active trypsin. Asterisk indicates active catalytic site of trypsin. S indicates the specificity pocket of the serene protease that selects the amino acid side chain for cleavage. Note that this molecule is a derivative of trypsin and SPINK1 to mimic chymotrypsin binding to stabilize the protein complex for crystallography. Image courtesy of William Furry

of acute pancreatitis and later chronic pancreatitis in families with these mutations. They are inherited in an autosomal dominant pattern with high penetrance (e.g., most individuals with the mutation get the disease phenotype and usually at an early age) and the disease is cause *hereditary pancreatitis*.

The autodigestion protective mechanism of the trypsin molecule has two steps: hydrolysis of the autolysis site at R122 by a second trypsin molecule and hydrolysis of the trypsin molecule at Leu81-Glu82 peptide bond by another trypsin-activated enzyme, the serine protease chymotrypsin C (CTRC) (Szabo and Sahin-Toth 2012). Loss-of-function mutations in *CTRC* also increase the risk of chronic pancreatitis

(Beer et al. 2012; Rosendahl et al. 2008; LaRusch et al. 2015), although the risk is much lower than the risk associated with *PRSS1* mutations and is usually pathogenic only in the context of additional risk factors. The trypsin autodigestion protective mechanism only works in solutions with low levels of calcium and therefore exists within the acinar cell under normal conditions, but not in the high calcium concentrations that exist in the pancreatic duct due to the action of calcium pumps transferring calcium from the acinar cell to the duct lumen. Fortunately, there are several other protective mechanisms if the trypsin-autodigestion protective mechanism fails (Whitcomb 2013).

The pancreatic acinar cells respond to acute inflammation with upregulation of a trypsin inhibitor: serine protease inhibitor Kazal type 1 gene (*SPINK1*), also known as pancreatic secretory trypsin inhibitor (PSTI) (Pfutzer et al. 2000; Witt et al. 2000). SPINK1 is a specific inhibitor of the two major isoforms of trypsin, PRSS1 (cationic trypsin) and PRSS2 (anionic trypsin). Pathogenic variants in *SPINK1* are loss-of-functions or loss of expression types. *SPINK1* mutations are associated with recurrent acute pancreatitis and/or chronic pancreatitis linked to multiple other etiologies (Aoun et al. 2008). SPINK1 is synthesized in the acinar cell with PRSS1 and PRSS2 so that the utility of SPINK1 as a trypsin inhibitor is seen both in the acinar cell and in the duct system.

The duct cells are important for the production of a bicarbonate-rich juice that flushes the zymogens out of the pancreas and into the duodenum where zymogens are normally activated and digest nutrients as they leave the stomach (Larusch and Whitcomb 2012; LaRusch et al. 2014). CFTR is critical to normal duct function. CFTR is a regulated ATP-binding cassette (ABC) transporter-type protein [ABCC7 (Lee et al. 2012)] localized to the apical plasma membrane of epithelial cells. It differs from other ABC transporters in that it acts as a regulated anion channel rather than a transporter (Bear et al. 1992). Loss of function of the CFTR represents the primary defect in ion channel/transport function that has been shown to be associated with pancreatic disease in humans. For more information about CFTR see Chaps. 15 and 16 of Volume 3.

This chapter focuses on cystic fibrosis of the pancreas and cystic-fibrosis-related disorders of the pancreas including new insights into the role of bicarbonate secretion through the CFTR molecule. Claudin-2, a tight-junction protein that functions as a cation–water channel between cells, is also discussed.

4.3 Pancreatic Duct Cell Biology

The pancreatic duct cells are polarized epithelial cells that form an arborizing tree of cells with slightly different functions extending from every acinus to the intestine. The intercalated ducts are immediately connected to the acinar cells and centroacinar cells and these cells have the highest level of CFTR expression by immunohistochemistry (Marino et al. 1991), while CFTR is expressed at high levels in the small, intercalated ducts and at lower levels in the interlobular ducts by in situ hybridization to CFTR mRNA (Strong et al. 1994).

The duct cells are linked by tight junctions that do not allow fluid or molecules to pass between the cells when the pancreas is at rest (Gunzel and Yu 2013). We believe, but it has not been proven, that there is an exchange of claudin-3 and/or claudin-4 for claudin-2 in the small ducts with stimulation so that sodium and water can pass between the cells and join bicarbonate to mediate sodium–bicarbonate secretion (Gunzel and Yu 2013; Amasheh et al. 2002; Rahner et al. 2001; Larusch and Whitcomb 2012). As the intercalated ducts from adjacent acini coalesce, they form interlobular ducts, then interlobular ducts, then interlobar ducts, and finally the main pancreatic duct. Progressively larger ducts display changes in anatomic features as well as epithelial cell function.

4.4 Ion Transport Across the Pancreatic Ductal Epithelial Cell

Ion transport across lipid membranes is facilitated by ion channels and exchangers and is dependent on the electrochemical gradient across the membrane. CFTR, as an anion-specific channel, specifically facilitates conductance of chloride (Cl^-), bicarbonate (HCO_3^-), and other anions (Bear et al. 1992). The direction of ion movement across the membrane is determined by the concentration of the ion on each side of the membrane and the electrical potential. From a physiologic perspective, ions can only be secreted or absorbed across epithelial cell barrier if there are channels or transporters capable of moving a specific ion across both the apical (luminal) and basolateral (plasma) membranes. Interestingly, the human body uses CFTR to *secrete* Cl^- in the lung, *secrete* HCO_3^- into the pancreatic duct, and *absorb* Cl^- in the sweat gland. The differences are all related to cell organization, membrane potential, and electrochemical gradients.

Potential mechanisms used by the pancreas to secrete bicarbonate have been tested using multiple experimental, theoretical, and mathematical models. The first important conceptual model was developed by Case and Argent [see Review (Case and Argent 1993)]. The model (Fig. 4.3) depicted entry of CO_2 into the duct cell from the blood by diffusion, transformation of CO_2 + H_2O into H^+ and HCO_3^- within the duct cell (facilitated by carbonic anhydrase; CA), and then secretion of HCO_3^- into the duct lumen by an antiporter in exchange for Cl^-. The intracellular Cl^- levels were maintained at a low level by opening of the CFTR channel to allow Cl^- to exit by diffusion. The electrical properties suggested a transcellular $V_t = 5$ mV (duct negative), a basolateral potential of $V_{bl} = -35$ mV, and an apical potential of $V_a = -35$ mV (intracellular negative). The model was developed as an adaptation of models in the urinary bladder (Stetson et al. 1985) and guinea pig gallbladder (Stewart et al. 1989), with the rationale that if there were a bicarbonate channel, the relative high permeability of the known channels to chloride, compared to

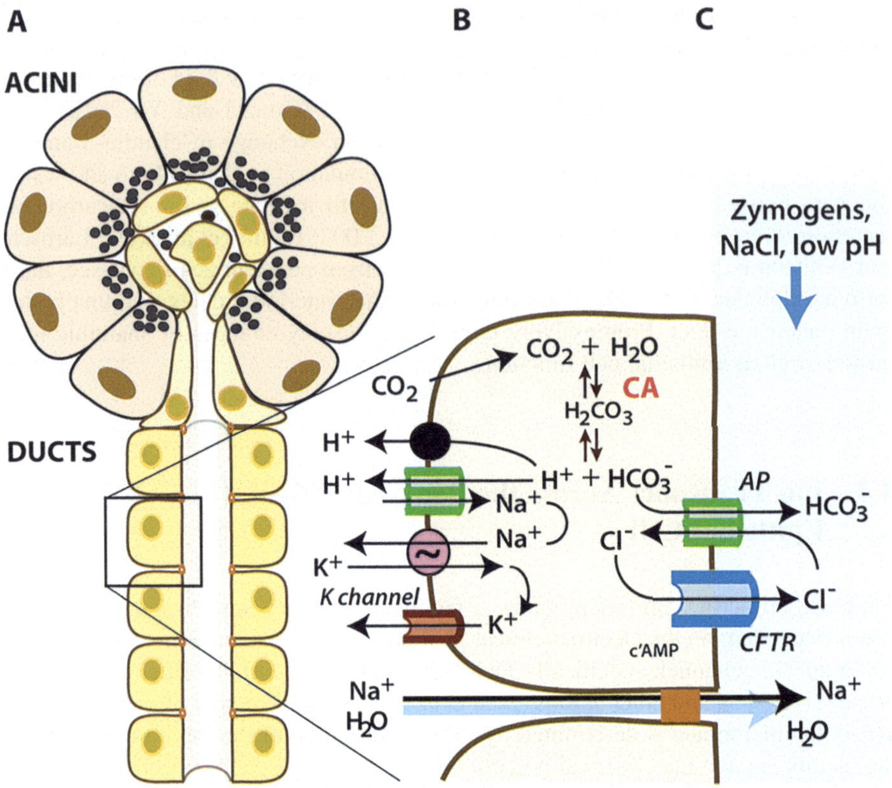

Fig. 4.3 The Case–Argent pancreatic duct model. One of the first and most influential hypothetical models of pancreatic duct physiology after the discovery of CFTR and before the sodium-bicarbonate cotransporter (NBC). CFTR was believed to be a selective chloride (Cl^-) channel (no significant HCO_3^- conductance), so bicarbonate (HCO_3^-) secretion was linked to a chloride-bicarbonate antiporter (AP). The role of CFTR was believed to actively conduct Cl^- out of the duct cell keeping intracellular chloride levels low enough to all Cl^- /HCO_3^- exchanger to occur—a hypothesis that influenced subsequent models. The HCO_3^- was believed to be generated within the duct cell through the effects of carbonic anhydrase (CA) (**a**) pancreatic acinus, (**b**) duct cell, (**c**) duct lumen

bicarbonate, would make any anion channel a chloride channel (Case and Argent 1993). However, upon experimental and mathematical model testing it was not capable of generating high luminal bicarbonate concentrations (Steward et al. 2005).

CFTR is clearly important to pancreatic duct bicarbonate secretion, but the channel is far more permeable to chloride than bicarbonate, indicating that the process of bicarbonate secretion is more complex than chloride secretion. Direct experimental measures of the dynamics of bicarbonate secretion are challenging, so alternate approaches, such as mathematical modeling, were used to answer the question of how the pancreas secretes a juice with high sodium-bicarbonate concentrations that increases with stimulation (Whitcomb and Ermentrout 2004).

The first major attempt at mathematical simulation of a pancreatic duct cell at rest and with stimulation was developed by Sohma et al. (1996). This mathematical model, built in collaboration with the Argent group, tested the Case–Argent hypothetical model. Variables included intracellular [Na^+], [K^+], [Cl^-], [HCO_3^-], membrane potential, and cell volume, plus extracellular fluids with or without a junctional pathway separating plasma from lumen. An equation for single diffusive pathway was included for Cl^- on the apical membrane (e.g., to include CFTR-related conductance), but it did not include conductance of HCO_3^- through CFTR. Dynamic simulation of fluid secretion indicated that it was not possible to generate high HCO_3^- concentrations in the duct lumen with the proposed conceptual model. Furthermore, HCO_3^- flux *decreased* as luminal HCO_3^- concentrations increased, pH was a problem, and the question of how opening of the CFTR channel resulted in brisk secretion of a sodium–bicarbonate fluid, as seen in vivo, was not adequately addressed.

The discovery of the pancreatic sodium–bicarbonate cotransporter (NBC_p) and the possibility of a HCO_3^- conductance pathway on the apical membrane resulted in another set of mathematical modeling experiment, published by Sohma et al. (2000). This elaborate model required a series of duct cells with different properties extending from the acinus toward the duct. Secretion began with Cl^- conductance through CFTR, and then the pancreatic juice was modified with Cl^- /HCO_3^- exchange, until a high HCO_3^- concentration was reached. However, simulation of fluid secretion with this model did not *generate* high luminal HCO_3^- concentrations and sustained high volume fluid secretion. Furthermore, the HCO_3^- would likely fall, rather than rise with CFTR channel opening and fluid volume increasing during stimulated secretion (Steward et al. 2005).

An alternative model was proposed that was conceptually different on several levels (Whitcomb 1999). The model was tested with the Whitcomb–Ermentrout equation-based mathematical modeling and secretion simulations (Whitcomb and Ermentrout 2004) and published in 2004. The Whitcomb–Ermentrout model addressed multiple conceptual differences from the Case–Argent models that were believed to be important in pancreatic physiology (Fig. 4.4). (1) The microanatomy of the pancreas included the duct with a small volume so that ion secretion could change the ion concentration in the duct. (2) The basolateral membrane needed to be impermeable to chloride to prevent transepithelial transport of chloride. (3) The CFTR channel was modeled with *variable permeability of bicarbonate* compared to chloride, initially 1:5. A chloride–bicarbonate antiporter (e.g., SLC26A6) was also included. (4) The membrane potential difference during secretion was predicted to be higher (e.g., −60 mV) than in the Case–Argent model (e.g., −35 mV), since the Nernst equation predicts that the direction of HCO_3^- flux would be from the lumen *into* the cell when luminal HCO_3^- was elevated. (5) The driving force for secretion was predicted to be the basolateral sodium–potassium pump which generated and maintained a large sodium electrochemical gradient. The gradient would transport bicarbonate into the cell against a bicarbonate electrical (+/− concentration) gradient via the sodium–bicarbonate co-transporter (NBC). The large membrane potential results in a disequilibrium for bicarbonate across the apical membrane, based on the

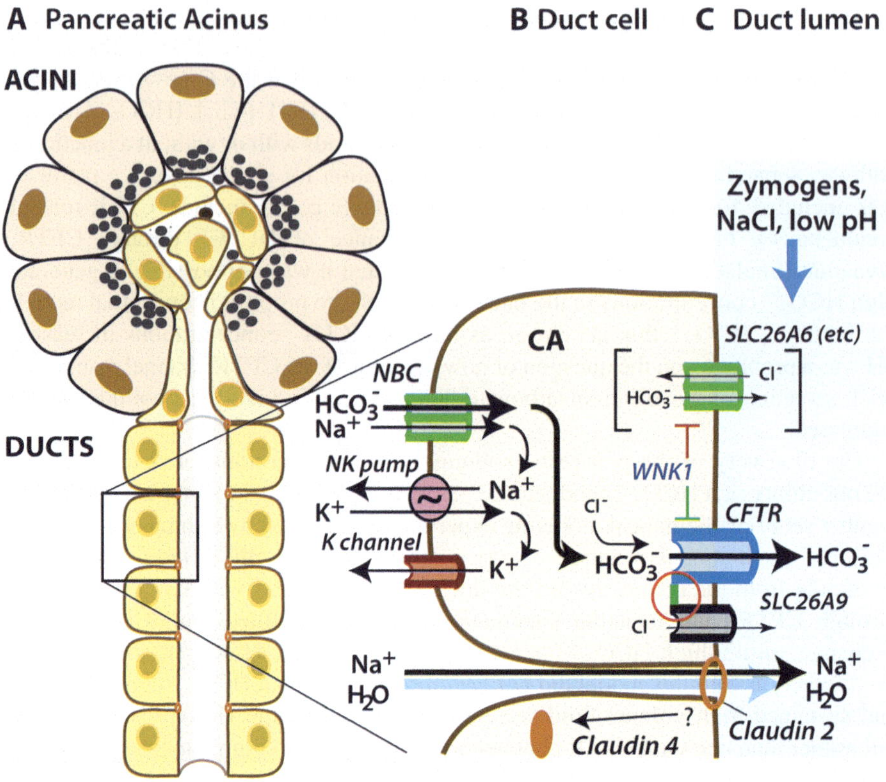

Fig. 4.4 Minimal model for a bicarbonate-secreting duct system. (**a**) Pancreatic acinus including the acinar cells and duct cells as a functional unit. (**b**) Proximal duct cell. (**c**) Luminal fluid beginning with acinar cell secretion of zymogens (pre-activated pancreatic digestive enzymes) a NaCl-rich fluid. Bicarbonate secretion is driven by a basolateral sodium–potassium pump generating a sodium gradient sufficient to bring bicarbonate into the cell via a sodium–bicarbonate cotransporter (NBD). Bicarbonate exits the duct cell down its electrochemical gradient through CFTR. Bicarbonate flux though CFTR is >>> chloride when (a) the basolateral membrane is impermeable to chloride so that equilibrates across the apical membrane, and (b) WNK1 acts on CFTR to increase bicarbonate permeability. The model predicts that sodium and water follow bicarbonate through a paracellular pathway facilitated by the insertion of claudin 2 and withdrawal of claudin 4 from the tight junction. With active bicarbonate secretion, the chloride–bicarbonate antiporters such as SLC26A6 are inhibited, but it becomes important when CFTR closes and the cell returns to equilibrium. SLC26A9 is also inhibited by WNK1, but may stabilize or modify CFTR function on the cell surface or provide some protection from dysfunctional CFTR in a yet to be determined mechanism. Carbonic anhydrase (CA) is important in protecting the duct cell from wide pH changes. Based on Whitcomb and Ermentrout (2004)

bicarbonate reversal potential, resulting in an efflux of bicarbonate until the concentration of bicarbonate in the lumen was >>150 mM. (6) The microanatomy of the duct is a cul-de-sac with a small volume. Efflux of bicarbonate would result in the generation a hypertonic solution within the duct lumen. As water is drawn into the duct by osmosis, it increases the hydrostatic pressure within the cul-de-sac, resulting in brisk flow of juice toward the duodenum (7). The continual influx of water into the small ducts would dilute the bicarbonate, so that an equilibrium concentration was never reached, and efflux of bicarbonate would continue from the duct cells. (8) The tight junction between duct cells would have needed to be permeable to both water and sodium, but not chloride, to facilitate sodium entry into the duct lumen following an electrical gradient generated by bicarbonate secretion. Water would follow the sodium and bicarbonate through the paracellular pathway and/or via a transcellular pathway linked to aquaporins. (9) The role of CFTR could be central to pancreatic bicarbonate secretion if it were permeable to both chloride and bicarbonate. The central question is *not the relative permeability* of CFTR to chloride or bicarbonate, but on the reversal potential (linked to the concentration of the anions on either side of the apical membrane) compared to the membrane potential during secretion, resulting in *anion flux*. If there were no pathway for chloride to cross the basolateral membrane, then with the opening of CFTR the chloride concentrations would equilibrate across the luminal membrane based on membrane potential. In contrast, bicarbonate is continually pulled into the duct cell via NBC, so that there is continued disequilibrium and outward anion flux. Thus, the CFTR would conduct bicarbonate rather than chloride since bicarbonate, but not chloride, has an outward driving force.

To test this conceptual model, pancreatic duct secretion was simulated using the Whitcomb–Ermentrout model (2004), applying the parameters of Sohma et al. (2000) with the exception of *CFTR* which was made permeable to bicarbonate in variable ratios to chloride. This simple model generated high concentrations of luminal bicarbonate (>140 mM) with high fluid output that was initiated by opening the CFTR channel and ended with closing the CFTR channel. The model also demonstrated that CFTR acted as a Cl^- and/or HCO_3^- channel depending on the conductance pathways on the basolateral membrane.

The Whitcomb–Ermentrout model also suggested that when conditions were optimized for maximal luminal bicarbonate secretion that several predictions could be made. (1) The basolateral membrane of duct cells from species that secreted high concentrations of bicarbonate in pancreatic fluid would be nearly impermeable to chloride. (2) The apical chloride–bicarbonate antiporter was not necessary for bicarbonate secretion. In fact, the apical chloride–bicarbonate antiporter actually ran *backward* (with low flux) from the predicted direction of anion flux required by the Case–Argent model (bicarbonate entering the cell in exchange for chloride). The model suggested that excluding the antiporter was only important after the CFTR channel was closed, since it played a role in the equilibration of chloride within the cell to resting levels (Fig. 4.4, SLC26A6 in parentheses). (3) Bicarbonate secretion was dependent on conductance through CFTR. When bicarbonate conductance but not chloride conductance was minimized, bicarbonate conductance and pancreatic fluid secretion rates dropped proportionally and independent of the antiporters. Thus, the model predicted that mutations that alter CFTR bicarbonate secretion but not chloride secretion would increase the risk of pancreatitis, since duct cell function and zymogen flushing are critical for protection of the pancreas. (4) There needed to be a paracellular pathway between the duct cells that was restricted to cations (e.g., sodium), but impermeable to chloride and bicarbonate to cause directional flow of cations and reduce the transcellular electrical potential (Fig. 4.4). (5) The model predicted that chloride would be washed out of the pancreas, and chloride concentrations within the duct and the lumen would drop to zero unless there were a small chloride leak or chloride entering the duct from upstream acinar cells.

The other major prediction of the model is that, if bicarbonate was secreted through CFTR, then the pancreatic model of bicarbonate secretion could be modified to focus on the proximal duct, rather than a series of duct cells that first secrete chloride proximally, then exchange chloride for bicarbonate distally (Sohma et al. 2000; Lee et al. 2012). The minimal model of Whitcomb and Ermentrout required testing, but over a decade would elapse before the critical information was available. The prediction that *CFTR* mutations selectively affecting bicarbonate would predispose to pancreatitis, and the unique features of claudin-2, found to be present in the small pancreatic ducts (Rahner et al. 2001; Lee et al. 2011; Larusch and Whitcomb 2012) are discussed below.

4.5 Duct Function and Pancreatitis

Pancreatic duct physiology is largely regulated dependent on CFTR functional. The details of the intracellular mechanism regulating CFTR opening are considered elsewhere (see Chap. 6 of Volume 1 and Chap. 16 of Volume 3). However, there appear to be several mechanisms that regulate the CFTR activity that is important in protecting the pancreas from pancreatitis.

There are multiple extracellular hormones, neurotransmitters, and receptors that are linked to duct function and specifically CFTR state. The first is secretin, the first hormone discovered in 1902 and the major challenge to the Pavlov perspective that the visceral organs were solely regulated by neural connections. Secretin is released from "S" cells in the duodenum mucosa into the blood stream following luminal acidification. The pancreatic duct cells express secretin receptors, as demonstrated by receptor autoradiography of rat ducts (Ulrich et al. 1998) and later human tissues (Korner et al. 2005). The possibility that secretin also acts on afferent vagal fibers or interneurons has not been excluded. Secretin receptors are seven transmembrane G-protein-coupled receptor family that increase intracellular cyclic AMP, as does the neuropeptide vasoactive intestinal peptide (VIP). Secretin stimulation of the duct results in CFTR-mediated fluid secretion.

Collection of human duct cells from fresh pancreas without disease remains a challenging obstacle to mechanistic research. Some insight can be gained by evaluation of well-differentiated duct cell lines that are capable of forming polarized monolayers with tight junctions or electrophysiology studies. Wang and Novak (2013) recently publish a series of studies using human-derived Capan-1 cells. They confirmed transepithelial anion transport in human pancreatic duct epithelium with Using chambers in response to secretin and VIP via cAMP, but also documented responses to acetylcholine (muscarinic receptors) and adenosine,

adenosine 5′-triphosphate (ATP), uridine 5′-triphosphate (UTP), and other pharmacological agents via purinergic P2 receptors and calcium signaling pathways (Wang and Novak 2013). These experiments demonstrate multiple pathways converging on the duct cell to stimulate secretion.

From the perspective of integrated physiology and systems biology, stimulation of the duct cells appears to occur under two conditions, meal stimulation and danger/inflammation. Multiple mechanisms are associated with the danger/inflammation cell activation. Extracellular ATP and UTP activate transepithelial anion secretion when applied to either the apical (luminal) or basolateral membrane (Wang and Novak 2013). ATP and UTP are normally intracellular molecules. These purines may be released by acinar cells during secretion, thus providing paracrine activation of local duct cell secretion (Novak 2008; Haanes et al. 2014) by binding to luminal receptors. Basolateral receptors may serve a different function. High concentrations are generated within the regional interstitial spaces with cell injury membrane disruption. Thus, ATP and UTP, by acting on the basolateral membrane of epithelial cells, on inflammatory cells, on neurons or other cells responding to injury represent DAMP molecules (Gombault et al. 2012).

The protease-activated receptors PAR2 (coagulation factor II (thrombin) receptor-like 1; F2RL1) and PAR4 (coagulation factor II (thrombin) receptor-like 3; F2RL3) are expressed in the pancreas and may be important for regulating duct secretion, inflammation, and pancreatic pain (Ceppa et al. 2011). PAR2 can be either protective or pathogenic in pancreatitis, depending on the model (Laukkarinen et al. 2008). PAR2 located on the apical membrane of the guinea pig and human duct cells is sensitive to trypsin activation and appeared to inhibit CFTR-medicated pancreatic secretion (Pallagi et al. 2011). The role of PARs in pancreatitis is likely important, but more studies are required to determine their role or effects under different circumstances.

The calcium sensing receptor (CaSR) is a member of the G-protein-coupled receptor (GPCR) superfamily (Brown et al. 1993). CaSR plays an important role in the body's overall calcium homeostasis and is expressed by cells of the parathyroid gland and renal tubules involved in the calcium metabolism. CaSR plays an important role in duodenal bicarbonate secretion by increasing intracellular calcium and opening CFTR (Xie et al. 2014). CaSR was identified in both human pancreatic acinar and ductal cells, as well as in various non-exocrine cells (Bruce et al. 1999). Mutations in the CaSR gene (*CASR)* that are classified as loss-of-function variants are associated with chronic pancreatitis in patients with *SPINK1* mutations (Felderbauer et al. 2003; Pidasheva et al. 2004; Muddana et al. 2008). This is a hypothesis-generating observation that CaSRs protect the pancreas from acute pancreatitis with high intraductal calcium concentrates favoring trypsin activation and protection from inhibition. *CASR* variants alone are not sufficient to cause chronic pancreatitis, but when combined with another major risk factor, such as a pathogenic *SPINK1* variant, the combined risk is sufficient to cause recurrent acute pancreatitis and eventually chronic pancreatitis.

4.6 Genetics of Human *CFTR* Variants

The biology of the CFTR channel is reviewed in Chap. 8 of Vol. 1 and Chap. 15 of Vol. 3. In humans, the primary organ affected by severe recessive *CFTR* mutations is the human pancreas. Indeed, the term "cystic fibrosis of the pancreas" refers to the chronic inflammation fibrosis and cyst that occur in infants who died of maldigestion from pancreatic exocrine failure and failure to thrive prior to enzyme replacement therapy. The fact that *CFTR* mutations have such devastating effects on the pancreas, and that no other disease of ion transporters affects the pancreas to cause a Mendelian genetic syndrome, demonstrates the central role of CFTR in pancreatic physiology.

Nearly 2000 mutations and/or sequence variants in *CFTR* have been identified in human subjects worldwide. The genetic variants are not equivalent in functional effect. The various consequences of pathogenic sequence variants on CFTR function have been divided into several classification systems. One of the clinical uses of classifications is to predict the consequence of the CFTR variant with complete loss of function resulting in a "severe" phenotype, reduced but residual function resulting in a "mild-variable" phenotype, and CFTR variants with no functional consequences being classified as "benign" (Zielenski 2000; Rowntree and Harris 2003). As an inherited disease, cystic fibrosis is an autosomal recessive genetic disorder and the severity or phenotype of the disease correlates with the amount of dysfunction in the *milder* mutation. In general, loss of a single *CFTR* allele does not result in human disease—up to 3% of the population may carry a single copy of the common *CFTR* F508del variant and be asymptomatic. However, as discussed below, even a 50% reduction in CFTR expression (e.g., one allele with a severe pathogenic *CFTR* variant and the other allele normal) makes the pancreas more susceptible to dysfunction when stressed. *CFTR* mutations classified as mild-variable variants typically retain at least 10% of normal function. In most cases, a 75–80% reduction in overall *CFTR* function (e.g., one allele with a severe mutation and the other with a mild-variable mutation) makes organs susceptible to injury from even moderate environmental or genetic stressors. With the retention of some function, the phenotype is milder than with no CFTR function, and some organs may function nearly normally while others become diseased. These milder clinical syndromes are classified as either atypical cystic fibrosis or as a CFTR-associated disorder (Kerem 2006; Bombieri et al. 2011). Patients without full-blown cystic fibrosis may have recurrent acute pancreatitis or chronic pancreatitis as part of the CFTR-related syndrome (Bombieri et al. 2011; Cohn et al. 1998; Sharer et al. 1998).

4.6.1 CFTR Bicarbonate-Defective Variants

The Whitcomb–Ermentrout model predicted that *CFTR* variants that specifically affected bicarbonate secretion would increase the risk of pancreatitis and disease in other organs using CFTR for bicarbonate conductance, but not the risk of disease in organs that used CFTR for chloride secretion, such as the sweat gland and lungs. This is a challenging prediction in humans, since *CFTR* functional testing is primarily done using sweat chloride testing, and the physician specialists caring for patients with cystic fibrosis are pulmonary physicians.

To test this hypothesis, we evaluated the CFTR gene in 984 patients from the North American Pancreatitis Study 2 [NAPS2 (Whitcomb et al. 2008)] in which patients underwent extensive phenotyping during ascertainment by experts from over 20 major pancreas centers (LaRusch et al. 2014). To improve the pre-test probability that any *CFTR* variant was truly associated with pancreatic disease, we conducted a literature review to identify all *CFTR* variants that have been reported at least twice in previous chronic pancreatitis case–control genetic studies, plus common severe, cystic fibrosis-causing *CFTR* variants ($CFTR^{CF}$). Using a panel of 81 *CFTR* variants we identified 43 variants in pancreatitis patients, including 9 seen multiple times that were previously classified as benign for classic cystic fibrosis (*CFTR* R74Q, R75Q, R117H, R170H, L967S, L997F, D1152H, S1235R, and D1270N) (LaRusch et al. 2014).

The nine candidate bicarbonate-defective *CFTR* variants ($CFTR^{BD}$) were individually cloned into normal CFTR vectors and expressed in HEK-293T cells for electrophysiological analysis. The goal was to test the hypothesis that the $CFTR^{BD}$ variants were specifically associated with impairment of the known WNK1-SPAK (Ste20-related proline/alanine-rich) pathway-stimulated increase in CFTR bicarbonate permeability (Park et al. 2010; LaRusch et al. 2014). WNK1 is a member with no lysine (K) kinases that serves as an intracellular sensor of osmolality, chloride concentration, and other factors and respond by activating additional kinases linked to a variety of ion channels and exchangers, including CFTR (Anselmo et al. 2006; Richardson and Alessi 2008; Park et al. 2010). Wild-type (WT) CFTR responds to WNK1-SPAK activation with an increase in CFTR HCO_3^- permeability and conductance to reach that of chloride (Park et al. 2010). In contrast, the candidate $CFTR^{BD}$ variants, which had normal chloride permeability and conductance, generally failed to respond to WNK1-SPAK activation with increased bicarbonate permeability or conductance (Fig. 4.5). Treatment with the CFTR inhibitor $CFTR_{inh}$-172 (20 μM) inhibited >90% of the HCO_3^- currents indicating that CFTR mediates most of the HCO_3^- currents rather than affecting other pathways.

The nine $CFTR^{BD}$ variants all result in amino acid substitutions, and the specific residues are not distributed equally across the gene. Three dimensional models revealed that the variant amino acids were clustered in 4 locations; L997F and D1152H projected into the lumen of the anion channel causing a reduction in functional diameter to block bicarbonate conductance, R74, R75, R170, L967, and R1162L [a candidate that did not fully meet the predefined criteria (LaRusch et al. 2014)] are at the hinge region that modulates the collective movements of the nucleotide-binding domains (NBDs) with respect to membrane-spanning domains (MSDs), D1270 and S1235 were on the apex surface of NBD2, and R117H projects into the extracellular space (Fig. 4.6a, b).

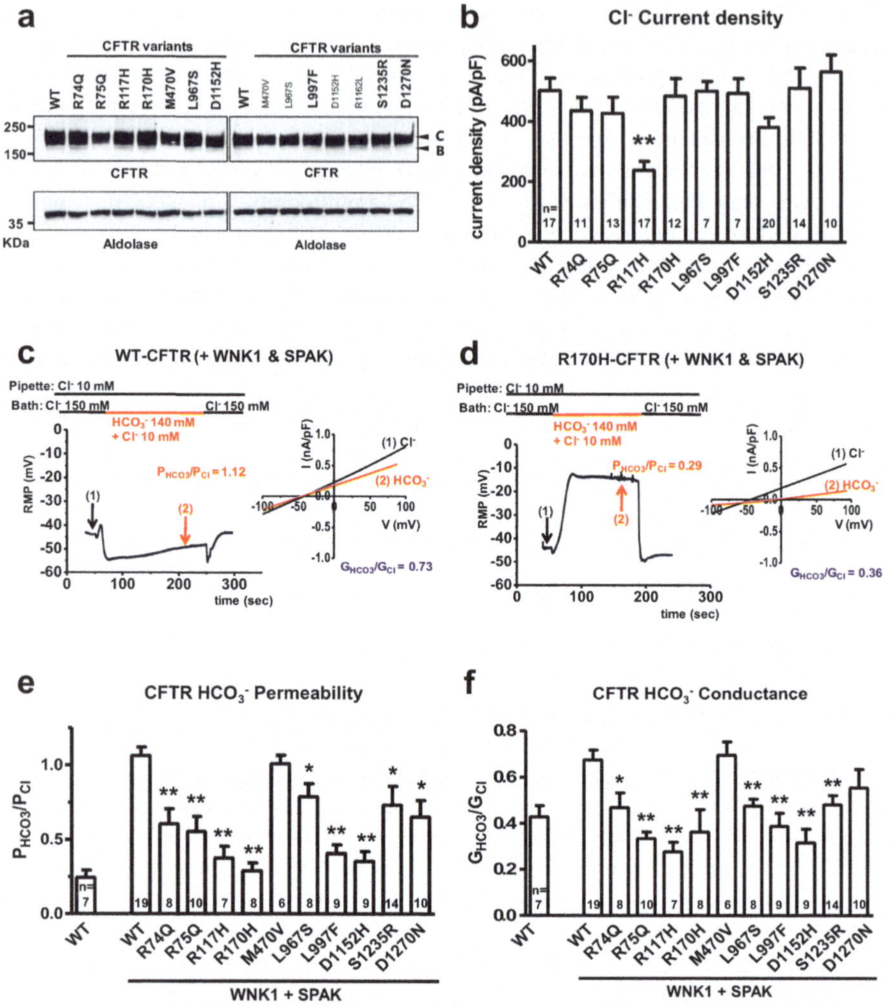

Fig. 4.5 Functional characteristics of the nine CFTR[BD] variants (LaRusch et al. 2014). Panel (**a**) Wild-type (WT) and variant CFTR proteins were expressed in HEK 293T cells and immunoblotted with anti-CFTR and anti-Aldolase antibodies. Replicate lanes are in small font. Band B, expected size of immature ER core-glycosylated CFTR; band C, mature complex-glycosylated CFTR. Panel (**b**) Whole-cell Cl2 currents were measured in WT and variant CFTR expressing HEK 293T cells, as described in Methods. Panel (**c**) Whole-cell currents of WT-CFTR were measured in HEK 293T cells co-expressed with WNK1 and SPAK using patch pipette contained a low concentration of Cl2 (10 mM). A representative trace of reversal potential measurement is shown in the left panel. The permeability ratio PHCO3/PCl was calculated according to the Goldman–Hodgkin–Katz equation. I–V relationships at the indicated points are presented in the accompanying graph. The conductance ratio GHCO3/GCl was calculated by measuring each outward current (i.e., slope between Erev and Erev+25 mV). RMP, resting membrane potential. Panel (**d**) Whole-cell currents of R170H-CFTR were measured in HEK 293T cells using the same protocol shown in panel (**c**). Panel (**e**) A summary of the PHCO3/PCl values obtained from WT-CFTR in the standard state (left) compared to WT-CFTR and the nine CFTRBD variants with WNK1 + SPAK activation (right, underlined). Panel (**f**) A summary of the GHCO3/GCl values in the standard state (left) with WNK1 + SPAK

Molecular dynamics (MD) simulations, based on homology-modeled structures of ABC transporters, were used to examine the effect of *CFTR*^{BD} variants on the structure and dynamics of the channel. The channel diameters of the WT and mutants L997F and D1152H (Figure 4.6c, d) demonstrate that the channel diameter is observed to narrow down from an average value of 10.3 Å to 7.5 Å (standard deviation, $\sigma = 0.5$ Å) at the pore region, near the L997F amino acid substitution (Fig. 4.6e), and from an average of 9.9 Å to 4.3 Å ($\sigma = 1.1$ Å) for the *CFTR*^{BD} mutant D1152H (Fig. 4.6f). On the other hand, L967, L997, D1152, and R1162 appear to serve as anchors for dynamic movements between NBD1 and NBD2, which affects channel gating (LaRusch et al. 2014).

The likely mechanism linking WNK1 to change in CFTR bicarbonate permeability has been discovered (Kim et al. 2020). Of the intracellular mechanisms to detect osmolarity and chloride concentrations, only WNK1 altered CFTR permeability to bicarbonate. Molecular analysis of WNK1 revealed that the WNK1 kinase domain is responsible for CFTR HCO3/Cl permeability regulation by direct association with CFTR at p.R74–75, while the surrounding N-terminal regions mediate the $[Cl^-]_i$-sensitivity of WNK1 (Fig. 4.7). Furthermore, the pancreatitis-causing R74Q and R75Q mutations in the elbow helix 1 of CFTR hampered WNK1-CFTR physical associations and reduced WNK1-mediated CFTR PHCO3/PCl regulation. Thus, the molecular biology and clinical observations are consistent and bring major insights into how CFTR and other regulated and selective ion channels work (LaRusch et al. 2014; Jun et al. 2016; Kim et al. 2020).

CFTR is utilized in multiple tissues for anion transport, including several organs in which CFTR is linked to bicarbonate secretion. To determine if *CFTR*^{BD} variants are associated with disease in other organs the prevalence of sinusitis and male infertility was compared between pancreatitis cases with and without *CFTR*^{BD} variants (LaRusch et al. 2014). Risk of sinusitis was significantly increased among carriers of *CFTR*^{BD} ($p = 0.001$; OR 2.60, CI 1.43–4.60), cystic fibrosis-associated variants (*CFTR*^{CF}) ($p = 0.01$; OR 2.47; CI 1.18–4.91), or either *CFTR*^{BD} or *CFTR*^{CF} variant allele ($p = 0.0001$; OR 2.55; CI 1.55–4.15) (LaRusch et al. 2014). Likewise, the risk of male infertility in patients with pancreatitis was similar to controls, but there significantly increased risk was associated with either *CFTR*^{BD} or *CFTR*^{CF} alleles ($p = 0.023$; OR 10.7; CI 1.03–536) or a recessive genotype ($p = 1.2 \times 10^{-7}$; OR 303; CI 23–15783) (LaRusch et al. 2014). Together, these data provide compelling evidence of disease in humans associated with defects in CFTR bicarbonate conductance.

Fig. 4.5 (continued) activation (right). Values throughout are mean ± SEM. **p*, 0.05, ***p*, 0.01: difference from WT in cells co-expressed with WNK1 and SPAK

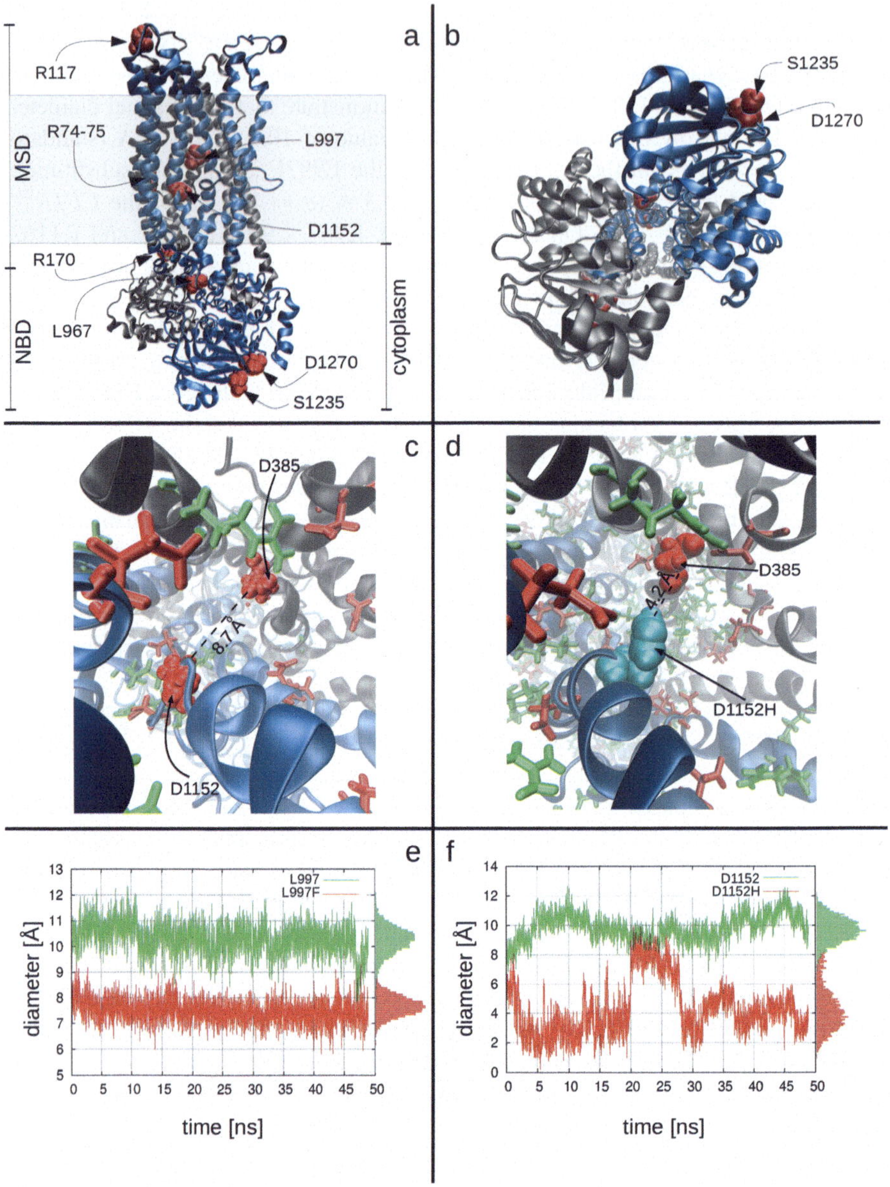

Fig. 4.6 Molecular modeling and simulations of CFTR WT and variants (LaRusch et al. 2014). Panels (**a**) and (**b**) display the side and bottom views, respectively, of the WT CFTR dimer, where the two nucleotide-binding domains and the two membrane-spanning domains are labeled as NBD and MSD. The shaded region indicates the location of the lipid bilayer. Color key: black, subunit 1 of CFTR, with residues 1–859; blue, subunit 2, residues 860–1480; red CFTR variants studied. Panel (**c**) shows the charge distribution around D1152H: this negatively charged residue (left; shown in red space-filling representation) is surrounded by several positively charged residues (green), especially on its side of the cavity, creating an attractive force that keeps the residue from extending into the cavity. Also shown are other negatively charged residues (red stick or space-filling representation), including D385, diametrically opposite to D1152. Panel (**d**) shows the

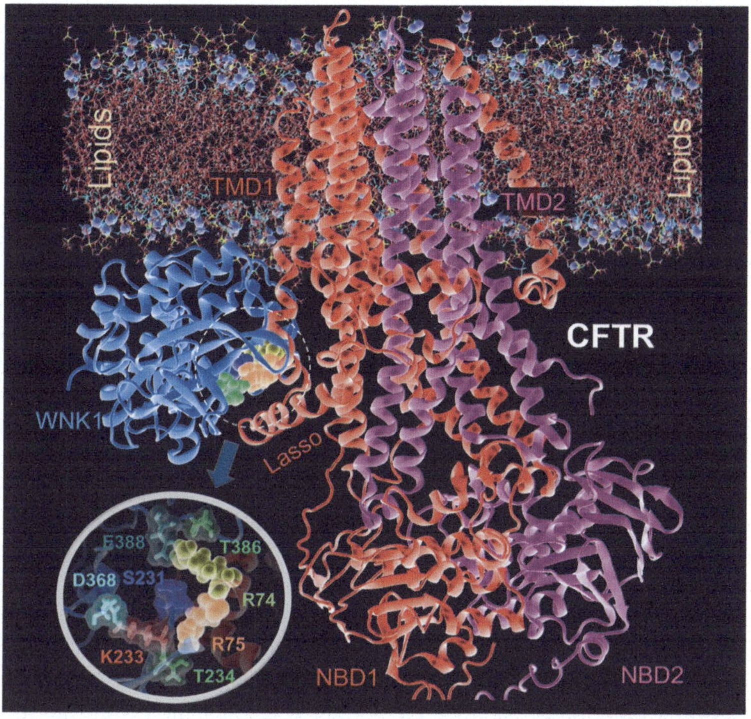

Fig. 4.7 Structural model for the MD equilibrated complex between hCFTR and WNK1 kinase domain in the presence of lipid bilayer. The R74 (purple balls) and R75 (blue balls) residues from hCFTR participate in the binding interface. The figure displays the hCFTRWNK1 complex predicted by ClusPro, after equilibration in the MD simulation system where it is embedded into the membrane lipids (lines with their phosphorus atoms shown in tan spheres) and solvated by 0.1 M NaCl solution. The snapshot was taken after 100-ns MD simulations. The inset figure shows a close-up view of interfacial interactions. WNK1 residues at the interface include S231, F232, K233, T234, and I384, G385, T386 and E388 (Modified from Kim et al. (2020) by Mary Hongying Cheng and Ivet Bahar)

4.6.2 Variant-Specific Rescue of CFTR Function

Several drugs have been developed to treat CF by modifying the CFTR protein. Ivacaftor increases the open probability (P_o) of CFTR, thereby permitting the

Fig. 4.6 (continued) corresponding scene for the variant residue, D1152H (cyan), which can move toward the center of the cavity, thus leading to a constriction in the channel diameter. Channel diameter at the location of variant residues: Panel (**e**) shows the diameter of the channel at the location of L997, as a function of time, both for the WT (L997, green curve) and the variant (F997, red curve), based on closest interatomic distance between L997/F997 and D385. On panel (**f**), the same information for the WT and variant D1152H is shown. In both plots, the pore diameter in the WT is larger than that stabilized in the mutants. The histograms of channel sizes are shown along the right ordinate

movement of additional anions through the channel and improving CFTR function (Van Goor et al. 2009; Eckford et al. 2012; Jih and Hwang 2013). Ivacaftor was first approved to treat CF in individuals with one or more copies of the variant that results in p.Gly551Asp, but has since been expanded to include eight variants that decrease CFTR P_o (Mosli et al. 2014; De Boeck et al. 2014). Another drug, lumacaftor, acts by correcting CFTR protein folding defects and was invented to treat individuals heterozygous for p.Phe508del (Van Goor et al. 2011; Lukacs and Verkman 2012). Lumacaftor alone is not clinically effective for p.Phe508del homozygotes (Clancy et al. 2012); however, p.Phe508del homozygotes see improved clinical outcomes when lumacaftor is used in conjunction with ivacaftor (Wainwright et al. 2015). Another drug that similarly corrects CFTR protein folding, tezacaftor, has also been found to be clinically effective in p.Phe508del homozygotes when used with ivacaftor (Wainwright et al. 2015). Ivacaftor in combination with lumacaftor or tezacaftor has also been demonstrated to be effective in CF individuals with one copy of p.Phe508del along with certain other CFTR functional variants (Rowe et al. 2017).

Recently, Han et al. tested the effectiveness of ivacaftor and lumacaftor in enhancing channel function in 63 CFTR variants which exhibit residual channel function expressed in CF bronchial epithelial cell line CFBE41o$^-$ (CFBE) cells and Fischer Rat Thyroid (FRT) cells (Han et al. 2018). Forty-five CFTR variants were expressed in CFBE cells and eighteen CFTR variants were expressed in FRT cells; four variants were expressed in both CFBE and FRT cells.

The residual function of each CFTR variant was calculated from the measured function of the CFTR variant by forskolin and the function of wild-type CFTR (Han et al. 2018). The CFTR variants were first tested for rescue of CFTR activity by 10 μM ivacaftor. Many of the cell lines responded to ivacaftor treatment with increased CFTR activity and the magnitude of response to ivacaftor treatment was largely dependent on the residual CFTR function of each variant. Four variants that exhibited increased CFTR activity greater than two standard deviations above the mean of all variants tested were designated as high-response, four variants were designated as intermediate-response, and all other variants were designated as moderate-response.

The CFTR variants were similarly treated with lumacaftor. Five variants with a lumacaftor response greater than two standard deviations above the mean were designated as high-response variants and four variants were designated as intermediate-response variants. Lumacaftor increased the total CFTR protein in all variants expressed in CFBE cells, this increase in total protein was not due to an increase in mRNA expression. Thus, in general the CFTR correctors increase CFTR function in general, with the magnitude generally directly proportional to the residual function (Han et al. 2018; Raraigh et al. 2018).

As ivacaftor in combination with lumacaftor has clinical benefit in homozygous CFTR F508del individuals, this drug combination was tested in the 59 CFTR variants (Han et al. 2018). Cells were incubated in lumacaftor (24 h for variants expressed in CFBE cells and 48 h for FRT cells) before the addition of ivacaftor. The response of 45 variants expressed in CFBE cells and 18 variants expressed in FRT cells to the combination of ivacaftor and lumacaftor was similar to CFTR function without the drugs. Five variants were high-response variants and four variants were intermediate-response variants. There was a robust correlation between forskolin-stimulated CFTR and CFTR that had been modulated with the combination of ivacaftor and lumacaftor across all CFTR variants, showing that the relationship is independent of mechanism or cell type. The response of the 45 variants expressed in CFBE cells to the ivacaftor/lumacaftor combination was compared to the response of the variants to each individual drug. Thirty of the 33 modest-response variants to ivacaftor and lumacaftor in combination demonstrated greater than or equal response to the drugs in combination than to either individual drug. Five rare CFTR variants had an intermediate response or high response to ivacaftor, and all five variants exhibited greater than or equal response to ivacaftor and lumacaftor in combination than the individual drugs. Six CFTR variants had an intermediate response or high response to lumacaftor, and all six variants exhibited greater than or equal response to ivacaftor and lumacaftor in combination than to the individual drugs.

These findings demonstrate that many CFTR variants, even those with low residual function, can be rescued using ivacaftor, lumacaftor, or the combination of the two drugs, and the expected improvement is proportional to residual function of the mutant CFTR protein. This demonstrates the potential usefulness of these drugs beyond the current list of approved CFTR variants. In addition to the potential expanded usefulness of ivacaftor and lumacaftor, the finding that ivacaftor and lumacaftor in combination restore CFTR function greater than or equal to either drug alone expands the potential usefulness of the drug combination beyond treating CFTR p.Phe508del homozygotes.

4.7 SLC26A9 and Cystic Fibrosis-Related Diabetes Mellitus

The solute carrier family 26 member 9 (SLC26A9) is expressed in a number of tissues including heart, lung, stomach, skin, kidney, thyroid, salivary gland, and prostate (Han et al. 2018; Chang et al. 2009; GTEx Consortium 2013; Xu et al. 2005; Lee et al. 2015; El Khouri and Toure 2014). In the airway, SLC26A9 is a cystic fibrosis disease modifier (Balazs and Mall 2018). SLC26A9 expression is governed by CFTR, and the two proteins bind to one another via an STAS domain and PDZ-binding motif (Chang et al. 2009; Bertrand et al. 2009; Ousingsawat et al. 2012).

Lam et al. (2020) examined 762 individuals homozygous for the CFTR variant p. Phe508del that were sequenced across the genomic region of *SLC26A9* including all *SLC26A9* variants associated with age-at-onset cystic fibrosis-related diabetes mellitus (CFRD) (Lam et al. 2020; Blackman et al. 2013). *SLC26A9* variants that were previously significantly associated with CFRD were also associated with CFRD in this sample ($p < 0.005$). SKAT-O (Fan et al. 2016) was used to test *SLC26A9* variants within a 5 kb sliding window in aggregate for associations with age-at-onset CFRD. Windows containing common variants (MAF $\geq$ 0.01) 5′ of

SLC26A9 and in intron 1 were significantly associated with age-at-onset CFRD ($p < 2.7E\text{-}4$). No windows containing only rare *SLC26A9* variants were significantly associated with age-at-onset CFRD (Lam et al. 2020).

Lam et al. then identified two *SLC26A9* haplotypes associated with CFRD; one associated with the later onset of CFRD and the other associated with earlier onset of CFRD (Lam et al. 2020). The former haplotype contains all alleles of variants associated with the later onset of CFRD from the GWAS, together, while the latter haplotype contains all alleles of variants found to be associated with the earlier onset CFRD. The incidence of CFRD was significantly different in individuals that were homozygous for the later onset CFRD haplotype compared to individuals that were homozygous for the earlier onset haplotype. PANC-1 and CFPAC-1 cells expressing variants from the later onset CFRD haplotype with a luciferase reporter exhibited 19% and 20%, respectively, greater expression than cells expressing variants from the earlier onset CFRD haplotype.

The data from Lam et al. (2020) demonstrated an association of *SLC26A9* with CFRD and how CFRD-associated *SLC26A9* haplotypes alter SLC26A9 expression. The *SLC26A9* haplotype associated with the later onset CFRD also exhibited greater SLC26A9 expression compared to the earlier onset CFRD-associated haplotype. This may indicate that greater SLC26A9 expression is protective against CFRD and this may be due to increased ion conductance by SLC26A9 in the presence of decreased CFTR function. If SLC26A9 is able to partially restore duct cell function, then it may decrease injury to the pancreas and delay the onset of CFRD.

4.8 Claudin-2 in the Pancreas

The claudins are family of at least 27 different tight junction proteins in mammals with a molecular range of 20–27 kDa and four membrane spanning domains (Angelow et al. 2008; Krause et al. 2008; Gunzel and Yu 2013). They differ from other tight junction proteins such as occludin and tricellulin as they form the "gasket" that seals the paracellular space between epithelial cells, separate the apical from basolateral membranes, mediate cell–cell adhesion, and polymerize to form tight junction fibrils (Angelow et al. 2008). The most extensive work on claudins has been performed in the kidney where various claudins are expressed in the different segments of the nephron. Although claudins have not been extensively studied in the pancreas, several claudins, including caludin-1, -2, -3, and -4, are expressed in mouse and human pancreas (Lee et al. 2011).

Genetic variants in the claudin-2 gene (*CLDN2*) locus confer a very high risk for pancreatitis, especially in alcoholics (Whitcomb et al. 2012). The finding was especially interesting since *CLDN2* is located on the X chromosome and these *CLDN2* variants exhibit X-linked recessive inheritance. Since men have one X chromosome, hemizygous males are susceptible to disease, whereas women, who have two X chromosomes, exhibit risk in a recessive pattern (Whitcomb et al. 2012). The risk variant was not in the coding region and appeared to increase inflammation response through aberrant localization on acinar cells. The functional consequences appear to reflect abnormal expression of claudin-2 on the acinar cell (or partially de-differentiated acinar cell) after injury. The gain-of-function could be affected and appear recessive in women as one of the two X chromosome alleles are silenced, but this has not yet been demonstrated. Although the pathobiology is not known, these findings brought attention to the potential role of the claudins in normal pancreatic biology.

Claudin-2 is a highly regulated tight junction protein normally expressed between cells of the pancreatic ducts, kidney proximal tubules, gall bladder hepatobiliary system, and intestinal epithelium (Lee et al. 2011; Aung et al. 2006). Claudin-2 differs from other claudins in the pancreas because it forms a low-resistance cation-selective ion and water channel (Van Itallie et al. 2008; Amasheh et al. 2002). In the pancreatic duct, claudin-2 may allow sodium and water to freely enter the duct lumen where it mixes with the bicarbonate that is actively secreted through the duct cells via a CFTR-dependent pathway to form pancreatic juice (Fig. 4.4). The expression of claudin-2 in human pancreas has recently been confirmed in pancreatic samples by quantitative PCR of cDNA, by Western blot, by immunofluorescence, and by immunohistochemistry (Whitcomb et al. 2012).

Claudin-2 and other tight junction proteins are continually cycled by insertion into the plasma membrane and removal by endocytosis (Dukes et al. 2012). The *CLDN2* promoter contains an NFκB binding site (Sakaguchi et al. 2002) and expression is upregulated by tumor necrosis factor-alpha (TNFa) (Mankertz et al. 2009), interleukin-6 (IL-6) (Suzuki et al. 2011), and the Wnt/LEF-1/beta-catenin pathway (Mankertz et al. 2004) and downregulated by epidermal growth factor (Ikari et al. 2011) by endocytosis linked to the Endosomal Sorting Complex Required for Transport (ESCRT) mechanism (Dukes et al. 2012). Expression of claudin-4, which is also expressed in the pancreas but differs from claudin-2 by forming high resistance barrier to ions and water (Van Itallie et al. 2001; Angelow et al. 2008), appears to be regulated in a reciprocate manner. In the gallbladder, for example, the immunohistochemical staining of claudin-2 was weak in normal tissue but became intense with inflammation, while claudin-4 staining shifted from strong to faint (Laurila et al. 2007). Madin-Darby canine kidney (MDCKII) cells respond to EGF with ERK1/2-mediated downregulation of claudin-2 and upregulation of claudin-1, -3, and -4 (Singh and Harris 2004), possibly through STAT3 and Src (Garcia-Hernandez et al. 2015). Both claudin-2 and -4 are phosphorylated by WNK4 (Yamauchi et al. 2004), and possibly WNK1 (Ohta et al. 2006), although the importance of this observation has not been applied to pancreatic duct cells. The dynamic changes may also be important during stimulated pancreatic secretion, which is not usually seen in human tissues since samples coming from humans are typically linked to surgical operations after the patient has been fasting for some time.

4.9 Summary and Conclusions

Diseases of the exocrine pancreas related to ion transporters are currently limited to cystic fibrosis and recurrent acute and chronic pancreatitis. While there may be other diseases, the clinical severity has not been sufficient to be fully recognized or appreciated. Clinical recognition of pathology is generally secondary to the increased susceptibility to intrapancreatic trypsin activation, which causes significant inflammatory responses that damage the pancreas. Ion transport dysfunction that reduces ductal fluid secretion and flushing of trypsin out of the duct results in pancreatic injury and disease. CFTR is used by the pancreas for bicarbonate secretion. A new finding of *CFTR* variants that specifically affect regulated bicarbonate conductance in response to WNK1 activation provides new insights into ion channel biology, models of pancreatic duct physiology, and have implications for multiple related organs systems.

Acknowledgements This work was supported by NIH DK063922 (DCW) and NIH DK061451 (DCW).

References

Amasheh S, Meiri N, Gitter AH, Schoneberg T, Mankertz J, Schulzke JD, Fromm M (2002) Claudin-2 expression induces cation-selective channels in tight junctions of epithelial cells. J Cell Sci 115:4969–4976

Angelow S, Ahlstrom R, Yu AS (2008) Biology of claudins. Am J Physiol Ren Physiol 295:F867–F876

Anselmo AN, Earnest S, Chen W, Juang YC, Kim SC, Zhao Y, Cobb MH (2006) WNK1 and OSR1 regulate the Na+, K+, 2Cl- cotransporter in HeLa cells. Proc Natl Acad Sci U S A 103:10883–10888

Aoun E, Chang CC, Greer JB, Papachristou GI, Barmada MM, Whitcomb DC (2008) Pathways to injury in chronic pancreatitis: decoding the role of the high-risk SPINK1 N34S haplotype using meta-analysis. PLoS One 3:e2003

Aung PP, Mitani Y, Sanada Y, Nakayama H, Matsusaki K, Yasui W (2006) Differential expression of claudin-2 in normal human tissues and gastrointestinal carcinomas. Virchows Archiv 448:428–434

Balazs A, Mall MA (2018) Role of the SLC26A9 chloride channel as disease modifier and potential therapeutic target in cystic fibrosis. Front Pharmacol 9:1112

Bear CE, Li CH, Kartner N, Bridges RJ, Jensen TJ, Ramjeesingh M, Riordan JR (1992) Purification and functional reconstitution of the cystic fibrosis transmembrane conductance regulator (CFTR). Cell 68:809–818

Beer S, Zhou J, Szabo A, Keiles S, Chandak GR, Witt H, Sahin-Toth M (2012) Comprehensive functional analysis of chymotrypsin C (CTRC) variants reveals distinct loss-of-function mechanisms associated with pancreatitis risk. Gut 62:1616–1624

Bertrand CA, Zhang R, Pilewski JM, Frizzell RA (2009) SLC26A9 is a constitutively active, CFTR-regulated anion conductance in human bronchial epithelia. J Gen Physiol 133:421–438

Blackman SM, Commander CW, Watson C, Arcara KM, Strug LJ, Stonebraker JR, Wright FA, Rommens JM, Sun L, Pace RG, Norris SA, Durie PR, Drumm ML, Knowles MR, Cutting GR (2013) Genetic modifiers of cystic fibrosis-related diabetes. Diabetes 62:3627–3635

Bombieri C, Claustres M, De Boeck K, Derichs N, Dodge J, Girodon E, Sermet I, Schwarz M, Tzetis M, Wilschanski M, Bareil C, Bilton D, Castellani C, Cuppens H, Cutting GR, Drevinek P, Farrell P, Elborn JS, Jarvi K, Kerem B, Kerem E, Knowles M, Macek M Jr, Munck A, Radojkovic D, Seia M, Sheppard DN, Southern KW, Stuhrmann M, Tullis E, Zielenski J, Pignatti PF, Ferec C (2011) Recommendations for the classification of diseases as CFTR-related disorders. J Cyst Fibros 10(Suppl 2):S86–S102

Brown EM, Gamba G, Riccardi D, Lombardi M, Butters R, Kifor O, Sun A, Hediger MA, Lytton J, Hebert SC (1993) Cloning and characterization of an extracellular Ca(2+)-sensing receptor from bovine parathyroid. Nature 366:575–580

Bruce JI, Yang X, Ferguson CJ, Elliott AC, Steward MC, Case RM, Riccardi D (1999) Molecular and functional identification of a Ca2+ (polyvalent cation)-sensing receptor in rat pancreas. J Biol Chem 274:20561–20568

Case RM, Argent BE (1993) Pancreatic duct cell secretion. Control and mechanism of transport. In: Go VLW et al (eds) The pancreas: biology, pathobiology and disease. Raven Press, New York

Ceppa EP, Lyo V, Grady EF, Knecht W, Grahn S, Peterson A, Bunnett NW, Kirkwood KS, Cattaruzza F (2011) Serine proteases mediate inflammatory pain in acute pancreatitis. Am J Physiol Gastrointest Liver Physiol 300:G1033–G1042

Chang MH, Plata C, Sindic A, Ranatunga WK, Chen AP, Zandi-Nejad K, Chan KW, Thompson J, Mount DB, Romero MF (2009) Slc26a9 is inhibited by the R-region of the cystic fibrosis transmembrane conductance regulator via the STAS domain. J Biol Chem 284:28306–28318

Chen JM, Ferec C (2012) Genetics and pathogenesis of chronic pancreatitis: The 2012 update. Clin Res Hepatol Gastroenterol 36(4):334–340

Clancy JP, Rowe SM, Accurso FJ, Aitken ML, Amin RS, Ashlock MA, Ballmann M, Boyle MP, Bronsveld I, Campbell PW, De Boeck K, Donaldson SH, Dorkin HL, Dunitz JM, Durie PR, Jain M, Leonard A, McCoy KS, Moss RB, Pilewski JM, Rosenbluth DB, Rubenstein RC, Schechter MS, Botfield M, Ordonez CL, Spencer-Green GT, Vernillet L, Wisseh S, Yen K, Konstan MW (2012) Results of a phase IIa study of VX-809, an investigational CFTR corrector compound, in subjects with cystic fibrosis homozygous for the F508del-CFTR mutation. Thorax 67:12–18

Cohn JA, Friedman KJ, Noone PG, Knowles MR, Silverman LM, Jowell PS (1998) Relation between mutations of the cystic fibrosis gene and idiopathic pancreatitis. N Engl J Med 339:653–658

De Boeck K, Munck A, Walker S, Faro A, Hiatt P, Gilmartin G, Higgins M (2014) Efficacy and safety of ivacaftor in patients with cystic fibrosis and a non-G551D gating mutation. J Cyst Fibros 13:674–680

Deng X, Whitcomb DC (1998) Neurohormonal control of the exocrine pancreas. Curr Opin Gastroenterol 14:362–368

Derikx MH, Kovacs P, Scholz M, Masson E, Chen JM, Ruffert C, Lichtner P, Te Morsche RH, Cavestro GM, Ferec C, Drenth JP, Witt H, Rosendahl J, Members PanEuropean Working group on Alcoholic Chronic Pancreatitis, and Collaborators (2015) Polymorphisms at PRSS1-PRSS2 and CLDN2-MORC4 loci associate with alcoholic and non-alcoholic chronic pancreatitis in a European replication study. Gut 64:1426–1433

Dukes JD, Whitley P, Chalmers AD (2012) The PIKfyve inhibitor YM201636 blocks the continuous recycling of the tight junction proteins claudin-1 and claudin-2 in MDCK cells. PLoS One 7:e28659

Eckford PD, Li C, Ramjeesingh M, Bear CE (2012) Cystic fibrosis transmembrane conductance regulator (CFTR) potentiator VX-770 (ivacaftor) opens the defective channel gate of mutant CFTR in a phosphorylation-dependent but ATP-independent manner. J Biol Chem 287:36639–36649

El Khouri E, Toure A (2014) Functional interaction of the cystic fibrosis transmembrane conductance regulator with members of the SLC26 family of anion transporters (SLC26A8 and SLC26A9): physiological and pathophysiological relevance. Int J Biochem Cell Biol 52:58–67

Fan R, Chiu CY, Jung J, Weeks DE, Wilson AF, Bailey-Wilson JE, Amos CI, Chen Z, Mills JL, Xiong M (2016) A comparison study of fixed and mixed effect models for gene level association studies of complex traits. Genet Epidemiol 40:702–721

Felderbauer P, Hoffmann P, Einwachter H, Bulut K, Ansorge N, Schmitz F, Schmidt WE (2003) A novel mutation of the calcium sensing receptor gene is associated with chronic pancreatitis in a family with heterozygous SPINK1 mutations. BMC Gastroenterol 3:34

Garcia-Hernandez V, Flores-Maldonado C, Rincon-Heredia R, Verdejo-Torres O, Bonilla-Delgado J, Meneses-Morales I, Gariglio P, Contreras RG (2015) EGF regulates claudin-2 and -4 expression through Src and STAT3 in MDCK cells. J Cell Physiol 230(1):105–115. PMID:24909426

Gombault A, Baron L, Couillin I (2012) ATP release and purinergic signaling in NLRP3 inflammasome activation. Front Immunol 3:414

Gorry MC, Gabbaizedeh D, Furey W, Gates LK Jr, Preston RA, Aston CE, Zhang Y, Ulrich C, Ehrlich GD, Whitcomb DC (1997) Mutations in the cationic trypsinogen gene are associated with recurrent acute and chronic pancreatitis. Gastroenterology 113:1063–1068

GTEx Consortium (2013) The Genotype-Tissue Expression (GTEx) project. Nat Genet 45:580–585

Gunzel D, Yu AS (2013) Claudins and the modulation of tight junction permeability. Physiol Rev 93:525–569

Haanes KA, Kowal JM, Arpino G, Lange SC, Moriyama Y, Pedersen PA, Novak I (2014) Role of vesicular nucleotide transporter VNUT (SLC17A9) in release of ATP from AR42J cells and mouse pancreatic acinar cells. Purinergic Signal 10:431–440

Han ST, Rab A, Pellicore MJ, Davis EF, McCague AF, Evans TA, Joynt AT, Lu Z, Cai Z, Raraigh KS, Hong JS, Sheppard DN, Sorscher EJ, Cutting GR (2018) Residual function of cystic fibrosis mutants predicts response to small molecule CFTR modulators. JCI Insight 3

Hoque R, Malik AF, Gorelick F, Mehal WZ (2012) Sterile inflammatory response in acute pancreatitis. Pancreas 41:353–357

Ikari A, Takiguchi A, Atomi K, Sugatani J (2011) Epidermal growth factor increases clathrin-dependent endocytosis and degradation of claudin-2 protein in MDCK II cells. J Cell Physiol 226:2448–2456

Jih KY, Hwang TC (2013) Vx-770 potentiates CFTR function by promoting decoupling between the gating cycle and ATP hydrolysis cycle. Proc Natl Acad Sci U S A 110:4404–4409

Jun I, Cheng MH, Sim E, Jung J, Suh BL, Kim Y, Son H, Park K, Kim CH, Yoon JH, Whitcomb DC, Bahar I, Lee MG (2016) Pore dilatation increases the bicarbonate permeability of CFTR, ANO1 and glycine receptor anion channels. J Physiol 594:2929–2955

Kerem E (2006) Atypical CF and CF related diseases. Paediatr Respir Rev 7(Suppl 1):S144–S146

Kim Y, Jun I, Shin DH, Yoon JG, Piao H, Jung J, Park HW, Cheng MH, Bahar I, Whitcomb DC, Lee MG (2020) Regulation of CFTR bicarbonate channel activity by WNK1: implications for pancreatitis and CFTR-related disorders. Cell Mol Gastroenterol Hepatol 9(1):79–103

Korner M, Hayes GM, Rehmann R, Zimmermann A, Friess H, Miller LJ, Reubi JC (2005) Secretin receptors in normal and diseased human pancreas: marked reduction of receptor binding in ductal neoplasia. Am J Pathol 167:959–968

Krause G, Winkler L, Mueller SL, Haseloff RF, Piontek J, Blasig IE (2008) Structure and function of claudins. Biochim Biophys Acta 1778:631–645

Lam AN, Aksit MA, Vecchio-Pagan B, Shelton CA, Osorio DL, Anzmann AF, Goff LA, Whitcomb DC, Blackman SM, Cutting GR (2020) Increased expression of anion transporter SLC26A9 delays diabetes onset in cystic fibrosis. J Clin Invest 130(1):272–286

Larusch J, Whitcomb DC (2012) Genetics of pancreatitis with a focus on the pancreatic ducts. Minerva Gastroenterol Dietol 58:299–308

LaRusch J, Jung J, General IJ, Lewis MD, Park HW, Brand RE, Gelrud A, Anderson MA, Banks PA, Conwell D, Lawrence C, Romagnuolo J, Baillie J, Alkaade S, Cote GA, Gardner TB, Amann ST, Slivka A, Sandhu B, Aloe A, Kienholz ML, Yadav D, Barmada MM, Bahar I, Lee MG, Whitcomb DC (2014) Mechanisms of CFTR functional variants that impair regulated

bicarbonate permeation and increase risk for pancreatitis but not for cystic fibrosis. PLoS Genet 10:e1004376
LaRusch J, Lozano-Leon A, Stello K, Moore A, Muddana V, O'Connell M, Diergaarde B, Yadav D, Whitcomb DC (2015) The common chymotrypsinogen C (CTRC) variant G60G (C.180T) increases risk of chronic pancreatitis but not recurrent acute pancreatitis in a north american population. Clin Transl Gastroenterol 6:e68
Laukkarinen JM, Weiss ER, van Acker GJ, Steer ML, Perides G (2008) Protease-activated receptor-2 exerts contrasting model-specific effects on acute experimental pancreatitis. J Biol Chem 283:20703–20712
Laurila JJ, Karttunen T, Koivukangas V, Laurila PA, Syrjala H, Saarnio J, Soini Y, Ala-Kokko TI (2007) Tight junction proteins in gallbladder epithelium: different expression in acute acalculous and calculous cholecystitis. J Histochem Cytochem 55:567–573
Lee JH, Kim KS, Kim TJ, Hong SP, Song SY, Chung JB, Park SW (2011) Immunohistochemical analysis of claudin expression in pancreatic cystic tumors. Oncol Rep 25:971–978
Lee MG, Ohana E, Park HW, Yang D, Muallem S (2012) Molecular mechanism of pancreatic and salivary gland fluid and HCO formula secretion. Physiol Rev 92:39–74
Lee HJ, Yoo JE, Namkung W, Cho HJ, Kim K, Kang JW, Yoon JH, Choi JY (2015) Thick airway surface liquid volume and weak mucin expression in pendrin-deficient human airway epithelia. Physiol Rep 3(8):e12480
Lukacs GL, Verkman AS (2012) CFTR: folding, misfolding and correcting the DeltaF508 conformational defect. Trends Mol Med 18:81–91
Mankertz J, Hillenbrand B, Tavalali S, Huber O, Fromm M, Schulzke JD (2004) Functional crosstalk between Wnt signaling and Cdx-related transcriptional activation in the regulation of the claudin-2 promoter activity. Biochem Biophys Res Commun 314:1001–1007
Mankertz J, Amasheh M, Krug SM, Fromm A, Amasheh S, Hillenbrand B, Tavalali S, Fromm M, Schulzke JD (2009) TNFalpha up-regulates claudin-2 expression in epithelial HT-29/B6 cells via phosphatidylinositol-3-kinase signaling. Cell Tissue Res 336:67–77
Marino CR, Matovcik LM, Gorelick FS, Cohn JA (1991) Localization of the cystic fibrosis transmembrane conductance regulator in pancreas. J Clin Investig 88:712–716
Masamune A, Nakano E, Hamada S, Kakuta Y, Kume K, Shimosegawa T (2015) Common variants at PRSS1-PRSS2 and CLDN2-MORC4 loci associate with chronic pancreatitis in Japan. Gut 64:1345–1346
Masamune A, Kotani H, Sorgel FL, Chen JM, Hamada S, Sakaguchi R, Masson E, Nakano E, Kakuta Y, Niihori T, Funayama R, Shirota M, Hirano T, Kawamoto T, Hosokoshi A, Kume K, Unger L, Ewers M, Laumen H, Bugert P, Mori MX, Tsvilovskyy V, Weissgerber P, Kriebs U, Fecher-Trost C, Freichel M, Diakopoulos KN, Berninger A, Lesina M, Ishii K, Itoi T, Ikeura T, Okazaki K, Kaune T, Rosendahl J, Nagasaki M, Uezono Y, Algul H, Nakayama K, Matsubara Y, Aoki Y, Ferec C, Mori Y, Witt H, Shimosegawa T (2020) Variants that affect function of calcium channel TRPV6 are associated with early-onset chronic pancreatitis. Gastroenterology 158(6):1626–1641.e8
Mosli MH, Sandborn WJ, Kim RB, Khanna R, Al-Judaibi B, Feagan BG (2014) Toward a personalized medicine approach to the management of inflammatory bowel disease. Am J Gastroenterol 109:994–1004
Muddana V, Lamb J, Greer JB, Elinoff B, Hawes RH, Cotton PB, Anderson MA, Brand RE, Slivka A, Whitcomb DC (2008) Association between calcium sensing receptor gene polymorphisms and chronic pancreatitis in a US population: role of serine protease inhibitor Kazal 1type and alcohol. World J Gastroenterol 14:4486–4491
Mussa BM, Verberne AJ (2013) The dorsal motor nucleus of the vagus and regulation of pancreatic secretory function. Exp Physiol 98:25–37
Noble MD, Liddle RA (2005) Neurohormonal control of exocrine pancreatic secretion. Curr Opin Gastroenterol 21:531–537
Novak I (2008) Purinergic receptors in the endocrine and exocrine pancreas. Purinergic Signal 4:237–253

Ohta A, Yang SS, Rai T, Chiga M, Sasaki S, Uchida S (2006) Overexpression of human WNK1 increases paracellular chloride permeability and phosphorylation of claudin-4 in MDCKII cells. Biochem Biophys Res Commun 349:804–808

Ousingsawat J, Schreiber R, Kunzelmann K (2012) Differential contribution of SLC26A9 to Cl(−) conductance in polarized and non-polarized epithelial cells. J Cell Physiol 227:2323–2329

Pallagi P, Venglovecz V, Rakonczay Z Jr, Borka K, Korompay A, Ozsvari B, Judak L, Sahin-Toth M, Geisz A, Schnur A, Maleth J, Takacs T, Gray MA, Argent BE, Mayerle J, Lerch MM, Wittmann T, Hegyi P (2011) Trypsin reduces pancreatic ductal bicarbonate secretion by inhibiting CFTR Cl(−) channels and luminal anion exchangers. Gastroenterology 141:2228–39 e6

Park HW, Nam JH, Kim JY, Namkung W, Yoon JS, Lee JS, Kim KS, Venglovecz V, Gray MA, Kim KH, Lee MG (2010) Dynamic regulation of CFTR bicarbonate permeability by [Cl-]i and its role in pancreatic bicarbonate secretion. Gastroenterology 139:620–631

Pfutzer RH, Barmada MM, Brunskill AP, Finch R, Hart PS, Neoptolemos J, Furey WF, Whitcomb DC (2000) SPINK1/PSTI polymorphisms act as disease modifiers in familial and idiopathic chronic pancreatitis. Gastroenterology 119:615–623

Pidasheva S, D'Souza-Li L, Canaff L, Cole DE, Hendy GN (2004) CASRdb: calcium-sensing receptor locus-specific database for mutations causing familial (benign) hypocalciuric hypercalcemia, neonatal severe hyperparathyroidism, and autosomal dominant hypocalcemia. Hum Mutat 24:107–111

Rahner C, Mitic LL, Anderson JM (2001) Heterogeneity in expression and subcellular localization of claudins 2, 3, 4, and 5 in the rat liver, pancreas, and gut. Gastroenterology 120:411–422

Raraigh KS, Han ST, Davis E, Evans TA, Pellicore MJ, McCague AF, Joynt AT, Lu Z, Atalar M, Sharma N, Sheridan MB, Sosnay PR, Cutting GR (2018) Functional assays are essential for interpretation of missense variants associated with variable expressivity. Am J Hum Genet 102 (6):1062–1077

Richardson C, Alessi DR (2008) The regulation of salt transport and blood pressure by the WNK-SPAK/OSR1 signalling pathway. J Cell Sci 121:3293–3304

Rosendahl J, Witt H, Szmola R, Bhatia E, Ozsvari B, Landt O, Schulz HU, Gress TM, Pfutzer R, Lohr M, Kovacs P, Bluher M, Stumvoll M, Choudhuri G, Hegyi P, te Morsche RH, Drenth JP, Truninger K, Macek M Jr, Puhl G, Witt U, Schmidt H, Buning C, Ockenga J, Kage A, Groneberg DA, Nickel R, Berg T, Wiedenmann B, Bodeker H, Keim V, Mossner J, Teich N, Sahin-Toth M (2008) Chymotrypsin C (CTRC) variants that diminish activity or secretion are associated with chronic pancreatitis. Nat Genet 40:78–82

Rosendahl J, Kirsten H, Hegyi E, Kovacs P, Weiss FU, Laumen H, Lichtner P, Ruffert C, Chen JM, Masson E, Beer S, Zimmer C, Seltsam K, Algul H, Buhler F, Bruno MJ, Bugert P, Burkhardt R, Cavestro GM, Cichoz-Lach H, Farre A, Frank J, Gambaro G, Gimpfl S, Grallert H, Griesmann H, Grutzmann R, Hellerbrand C, Hegyi P, Hollenbach M, Iordache S, Jurkowska G, Keim V, Kiefer F, Krug S, Landt O, Leo MD, Lerch MM, Levy P, Loffler M, Lohr M, Ludwig M, Macek M, Malats N, Malecka-Panas E, Malerba G, Mann K, Mayerle J, Mohr S, Te Morsche RHM, Motyka M, Mueller S, Muller T, Nothen MM, Pedrazzoli S, Pereira SP, Peters A, Pfutzer R, Real FX, Rebours V, Ridinger M, Rietschel M, Rosmann E, Saftoiu A, Schneider A, Schulz HU, Soranzo N, Soyka M, Simon P, Skipworth J, Stickel F, Strauch K, Stumvoll M, Testoni PA, Tonjes A, Werner L, Werner J, Wodarz N, Ziegler M, Masamune A, Mossner J, Ferec C, Michl P, Drenth JPH, Witt H, Scholz M, Sahin-Toth M, and All members of the PanEuropean Working group on ACP (2017) Genome-wide association study identifies inversion in the CTRB1-CTRB2 locus to modify risk for alcoholic and non-alcoholic chronic pancreatitis. Gut 67(10):1855–1863

Rowe SM, Daines C, Ringshausen FC, Kerem E, Wilson J, Tullis E, Nair N, Simard C, Han L, Ingenito EP, McKee C, Lekstrom-Himes J, Davies JC (2017) Tezacaftor-ivacaftor in residual-function heterozygotes with cystic fibrosis. N Engl J Med 377:2024–2035

Rowntree RK, Harris A (2003) The phenotypic consequences of CFTR mutations. Ann Hum Genet 67:471–485

Sakaguchi T, Gu X, Golden HM, Suh E, Rhoads DB, Reinecker HC (2002) Cloning of the human claudin-2 5′-flanking region revealed a TATA-less promoter with conserved binding sites in mouse and human for caudal-related homeodomain proteins and hepatocyte nuclear factor-1alpha. J Biol Chem 277:21361–21370

Sarles H, Adler G, Dani R, Frey C, Gullo L, Harada H, Martin E, Norohna M, Scuro LA (1989) The pancreatitis classification of Marseilles, Rome 1988. Scand J Gastroenterol 24:641–641

Sarner M, Cotton PB (1984) Definitions of acute and chronic pancreatitis. Clin Gastroenterol 13:865

Sharer N, Schwarz M, Malone G, Howarth A, Painter J, Super M, Braganza J (1998) Mutations of the cystic fibrosis gene in patients with chronic pancreatitis. N Engl J Med 339:645–652

Singh AB, Harris RC (2004) Epidermal growth factor receptor activation differentially regulates claudin expression and enhances transepithelial resistance in Madin-Darby canine kidney cells. J Biol Chem 279:3543–3552

Sohma Y, Gray MA, Imai Y, Argent BE (1996) A mathematical model of the pancreatic ductal epithelium. J Membr Biol 154:53–67

Sohma Y, Gray MA, Imai Y, Argent BE (2000) HCO3 transport in a mathematical model of the pancreatic ductal epithelium. J Membr Biol 176:77–100

Steer ML, Waxman I, Freedman S (1995) Chronic pancreatitis. N Engl J Med 332:1482–1490

Stetson DL, Beauwens R, Palmisano J, Mitchell PP, Steinmetz PR (1985) A double-membrane model for urinary bicarbonate secretion. Am J Phys 249:F546–F552

Steward MC, Ishiguro H, Case RM (2005) Mechanisms of bicarbonate secretion in the pancreatic duct. Annu Rev Physiol 67:377–409

Stewart CP, Winterhager JM, Heintze K, Petersen KU (1989) Electrogenic bicarbonate secretion by guinea pig gallbladder epithelium: apical membrane exit. Am J Phys 256:C736–C749

Strong TV, Boehm K, Collins FS (1994) Localization of cystic fibrosis transmembrane conductance regulator mRNA in the human gastrointestinal tract by in situ hybridization. J Clin Invest 93:347–354

Suzuki T, Yoshinaga N, Tanabe S (2011) Interleukin-6 (IL-6) regulates claudin-2 expression and tight junction permeability in intestinal epithelium. J Biol Chem 286:31263–31271

Szabo A, Sahin-Toth M (2012) Determinants of chymotrypsin C cleavage specificity in the calcium-binding loop of human cationic trypsinogen. FEBS J 279:4283–4292

Ulrich CD II, Wood P, Hadac EM, Kopras E, Whitcomb DC, Miller LJ (1998) Cellular distribution of secretin receptor expression in rat pancreas. Am J Phys 275:G1437–G1444

Van Goor F, Hadida S, Grootenhuis PD, Burton B, Cao D, Neuberger T, Turnbull A, Singh A, Joubran J, Hazlewood A, Zhou J, McCartney J, Arumugam V, Decker C, Yang J, Young C, Olson ER, Wine JJ, Frizzell RA, Ashlock M, Negulescu P (2009) Rescue of CF airway epithelial cell function in vitro by a CFTR potentiator, VX-770. Proc Natl Acad Sci U S A 106:18825–18830

Van Goor F, Hadida S, Grootenhuis PD, Burton B, Stack JH, Straley KS, Decker CJ, Miller M, McCartney J, Olson ER, Wine JJ, Frizzell RA, Ashlock M, Negulescu PA (2011) Correction of the F508del-CFTR protein processing defect in vitro by the investigational drug VX-809. Proc Natl Acad Sci U S A 108:18843–18848

Van Itallie C, Rahner C, Anderson JM (2001) Regulated expression of claudin-4 decreases paracellular conductance through a selective decrease in sodium permeability. J Clin Invest 107:1319–1327

Van Itallie CM, Holmes J, Bridges A, Gookin JL, Coccaro MR, Proctor W, Colegio OR, Anderson JM (2008) The density of small tight junction pores varies among cell types and is increased by expression of claudin-2. J Cell Sci 121:298–305

Wainwright CE, Elborn JS, Ramsey BW, Marigowda G, Huang X, Cipolli M, Colombo C, Davies JC, De Boeck K, Flume PA, Konstan MW, McColley SA, McCoy K, McKone EF, Munck A, Ratjen F, Rowe SM, Waltz D, Boyle MP, Traffic Study Group, and Transport Study Group (2015) Lumacaftor-ivacaftor in patients with cystic fibrosis homozygous for Phe508del CFTR. N Engl J Med 373:220–231

Wang J, Novak I (2013) Ion transport in human pancreatic duct epithelium, Capan-1 cells, is regulated by secretin, VIP, acetylcholine, and purinergic receptors. Pancreas 42:452–460

Whitcomb DC (1999) Pancreatic bicarbonate secretion: the role of CFTR and the sodium-bicarbonate cotransporter. Gastroenterology 117:275–277

Whitcomb DC (2013) Genetic risk factors for pancreatic disorders. Gastroenterology 144:1292–1302

Whitcomb DC, Ermentrout GB (2004) A mathematical model of the pancreatic duct cell generating high bicarbonate concentrations in pancreatic juice. Pancreas 29:E30–E40

Whitcomb DC, Gorry MC, Preston RA, Furey W, Sossenheimer MJ, Ulrich CD, Martin SP, Gates LK Jr, Amann ST, Toskes PP, Liddle R, McGrath K, Uomo G, Post JC, Ehrlich GD (1996) Hereditary pancreatitis is caused by a mutation in the cationic trypsinogen gene. Nat Genet 14:141–145

Whitcomb DC, Yadav D, Adam S, Hawes RH, Brand RE, Anderson MA, Money ME, Banks PA, Bishop MD, Baillie J, Sherman S, DiSario J, Burton FR, Gardner TB, Amann ST, Gelrud A, Lo SK, DeMeo MT, Steinberg WM, Kochman ML, Etemad B, Forsmark CE, Elinoff B, Greer JB, O'Connell M, Lamb J, Barmada MM, Group North American Pancreatic Study (2008) Multicenter approach to recurrent acute and chronic pancreatitis in the United States: the North American Pancreatitis Study 2 (NAPS2). Pancreatology 8:520–531

Whitcomb DC, Larusch J, Krasinskas AM, Klei L, Smith JP, Brand RE, Neoptolemos JP, Lerch MM, Tector M, Sandhu BS, Guda NM, Orlichenko L, Albert MS, Albin RL, Apostolova LG, Arnold SE, Baldwin CT, Barber R, Barnes LL, Beach TG, Beecham GW, Beekly D, Bennett DA, Bigio EH, Bird TD, Blacker D, Boxer A, Burke JR, Buxbaum JD, Cairns NJ, Cantwell LB, Cao C, Carney RM, Carroll SL, Chui HC, Clark DG, Cribbs DH, Crocco EA, Cruchaga C, Decarli C, Demirci FY, Dick M, Dickson DW, Duara R, Ertekin-Taner N, Faber KM, Fallon KB, Farlow MR, Ferris S, Foroud TM, Frosch MP, Galasko DR, Ganguli M, Gearing M, Geschwind DH, Ghetti B, Gilbert JR, Gilman S, Glass JD, Goate AM, Graff-Radford NR, Green RC, Growdon JH, Hakonarson H, Hamilton-Nelson KL, Hamilton RL, Harrell LE, Head E, Honig LS, Hulette CM, Hyman BT, Jicha GA, Jin LW, Jun G, Kamboh MI, Karydas A, Kaye JA, Kim R, Koo EH, Kowall NW, Kramer JH, Kramer P, Kukull WA, Laferla FM, Lah JJ, Leverenz JB, Levey AI, Li G, Lin CF, Lieberman AP, Lopez OL, Lunetta KL, Lyketsos CG, Mack WJ, Marson DC, Martin ER, Martiniuk F, Mash DC, Masliah E, McKee AC, Mesulam M, Miller BL, Miller CA, Miller JW, Montine TJ, Morris JC, Murrell JR, Naj AC, Olichney JM, Parisi JE, Peskind E, Petersen RC, Pierce A, Poon WW, Potter H, Quinn JF, Raj A, Raskind M, Reiman EM, Reisberg B, Reitz C, Ringman JM, Roberson ED, Rosen HJ, Rosenberg RN, Sano M, Saykin AJ, Schneider JA, Schneider LS, Seeley WW, Smith AG, Sonnen JA, Spina S, Stern RA, Tanzi RE, Trojanowski JQ, Troncoso JC, Tsuang DW, Valladares O, Van Deerlin VM, Van Eldik LJ, Vardarajan BN, Vinters HV, Vonsattel JP, Wang LS, Weintraub S, Welsh-Bohmer KA, Williamson J, Woltjer RL, Wright CB, Younkin SG, Yu CE, Yu L, Alkaade S, Amann ST, Anderson MA, Baillie J, Banks PA, Conwell D, Cote GA, Cotton PB, Disario J, Farrer LA, Forsmark CE, Johnstone M, Gardner TB, Gelrud A, Greenhalf W, Haines JL, Hartman DJ, Hawes RA, Lawrence C, Lewis M, Mayerle J, Mayeux R, Melhem NM, Money ME, Muniraj T, Papachristou GI, Pericak-Vance MA, Romagnuolo J, Schellenberg GD, Sherman S, Simon P, Singh VP, Slivka A, Stolz D, Sutton R, Weiss FU, Wilcox CM, Zarnescu NO, Wisniewski SR, O'Connell MR, Kienholz ML, Roeder K, Barmada MM, Yadav D, Devlin B (2012) Common genetic variants in the CLDN2 and PRSS1-PRSS2 loci alter risk for alcohol-related and sporadic pancreatitis. Nat Genet 44:1349–1354

Witt H, Luck W, Hennies HC, Classen M, Kage A, Lass U, Landt O, Becker M (2000) Mutations in the gene encoding the serine protease inhibitor, Kazal type 1 are associated with chronic pancreatitis. Nat Genet 25:213–216

Witt H, Apte MV, Keim V, Wilson JS (2007) Chronic pancreatitis: challenges and advances in pathogenesis, genetics, diagnosis, and therapy. Gastroenterology 132:1557–1573

Witt H, Beer S, Rosendahl J, Chen JM, Chandak GR, Masamune A, Bence M, Szmola R, Oracz G, Macek M Jr, Bhatia E, Steigenberger S, Lasher D, Buhler F, Delaporte C, Tebbing J, Ludwig M,

Pilsak C, Saum K, Bugert P, Masson E, Paliwal S, Bhaskar S, Sobczynska-Tomaszewska A, Bak D, Balascak I, Choudhuri G, Nageshwar Reddy D, Rao GV, Thomas V, Kume K, Nakano E, Kakuta Y, Shimosegawa T, Durko L, Szabo A, Schnur A, Hegyi P, Rakonczay Z Jr, Pfutzer R, Schneider A, Groneberg DA, Braun M, Schmidt H, Witt U, Friess H, Algul H, Landt O, Schuelke M, Kruger R, Wiedenmann B, Schmidt F, Zimmer KP, Kovacs P, Stumvoll M, Bluher M, Muller T, Janecke A, Teich N, Grutzmann R, Schulz HU, Mossner J, Keim V, Lohr M, Ferec C, Sahin-Toth M (2013) Variants in CPA1 are strongly associated with early onset chronic pancreatitis. Nat Genet 45:1216–1220

Xie R, Dong X, Wong C, Vallon V, Tang B, Sun J, Yang S, Dong H (2014) Molecular mechanisms of calcium-sensing receptor-mediated calcium signaling in the modulation of epithelial ion transport and bicarbonate secretion. J Biol Chem 289:34642–34653

Xu J, Henriksnas J, Barone S, Witte D, Shull GE, Forte JG, Holm L, Soleimani M (2005) SLC26A9 is expressed in gastric surface epithelial cells, mediates Cl-/HCO3- exchange, and is inhibited by NH4+. Am J Physiol Cell Physiol 289:C493–C505

Yamauchi K, Rai T, Kobayashi K, Sohara E, Suzuki T, Itoh T, Suda S, Hayama A, Sasaki S, Uchida S (2004) Disease-causing mutant WNK4 increases paracellular chloride permeability and phosphorylates claudins. Proc Natl Acad Sci U S A 101:4690–4694

Zielenski J (2000) Genotype and phenotype in cystic fibrosis. Respiration 67:117–133

Chapter 5
Fundamentals of Ion Transport Across Human Sweat Gland in Health and Disease

M. M. Reddy

Abstract The primary function of the sweat gland is to regulate body temperature through evaporative cooling during heat exposure or physical activity. Understanding the fundamentals of electrolyte transport by the sweat glands is essential due to their pivotal role in thermoregulation. Furthermore, knowledge gained in the process will significantly enhance our understanding of the two major epithelial functions that are common to most exocrine glands, namely "absorption" and "secretion" of electrolyte fluid that play a critical role in numerous exocrine functions. In addition, epithelial cells of the sweat gland employ transport elements such as Na^+/K^+ pump, basolateral K^+ channels, amiloride-sensitive epithelial Na^+ channels (ENaC), the cystic fibrosis transmembrane conductance regulator (CFTR) anion channel, and the $Na^+/K^+/2Cl^-$ cotransporter that find common expression among relatively inaccessible internal epithelial organs such as the kidneys, airways, pancreas, intestine, and salivary glands. Therefore, the sweat glands provide an important model system for investigating the electrophysiological principles involved in epithelial ion-transport processes in both secretion and absorption. This chapter is primarily focused on the fundamentals of ion-transport mechanisms involved in both secretion and absorption of electrolyte fluid by two morphologically distinct regions of the sweat gland, namely the secretory coil and the reabsorptive duct, in health and disease.

Keywords Sweat gland · Ion transport · CFTR · ENaC · ATP · cAMP · Glutamate · CF · PHA-1

M. M. Reddy (✉)
Department of Pediatrics, UCSD School of Medicine, University of California, San Diego, La Jolla, CA, USA
e-mail: rmadireddi@ucsd.edu

K. L. Hamilton, D. C. Devor (eds.), *Ion Transport Across Epithelial Tissues and Disease*, Physiology in Health and Disease,
https://doi.org/10.1007/978-3-030-55310-4_5

5.1 Introduction

The eccrine sweat gland is an important exocrine gland of the skin with about 1.6 to 4.0 million glands distributed nearly all over the body (Sato et al. 1989). The primary function of the sweat gland is thermoregulation during heat exposure or physical activity through evaporative cooling following hypotonic sweat secretion on to the skin surface. Hyperthermia (heat stroke), failure to regulate body temperature and loss of body electrolytes in secreted sweat during heat waves is one of the major causes of death around the world. It is important to understand the fundamentals of electrolyte transport by the sweat gland epithelial cells based on the following rationale. Sweat glands play a pivotal role in thermoregulation. They are involved in two major epithelial functions that are common to most exocrine glands, namely "absorption" and "secretion" of electrolyte fluid. Depending on the exocrine organ, secretion and absorption of electrolyte fluid serve distinct physiological functions such as hydration of outward facing epithelial surfaces, conservation of body volume, and providing optimal aqueous phase for carrying out important biological reactions. Furthermore, unlike in the case of many internal exocrine glands, secretory and absorptive functions of the sweat gland are performed by two morphologically distinct regions, the "secretory coil" and the "reabsorptive duct" so that these functions can be studied in relative isolation. In addition, the epithelial cells of the sweat gland share similar transport elements and properties of regulation with many internal epithelial organs such as the kidneys, airways, pancreas, intestine, and salivary glands. Most of these internal organs are morphologically complex and are relatively inaccessible for electrophysiological investigation. In contrast, freshly isolated human sweat glands are relatively easy to get access, and therefore, provide an important model system for studying the electrophysiological mechanisms of epithelial ion-transport processes in heath and disease (Quinton 1999a; Reddy and Stutts 2013). Herein, we will describe the fundamentals of ion-transport mechanisms involved in both secretion and absorption of electrolyte fluid by epithelial cells of the sweat glands.

5.2 The Secretory Coil

The secretory coil is comprised of three distinct cell types, the clear cells, the dark cells, and the myoepithelial (ME) cells (Quinton and Reddy 1990; Reddy and Bell 1996; Reddy et al. 1992, 1997; Sato 1977b) (Fig. 5.1). The clear cells are also called β-S cells (β-adrenergic-sensitive cells) and are generally believed to be responsible for fluid secretion due to: (a) an abundance of mitochondria in proximity to the basolateral membrane to support Na^+/K^+ pump activity that play a critical role in maintaining transepithelial salt secretion; (b) intercellular canaliculi that expand the surface area needed for ion-transport activity; (c) β-adrenergic receptors in the basolateral membrane; and (d) CFTR channels responsible for Cl^- secretion

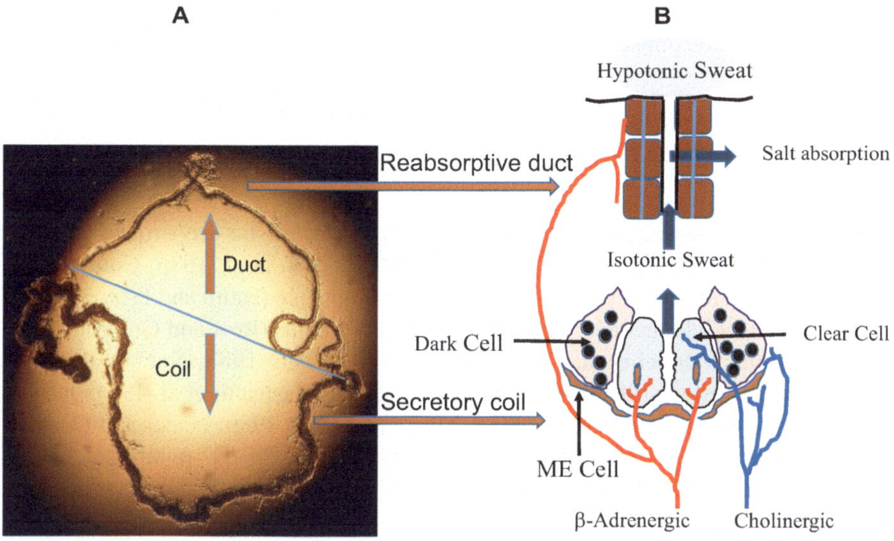

Fig. 5.1 (**a**) An intact freshly dissected human sweat gland: The gland is uncoiled after collagenase treatment. Sweat glands have two morphologically distinct regions. A secretory coil which is responsible for secreting isotonic sweat and a reabsorptive sweat duct, which is responsible absorbing salt from the primary sweat so that only hypotonic sweat reaches the skin surface for evaporative cooling during heat exposure and physical activity (Quinton 1999b). (**b**) Lower panel: Graphic representation of cell types and innervation of sweat gland secretory coil and reabsorptive sweat duct. Notice that there are three distinct cell types in the secretory coil: (1) the clear or β-S cells that respond to both cholinergic and β-adrenergic stimulation; (2) the dark or β-I cell that only respond to cholinergic stimulation and; and (3) the contractile myoepithelial cells (ME-cells) that only respond to cholinergic stimulation. All three cell types exhibit distinct electrophysiological signatures that clearly distinguish them from each other. (**b**) Upper panel: The reabsorptive sweat duct is a syncytium of two cell layers connected with gap junctions (between the cells) and tight junctions (facing the luminal surface). The reabsorptive sweat duct cells primarily respond to β-adrenergic stimulation (Reddy and Bell 1996; Sato and Sato 2000, 1981, 1984; Uno and Montagna 1975)

(Reddy and Bell 1996; Reddy et al. 1992, 1997). The second secretory cell types are called the dark cells, which are also known as the β-I cell (β-adrenergic-insensitive cells). These cells are generally considered to be responsible for macromolecular secretion as they are rich in periodic acid Schiff positive cytoplasmic granules, insensitive to β-adrenergic stimulation due to the absence of CFTR anion channels in the apical plasma membrane and they have cholinergic receptors responsible for secreting a significant macromolecular content (Sato 1977b; Sato et al. 1989). The third cell type is the myoepithelial cell (ME cell). These are spindle-shaped cells that surround the secretory coil. These cells are primarily responsible for squeezing the primary sweat out of the lumen and on to the skin surface by their contractile activity (Sato 1977a, b, 1980; Sato et al. 1979). The contractile activity of ME-cells is initiated by cholinergic stimulation. These three cell types can be identified by their distinct electrophysiological signatures with characteristic cell potentials and pharmacological responses (Reddy and Bell 1996; Reddy et al. 1992, 1997).

5.2.1 Properties of Clear Cells

The sweat gland secretory cells represent many characteristic properties of a fluid secreting epithelial cells such as the airways (Bedrossian et al. 1976; Cotton et al. 1988; Lethem et al. 1993; Widdicombe 1994), the intestine (Halm and Frizzell 1990; Hogan et al. 1997; Hogenauer et al. 2000), salivary glands (Best and Quinton 2005; Ma et al. 1999; Mandel et al. 1967; Martinez and Cassity 1983; Noel et al. 2008; Sorscher and Breslow 1982), and submucosal glands (Basbaum et al. 1981; Jeffrey and Brain 1988; Lee and Foskett 2014; Marin and Culp 1986; Nadel and German 1980; Shimura et al. 1986). The basolateral electrogenic Na^+/K^+ pump that pumps 3 Na^+ ions out of the cell in exchange for two K^+ ions simultaneously accomplishes three important functions. First, it helps establish the Na^+ electrochemical gradient across the cell membranes that is essential for transepithelial Na^+ transport. Second, it provides the chemical driving force for Na^+-dependent co- and counter-ion transport processes such as the $Na^+/K^+/2Cl^-$ cotransporter, the Na^+/HCO_3^- cotransporter, and the Na^+/H^+ exchanger. These carrier-mediated transporters play a critical role in transepithelial Cl^- and HCO_3^- transport as well as in controlling cytosolic pH. Third, the Na^+/K^+ pump helps maintain the transmembrane electrochemical potential gradient for K^+, which is essential for establishing resting membrane potential across the predominantly K^+-selective basolateral membrane and also provides transepithelial driving force for the movement of diffusible ions such as Cl^- and HCO_3^- from serosa to mucosa during secretion. In addition to the Na^+/K^+ pump, the $Na^+/K^+/2Cl^-$ cotransporter and at least two types of K^+ channels reside in the basolateral membrane that act in parallel with the Na^+/K^+ pump activity. In contrast, the apical membrane is predominantly anion selective (mainly Cl^-). They include, cAMP-activated CFTR anion channels and Ca^{2+}-activated Cl^- channels (Fig. 5.2).

5.2.1.1 Mechanism of Cholinergic Sweat Secretion

One of the crucial steps in the secretory process is the accumulation of intracellular Cl^- above the electrochemical equilibrium potential across the apical membrane of the clear cells by the $Na^+/K^+/2Cl^-$ cotransporter activity in the basolateral membrane. The critical role of $Na^+/K^+/2Cl^-$ activity in the overall secretory process was shown by the fact that sweat secretion by the sweat glands is inhibited by the $Na^+/K^+/2Cl^-$ inhibitor bumetanide (Quinton 1981, 1987, 1999a; Saga et al. 1988; Saga and Sato 1989; Sato 1977b). The basolateral membranes of clear cells were shown to abundantly express $Na^+/K^+/2Cl^-$ transporters (Bovell and Quinton 2002; Nejsum et al. 2005; Quinton 1999a). This action of $Na^+/K^+/2Cl^-$ facilitates passive diffusion

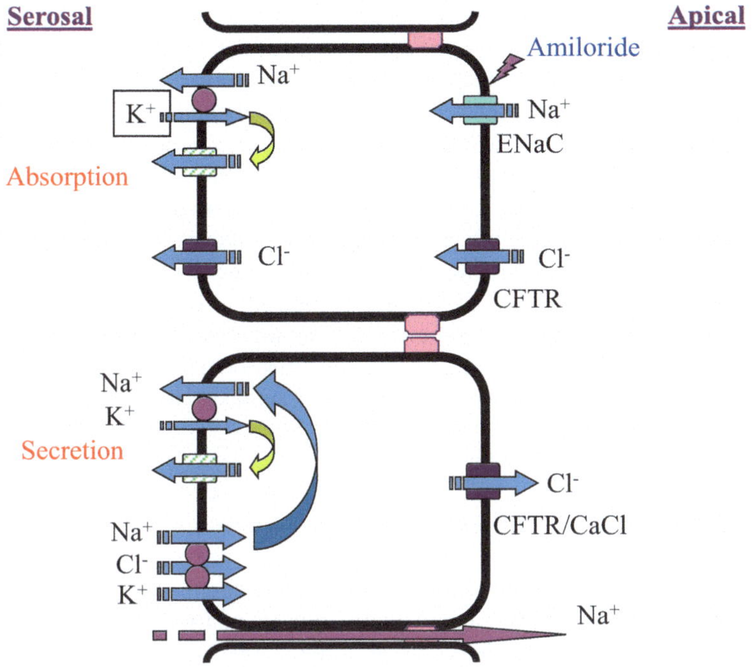

Fig. 5.2 Basic mechanisms of isotonic sweat secretion by the secretory coil and reabsorption of salt by the sweat duct. The Na^+/K^+ pump in the basolateral membrane is responsible for establishing electrochemical gradients for an inwardly directed Na^+ gradient and an outwardly directed K^+ gradient across the basolateral membrane of both the secretory as well as the reabsorptive duct cells. The $N^+/K^+/2Cl^-$ cotransporter in the basolateral membrane of secretory cells exploit the chemical gradients established by the pump and helps build intracellular Cl^- concentrations above the Nernstian equilibrium potential across the apical membrane. Activation of either apical CFTR or the Ca^{2+}-activated Cl^- channels by β-adrenergic and cholinergic agonists, respectively, results in the secretion of Cl^- into the lumen causing hyperpolarization of the lumen relative to the serosal side. Luminal negative potential provides the driving force for positively charged Na^+ ions to diffuse into the lumen across the tight junctions. Thus, accumulation of NaCl in the lumen provides the osmatic gradient for water to enter the lumen that forms the isotonic primary sweat (*lower panel*). As the isotonic sweat enters the ductal lumen, NaCl is reabsorbed by passive diffusion into the duct cells via CFTR and ENaC channels following favorable electrochemical gradients for both ions. Once entering the cells, Cl^- diffuses out of the cell via CFTR Cl^- channels in the basolateral membrane. The Na^+ is actively transported out of the cells by the pump. Excess K^+ accumulated during pump activity exits the cell through basolateral membrane K^+ channels (*upper panel*) (Reddy and Bell 1996; Reddy et al. 1992)

of Cl^- down the electrochemical gradient across the apical membrane following cholinergic stimulation (Reddy and Bell 1996). Cholinergic stimulation triggers a series of events in the ion-transport properties of both the apical and the basolateral membranes culminating in the secretion of isotonic sweat into the lumen. These events include: (1) increase in intracellular Ca^{2+} due to mobilization of both extracellular as well as intracellular Ca^{2+} stores, (2) activation of Ca^{2+}-dependent K^+ channels in the basolateral membrane that cause cell membrane hyperpolarization, which increases driving force for Cl^- exit across the apical membrane, (3) activation of Ca^{2+}-dependent Cl^- channels in the apical membrane that allow passive diffusion of Cl^- down the electrochemical gradient, (4) hyperpolarization of the luminal potential as a consequence of the secretion of negatively charged Cl^- ions, (5) passive diffusion of Na^+ ions through the para-cellular pathway down the electrochemical gradient created by the luminal negative transepithelial potential that leads to a transepithelial osmatic gradient, and finally (6) secretion of water following the osmotic gradient created by the aforementioned sequence of events. It is also important to point out that cholinergic stimulation does not activate CFTR anion channels. Furthermore, there are two pharmacologically distinct Ca^{2+}-dependent K^+ channels in the basolateral membrane, which include a Ba^{2+}-sensitive K^+ channel that is responsible for maintaining the resting membrane potential and a Ba^{2+}-insensitive, K^+ channel activated by cholinergic agonists (Reddy and Bell 1996; Reddy et al. 1992, 1997).

5.2.1.2 Mechanism of β-Adrenergic Sweat Secretion

The electrophysiological mechanisms of β-adrenergic agonist-induced sweat secretion are distinctly different from that of cholinergic agonist-induced sweat secretion. Unlike in the case of cholinergic secretion, β-adrenergic stimulation involves only the activation of cAMP-dependent CFTR anion channels but not the Ca^{2+}-dependent Cl^- channels. Furthermore, β-adrenergic agonist stimulation is not accompanied by the activation of basolateral K^+ channels (Reddy and Bell 1996). As a consequence, the electrical driving force for Cl^- secretion is significantly smaller compared to the driving force for Cl^- secretion during cholinergic stimulation. This difference in the magnitude of electrochemical driving force for Cl^- secretion between cholinergic and β-adrenergic agonist stimulations reflects in relatively smaller volumes of sweat secretion during β-adrenergic stimulation as compared to that induced by cholinergic agonist stimulation (Bijman and Quinton 1984a, b, c; Dobson and Sato 1972; Quinton 1979, 1984; Reddy and Bell 1996; Sato 1977b).

5.2.2 Properties of Dark Cells

Also known as β-adrenergic-insensitive cells (β-I cell) (Reddy and Bell 1996; Reddy et al. 1992; Reddy et al. 1992), the dark cells do not express CFTR as indicated by the lack of an electrophysiological response to stimulation by β-adrenergic agonists such as isoproterenol and cAMP (Reddy et al. 1992). These cells acutely respond to cholinergic stimulation as reflected in changes in intracellular ion concentrations (Saga and Sato 1988, 1989) and secretion of periodic acid Schiff positive material (Yanagawa et al. 1986). These cells respond to cholinergic stimulation that is dependent on Ca^{2+}. Cholinergic receptor activation is characterized by cell membrane hyperpolarization due to activation of Ca^{2+}-dependent K^+ channels in the

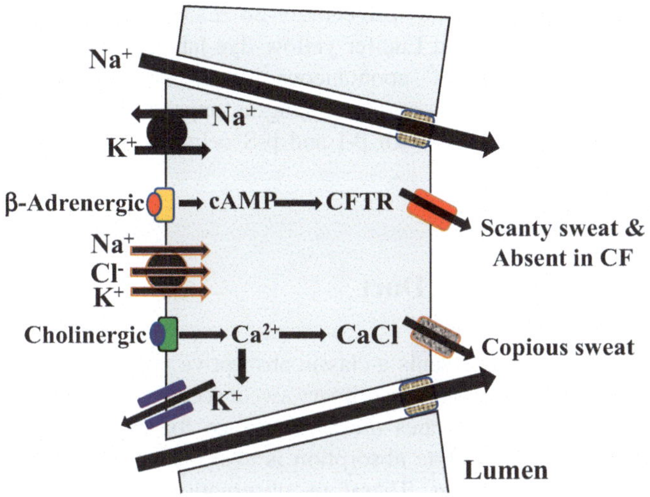

Fig. 5.3 Mechanisms of cholinergic and β-adrenergic sweat secretion by the clear (β-S) cells: Basic transport elements including the N^+/K^+ pump and the $Na^+/K^+/2Cl^-$ cotransporters are common to both adrenergic and cholinergic mechanisms of sweat secretion. However, cholinergic stimulation simultaneously activates Ca^{2+}-dependent Cl^- channels in the apical membrane and K^+ channels in the basolateral membrane. In contrast, β-adrenergic stimulation activates only CFTR Cl^- channels without causing simultaneous activation of basolateral K^+ conductance. Therefore, cholinergic stimulation results in a much greater electrochemical driving force for Cl^- secretion due to hyperpolarization of cell potential following activation of basolateral K^+ conductance. In contrast, β-adrenergic stimulation results in much smaller electrochemical driving force for Cl^- secretion due to lack of simultaneous activation of basolateral K^+ conductance. Hence, cholinergic stimulation results in a significantly larger volume of sweat secretion compared to scanty and often transient β-adrenergic sweat secretion. Furthermore, the absence of CFTR Cl^- channels in the CF β-cells results in the complete absence of β-adrenergic sweat secretion (Reddy and Bell 1996; Reddy et al. 1992)

basolateral membrane (Fig. 5.3). Two main electrical signatures distinguish β-I Cells from β-S cells of the secretory coil. The intracellular potentials of β-I cells are significantly less hyperpolarized (less negative) as compared to those of β-S cells (Reddy and Bell 1996; Reddy et al. 1992) and the cholinergic agonist-stimulated Ca^{2+}-dependent K^+ channels can be acutely blocked by Ba^{2+} only in β-I cells but not in β-S cells (Reddy and Bell 1996).

5.2.3 Properties of ME Cells

The ME cells are spindle-shaped contractile cells surrounding the secretory coil (Sato 1977a, b, 1980). These cells can be further distinguished from the secretory cells by their characteristic electrical signatures that include: (1) larger cell membrane potentials (-77 ± 4.0 mV) as compared to both clear cells (β-S,

−57 ± 4.0 mV) and dark cells (β-I cells, −28 ± 3.0 mV) as shown by both intracellular microelectrode and Lucifer yellow dye-labeling studies (Reddy et al. 1992), (2) the presence of spontaneous electrical activity (spontaneous depolarizing potentials), and (3) a depolarizing as opposed to the hyperpolarizing response of the secretory cells (both β-I and β-S cells) to cholinergic stimulation (Reddy et al. 1992).

5.3 Reabsorptive Sweat Duct

Reabsorptive sweat duct represents a classic absorptive epithelium involved in the reabsorption of salt from the isotonic primary sweat secreted by the secretory coil so that only hypotonic sweat reaches the skin surface for evaporative cooling. As previously mentioned, electrolyte absorption is also one of the basic functions of several internal exocrine organs. Therefore, observations made on this tissue can shed light on the electrolyte absorptive process in otherwise inaccessible internal exocrine glands such as airways of the lung, the intestine, and the kidneys (Quinton 1990, 1994, 1999a; Reddy and Stutts 2013). The ion channels such as ENaC and CFTR channels are widely distributed among major epithelial organs that are involved in numerous diseases including CF, pseudohypoaldosteronism (PHA), Liddle's syndrome, and hypertension. Observations first reported on this tissue such as Cl^- impermeability (Quinton 1983) and defective β-adrenergic secretions in CF (Quinton 2007; Sato and Sato 1984), CFTR regulation by ATP and cellular energy charge (Quinton and Reddy 1992) served as early pointers to similar observations in other exocrine glands affected by CF including airways, intestine, and pancreas (Anderson et al. 1991a, 1992; Anderson and Welsh 1992; Gray et al. 1989; Quinton 2007). Sweat duct is morphologically and functionally among the simplest models of native human tissue to study and interpret. The sweat duct is the most available source of CF-affected intact epithelia inherently expressing high levels of CFTR (Quinton 2007; Reddy et al. 1999, 1996; Reddy and Quinton 1989b, 1992a, 2003b). Therefore, understanding the physiology of electrolyte transport in the sweat duct leads to a better understanding of the mechanism of electrolyte absorption by epithelial cells not only in sweat duct but also in epithelial cells of other exocrine organs in general.

5.3.1 Morphology

The sweat duct is comprised of a syncytium of double layers of polarized epithelial cells with distinct apical and basolateral membranes joined through gap junctions (between the cell layers) and tight junctions separating the lumen from the serosal surface (Bijman and Quinton 1987; Reddy and Quinton 1987a, b; Sato 1977b). It is a tubular epithelium of ~10 μM in diameter and ~2000 μM in length. The duct has coiled portion intertwined with the secretory coil and a straight portion that reaches the skin surface (Sato 1977b).

5.3.2 Electrical Properties

In comparison to most epithelial cells (Frizzell et al. 1979; Fromter and Gebler 1977; Reuss and Finn 1975; Schettino et al. 1985; van Os and Slegers 1975), sweat duct cells have low membrane potentials (a basolateral membrane potential of about −34 mV) (Reddy and Quinton 1987b) even though such low intracellular potentials are not without precedent (Schettino et al. 1985; Willumsen et al. 1989a, b). The epithelial cells with significantly larger negative cell membrane potentials as compared to sweat duct cells are generally characterized by the presence of coupled carriers (e.g. Na^+/Cl^-, Na^+/H^+, $N^+/K^+/2Cl^-$ cotransporters, etc.,) such that ENaC-mediated conductance of these cells is relatively low and Na^+ transport is electrically silent (Frizzell et al. 1979; Reuss and Finn 1975; Shindo and Spring 1981; van Os and Slegers 1975). Furthermore, even in epithelia with significant ENaC conductance such as frog skin (Harvey and Kernan 1984) and urinary bladder (Fromter and Gebler 1977), they retain larger cell potentials compared to the sweat duct cells because they retain significantly larger transepithelial electrical resistance. Sweat duct is somewhat unique in that it combines an electro-diffusive Na^+ permeability mediated by ENaC in the apical membrane and a very large transepithelial electrical conductance that serves as an electrical shunt to reduce cell potentials. Large transepithelial electrical conductance of sweat duct (~100 mS) is almost entirely due to CFTR-mediated Cl^- conductance located in both the apical and basolateral membranes (Bijman and Fromter 1986; Quinton 1986; Reddy and Quinton 1987a, 1989a, b, 1992a). As a consequence of differential distribution of ENaC, CFTR (in the apical membrane) and CFTR, K^+ channels, and Na^+/K^+ pump (in the basolateral membrane), the apical membrane potential (V_a) is relatively less negative ($V_a = -24$ mV) as compared to the basolateral membrane potential (V_b) ($V_b = -34$ mV). Hence, there is a −10 mV lumen negative transepithelial potential (V_t) that provides a net electrical driving force for salt absorption in the human sweat duct (Quinton and Reddy 1990, 1991; Reddy and Quinton 1989a).

5.3.3 Intracellular Ion Activities

Using double-barreled intracellular microelectrodes (ISME), intracellular K^+ and Cl^- activities along with V_a, V_b, and V_t potentials were measured in isolated microperfused sweat ducts to determine the electrochemical driving forces for transepithelial NaCl absorption (Reddy and Quinton 1991, 1994a).

5.3.3.1 Intracellular Cl^- Activity

Based on the data shown in Table 5.1, it is evident that there is a significant driving force for passive diffusion of Cl^- across both apical and basolateral membranes through CFTR anion channels. The intracellular Cl^- activity favors downhill influx of Cl^- across the apical membrane with an electromotive force (EMF) of 9 mV (the difference between the Cl^- equilibrium potential (ECl^-) and the V_a). Similarly, Cl^- can passively exit the basolateral membrane with a downhill EMF of ~3 mV (Reddy and Quinton 1994a). These data also indicated that both apical and basolateral membrane potentials are closer to the Cl^- equilibrium potential than to that of K^+ ($EK^+ = 72$ mV). This observation is consistent with an early report that human sweat duct cell membranes are predominantly Cl^- selective (Reddy and Quinton 1994a), and therefore, the low cell membrane potentials reported previously are not due to membrane damage during microelectrode impalement but rather they reflect the ion-selectivity properties of the absorptive epithelium (Reddy and Quinton 1987b).

5.3.3.2 Intracellular K^+ Activity

Intracellular K^+ activity with a Nernstian equilibrium potential of ~−72 mV is significantly above the electrochemical equilibrium distribution across both the apical and basolateral membranes (Table 5.1) with a large driving force favoring the downhill diffusion across both cell membranes. Only the basolateral membrane has a Ba^{2+}-sensitive K^+ conductance that seems to change in parallel with Na^+/K^+ pump activity in that membrane (Reddy and Quinton 1991). Contribution of basolateral membrane K^+ conductance is less than 10% of the total transepithelial electrical conductance of the sweat duct. Early studies indicated the basolateral K^+ conductance is acutely activated by increases in the activities of intracellular $[K^+]_i$, intracellular $[Ca^{2+}]_i$, Na^+/K^+ pump in the basolateral membrane and ENaC in the apical membrane (Davis and Finn 1982b; Reddy and Quinton 1991). In fact, a close cross-talk between the basolateral K^+ channels, Na^+/K^+ pump, and the apical ENaC channels has been observed in other epithelial cells (Davis and Finn 1982b). It is significant to note that a coordination between the activities of K^+ and Na^+ channels relative to the Na^+/K^+ pump activity seems critical for the smooth flow of transcellular ion transport without disrupting the cell volume (Davis and Finn 1982b; Reddy and Quinton 1991, 2006).

5.3.3.3 Intracellular Na^+ Activity

There are no direct measurements of intracellular Na^+ activity in the sweat duct. However, based on: (1) the assumption that the activities $[Na^+]_i + [K^+]_i$ in the intracellular and extracellular compartments remain constant at ~150 mM, and (2) the previously measured $[K^+]_i$ activity of ~80 mM (Reddy and Quinton 1991,

Table 5.1 Spontaneous intracellular chloride (Cl^-), potassium (K+) activities and the corresponding transcellular electrochemical driving forces in the human re-absorptive sweat duct when lumen and bath were perfused with 150 mM NaCl containing Ringer's solution

ISME	V_t (mV)	V_a (mV)	V_b (mV)	$[Ion]_I$(mM)	E[Ion](mM)	ΔE_{Va}(mV)	ΔE_{Vb}(mV)
Cl^- ($n = 33$)	-12 ± 1	-21 ± 2	-33 ± 1.5	53 ± 4	−30	9 (influx)	3.0 (efflux)
K^+($n = 39$)	-9 ± 0.8	-19 ± 1	-28 ± 0.9	80 ± 4	−72	53 (efflux)	45 (efflux)

Notice that there is a significant electrochemical driving force for passive Cl^- absorption across both the apical and the basolateral membranes. Furthermore, intracellular K+ is significantly above it's electrochemical equilibrium potential so that an increase in basolateral K+ conductance during absorptive process provides enhanced driving force for passive Cl^- as well as Na+ diffusion across the apical membrane

n = number of intracellular measurements from as many individual ducts from a minimum of 10 human subjects. ΔE_{Va}(mV) = Electrochemical driving force across the apical membrane, ΔE_{Vb}(mV) = Electrochemical driving force across the basolateral membrane membrane, which is calculated as the difference between respective ion Nernstian equilibrium potential (E[ion]) and the respective membrane potentials (e.g., ΔEva = E[ion] − Va). Influx and efflux represent the direction of electrochemical driving force favoring into and out of cell, respectively. The ion-measurements were made while perfusing the bath and lumen with 150 mM NaCl and 5 mM K^+ containing Ringer's solution

1994a), the calculated $[Na^+]_i$ would be ~70 mM. Therefore, the estimated value of Na^+ electromotive force (ENa) would be ~ +20 mV favoring passive diffusion of Na^+ into the cells via ENaC across the apical membrane (Reddy and Quinton 1989a, b, 1991, 1994a).

5.3.4 Mechanism of Salt Absorption

The primary function of sweat duct is to reabsorb salt from the isotonic sweat secreted by the coil. A major portion of the primary sweat is composed of among others (such as macromolecules) in mM: Na^+(145), K^+(5), Cl^- (125), and HCO_3^- (20). Based on the transepithelial electrical potential profile, there are two main barriers for these ion to cross the apical and basolateral membranes. There is a passive electrochemical gradient for both Na^+ and Cl^- influx into the cell across the apical membrane down the favorable electrochemical gradient. However, only Cl^- can move passively across the basolateral membrane as there is significant electrochemical driving force and an anion channel (CFTR) to support such diffusion. Na^+ is actively pumped out of the cell (Reddy and Quinton 1987b, 1989a, b, 1992a, 1994a; Reddy and Stutts 2013) (Fig. 5.2).

5.3.4.1 Regulation of CFTR

CFTR plays a crucial role in both absorption and secretion of salt by numerous epithelial organs (Quinton 1999a). CFTR is also unique among ion channels with respect to its complex molecular structure (Riordan et al. 1989; Welsh et al. 1992), the multiple regulatory controls involving different kinases (86–94), protein phosphatases (Berger et al. 1993; Fischer et al. 1998; Gadsby and Nairn 1999; Luo et al. 1998), and its multifunctional behavior involving the control of other ion channels including ENaC (Schwiebert et al. 1999). Structurally, it is distinguished from most ion channels by the presence of a regulatory domain (R-domain) with multiple consensus phosphorylation sites and two nucleotide-binding domains (NBDs) (Riordan et al. 1989; Welsh et al. 1992). Even though early studies implicated several kinases and phosphatases in the regulation of the CFTR channel, we do not know whether these regulatory processes are tissue specific or coexist within most epithelial cells. Even if these control mechanisms do exist, the purpose of such complex regulation remains puzzling. Another significant factor in CFTR regulation involves both hydrolytic and nonhydrolytic ATP binding to NBDs (Anderson et al. 1991a; Ko and Pedersen 1995; Quinton and Reddy 1991; Randak et al. 1997; Reddy and Quinton 1996b; Sheppard and Welsh 1999; Winter et al. 1994). Nonhydrolytic ATP binding may be required for coupling transepithelial electrolyte transport to cellular energy charge (Quinton and Reddy 1991; Wine and Silverstein 1992). However, it is intriguing that unlike most ion channels that allow passive diffusion of ions (as opposed to energy-dependent active transport), CFTR utilizes ATP energy during the transport process. Recent studies have indicated that these multiple controls may play a role in determining the anion selectivity of the CFTR channel to HCO_3^- and Cl^- (Reddy and Quinton 2003a). Thus, defining the regulatory controls over CFTR is essential not only to better understand how different mutations selectively affect epithelial Cl^- and HCO_3^- conductance's in CF (Reddy and Quinton 2003a), but also to develop effective therapeutic strategies to treat diseases such as CF. Therefore, understanding the mechanisms of regulation and its interaction with other transport elements sheds light on several physiological functions of internal organs in health and disease. Early studies have indicated that CFTR in the sweat duct is subject to regulation by phosphorylation by different kinases, cytosolic glutamate, ATP, and intracellular ions including Na^+, K^+, and pH. These regulatory processes are discussed in greater detail below (Reddy et al. 1998, 2001; Reddy and Quinton 1992a, 1994b, 1996a, b, 1998, 2001a, 2003a, 2006, 2009).

5.3.4.1.1 Phosphorylation Activation

Several studies suggested that CFTR may be controlled by kinases involving PKC (Jia et al. 1997), PKA (Berger et al. 1993; Cheng et al. 1991; Dulhanty and Riordan 1994), PKG (French et al. 1995; Picciotto et al. 1992; Vaandrager et al. 1996, 1998), Ca^{2+} calmodulin-dependent protein kinases (Picciotto et al. 1992), and protein tyrosine kinases (Gadsby and Nairn 1999), as well as different protein phosphatases (Berger et al. 1993; Fischer et al. 1998; Gadsby and Nairn 1999; Luo et al. 1998). CFTR is known to have several phosphorylation sites that can be targeted by protein kinase A (PKA), protein kinase C (PKC), protein kinase G (PKG), G-protein-dependent kinase, Ca^{2+} calmodulin-dependent protein kinases, and protein tyrosine kinases (Bradbury 1999; Chappe et al. 2003; Hanrahan et al. 1996; Jia et al. 1997; Luo et al. 1998, 2000; Mathews et al. 1998; Seibert et al. 1999; Sheppard and Welsh 1999; Welsh et al. 1992; Winter et al. 1994). Even though the aforementioned kinases were shown to phosphorylate and activate CFTR in different model systems, only cAMP, cGMP, and G-protein-dependent kinases were shown to have a significant control over CFTR activity in the native human sweat duct. It was also shown that these kinases act independently of each other with specific characteristic properties that distinguish them (Reddy and Quinton 2009). For example, protein phosphatase 2A (PP2A) can dephosphorylate PKA-phosphorylated but not PKG-phosphorylated CFTR (Reddy and Quinton 2009). Furthermore, in contrast to cAMP dependence of PKA phosphorylation, G-protein-dependent kinase does not depend on intracellular cAMP for activating CFTR (Reddy and Quinton 2001a). In addition, the CFTR activity in human sweat duct is independent of changes in intracellular Ca^{2+} indicating that PKC and Ca^{2+} calmodulin-dependent kinases do not seem to have a physiological role in regulating CFTR in this native human tissue.

5.3.4.1.2 Phosphorylation-Independent Regulation

CFTR activity is also dependent on intracellular changes in ATP (Reddy and Quinton 1996b), glutamate, and its metabolite α-ketoglutarate (Reddy and Quinton 2003a). While ATP participates in phosphorylation of CFTR as a substrate, it is also involved in CFTR regulation through allosteric nonhydrolytic binding to CFTR thereby acting as a sensor of cellular energy charge (Quinton and Reddy 1992, 1994). Thus, it appears that by coupling the availability of cellular energy to CFTR-dependent transepithelial transport activity by feedback mechanism, the epithelial cells seem to be protecting themselves from excessive fluctuations in cell volume during bulk transcellular ion transport. Similarly, cytosolic glutamate and α-ketoglutarate that are associated with the Krebs cycle, and hence, ATP production also seems to be coupled to CFTR activity by an as yet unknown mechanism that is independent of ATP (Reddy and Quinton 2001b, 2003a).

In addition, cytosolic glutamate regulation of CFTR is unique with distinct regulatory characteristics that distinguish it from kinase- and ATP-dependent regulation. Glutamate acutely activates CFTR-mediated Cl^- conductance in the complete absence of ATP. Furthermore, glutamate activation of CFTR Cl^- channels is insensitive to the presence of general kinase blocker staurosporine and removal of Mg^{2+}, which is necessary for phosphorylation and ATP hydrolysis (Reddy and Quinton 2003a). These observations indicated that glutamate regulation of CFTR is independent of phosphorylation. In addition, kinase-dependent activation of CFTR Cl^- conductance was completely inhibited by cytosolic Na^+ due to activation of dephosphorylation by endogenous phosphatases (Reddy and Quinton 2006, 2009; Reddy and Stutts 2013). In contrast, the glutamate activation of CFTR is in fact potentiated by cytosolic Na^+ providing additional evidence that glutamate regulation of CFTR is distinctly different from either ATP- or kinase-dependent regulatory processes. Furthermore, early studies have shown that cytosolic glutamate plays a role in changing the CFTR Cl^-/HCO_3^- selectivity. While cytosolic glutamate alone can stimulate CFTR-mediated Cl^- conductance, the CFTR-mediated HCO_3^- conductance by glutamate involves ATP hydrolysis. These observations indicate that unlike most ion channels, the CFTR ion selectivity is not fixed but it is dynamic, which can change depending upon the cytosolic regulatory conditions (Reddy and Quinton 2003b; Reddy et al. 2018).

5.3.4.1.2.1 Regulation by Intracellular Ions

Intracellular ion composition changes as a function of transport activity involving CFTR, ENaC, Na^+/K^+ pump, Na^+/H^+ exchange, and the basolateral K^+ channels (Reddy and Quinton 1991, 1994a, 2006; Reddy et al. 2008). Available evidence indicates the activities of these transporters have a significant role in regulating the CFTR anion channel function, which is predominantly mediated by changes in the cytosolic ion composition.

5.3.4.2 Effect of $[Na^+]_i$ and $[K^+]_i$ on CFTR Activity

Following Na^+ entry into the cells, via ENaC across the apical membrane, the intracellular Na^+ activity ($[Na^+]_i$) increases that seems to trigger a series of events including increases in Na^+/K^+ pump activity, basolateral K^+ conductance (gK), and K^+ efflux down the electrochemical gradient across the basolateral membrane (Davis and Finn 1982b; DeLong and Civan 1978; Reddy and Quinton 1991). Such cross-talk between the transport activity at the apical membrane and the basolateral membrane has been widely reported among epithelia involved in transepithelial ion transport (Civan 1978; Davis and Finn 1982a, b; DeLong and Civan 1978). Coordination between the apical and basolateral membrane transport activity seems essential to maintain a stable cell volume by matching the entry and exit of salt into and out of the epithelial cells (Davis and Finn 1982a, b; Reddy et al. 1999; Reddy and Quinton 1991, 2003b, 2006). One of the strategies employed by the sweat duct epithelial cells appears to be aimed at controlling CFTR-mediated Cl^- influx into the cells across the apical membrane. It has been shown that loss of $[K^+]_i$ during transport activity results in the activation of protein phosphatase 2A (PP-2A) activity that is responsible for dephosphorylation and hence deactivation of CFTR (Reddy and Quinton 1996a). Thus, restricting Cl^- entry by deactivating CFTR results in the depolarization of the apical membrane and reduction of driving force for Na^+ influx and reduced Na^+ absorption. Parallel inhibition of Na^+ absorption in the absence of CFTR activity is clearly demonstrated in CF sweat ducts (Fig. 5.4).

5.3.4.3 Control of CFTR Activity by pH_i

Sweat duct shows the significant presence of Na^+/H^+ exchanger activity that is energized by the Na^+ chemical gradient (Granger et al. 2003; Li et al. 2014; Reddy et al. 2008). As previously mentioned, influx of Na^+ gradually reduces the Na^+ chemical gradient thereby lowering Na^+/H^+ (NHE) activity and pH_i (Li et al. 2014; Reddy et al. 1998, 2008). Studies on basolaterally α-toxin-permeabilized microperfused sweat ducts showed that the luminal pH (5.0–8.5) had little effect on the cAMP/ATP-activated CFTR Cl^- conductance (Reddy et al. 1998), indicating that channel activity is maintained over a broad range of luminal pH in which it functions physiologically. However, phosphorylation activation of CFTR Cl^- conductance was shown to be sensitive to changes in intracellular pH (Reddy et al. 1998). That is, in the presence of cAMP and ATP, CFTR could be phosphorylated at physiological pH (6.8) but not at low pH (approximately 5.5). On the other hand, basic pH prevented endogenous okadaic acid-sensitive phosphatase (PP-2A) from dephosphorylating CFTR. After phosphorylation of CFTR with cAMP and ATP, CFTR is normally deactivated within 1 min after removal of cAMP, even in the presence of 5 mM ATP. This deactivation is due to an increase in endogenous phosphatase activity relative to kinase activity, since it is reversed by the reapplication of cAMP. However, increasing cytoplasmic pH (7.5–8.5) significantly

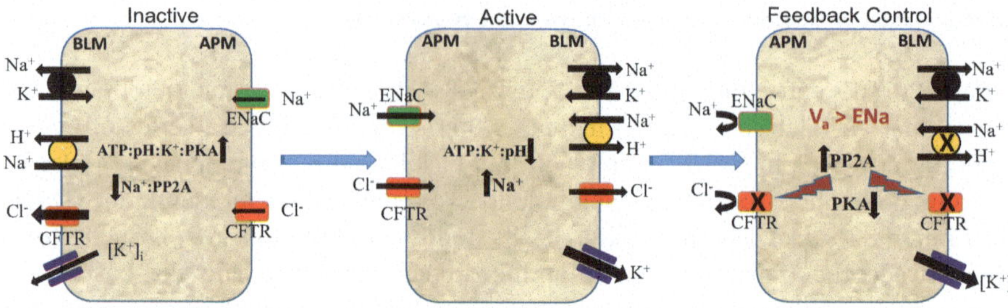

Fig. 5.4 Schematic representation of different ion-transport states during transepithelial salt absorption in the human sweat duct: Inactive state: represents the period of minimal sweat secretion. In this state, the ATP levels are high when little energy is spent on transport activity. As there is little Na^+ in the lumen to be absorbed, $[Na^+]_i$ and $[K^+]_i$ are at their lowest and highest values, respectively. Low $[Na^+]_i$ provides a favorable chemical energy gradient for the Na^+/H^+ exchanger in the basolateral membrane raising the $[pH]_i$. Elevated ATP, pH_i, and $[K^+]_i$ create favorable conditions for ATP/PKA activation of CFTR. In addition, higher pH and low $[Na^+]_i$ provide ideal conditions for ENaC activity. *Active state*: Primary sweat enters the ductal lumen. NaCl enters the cell through ENaC and CFTR channels, respectively. Elevated $[Na^+]_i$ during trans-cellular transport, levels of ATP, $[K^+]_i$, and $[pH]_i$ decrease. *Feed-back control*: Decrease in both $[K^+]_i$ and pH_i (as a consequence of elevated $[Na^+]_i$ due to transport activity) results in gradual deactivation of CFTR by reciprocal activity of PKA and PP2-A (Reddy and Quinton 2003b, 2009). Deactivation of CFTR and Acidic pH cause simultaneous deactivation of ENaC (Reddy et al. 1998; Reddy and Quinton 1996a). Thus, limiting salt influx, allowing the cells to restore cellular energy charge and protecting the cells from abrupt changes in the cell volume (Reddy and Quinton 2009; Reddy et al. 2008)

delayed the deactivation of CFTR Cl^- conductance in a dose-dependent manner, indicating inhibition of dephosphorylation. These results indicated that CFTR Cl^- conductance is regulated via shifts in cytoplasmic pH that mediate reciprocal control of endogenous kinase and phosphatase activities (Reddy et al. 1998; Reddy and Stutts 2013) (Fig. 5.4).

5.3.4.4 Regulation of ENaC

In spite of the central role played by ENaC in a number of life-threatening diseases including CF, it is surprising that we know so little of the regulation of ENaC in native human tissues. Point mutation studies have implicated intracellular domains of CFTR as possible regions that are significant for ENaC regulation (Kunzelmann et al. 1997). Early evidence suggested a possible role for cytoskeletal elements (Dinudom et al. 1995; Ismailov et al. 1997a, b), G-proteins (Dinudom et al. 1995), and changes in intracellular and or extracellular ions including Na^+ and Cl^- concentrations in regulating ENaC in ex vivo systems (Cook et al. 1998; Dinudom et al. 1998). Even though PKA phosphorylation was shown to activate ENaC in different cells (Awayda et al. 1996), recent studies on the sweat duct suggested that conditions sufficient for PKA phosphorylation cannot activate ENaC in this native tissue (Reddy et al. 1999). Some of the mechanisms of regulating ENaC activity in the sweat duct include: (1) CFTR activity (Reddy et al. 1999), (2) cytosolic pH (Reddy et al. 2008), and (3) cytosolic Cl^- activity (Reddy and Quinton 2003b).

5.3.4.4.1 CFTR Activity Regulates ENaC

Early studies indicated that CFTR activation causes parallel activation of ENaC in the native human sweat duct (Reddy et al. 1999; Reddy and Quinton 2005). However, it was not clear whether phosphorylated CFTR, phosphorylated ENaC, or only Cl^- channel function is required for ENaC activation. Subsequent investigations were focused on determining whether ENaC activation depends only on Cl^- channel function and not on phosphorylation of either CFTR or ENaC. As previously discussed, CFTR is activated by PKA in the presence of ATP, but cytosolic glutamate activation of CFTR is independent of ATP and phosphorylation (Reddy et al. 1999; Reddy and Quinton 2003b). It was shown that both phosphorylation-dependent (PKA) and phosphorylation-independent (glutamate) activation of CFTR Cl^- channel function support ENaC activation (Reddy et al. 1999; Reddy and Quinton 2005). It was further examined whether cytosolic application of 5 mM ATP alone, phosphorylation by cAMP-, cGMP-, and G-protein-dependent kinases (all in the presence of 100 μM ATP), or glutamate could support ENaC activation in the absence of CFTR Cl^- conductance. Those studies have revealed that none of those agonists activated ENaC by themselves when Cl^- current through CFTR was blocked by: (1) Cl^- removal, (2) DIDS inhibition, (3) lowering the ATP concentration to 100 μM (instead of 5 mM required to support CFTR channel function), or (4) CFTR mutation (homozygous F508del CF ducts). However, Cl^- gradients in the direction of absorption supported, while Cl^- gradients in the direction of secretion prevented ENaC activation. It was concluded that the interaction between CFTR and ENaC is dependent on activated Cl^- current through CFTR in the direction of absorption (Cl^- gradient from lumen to cell). However, such activation of ENaC is independent of phosphorylation and ATP. However, reversing Cl^- current through CFTR in the direction of secretion (Cl^- gradient from cell to lumen) prevents ENaC activation even in the presence of Cl^- current through CFTR (Reddy and Quinton 2003b, 2005). The findings that ENaC and CFTR activation are closely coupled, apparently independent of simultaneous phosphorylation by PKA, suggest possible molecular interactions between ENaC and CFTR (Kunzelmann 2001; Kunzelmann et al. 2001; Reddy and Stutts 2013).

5.3.4.4.2 Regulation of ENaC by Cytosolic Ions

ENaC regulation by intracellular ions has been widely reported in early studies using ex vivo model systems (Cook et al. 1998; Dinudom et al. 1998, 1995). However, much of the direct evidence of ENaC regulation by cytosolic ions in the native human sweat duct is derived from recent studies on α-toxin-permeabilized microperfused sweat duct (Reddy et al. 1999, 2008; Reddy and Quinton 2003b, 2005).

5.3.4.4.2.1 Control of ENaC by pH_i

As previously mentioned, the activities of CFTR and ENaC are acutely coordinated in the sweat duct. Previous studies indicated that luminal pH of sweat ducts varies over 3 pH units and that the cytoplasmic pH affects both CFTR and ENaC (Quinton and Reddy 1989; Reddy et al. 1998; Reddy and Quinton 2000). Therefore, using basolaterally α-toxin-permeabilized apical membrane preparations of sweat ducts as an experimental system, it was investigated whether changes in the cytosolic pH could mediate the cross-talk between CFTR and ENaC. The results showed that while luminal pH had no effect, cytosolic pH acutely affected ENaC activity. That is, acidic pH inhibited, while basic pH activated, ENaC. pH regulation of ENaC appears to be independent of CFTR or endogenous kinase activities because basic pH independently stimulated ENaC: (1) in normal ducts even when CFTR was deactivated, (2) in CF ducts that lack CFTR in the plasma membranes, and (3) after blocking endogenous kinase activity with staurosporine. Considering the evidence of Na^+//H^+ exchange activity as shown by the expression of mRNA and function of NHE in the basolateral membrane of the sweat duct, it seems likely that the increase in cytosolic $[Na^+]_i$ during salt absorption may lower cytosolic pH. Such a decrease in pH_i appears to cause a feedback inhibition of ENaC that is designed to prevent drastic changes in the cell volume following massive influx of salt (Reddy and Quinton 2000; Reddy and Stutts 2013).

5.3.4.4.2.2 Control of ENaC by $[Cl]_i$

Understanding the regulation of ENaC by $[Cl]_i$ is essential for the following reasons. CFTR and ENaC co-exist in the apical membranes of several important epithelial organs including the kidneys, colonic epithelium, and the airways (Boucherot et al. 2001; Reddy and Stutts 2013). Interaction between CFTR and ENaC activities has been widely reported (Boucherot et al. 2001; Donaldson et al. 2002; Kunzelmann et al. 2001; Reddy and Stutts 2013; Schwiebert et al. 1999). For example, it was shown that CFTR inhibited ENaC activity (Donaldson et al. 2002; Reddy and Stutts 2013). In contrast, parallel activation of ENaC following CFTR activation was observed in the sweat duct indicating that the interaction between CFTR and ENaC could be tissue specific (Reddy et al. 1999; Reddy and Quinton 2003b, 2005; Reddy and Stutts 2013). How to reconcile with these seemingly contradictory observations regarding the functional interactions between CFTR and ENaC is an important physiological question. Understanding the role of intracellular Cl^- in mediating the functional interaction between CFTR and ENaC seems quite relevant in solving this riddle. It is important to note that increase in $[Cl^-]_i$ as a function of CFTR activity in the apical membrane results in deactivation of ENaC activity in the sweat duct (Reddy and Quinton 2003b). It was shown that ENaC activity was significantly inhibited when $[Cl^-]_i$ was raised to 50 mM in basolaterally α-toxin-permeabilized ducts (Reddy and Quinton 2003b). Studies using intracellular Cl^--sensitive microelectrodes revealed that $[Cl^-]_i$ varies between 20 and 80 mM (Reddy and Quinton 1994a). These observations indicated that the cross-talk between CFTR and ENaC occurs at different levels involving secondary changes in intracellular ion activities (Reddy and Stutts 2013), i.e., activation of CFTR results in a gradual buildup of $[Cl^-]_i$ and $[Na^+]_i$ along with it. Uncontrolled influx of NaCl can result in a mismatch between the salt influx through passive diffusion across the apical membrane and active Na^+ pump across the basolateral membrane. Therefore, it appears that increased $[Cl^-]_i$ acts as a feedback inhibitor of ENaC activity in order to prevent further influx of salt thereby preventing drastic changes in the cell volume. These observations in the sweat duct, while being consistent with ex vivo studies showing a similar inhibition of ENaC by intracellular Cl^-, offer at least one plausible explanation of how an increase in $[Cl^-]_i$, following CFTR activation, could potentially cause simultaneous inhibition of ENaC as observed in some CF-affected epithelia (Boucherot et al. 2001; Konig et al. 2001; Kunzelmann et al. 2001; Reddy and Stutts 2013).

5.4 Abnormal Ion Transport by Sweat Gland in Disease Conditions

CFTR and ENaC play a central role in major diseases associated with electrolyte transport as previously mentioned (Quinton 1999a). These two significant ion channels play a critical role in maintaining several vital physiological functions including blood pressure, air way surface liquid volume and composition, and maintain electrolyte balance during heat exposure (Di Sant'Agnese et al. 1953a, b; Quinton 1999b). While CFTR is primarily responsible for abnormal secretion and absorption of electrolyte fluid by exocrine glands in CF, ENaC plays a significant role in several disease conditions including the pathology of PHA (Furgeson and Linas 2010; Reddy et al. 2005) and Liddle's syndrome (Botero-Velez et al. 1994; Tetti et al. 2018). Therefore, understanding the role of these principle epithelial ion channels in the secretory coil and in the absorptive sweat duct can provide important insights into the physiological basis of certain disease conditions.

5.4.1 Secretory Coil

5.4.1.1 Abnormal CFTR Function and Sweat Secretion in CF

Sweat glands played a significant role in determining the basic physiological defect in CF. It was shown for the first time that the β-adrenergic secretion was specifically affected in CF (Behm et al. 1987; Johnson et al. 1991; Sato and Sato 1984). Immediately following this discovery, a significant observation was made by two research teams, which showed that the culture cells derived from CF airways showed Cl^- channels (outward rectifying Cl^- channels or ORDIC channels) in both normal and CF tissues but those channels could not be activated by PKA in CF tissue. Taken together, with these observations on sweat secretion by (Sato and Sato 1984), it was suspected that the regulation of an anion channel may be the primary defect in CF (Schoumacher et al. 1987; Welsh 1986). The identification of the CF gene and the gene product finally resolved this crucial issue. The CF gene product was subsequently identified as a low-conductance, linear, ATP-sensitive, and PKA-activated Cl^- channel, which was shown to be expressed in all the CF-affected organs (Anderson et al. 1991b, 1991c; Riordan et al. 1989; Wine et al. 1991). It was subsequently revealed that the PKA regulation of ORDIC channel activity itself is dependent on the presence of CFTR, and therefore, lack of CFTR in CF-affected cells resulted in the apparent lack of PKA regulation in CF cells (Gabriel et al. 1993). Further studies on the mechanisms of β-adrenergic sweat secretion by the human sweat glands revealed that the β-adrenergic sweat secretion is exclusively depending on CFTR. Therefore, the absence of CFTR activity directly affected β-adrenergic sweat secretion (Fig. 5.3). However, the cholinergic secretion is unaffected in CF because it exclusively depends on the activation of Ca^{2+}-activated Cl^- channels that are not affected, in CF. These observations provide a cellular basis of the defective β-adrenergic secretion in CF (Reddy and Bell 1996; Reddy et al. 1992). Since β-adrenergic sweat is specifically affected in CF, it seems appropriate to use this basic defect as an endpoint biomarker in testing the therapeutic efficiency of potential CF drugs (Behm et al. 1987; Johnson et al. 1991; Shamsuddin et al. 2008). Therefore, as previously mentioned, relatively easy access of the sweat gland tissue lends itself as a desirable choice for early screening of CF drugs.

5.4.1.2 Which Secretory Cell Type(s) Is/Are Affected in CF?

Exocrine organs such as intestine, pancreas, and airways have multiple cell types such as goblet, serous, ciliated, and myoepithelial cells, which perform specific functions. Determining which cell type(s) normally express CFTR is essential to understand the potential impact of CF on specific organ functions. Studies on the secretory coil provided one of the earliest indications that the expression of CFTR Cl^- channels appears to be selective and restricted to specific cell types of epithelial organs (Reddy et al. 1992). Sweat gland secretory coil with just three epithelial cell types and the easy access of the tissue for experimentation permitted examination of this issue using electrophysiological techniques (Reddy and Quinton 1992b). Using intracellular microelectrodes, it was shown that the clear cells (β-S cells), which are the only cell type that normally expresses Camp-activated CFTR Cl^- channels, are clearly affected in CF as indicated by a lack of the β-adrenergic electrophysiological response (Reddy et al. 1992; Figs. 5.1 and 5.3). These observations have significant implications for choosing appropriate therapeutic choices. For example, the consequences of inserting CFTR by gene therapy into cells that normally do not express CFTR may lead to unintended consequences (Reddy et al. 1997; Strong et al. 1994).

5.4.2 *Sweat Duct*

5.4.2.1 Cl^- Impermeability in CF Sweat Duct

Elevated sweat salt concentration in CF has been the hall mark of CF diagnosis (Di Sant'Agnese et al. 1953a, b). Early measurements of transepithelial potentials in the reabsorptive sweat duct by Dr. Paul Quinton revealed that the transepithelial potential in CF ducts is an order of magnitude more negative as compared to the values from normal ducts. Furthermore, CF sweat ducts showed little or no Cl^- diffusion potential as compared to non-CF control ducts. That is, substituting 150 mM luminal Cl^- with the impermeant anion gluconate, in the presence of 150 mM NaCl Ringer's solution in the contra-luminal bath, induced about −100 mV transepithelial Cl^- diffusion potential in normal ducts, whereas similar manipulation resulted in little change in CF ducts (Quinton 1983). Furthermore, measurement of transepithelial electrical conductance revealed that the Cl^- conductance of CF ducts is an order of magnitude smaller than that of normal ducts (Quinton 1986).

Furthermore, while CFTR is expressed in the apical membrane of most epithelial cells (Reddy and Quinton 1992b), this anion channel is expressed in both the apical and the basolateral membranes of sweat duct (Fig. 5.2) based on the following evidence. The Cl^- conductance is activated by cAMP in both cell membranes. Cl^- conductance in both membranes is affected in CF. The apical membrane is depolarized toward Na^+ electromotive force (EMF) and the basolateral membrane is hyperpolarized toward K^+ EMF in the CF ducts indicating the absence of Cl^- conductance in respective membranes (Reddy and Quinton 1989a, b). The changes in the apical voltage divider ratios (ratio of apical to basolateral membrane resistances) in CF ducts as compared to that of normal ducts is consistent with the absence of Cl^- conductance in both cell membranes (Reddy and Quinton 1989b). While ~80% of the transepithelial conductance in normal sweat duct is due to Cl^-, this conductance is almost completely lost in CF sweat ducts (Quinton 1986). Even though four types of anion channels were found in the apical membrane of culture cells from whole sweat gland (Krouse et al. 1989), only one type of anion channel is found in both the apical and basolateral membranes of native human sweat duct that appears similar to CFTR (Lew and Krasne 1991). The anion selectivity sequence of the Cl^- conductance in both cell membranes is identical and similar to that of CFTR (Reddy and Quinton 1992b).

5.4.2.2 Cellular Basis of Abnormal Salt Transport in CF

The electrical potential profile of normal and CF duct reveals that the absence of CFTR in the apical membrane of the CF ducts significantly depolarized the apical membrane closer to the Na^+ EMF across that membrane. Such depolarization almost completely eliminated the driving force for Na^+ influx into the cell even though there

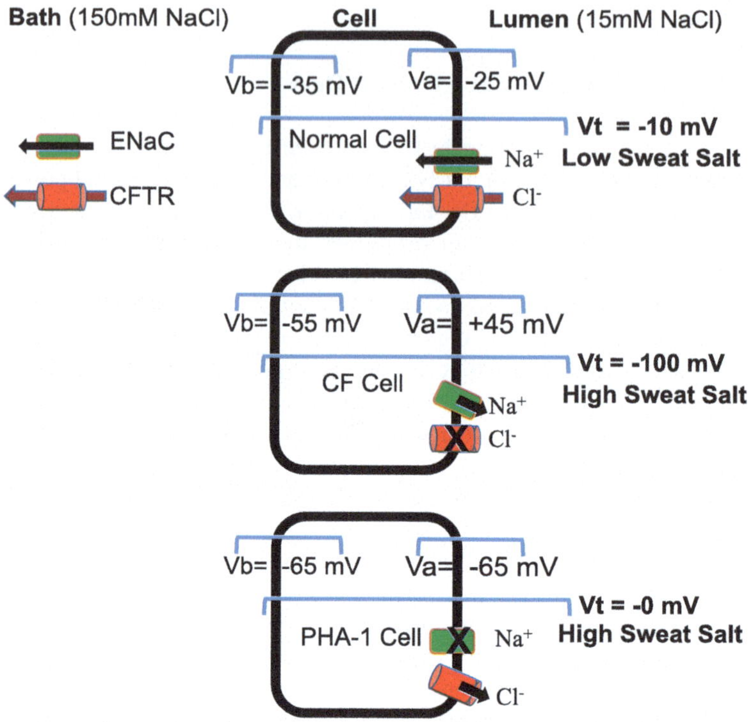

Fig. 5.5 Abnormal salt transport by sweat duct in health and disease: In normal cells (*upper panel*), there is a favorable transcellular driving force for NaCl absorption when both CFTR and ENaC are functioning normally. However, in CF cells (*middle panel*), the absence of CFTR in the apical membrane not only blocked Cl^- but also Na^+ absorption due to reversal of electrical driving force for Na^+ entry. Notice that in CF, the apical ($V_a = +45$ mV) and transepithelial ($V_t = -100$ mV) potentials turned significantly positive and negative, respectively, as compared to their respective controls from normal ducts (Reddy and Quinton 1989a). In PHA-1 cells (*Lower Panel*), the absence of ENaC not only blocked Na^+ but also Cl^- absorption due to abolition of driving force for Cl^- entry into the cells. Notice that the apical ($V_a \sim -65$ mV) is more hyperpolarized and the transepithelial potential ($V_t \sim 0$ mV) is abolished. It is important to note that blocking either CFTR (as in CF) or ENaC (as in PHA-1) results in elevated sweat salt concentration as compared to the normal subjects. One way to distinguish between the sweat glands from CF and PHA-1 patients is by comparing the dramatically different transepithelial potentials (CF $V_t \sim -100$ mV vs PHA-1 $V_t \sim 0$ mV) (Reddy et al. 2005). Note: Value of intracellular potentials from normal and CF subjects actually reflects direct measurements with microelectrodes (Reddy and Quinton 1989a). The intracellular potentials shown in the figure for PHA-1 ducts were not directly measured but they reflect values from normal ducts after blocking ENaC with amiloride (Reddy and Quinton 1987b; Reddy et al. 2005) to mimic PHA-1. The values of transepithelial potentials from all three groups are directly measured by the microperfusion technique

is a significant ENaC activity (Reddy and Quinton 1989a, b, 2003b, 2005) (Fig. 5.5). Therefore, the absence of CFTR not only effected Cl^- but also prevented Na^+ absorption regardless of the status of Na^+/K^+ pump and ENaC activities as indicated by elevated sweat salt concentrations in CF (Reddy and Stutts 2013). Furthermore, it is important to note that potential secondary changes in intracellular ion activities such as for Na^+ and Cl^- could have a secondary influence on other carrier-mediated ion exchangers (e.g., Cl^-/HCO_3^- exchanger) because of their dependence on the chemical driving forces provided by the transmembrane Na^+ and Cl^- distributions (Reddy and Stutts 2013). In fact, secondary changes in the carrier transport functions may have significant role in the acidification of exocrine secretions in the CF-affected organs such as the pancreas, airways, and the sweat gland (Abou Alaiwa et al. 2018; Park et al. 2010; Quinton and Reddy 1989; Sohma et al. 1996).

5.4.2.3 Defective CFTR HCO_3^- Conductance in CF Sweat Duct

As previously discussed, CFTR is an anion channel that conducts both Cl^- and HCO_3^-. It was shown for the first time in the human sweat duct that the selectivity of CFTR to Cl^- vs HCO_3^- is not static but shows dynamic change as a function of intracellular regulatory environment involving cAMP, cGMP, glutamate metabolites, and ATP (Reddy and Quinton 2003b). It was also revealed that certain CF mutations selectively affect Cl^-/HCO_3^- selectivity. For example, sweat ducts from R117H/F508del heterozygote CF patients showed significant HCO_3^- conductance in the face of a dramatically reduced Cl^- conductance (Choi et al. 2001; Reddy and Quinton 2003b). Early data from F508del heterozygote CF ducts carrying G542X and W1282X nonsense CF mutations that do not produce functional CFTR protein showed little HCO_3^- conductance even after correcting F508del CFTR with the corrector compound Vx-809 (Van Goor et al. 2009, 2011), which is indicated by a significant presence of CFTR-mediated Cl^- conductance after correction (Reddy et al. 2017). These observations in freshly isolated human sweat duct are consistent with early evidence showing that CF mutations also affected HCO_3^- permeability function of CFTR and that HCO_3^- secretion is poor in CF-affected organs (DiMagno et al. 1977; Johansen et al. 1968; Ko et al. 2002; Kopelman et al. 1988; Luckie et al. 2001; Wine 2001). In addition, the disease severity is most likely correlated with the aberrant HCO_3^- transport caused by CFTR mutations. For example, the R117H mutation, which causes a pancreatic sufficient and milder form of CF disease showed little Cl^- conductance but retained 73% of wild type HCO_3^- transport in heterologous systems (Choi et al. 2001). The E193K, D648V, H949Y, and R1070Q mutations, all associated with CF had no effect on Cl^- transport but reduced HCO_3^- transport by 50–65% (Choi et al. 2001). More recent evidence indicates that abnormal CFTR HCO_3^- permeability affects mucus release into the lumen of the gut (Garcia et al. 2009), reproductive tract (Muchekehu and Quinton 2010), and in bronchioles (Shamsuddin and Quinton 2012). Furthermore, it is known that ASL in CF is relatively acidic (Coakley et al. 2003; Shah et al. 2016; Shamsuddin and Quinton 2012), HCO_3^- affects the viscoelastic properties of airway mucus, neutrophil responses, and bacterial viability. Mucus secretion is closely associated with HCO_3^- secretion (Garcia et al. 2009; Joo et al. 2001; Livingston et al. 1995; Muchekehu and Quinton 2010; Quinton 2008; Trout et al. 1998). CF airways lacking HCO_3^- transport are not able to neutralize H^+ airway surface acid

(Coakley et al. 2003), and if HCO_3^- is not secreted (Welsh and Smith 2001) fluid for displacing debris may be compromised (Joo et al. 2001; Quinton 1999a; Smith and Welsh 1992) and clearance slowed or remain inactive (Matsui et al. 1998). These observations in the sweat duct and other CF-affected tissues emphasize the need for new therapeutic strategies aimed at correcting aberrant CFTR-mediated HCO_3^- transport in CF patients.

5.4.2.4 ENaC and CFTR Activities in Sweat Ducts Affected by PHA-1

As previously discussed, the CFTR Cl^- channel function is required for activating ENaC in sweat duct (Reddy et al. 1999; Reddy and Quinton 2003b). Since, ENaC and CFTR must work in concert during salt absorption, the question arises whether ENaC activity also influence the CFTR function. In fact, studies on heterologous systems indicated that ENaC expression significantly stimulated CFTR Cl^- currents in *Xenopus* oocytes (Ji et al. 2000). The sweat ducts from type-1 PHA patients offered significant insights regarding the role of ENaC in regulating CFTR activity in a native epithelium. The autosomal recessive PHA-1 is caused by loss-of-function mutations in ENaC (Chang et al. 1996; Firsov et al. 1998, 1999; Hummler and Horisberger 1999; Schafer 2002; Strautnieks et al. 1996). These studies on sweat ducts from PHA-1 patients showed that the ENaC channel function is not required for CFTR activation (Reddy et al. 2005). The PHA-1 ducts completely lacked spontaneous ENaC conductance. In contrast, the normal ducts showed large spontaneous ENaC conductance (46 ± 10 mS, mean ± SE). cAMP activation of CFTR Cl^- conductance or alkalinization of cytosolic pH (6.8–8.5)-stimulated ENaC conductance of normal but not PHA-1 ducts. In contrast, both spontaneous CFTR Cl^- conductance in intact ducts and cAMP-activated CFTR conductance in permeabilized ducts appeared to be similar in normal and PHA-1 subjects (Reddy et al. 2005). These observations indicated that virtual lack of ENaC in PHA-1 ducts had little effect on CFTR activity. Furthermore, the lack of ENaC in PHA-1 ducts completely blocked salt absorption and caused a dramatic increase in the sweat NaCl concentration. These studies provided strong evidence that normal salt absorption requires normal functioning of both CFTR and ENaC. Abnormality in either CFTR function as in CF or ENaC function as in PHA-1 has a similar functional consequence of elevated sweat NaCl in both cases, and therefore, cannot serve as a diagnostic tool to distinguish between CF and PHA-1 patients (Fig. 5.5). However, dramatically different transepithelial potentials between CF ($V_t = \sim -100$ mV) and PHA-1 ($V_t = \sim 0$ mV) sweat ducts clearly distinguish between these two groups (Fig. 5.5).

5.4.2.5 Summary and Future Directions

The human sweat glands perform two of the most significant and commonly shared exocrine functions, i.e., the “secretion” and “absorption” of electrolyte fluid. These glands also express two of the most significant transport elements, CFTR and ENaC, that play a central role in transepithelial ion transport by most exocrine glands. Furthermore, unlike most internal exocrine organs such as the intestine and airways, the sweat glands are relatively accessible for experimentation. They are morphologically as well as functionally among the simplest models of human exocrine tissues to study and interpret. Therefore, the human sweat gland can serve as an ideal and clinically relevant model system for studying the physiological and molecular mechanisms of ion transport by the epithelial cells in health and disease. Hence, future research involving sweat glands can be focused on defining the: (1) molecular mechanisms of regulating transepithelial ion transport involved in electrolyte-absorptive and -secretory processes by exocrine glands, (2) effects of potential therapeutic drugs aimed at correcting the defective CFTR anion channel function in disease conditions, (3) relative effects of different mutations on CFTR-mediated anion transport in CF, and (4) properties of regulation of ENaC in health and disease conditions such as PHA, Liddle's syndrome, and CF.

Acknowledgments The author is immensely thankful to Prof. Paul Quinton for useful discussions and valuable suggestions. The author is also thankful to Kirk Taylor and Sucharitha Reddy for expert technical assistance. This work was funded in part by the USPHS-NIH, The USCF Foundation, The Cystic Fibrosis Therapeutics, Inc. (CFFT), and The Cystic Fibrosis Research, Inc. (CFRI).

References

Abou Alaiwa MH, Launspach JL, Grogan B, Carter S, Zabner J, Stoltz DA, Singh PK, McKone EF, Welsh MJ (2018) Ivacaftor-induced sweat chloride reductions correlate with increases in airway surface liquid pH in cystic fibrosis. JCI Insight 3(15)

Anderson MP, Welsh MJ (1992) Regulation by ATP and ADP of CFTR chloride channels that contain mutant nucleotide-binding domains. Science 257(5077):1701–1704

Anderson MP, Berger HA, Rich DP, Gregory RJ, Smith AE, Welsh MJ (1991a) Nucleoside triphosphates are required to open the CFTR chloride channel. Cell 67(4):775–784

Anderson MP, Gregory RJ, Thompson S, Souza DW, Paul S, Mulligan RC, Smith AE, Welsh MJ (1991b) Demonstration that CFTR is a chloride channel by alteration of its anion selectivity. Science 253(5016):202–205

Anderson MP, Rich DP, Gregory RJ, Smith AE, Welsh MJ (1991c) Generation of cAMP-activated chloride currents by expression of CFTR. Science 251(4994):679–682

Anderson MP, Sheppard DN, Berger HA, Welsh MJ (1992) Chloride channels in the apical membrane of normal and cystic fibrosis airway and intestinal epithelia. Am J Physiol 263:L1–L14

Awayda MS, Ismailov II, Berdiev BK, Fuller CM, Benos DJ (1996) Protein kinase regulation of a cloned epithelial Na+ channel. J Gen Physiol 108(1):49–65

Basbaum CB, Ueki I, Brezina L, Nadel JA (1981) Tracheal submucosal gland serous cells stimulated in vitro with adrenergic and cholinergic agonists. A morphometric study. Cell Tissue Res 220:481–498

Bedrossian CW, Greenberg SD, Singer DB, Hansen JJ, Rosenberg HS (1976) The lung in cystic fibrosis. A quantitative study including prevalence of pathologic findings among different age groups. Hum Pathol 7:195–204

Behm JK, Hagiwara G, Lewiston NJ, Quinton PM, Wine JJ (1987) Hyposecretion of beta-adrenergically induced sweating in cystic fibrosis heterozygotes. Pediatr Res 22(3):271–276

Berger HA, Travis SM, Welsh MJ (1993) Regulation of the cystic fibrosis transmembrane conductance regulator Cl^- channel by specific protein kinases and protein phosphatases. J Biol Chem 268:2037–2047

Best JA, Quinton PM (2005) Salivary secretion assay for drug efficacy for cystic fibrosis in mice. Exp Physiol 90(2):189–193

Bijman J, Fromter E (1986) Direct demonstration of high transepithelial chloride-conductance in normal sweat duct which is absent in cystic fibrosis. Pflugers Arch 407(Suppl 2):S123–S127

Bijman J, Quinton PM (1984a) Influence of abnormal Cl- impermeability on sweating in cystic fibrosis. Am J Phys 247(1 Pt 1):C3–C9

Bijman J, Quinton PM (1984b) Influence of calcium and cyclic nucleotides on beta-adrenergic sweat secretion in equine sweat glands. Am J Phys 247(1 Pt 1):C10–C13

Bijman J, Quinton PM (1984c) Properties of equine sweat glands stimulated in vitro. In: Case M, Lingard J, Young J (eds) Secretion: mechanisms and control. Manchester University Press, Manchester, pp 253–258

Bijman J, Quinton PM (1987) Permeability properties of cell membranes and tight junctions of normal and cystic fibrosis sweat ducts. Pflugers Arch 408(5):505–510

Botero-Velez M, Curtis JJ, Warnock DG (1994) Brief report: Liddle's syndrome revisited--a disorder of sodium reabsorption in the distal tubule. N Engl J Med 330(3):178–181

Boucherot A, Schreiber R, Kunzelmann K (2001) Role of CFTR's PDZ1-binding domain, NBF1 and Cl(−) conductance in inhibition of epithelial Na(+) channels in Xenopus oocytes. Biochim Biophys Acta 1515(1):64–71

Bovell DL, Quinton PM (2002) The immunocytochemical localisation of anion exchangers (AE) in the reabsorptive duct of the human eccrine sweat gland. J Physiol Lond 544P:107P

Bradbury NA (1999) Intracellular CFTR: localization and function. Physiol Rev 79(1 Suppl):S175–S191

Chang SS, Grunder S, Hanukoglu A, Rosler A, Mathew PM, Hanukoglu I, Schild L, Lu Y, Shimkets RA, Nelson-Williams C et al (1996) Mutations in subunits of the epithelial sodium channel cause salt wasting with hyperkalaemic acidosis, pseudohypoaldosteronism type 1. Nat Genet 12(3):248–253

Chappe V, Hinkson DA, Zhu T, Chang XB, Riordan JR, Hanrahan JW (2003) Phosphorylation of protein kinase C sites in NBD1 and the R domain control CFTR channel activation by PKA. J Physiol 548(Pt 1):39–52

Cheng SH, Rich DP, Marshall J, Gregory RJ, Welsh MJ, Smith AE (1991) Phosphorylation of the R domain by cAMP-dependent protein kinase regulates the CFTR chloride channel. Cell 66 (5):1027–1036

Choi JY, Muallem D, Kiselyov K, Lee MG, Thomas PJ, Muallem S (2001) Aberrant CFTR-dependent HCO3- transport in mutations associated with cystic fibrosis. Nature 410 (6824):94–97

Civan MM (1978) Intracellular activities of sodium and potassium. Am J Phys 234(4):F261–F269

Coakley RD, Grubb BR, Paradiso AM, Gatzy JT, Johnson LG, Kreda SM, O'Neal WK, Boucher RC (2003) Abnormal surface liquid pH regulation by cultured cystic fibrosis bronchial epithelium. Proc Natl Acad Sci U S A 100(26):16083–16088

Cook DI, Dinudom A, Komwatana P, Young JA (1998) Control of Na+ transport in salivary duct epithelial cells by cytosolic Cl- and Na+. Eur J Morphol 36(Suppl):67–73

Cotton CU, Boucher RC, Gatzy JT (1988) Bioelectric properties and ion transport across excised canine fetal and neonatal airways. J Appl Physiol 65(6):2367–2375

Davis CW, Finn AL (1982a) Sodium transport effects on the basolateral membrane in toad urinary bladder. J Gen Physiol 80(5):733–751

Davis CW, Finn AL (1982b) Sodium transport inhibition by amiloride reduces basolateral membrane potassium conductance in tight epithelia. Science 216:525–527

DeLong J, Civan MM (1978) Dissociation of cellular K+ accumulation from net Na+ transport by toad urinary bladder. J Membr Biol 42(1):19–43
Di Sant'Agnese PA, Darling RC, Perera GA, Shea E (1953a) Abnormal electrolyte composition of sweat in cystic fibrosis of the pancreas; clinical significance and relationship to the disease. Pediatrics 12(5):549–563
Di Sant'Agnese PA, Darling RC, Perera GA, Shea E (1953b) Sweat electrolyte disturbances associated with childhood pancreatic disease. Am J Med 15(6):777–784
DiMagno EP, Malagelada JR, Go VLM (1977) Fate of orally ingested enzymes in pancreatic insufficiency: comparison of two dosage schedules. N Engl J Med 296:1318
Dinudom A, Komwatana P, Young JA, Cook DI (1995) Control of the amiloride-sensitive Na+ current in mouse salivary ducts by intracellular anions is mediated by a G protein. J Physiol 487 (Pt 3):549–555
Dinudom A, Harvey KF, Komwatana P, Young JA, Kumar S, Cook DI (1998) Nedd4 mediates control of an epithelial Na+ channel in salivary duct cells by cytosolic Na+. Proc Natl Acad Sci U S A 95(12):7169–7173
Dobson RL, Sato K (1972) The secretion of salt and water by the eccrine sweat gland. Arch Dermatol 105:366–370
Donaldson SH, Poligone EG, Stutts MJ (2002) CFTR regulation of ENaC. Methods Mol Med 70:343–364
Dulhanty AM, Riordan JR (1994) Phosphorylation by cAMP-dependent protein kinase causes a conformational change in the R domain of the cystic fibrosis transmembrane conductance regulator. Biochemistry 33:4072–4079
Firsov D, Gautschi I, Merillat AM, Rossier BC, Schild L (1998) The heterotetrameric architecture of the epithelial sodium channel (ENaC). EMBO J 17(2):344–352
Firsov D, Robert-Nicoud M, Gruender S, Schild L, Rossier BC (1999) Mutational analysis of cysteine-rich domains of the epithelium sodium channel (ENaC). Identification of cysteines essential for channel expression at the cell surface. J Biol Chem 274(5):2743–2749
Fischer H, Illek B, Machen TE (1998) Regulation of CFTR by protein phosphatase 2B and protein kinase C. Pflugers Arch 436(2):175–181
French PJ, Bijman J, Edixhoven M, Vaandrager AB, Scholte BJ, Lohmann SM, Nairn AC, De Jonge HR (1995) Isotype-specific activation of cystic fibrosis transmembrane conductance regulator-chloride channels by cGMP-dependent protein kinase II. J Biol Chem 270:26626–26631
Frizzell RA, Field M, Schultz SG (1979) Sodium-coupled chloride transport by epithelial tissues. Am J Physiol 5:F1–F8
Fromter E, Gebler B (1977) Electrical properties of amphibian urinary bladder epithelia III. The cell membrane resistances and the effect oif amiloride. Eur J Phys 371:99–108
Furgeson SB, Linas S (2010) Mechanisms of type I and type II pseudohypoaldosteronism. J Am Soc Nephrol 21(11):1842–1845
Gabriel SE, Clarke LL, Boucher RC, Stutts MJ (1993) CFTR and outward rectifying chloride channels are distinct proteins with a regulatory relationship. Nature 363(6426):263–268
Gadsby DC, Nairn AC (1999) Control of CFTR channel gating by phosphorylation and nucleotide hydrolysis. Physiol Rev 79(1 Suppl):S77–S107
Garcia MA, Yang N, Quinton PM (2009) Normal mouse intestinal mucus release requires cystic fibrosis transmembrane regulator-dependent bicarbonate secretion. J Clin Invest 119 (9):2613–2622
Granger D, Marsolais M, Burry J, Laprade R (2003) Na+/H+ exchangers in the human eccrine sweat duct. Am J Physiol Cell Physiol 285(5):C1047–C1058
Gray MA, Harris A, Coleman L, Greenwell JR, Argent BE (1989) Two types of chloride channel on duct cells cultured from human fetal pancreas. Am J Phys 257(2 Pt 1):C240–C251
Halm DR, Frizzell RA (1990) Intestinal Cl^- secretion. In: Lebenthal E, Duffy ME (eds) Textbook in secretory diarrhea. Raven Press, New York, pp 47–58

Hanrahan JW, Mathews CJ, Grygorczyk R, Tabcharani JA, Grzelczak Z, Chang XB, Riordan JR (1996) Regulation of the CFTR chloride channel from humans and sharks. J Exp Zool 275 (4):283–291

Harvey BJ, Kernan RP (1984) Intracellular ion activities in frog skin in relation to external sodium and effects of amiloride and/or ouabain. J Physiol 349:501–517

Hogan DL, Crombie DL, Isenberg JI, Svendsen P, Schaffalitzky de Muckadell OB, Ainsworth MA (1997) CFTR mediates cAMP- and Ca2+−activated duodenal epithelial HCO3- secretion. Am J Physiol 272:G872–G878

Hogenauer C, Santa Ana CA, Porter JL, Millard M, Gelfand A, Rosenblatt RL, Prestidge CB, Fordtran JS (2000) Active intestinal chloride secretion in human carriers of cystic fibrosis mutations: an evaluation of the hypothesis that heterozygotes have subnormal active intestinal chloride secretion. Am J Hum Genet 67(6):1422–1427

Hummler E, Horisberger JD (1999) Genetic disorders of membrane transport. V The epithelial sodium channel and its implication in human diseases. Am J Physiol 276(3 Pt 1):G567–G571

Ismailov II, Berdiev BK, Shlyonsky VG, Benos DJ (1997a) Mechanosensitivity of an epithelial Na + channel in planar lipid bilayers: release from Ca2+ block. Biophys J 72(3):1182–1192

Ismailov II, Berdiev BK, Shlyonsky VG, Fuller CM, Prat AG, Jovov B, Cantiello HF, Ausiello DA, Benos DJ (1997b) Role of actin in regulation of epithelial sodium channels by CFTR. Am J Phys 272(4 Pt 1):C1077–C1086

Jeffrey PK, Brain APR (1988) Surface morphology of human airway mucosa: normal, carcinoma or cystic fibrosis. Scanning Microsc 2(1):553–560

Ji HL, Chalfant ML, Jovov B, Lockhart JP, Parker SB, Fuller CM, Stanton BA, Benos DJ (2000) The cytosolic termini of the beta- and gamma-ENaC subunits are involved in the functional interactions between cystic fibrosis transmembrane conductance regulator and epithelial sodium channel. J Biol Chem 275(36):27947–27956

Jia Y, Mathews CJ, Hanrahan JW (1997) Phosphorylation by protein kinase C is required for acute activation of cystic fibrosis transmembrane conductance regulator by protein kinase A. J Biol Chem 272(8):4978–4984

Johansen PG, Anderson CM, Hadorn B (1968) Cystic fibrosis o the pancreas. A generalised disturbance of water and electrolyte movement in exocrine tissues. Lancet 1(7540):455–460

Johnson JP, Louie E, Lewiston NJ, Wine JJ (1991) b-adrenergic sweat responses in cystic fibrosis heterozygotes with and without the D F508 allele. Pediatr Res 29(6):525–528

Joo NS, Krouse ME, Wu JV, Saenz Y, Jayaraman S, Verkman AS, Wine JJ (2001) HCO3- transport in relation to mucus secretion from submucosal glands. JOP 2(4 Suppl):280–284

Ko YH, Pedersen PL (1995) The first nucleotide binding fold of the cystic fibrosis transmembrane conductance regulator can function as an active ATPase. J Biol Chem 270(38):22093–22096

Ko SB, Shcheynikov N, Choi JY, Luo X, Ishibashi K, Thomas PJ, Kim JY, Kim KH, Lee MG, Naruse S et al (2002) A molecular mechanism for aberrant CFTR-dependent HCO(3)(−) transport in cystic fibrosis. EMBO J 21(21):5662–5672

Konig J, Schreiber R, Voelcker T, Mall M, Kunzelmann K (2001) The cystic fibrosis transmembrane conductance regulator (CFTR) inhibits ENaC through an increase in the intracellular Cl-concentration. EMBO Rep 2(11):1047–1051

Kopelman H, Corey M, Gaskin K, Durie P, Weizman Z, Forstner G (1988) Impaired chloride secretion, as well as bicarbonate secretion, underlies the fluid secretory defect in the cystic fibrosis pancreas. Gastroenterology 95(2):349–355

Krouse ME, Hagiwara G, Chen J, Lewiston NJ, Wine JJ (1989) Ion channels in normal human and cystic fibrosis sweat gland cells. Am J Physiol 257:C129–C140

Kunzelmann K (2001) CFTR: interacting with everything? News Physiol Sci 16:167–170

Kunzelmann K, Kiser GL, Schreiber R, Riordan JR (1997) Inhibition of epithelial Na+ currents by intracellular domains of the cystic fibrosis transmembrane conductance regulator. FEBS Lett 400(3):341–344

Kunzelmann K, Schreiber R, Boucherot A (2001) Mechanisms of the inhibition of epithelial Na(+) channels by CFTR and purinergic stimulation. Kidney Int 60(2):455–461

Lee RJ, Foskett JK (2014) Ca(2)(+) signaling and fluid secretion by secretory cells of the airway epithelium. Cell Calcium 55(6):325–336
Lethem MI, Dowell ML, Van Scott M, Yankaskas JR, Egan T, Boucher RC, Davis CW (1993) Nucleotide regulation of goblet cells in human airway epithelial explants: normal exocytosis in cystic fibrosis. Am J Respir Cell Mol Biol 9(3):315–322
Lew TS, Krasne S (1991) The only anion-selective channel measured in intact sweat ducts has a unit conductance of 8 pS. Pediatric Pulmonology Suppl 6:251
Li H, Zhang X, Zeng S, Chen L, Li X, Lin C, Zhang M, Shu S, Xie S, He Y et al (2014) The cellular localization of Na(+)/H(+) exchanger 1, cystic fibrosis transmembrane conductance regulator, potassium channel, epithelial sodium channel gamma and vacuolar-type H+-ATPase in human eccrine sweat glands. Acta Histochem 116(8):1237–1243
Livingston EH, Miller J, Engel E (1995) Bicarbonate diffusion through mucus. Am J Phys 269(3 Pt 1):G453–G457
Luckie DB, Singh CN, Wine JJ, Wilterding JH (2001) CFTR activation raises extracellular pH of NIH/3T3 mouse fibroblasts and C127 epithelial cells. J Membr Biol 179(3):275–284
Luo J, Pato MD, Riordan JR, Hanrahan JW (1998) Differential regulation of single CFTR channels by PP2C, PP2A, and other phosphatases. Am J Phys 274(5 Pt 1):C1397–C1410
Luo J, Zhu T, Evagelidis A, Pato MD, Hanrahan JW (2000) Role of protein phosphatases in the activation of CFTR (ABCC7) by genistein and bromotetramisole. Am J Physiol Cell Physiol 279(1):C108–C119
Ma T, Song Y, Gillespie A, Carlson EJ, Epstein CJ, Verkman AS (1999) Defective secretion of saliva in transgenic mice lacking aquaporin-5 water channels. J Biol Chem 274 (29):20071–20074
Mandel ID, Kutscher A, Denning CR, Thompson RH, Zegarelli EV (1967) Salivary studies in cystic fibrosis. Am J Dis Child 113:431–438
Marin MG, Culp DJ (1986) Isolation and culture of submucosal gland cells. Clin Chest Med 7 (2):239–245
Martinez JR, Cassity N (1983) Salivary secretion induced from isolated, perfused rat submandibular glands by sympathomimetic agents. Arch Oral Biol 28(12):1101–1108
Mathews CJ, Tabcharani JA, Hanrahan JW (1998) The CFTR chloride channel: nucleotide interactions and temperature-dependent gating. J Membr Biol 163:55–66
Matsui H, Grubb BR, Tarran R, Randell SH, Gatzy JT, Davis CW, Boucher RC (1998) Evidence for periciliary liquid layer depletion, not abnormal ion composition, in the pathogenesis of cystic fibrosis airways disease. Cell 95(7):1005–1015
Muchekehu R, Quinton PM (2010) Cystic fibrosis: a lesson in chemistry. Physiol News 81:37–39
Nadel JA, German VF (1980) Micropipette measurement of airway submucosal gland secretion: autonomic effects. Am Rev Respir Dis 121:351–357
Nejsum LN, Praetorius J, Nielsen S (2005) NKCC1 and NHE1 are abundantly expressed in the basolateral plasma membrane of secretory coil cells in rat, mouse, and human sweat glands. Am J Physiol Cell Physiol 289(2):C333–C340
Noel S, Strale PO, Dannhoffer L, Wilke M, DeJonge H, Rogier C, Mettey Y, Becq F (2008) Stimulation of salivary secretion in vivo by CFTR potentiators in Cftr+/+ and Cftr−/− mice. J Cyst Fibros 7(2):128–133
Park HW, Nam JH, Kim JY, Namkung W, Yoon JS, Lee JS, Kim KS, Venglovecz V, Gray MA, Kim KH et al (2010) Dynamic regulation of CFTR bicarbonate permeability by [Cl-]i and its role in pancreatic bicarbonate secretion. Gastroenterology 139(2):620–631
Picciotto MR, Cohn JA, Bertuzzi G, Greengard P, Nairn AC (1992) Phosphorylation of the cystic fibrosis transmembrane conductance regulator. J Biol Chem 267(18):12742–12752
Quinton PM (1979) Composition and control of secretions from tracheal bronchial submucosal glands. Nature 279(5713):551–552
Quinton PM (1981) Effects of some ion transport inhibitors on secretion and reabsorption in intact and perfused single human sweat glands. Pflugers Arch 391(4):309–313
Quinton PM (1983) Chloride impermeability in cystic fibrosis. Nature 301(5899):421–422

Quinton PM (1984) The pathology of secretion in cystic fibrosis. Secretion: mechanisms and control. Manchester University Press, Manchester, pp 282–293
Quinton PM (1986) Missing Cl conductance in cystic fibrosis. Am J Phys 251(4 Pt 1):C649–C652
Quinton PM (1987) Physiology of sweat secretion. Kidney Int Suppl 21:S102–S108
Quinton PM (1990) Cystic fibrosis: a disease in electrolyte transport. FASEB J 4(10):2709–2717
Quinton PM (1994) Cystic fibrosis: old questions, new answers. The Second Joseph Levy Memorial Lecture, XIXth European Cystic Fibrosis Conference. Cystic Fibrosis Trust, Bromley, Kent, pp 1–35
Quinton PM (1999a) Physiological basis of cystic fibrosis: a historical perspective. Physiol Rev 79 (1 Suppl):S3–S22
Quinton PM (1999b) The sweat gland. In: Yankaskas JR, Knowles MR (eds) Cystic fibrosis in adults. Lipencott-Raven, Philadelphia, PA, pp 419–438
Quinton PM (2007) Cystic fibrosis: lessons from the sweat gland. Physiology (Bethesda) 22 (3):212–225
Quinton PM (2008) Cystic fibrosis: impaired bicarbonate secretion and mucoviscidosis. Lancet 372 (9636):415–417
Quinton PM, Reddy MM (1989) Cl- conductance and acid secretion in the human sweat duct. Ann N Y Acad Sci 574:438–446
Quinton PM, Reddy MM (1990) The sweat gland in cystic fibrosis. In: Davis P (ed) Cystic fibrosis. Marcel Dekker, New York
Quinton PM, Reddy MM (1991) Regulation of absorption in the human sweat duct. Adv Exp Med Biol 290:159–170. discussion 170–152
Quinton PM, Reddy MM (1992) Control of CFTR chloride conductance by ATP levels through non-hydrolytic binding. Nature 360(6399):79–81
Quinton PM, Reddy MM (1994) Regulation of absorption by phosphorylation of CFTR. Jpn J Physiol 44(Suppl 2):S207–S213
Randak C, Neth P, Auerswald EA, Eckerskorn C, Assfalg-Machleidt I, Machleidt W (1997) A recombinant polypeptide model of the second nucleotide-binding fold of the cystic fibrosis transmembrane conductance regulator functions as an active ATPase, GTPase and adenylate kinase. FEBS Lett 410(2–3):180–186
Reddy MM, Bell CL (1996) Distinct cellular mechanisms of cholinergic and beta-adrenergic sweat secretion. Am J Phys 271(2 Pt 1):C486–C494
Reddy MM, Light M, Quinton PM (1999) Activation of epithelial Na+ channel (ENaC) requires CFTR Cl- channel function. Nature 402:301–304
Reddy MM, Quinton PM (1987a) Cl- permeability of sweat duct cell membranes: intracellular microelectrode analysis. Prog Clin Biol Res 254:45–57
Reddy MM, Quinton PM (1987b) Intracellular potentials of microperfused human sweat duct cells. Pflugers Arch 410(4–5):471–475
Reddy MM, Quinton PM (1989a) Altered electrical potential profile of human reabsorptive sweat duct cells in cystic fibrosis. Am J Phys 257(4 Pt 1):C722–C726
Reddy MM, Quinton PM (1989b) Localization of Cl- conductance in normal and Cl- impermeability in cystic fibrosis sweat duct epithelium. Am J Phys 257(4 Pt 1):C727–C735
Reddy MM, Quinton PM (1991) Intracellular potassium activity and the role of potassium in transepithelial salt transport in the human reabsorptive sweat duct. J Membr Biol 119 (3):199–210
Reddy MM, Quinton PM (1992a) cAMP activation of CF-affected Cl- conductance in both cell membranes of an absorptive epithelium. J Membr Biol 130(1):49–62
Reddy MM, Quinton PM (1992b) Electrophysiologically distinct cell types in human sweat gland secretory coil. Am J Phys 262(2 Pt 1):C287–C292
Reddy MM, Quinton PM (1994a) Intracellular Cl activity: evidence of dual mechanisms of cl absorption in sweat duct. Am J Phys 267(4 Pt 1):C1136–C1144
Reddy MM, Quinton PM (1994b) Rapid regulation of electrolyte absorption in sweat duct. J Membr Biol 140(1):57–67

Reddy MM, Quinton PM (1996a) Deactivation of CFTR-Cl conductance by endogenous phosphatases in the native sweat duct. Am J Phys 270(2 Pt 1):C474–C480
Reddy MM, Quinton PM (1996b) Hydrolytic and nonhydrolytic interactions in the ATP regulation of CFTR Cl- conductance. Am J Phys 271(1 Pt 1):C35–C42
Reddy MM, Quinton PM (1998) G proteins activate CFTR GCl in the Native sweat duct. Ped Pulmon 17:208 A–208 200
Reddy MM, Quinton PM (2000) CFTR independent cytosolic basic pH activation of ENaC in the native sweat duct. Ped Pulmonol Suppl 20:131
Reddy MM, Quinton PM (2001a) cAMP-independent phosphorylation activation of CFTR by G proteins in native human sweat duct. Am J Physiol Cell Physiol 280(3):C604–C613
Reddy MM, Quinton PM (2001b) Selective activation of cystic fibrosis transmembrane conductance regulator Cl- and HCO3- conductances. JOP 2(4 Suppl):212–218
Reddy MM, Quinton PM (2003a) Control of dynamic CFTR selectivity by glutamate and ATP in epithelial cells. Nature 423:756–760
Reddy MM, Quinton PM (2003b) Functional interaction of CFTR and ENaC in sweat glands. Pflugers Arch 445(4):499–503
Reddy MM, Quinton PM (2005) ENaC activity requires CFTR channel function independently of phosphorylation in sweat duct. J Membr Biol 207(1):23–33
Reddy MM, Quinton PM (2006) Cytosolic potassium controls CFTR deactivation in human sweat duct. Am J Physiol Cell Physiol 291(1):C122–C129
Reddy MM, Quinton PM (2009) PKA mediates constitutive activation of CFTR in human sweat duct. J Membr Biol 231(2–3):65–78
Reddy MM, Stutts MJ (2013) Status of fluid and electrolyte absorption in cystic fibrosis. Cold Spring Harb Perspect Med 3(1):a009555
Reddy MM, Bell CL, Quinton PM (1992) Evidence of two distinct epithelial cell types in primary cultures from human sweat gland secretory coil. Am J Phys 262(4 Pt 1):C891–C898
Reddy MM, Quinton PM, Haws C, Wine JJ, Grygorczyk R, Tabcharani JA, Hanrahan JW, Gunderson KL, Kopito RR (1996) Failure of the cystic fibrosis transmembrane conductance regulator to conduct ATP. Science 271(5257):1876–1879
Reddy MM, Bell CL, Quinton PM (1997) Cystic fibrosis affects specific cell type in sweat gland secretory coil. Am J Phys 273(2 Pt 1):C426–C433
Reddy MM, Kopito RR, Quinton PM (1998) Cytosolic pH regulates GCl through control of phosphorylation states of CFTR. Am J Phys 275(4 Pt 1):C1040–C1047
Reddy MM, Light MJ, Quinton PM (1999) Activation of the epithelial Na+ channel (ENaC) requires CFTR Cl- channel function. Nature 402(6759):301–304
Reddy MM, Sun D, Quinton PM (2001) Apical heterotrimeric G-proteins activate CFTR in the native sweat duct. J Membr Biol 179(1):51–61
Reddy MM, Wang XF, Gottschalk M, Jones K, Quinton PM (2005) Normal CFTR activity and reversed skin potentials in pseudohypoaldosteronism. J Membr Biol 203(3):151–159
Reddy MM, Wang XF, Quinton PM (2008) Effect of cytosolic pH on epithelial Na+ channel in normal and cystic fibrosis sweat ducts. J Membr Biol 225(1–3):1–11
Reddy MM, Samnsudhin AK, Flores-Delgado G, Quinton PM (2017) VX-770 stimulates CFTR-HCO3- but not in the CF sweat ducts. Pediatr Pulmonol 52(S2):1
Reddy MM, Samsudhin AK, Quinton PM (2018) Glutamate activation of aberrant CFTR HCO3-conductance in CF sweat ducts. Pediatr Pulmonol 52(S47):1
Reuss L, Finn AL (1975) Electrical properties of the cellular transepithelial pathway in Necturus gallbladder II. Ionic permeability of the apical cell membrane. J Membr Biol 25:141–161
Riordan JR, Rommens JM, Kerem B, Alon N, Rozmahel R, Grzelczak Z, Zielenski J, Lok S, Plavsic N, Chou JL et al (1989) Identification of the cystic fibrosis gene: cloning and characterization of complementary DNA. Science 245(4922):1066–1073
Saga K, Sato K (1988) Ultrastructural localization of ouabain-sensitive, K-dependent p-nitrophenyl phosphatase activity in monkey eccrine sweat gland. J Histochem Cytochem 36(8):1023–1030

Saga K, Sato K (1989) Electron probe X-ray microanalysis of cellular ions in the eccrine secretory coil cells during methacholine stimulation. J Membr Biol 107(1):13–24
Saga K, Sato F, Sato K (1988) K+ efflux from the monkey eccrine secretory coil during the transient of stimulation with agonists. J Physiol 405:205–217
Sato K (1977a) Pharmacology and function of the myoepithelial cell in the eccrine sweat gland. Experientia 33(5):631–633
Sato K (1977b) The physiology, pharmacology, and biochemistry of the eccrine sweat gland. Rev Physiol Biochem Pharmacol 79:51–131
Sato K (1980) Pharmacological responsiveness of the myoepithelium of the isolated human axillary apocrine sweat gland. Br J Dermatol 103(3):235–243
Sato K, Sato F (1981) Pharmacologic responsiveness of isolated single eccrine sweat glands. Am J Phys 240(1):R44–R51
Sato K, Sato F (1984) Defective beta adrenergic response of cystic fibrosis sweat glands in vivo and in vitro. J Clin Invest 73(6):1763–1771
Sato F, Sato K (2000) cAMP-dependent Cl(−) channel protein (CFTR) and its mRNA are expressed in the secretory portion of human eccrine sweat gland. J Histochem Cytochem 48 (3):345–354
Sato K, Nishiyama A, Kobayashi M (1979) Mechanical properties and functions of the myoepithelium in the eccrine sweat gland. Am J Phys 237(3):C177–C184
Sato K, Kang WH, Saga K, Sato KT (1989) Biology of sweat glands and their disorders. I. Normal sweat gland function. J Am Acad Dermatol 20(4):537–563
Schafer JA (2002) Abnormal regulation of ENaC: syndromes of salt retention and salt wasting by the collecting duct. Am J Physiol Renal Physiol 283(2):F221–F235
Schettino T, Kohler M, Fromter E (1985) Membrane potentials of individual cells of isolated gastric glands of rabbit. Pflugers Arch 405(1):58–65
Schoumacher RA, Shoemaker RL, Halm DR, Tallant EA, Wallace RW, Frizzell RA (1987) Phosphorylation fails to activate chloride channels from cystic fibrosis airway cells. Nature 330(6150):752–754
Schwiebert EM, Benos DJ, Egan ME, Stutts MJ, Guggino WB (1999) CFTR is a conductance regulator as well as a chloride channel. Physiol Rev 79(1 Suppl):S145–S166
Seibert FS, Chang XB, Aleksandrov AA, Clarke DM, Hanrahan JW, Riordan JR (1999) Influence of phosphorylation by protein kinase A on CFTR at the cell surface and endoplasmic reticulum. Biochim Biophys Acta 1461:275–283
Shah VS, Meyerholz DK, Tang XX, Reznikov L, Abou Alaiwa M, Ernst SE, Karp PH, Wohlford-Lenane CL, Heilmann KP, Leidinger MR et al (2016) Airway acidification initiates host defense abnormalities in cystic fibrosis mice. Science 351(6272):503–507
Shamsuddin AK, Quinton PM (2012) Surface fluid absorption and secretion in small airways. J Physiol 590(Pt 15):3561–3574
Shamsuddin AK, Reddy MM, Quinton PM (2008) Iontophoretic beta-adrenergic stimulation of human sweat glands: possible assay for cystic fibrosis transmembrane conductance regulator activity in vivo. Exp Physiol 93(8):969–981
Sheppard DN, Welsh MJ (1999) Structure and function of the CFTR chloride channel. Physiol Rev 79:S23–S45
Shimura S, Sasaki T, Sasaki H, Takishima T (1986) Contractility of isolated single submucosal gland from trachea. J Appl Physiol 60(4):1237–1247
Shindo T, Spring KR (1981) Chloride movement across the basolateral membrane of proximal tubule cells. J Membr Biol 58(1):35–42
Smith JJ, Welsh MJ (1992) cAMP stimulates bicarbonate secretion across normal, but not cystic fibrosis airway epithelia. J Clin Invest 89(4):1148–1153
Sohma Y, Gray MA, Imai Y, Argent BE (1996) A mathematical model of the pancreatic ductal epithelium. J Membr Biol 154(1):53–67
Sorscher EJ, Breslow JL (1982) Cystic fibrosis: a disorder of calcium-stimulated secretion and transepithelial sodium transport? Lancet 1:368–370

Strautnieks SS, Thompson RJ, Gardiner RM, Chung E (1996) A novel splice-site mutation in the gamma subunit of the epithelial sodium channel gene in three pseudohypoaldosteronism type 1 families. Nat Genet 13(2):248–250
Strong TV, Boehm K, Collins FS (1994) Localization of cystic fibrosis transmembrane conductance regulator mRNA in the human gastrointestinal tract by in situ hybridization. J Clin Invest 93 (1):347–354
Tetti M, Monticone S, Burrello J, Matarazzo P, Veglio F, Pasini B, Jeunemaitre X, Mulatero P (2018) Liddle syndrome: review of the literature and description of a new case. Int J Mol Sci 19 (3):812
Trout L, King M, Feng W, Inglis SK, Ballard ST (1998) Inhibition of airway liquid secretion and its effect on the physical properties of airway mucus. Am J Phys 274(2 Pt 1):L258–L263
Uno H, Montagna W (1975) Catecholamine-containing nerve terminals of the eccrine sweat glands of macaques. Cell Tissue Res 158:1–13
Vaandrager AB, Ehlert EM, Jarchau T, Lohmann SM, de Jonge HR (1996) N-terminal myristoylation is required for membrane localization of cGMP-dependent protein kinase type II. J Biol Chem 271(12):7025–7029
Vaandrager AB, Smolenski A, Tilly BC, Houtsmuller AB, Ehlert EM, Bot AG, Edixhoven M, Boomaars WE, Lohmann SM, de Jonge HR (1998) Membrane targeting of cGMP-dependent protein kinase is required for cystic fibrosis transmembrane conductance regulator Cl- channel activation. Proc Natl Acad Sci U S A 95(4):1466–1471
Van Goor F, Hadida S, Grootenhuis PD, Burton B, Cao D, Neuberger T, Turnbull A, Singh A, Joubran J, Hazlewood A et al (2009) Rescue of CF airway epithelial cell function in vitro by a CFTR potentiator, VX-770. Proc Natl Acad Sci U S A 106(44):18825–18830
Van Goor F, Hadida S, Grootenhuis PD, Burton B, Stack JH, Straley KS, Decker CJ, Miller M, McCartney J, Olson ER et al (2011) Correction of the F508del-CFTR protein processing defect in vitro by the investigational drug VX-809. Proc Natl Acad Sci U S A 108(46):18843–18848
van Os CH, Slegers JF (1975) The electrical potential profile of gallbladder epithelium. J Membr Biol 24(3–4):341–363
Welsh MJ (1986) An apical-membrane chloride channel in human tracheal epithelium. Science 232 (4758):1648–1650
Welsh MJ, Smith JJ (2001) cAMP stimulation of HCO3- secretion across airway epithelia. JOP 2 (4 Suppl):291–293
Welsh MJ, Anderson MP, Rich DP, Berger HA, Denning GM, Ostedgaard LS, Sheppard DN, Cheng SH, Gregory RJ, Smith AE (1992) Cystic fibrosis transmembrane conductance regulator: a chloride channel with novel regulation. Neuron 8:821–829
Widdicombe JH (1994) Accumulation of airway mucus in cystic fibrosis. Pulm Pharmacol 7:225–233
Willumsen NJ, Davis CW, Boucher RC (1989a) Cellular Cl- transport in cultured cystic fibrosis airway epithelium. Am J Phys 256(5 Pt 1):C1045–C1053
Willumsen NJ, Davis CW, Boucher RC (1989b) Intracellular Cl- activity and cellular Cl- pathways in cultured human airway epithelium. Am J Phys 256(5 Pt 1):C1033–C1044
Wine JJ (2001) Cystic fibrosis: the 'bicarbonate before chloride' hypothesis. Curr Biol 11(12): R463–R466
Wine JJ, Silverstein SC (1992) Cystic fibrosis. ATP and chloride conductance. Nature 360 (6399):18
Wine JJ, Brayden DJ, Hagiwara G, Krouse ME, Law TC, Muller UJ, Solc CK, Ward CL, Widdicombe JH, Xia Y (1991) Cystic fibrosis, the CFTR, and rectifying Cl- channels. Adv Exp Med Biol 290:253–269. discussion 269–272
Winter MC, Sheppard DN, Carson MR, Welsh MJ (1994) Effect of ATP concentration on CFTR Cl- channels: a kinetic analysis of channel regulation. Biophys J 66(5):1398–1403
Yanagawa S, Yokozeki H, Sato K (1986) Origin of periodic acid-Schiff-reactive glycoprotein in human eccrine sweat. J Appl Physiol 60(5):1615–1622

Chapter 6
Transporters in the Lactating Mammary Epithelium

Margaret C. Neville, Akihiro Kamikawa, Patricia Webb, and Palaniappian Ramanathan

Abstract Membrane transporters are essential for the synthesis and secretion of milk. We are at a transition point; in the twentieth century, most transport studies were carried out by following the transported molecule into the cell and examining all the ways that transport could be perturbed by the environment, allowing functional descriptions of most transport processes. We now have powerful molecular tools so that the molecular nature of the transporters as well as the mRNAs from which they are transcribed can be studied; in addition, antibodies to the proteins allow the transporters to be localized within the cell so that real cellular pathways can be discerned. So that our current knowledge can be of most use to the scientific community we use mRNA expression levels from two datasets representing complete catalogues of expressed genes in the lactating mammary gland of the human and the mouse. These expression levels are compared with expression levels in the colostral and transitional milk phases in the case of human milk and with midpregnancy expression levels in the mouse. The data provide a molecular framework, which should greatly facilitate future elucidation of transporters and their function in the cellular compartments involved in the secretion of milk.

M. C. Neville (✉)
Departments of Obstetrics and Gynecology and Physiology and Biophysics, University of Colorado School of Medicine, Aurora, CO, USA
e-mail: Peggy.Neville@cuanschutz.edu

A. Kamikawa
Department of Veterinary Medicine, Obihiro University of Agriculture and Veterinary Medicine, Obihiro, Hokkaido, Japan
e-mail: akami@obihiro.ac.jp

P. Webb
Department of Obstetrics and Gynecology, University of Colorado School of Medicine, Aurora, CO, USA
e-mail: patricia.webb@cuanschutz.edu

P. Ramanathan
Department of Pathology, University of Texas Medical Branch, Galveston, TX, USA
e-mail: paramana@utmb.edu

K. L. Hamilton, D. C. Devor (eds.), *Ion Transport Across Epithelial Tissues and Disease*, Physiology in Health and Disease,
https://doi.org/10.1007/978-3-030-55310-4_6

Keywords Membrane transport · Milk secretion · Plasma membranes · Endoplasmic reticulum · Golgi compartment · Secretory vesicle · Lactation transcriptome

6.1 The Cell Biology of the Lactating Mammary Epithelial Cell

The lactating mammary epithelial cell has been known for more than 5 decades as a complex structure responsible for the secretion of all milk components during full lactation including water, ions, sugars, proteins, lipids, as well as certain substrates such as amino acids and nucleotides (Fig. 6.1, Linzell and Peaker 1971b). Organic substrates are transported across the basolateral surface of the cell and utilized in the various intracellular compartments to synthesize the final components in milk or in some cases, like immunoglobulins and albumin, transported intact from the interstitial space to the milk lumen. Complex and precise systems, best understood for Ca^{2+}, Zn^{2+}, and, surprisingly, Cu^{2+} are utilized to be sure that appropriate ionic concentrations are present in the various intercellular compartments and in milk. Before embarking on a journey into the molecular identity and localization of the transporters and channels involved in all these compartments as well as the plasma membrane, we must review the cell biology of the lactating mammary cell including the internal membrane-bound compartments. The actual discussion of transporters begins with Sect. 6.2.

6.1.1 Secretory Pathways

Milk components are secreted into the lumen utilizing the four classical pathways as well as a paracellular pathway as shown in Fig. 6.2 (Linzell and Peaker 1971b).

6.1.1.1 Pathway I. Vesicular Secretion

Proteins destined for secretion or, in some cases, for localization on the apical membrane carry an endoplasmic reticulum (ER) signal sequence, generally on their N-terminus, which directs them into the sec61 pore on the rough endoplasmic reticulum as they are synthesized (Hirschberg 2018). Once protein folding is complete, the proteins are packaged into transport vesicles, which ferry them to the Golgi complex where they are subjected to additional processing and packaged into secretory vesicles along with lactose and oligosaccharides, both synthesized in the Golgi complex. Major proteins secreted by this pathway include casein, α-lactoglobulin, lactoferrin, and, in ruminants, β-lactoglobulin. Other proteins like

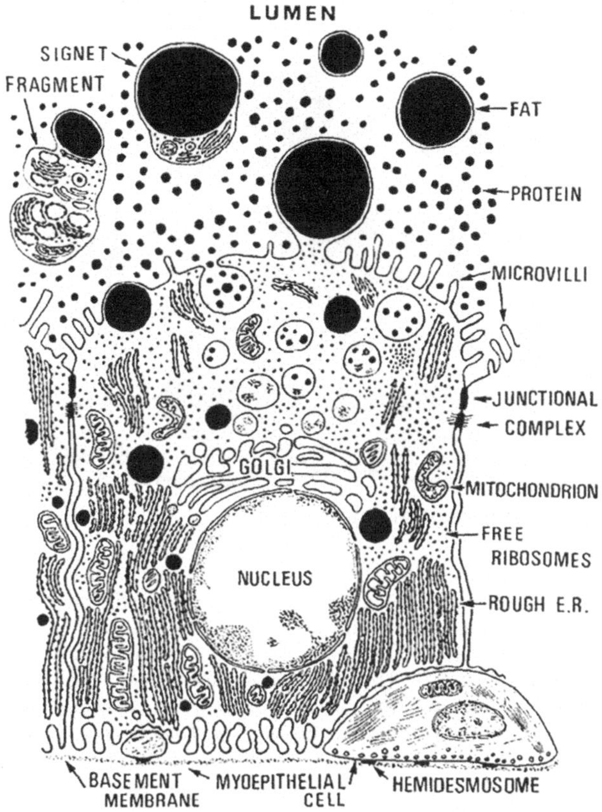

Fig. 6.1 Diagram of compartments in the lactating mammary alveolar cell. This diagram was drawn by Jim Linzell from electron micrographs available in the late 1960s. The major membrane-bound compartments are the cell itself bound by plasma membranes that define interactions of the cell with both the apical lumen and the basolateral space, the nucleus, the rough endoplasmic reticulum (ROUGH E.R.), the mitochondria, the Golgi compartment, and the secretory vesicles (round compartments containing puncta of casein). Black filled-circles represent milk fat droplets in the cytoplasm and the membrane-bound milk-fat globules in the lumen. Black dot-labeled PROTEIN represent casein micelles. Note that some of the luminal milk fat globules carry a fragment of cytoplasm. Myoepithelial cells, derived from basal cells in the virgin gland, are localized between the epithelial cell and the basement membrane. From Linzell and Peaker (1971b) used by permission of Malcolm Peaker

β1–4-galactosyltransferase-I (B4GALT1) remain in the Golgi compartment for the synthesis of lactose and oligosaccharides.

6.1.1.2 Pathway II. Secretion of the Milk Fat Globule

This exocytotic process is unique to the mammary gland among vertebrate species. Milk triglycerides are synthesized in the ER and travel to the apical membrane as

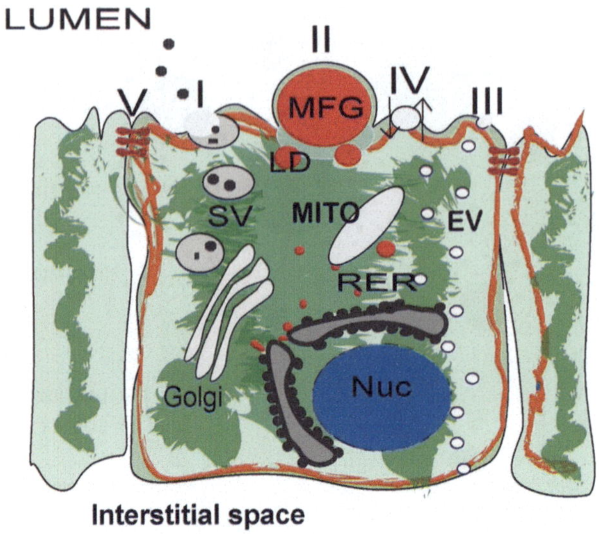

Fig. 6.2 Secretory pathways in the lactating mammary epithelium. Secretory pathways include I. The secretory vesicle pathway. Golgi-derived vesicles open directly into the mammary lumen releasing their contents, which include casein micelles, shown as black dots within the vesicles, and other proteins including alpha-lactalbumin, beta-lactoglobulin, lactoferrin, and many others as well as lactose and oligosaccharides. II. The milk fat globule whose secretion is described by McManaman Chap. 7 in this volume. III. Transcytotic vesicles that transfer immunoglobulins and other compounds from the interstitial space to the milk. IV. Direct transporters and channels that ferry ions like chloride, potassium, calcium, and sodium and possibly organic molecules like glucose directly into the milk. V. The paracellular pathway providing direct access between the interstitial and the milk space. This pathway is closed by junctional complexes in the lactating gland (Nguyen and Neville 1998) and is not dealt with here. *Nuc*, nucleus; *RER*, rough endoplasmic reticulum; *EV*, endocytic vesicles; *MITO*, mitochondrion; *SV*, secretory vesicles; *LD*, lipid droplets; and *MFG*, milk fat globule

small lipid droplets recently described in an elegant article by Mather and colleagues (2019). These droplets begin to attach to the plasma membrane where they aggregate into larger droplets and are eventually engulfed by the apical membrane to form the milk fat globule, surrounded by a membrane bilayer and containing a specialized protein population that includes butyrophilin, xanthine oxidase, mucins, MFGE8 (lactadherin), and many others (Liao et al. 2011; McManaman 2014). The transport processes associated with lipid synthesis and formation of the milk fat droplet are dealt with in McManaman Chap. 7 of this volume and will not be discussed here. However, one important aspect of the milk fat globule from human milk as well as that of many other species is that the process of engulfment of the lipid droplet also engulfs substantial cytoplasm (see FRAGMENT, Fig. 6.1). The amount of cytoplasm is often sufficient so that mRNAs associated with the mammary epithelial cell can be extracted from the milk fat globule and used to accurately determine the transcriptome of the lactating mammary alveolar cell (Maningat et al. 2007, 2009; Mohammad et al. 2012; Lemay et al. 2013a, b; Chen et al. 2016).

6.1.1.3 Pathway III. Direct Transfer of Proteins and Other Components from the Extracellular Space to the Milk Lumen

A direct vesicular pathway for transfer of certain proteins was first discovered as the mechanism by which immunoglobulins are transferred from the interstitial space, where they are manufactured by resident B cells, to the milk lumen (Mostov et al. 1980; Solari and Kraehenbuhl 1984). While the pathway does not make use of classical membrane transporters, it does provide a unique mechanism for transfer of proteins and possibly other molecules from the interstitial space to the milk lumen. It is not clear what membrane transporters are involved, and so the pathway will not be examined in this article.

6.1.1.4 Pathway IV. Transfer of Ions and Nutrients Across the Basolateral and Apical Membranes

Ion and nutrient concentrations must be regulated in the cytoplasm, in the membrane-bound cellular compartments of the lactating cell, and in the milk. Elaborate regulated transport mechanisms have evolved to do this job for monovalent ions such as sodium, potassium, and chloride ions, for divalent ions such as calcium, zinc, magnesium, and iron, and for the major nutrients involved in synthesis of milk carbohydrates, proteins, and lipids. The current status of transporters for all these entities except lipids will be outlined in this article.

6.1.1.5 Pathway V. Paracellular Transfer of Molecules from the Interstitial Space to the Milk Lumen and from the Milk Lumen to the Interstitial Space

Since specialized transport mechanisms are not involved in paracellular transport, where substances as large as albumin pass through the intercellular spaces in response to a concentration gradient (Monks and Neville 2004), we will not discuss this process in any detail here. Suffice it to say that these spaces are open during late pregnancy and close with the onset of lactation (Nguyen et al. 2001a, b). It is thought that changes in paracellular permeability also occur during mastitis and involution, but the mechanisms involved are poorly understood.

6.1.1.6 A Currently Undefined Pathway for Extracellular Vesicles

Milk contains a large population of small (~100 nM) and extra-small vesicles denoted as extracellular vesicles; the larger ones are often called exosomes, but the categories are not well defined. Provost and colleagues (Benmoussa et al. 2019) examined the proteome of two categories of vesicles pelleted from skim milk at

100,000 g and 25,000 g and found a number of specific proteins; however, the protein content did not contain any obvious transport molecules, and so we will not deal with this pathway in this article.

6.1.2 Methods for Studying Membrane Transport in the Lactating Mammary Epithelium

Transport of substrates into the mammary gland was first studied in the 1960s and 1970s in the lactating goat by measuring the arteriovenous difference in the concentration of these substrates or their isotopes in the blood (Linzell and Peaker 1971b; West et al. 1972; Linzell 1974a, b). The permeability of the apical membrane was assessed by injecting isotopes for the desired substances up the teat into the milk space and determining recovery sometime later. Up to this time, a great deal of progress had been made defining the major proteins present in milk and their sites of origin in the mammary gland. For example, work summarized by Brew defined the site of synthesis of lactose as the Golgi compartment (Brew 1969) on the basis of finding both constituents of lactose synthase, beta-1,4-galactosyltransferase and alpha-lactalbumin associated with the Golgi compartment. While transport of many substances has been functionally studied for decades (Shennan and Peaker 2000), the molecular identity of the transporter molecules has often not been precisely clear. For bivalent ions like Ca^{2+} and Zn^{2+}, this problem has largely been solved (see Sect. 6.2); for many others, much research is still required. An exemplary research project on chloride ion transport makes use of the many techniques available to study ion transport including patch clamp of the apical membrane, real-time PCR analysis of gene expression, and immunohistochemistry to determine the identity, functional properties, and localization of a calcium-activated chloride channel, anoctamin, in the apical membrane of the mouse mammary epithelial cell (Kamikawa et al. 2016).

In this review, we rely heavily on transcriptomic data available from two sources to help define the molecular identity of transporters for both charged and uncharged molecules in the lactating mammary epithelium. For human lactation, we rely on transcriptomic data from the analysis of mRNA extracted from the milk fat globule of 16 women at the colostral, transitional, and mature milk stages of lactation referred to here as the Lemay study (Lemay et al. 2013a). The dataset contains expression data for nearly 25,000 unique mRNAs. The use of milk fat globule RNA to analyze the lactation transcriptome was validated by a comparison of the transcriptome from the milk fat globule of rhesus monkeys compared to that of epithelial cells from the same population (Lemay et al. 2013b). It appears to provide a reasonable picture of the transporters actually present in the various compartments of the lactating mammary alveolar cell in the lactating woman. Transcriptomic data available from the Hadsell laboratory later in lactation (Maningat et al. 2007, 2009; Mohammad et al. 2012) correlates well with that from the Lemay study. It should be noted that the study from the Hadsell laboratory contains comprehensive longitudinal data on gene expression from seven women collected at close intervals during the onset of lactation and could be an excellent resource for developing a detailed understanding of the changes in gene expression, which underlie the process of secretory activation (formerly known as lactogenesis II) in women. A microarray study from the Neville laboratory examining the transcriptome of mammary cells from day 14 pregnant mice and day 2 lactating mice provides a guide to transporter and channel expression in the mouse mammary gland (Heinz et al. 2016). This dataset, which contains nearly 30,000 mRNAs, is referred to as the Ramanathan dataset here.

The caveats to the use of these large mRNA datasets without backup from proteomics, immunohistochemistry, and functional data were well-outlined by Lippolis and colleagues and are recognized here (Lippolis et al. 2018). The data presented here should therefore be regarded as a way station on a long adventure that will help us understand how small molecules enter the mammary cells and are distributed throughout its compartments prior to being secreted into milk, either as the small molecules themselves or as part of larger molecules.

6.1.3 Membrane Bound Compartments in the Lactating Mammary Alveolar Cell

Secretory epithelia are characterized by strong polarity as well as cell-cell junctions that control both paracellular permeability and the morphology of the epithelial structure as a whole. At the termination of embryogenesis, the rudimentary mammary gland already consists of a system of tubules lined by a two-layer epithelium with a luminal epithelial layer and a basal layer (Huebner et al. 2014). The hormones of puberty and pregnancy bring about growth and differentiation into a layer of presecretory or luminal epithelial cells surrounded by basal cells. During pregnancy, the gland differentiates so that the luminal cells acquire many characteristics of the secretory alveolar cells and the basal cells differentiate into myoepithelial cells that will respond to oxytocin to cause milk ejection from the lactating gland. Lactation itself turns on with a fall in progesterone along with an increase in prolactin signaling around parturition. The lactating alveolar cells have a number of membrane-bound compartments that are important to consider in defining their transport activities.

6.1.3.1 Basolateral Membrane

The basolateral membrane represents the interface of the alveolar cells with the extracellular environment; for this reason, the ions and substrates involved in milk secretion must all transit this barrier. Transporters for major substrates like amino acids, glucose, fatty acids, and mono- and divalent ions must reside here along with receptors and transporters for hormones and drugs, which we will not be discussing here.

6.1.3.2 Apical Membrane

Transporters in the apical membrane have not been as well characterized as those in the basolateral membrane although, as mentioned above, anoctamin, which transfers Cl^- from the cytoplasm to the milk, is the subject of an exemplary study (Kamikawa et al. 2016) and the plasma membrane Ca^{2+} ATPase PMCA has received considerable attention (Reinhardt et al. 2004; VanHouten et al. 2007). Other molecules that are found in both milk and basal extracellular space include glucose, certain amino acids, other monovalent and divalent cations, and others. We will suggest some candidates for transport of these molecules across the apical membrane, but further studies are needed.

6.1.3.3 Endoplasmic Reticulum

The endoplasmic reticulum is the location of folding of proteins destined for both the apical plasma membrane and apical secretion. It is also a storage reservoir for Ca^{2+} in most cells; a specific Ca^{2+}ATPase, SERCA, transports Ca^{2+} from the cytoplasm into the ER. Transporters for other molecules are present but have not been well characterized in the lactating alveolar cell.

6.1.3.4 Golgi Vesicles

Illustrating the localization of secretory vesicle Ca^{++} ATPase (SPCA1) in the mouse mammary gland, Cross and associates (2013) published the image in Fig. 6.3, showing colocalization of a Golgi vesicle marker, GM130, and SPCA1 in the lactating mouse mammary epithelium. We will discuss the calcium transport activities of SPCA1 in the section on calcium transport. Here, we point out the localization of the Golgi compartment near the apical pole of the secreting luminal cell and note that it occupies a fair proportion of the cytoplasmic compartment of this mammary secretory cell. Because we do not know the concentration of most ions and other small molecules in the Golgi-secretory vesicle complex, it could be that sequestration can lead to misestimation of cytoplasmic ion and nutrient concentrations if the estimates are not carried out with appropriate sensing dyes or electrodes.

6.1.3.5 Lysosome and Endocytic Vesicles

Very little is known about this compartment in the cytoplasm of lactating epithelial cells, although lysosomal uptake of milk fat globules is thought to mediate cell death

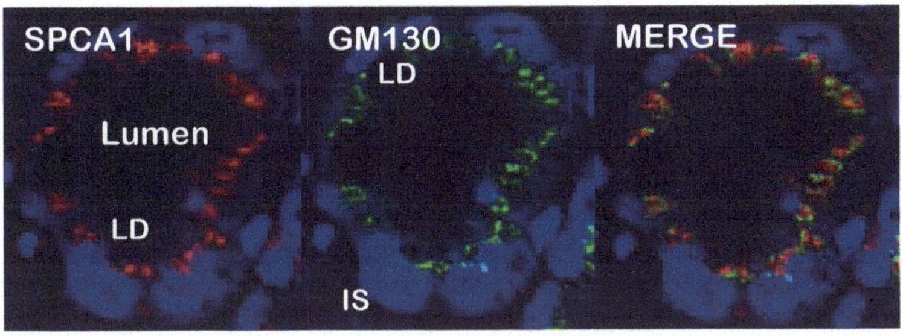

Fig. 6.3 Golgi compartment and SPCA1 positioning in lactating mammary alveolar cells. Image of a lactating mammary alveolus with cells stained for nuclei (DAPI, blue), transporter SPCA1 (red), and Golgi marker GM130 (green). Both SPCA1 and GM130 are localized near the apical plasma membrane of most cells. The two are not completely colocalized, suggesting that SPCA is confined to secretory vesicles. LD, lipid droplet; IS, interstitial space. We thank T.A. Reinhardt for providing these images modified from Cross et al. (2013)

during involution (Sargeant et al. 2014). The transcytotic transfer of immunoglobulins and other molecules is thought to involve the endocytic compartment although the pathway through the cell is unclear. Transporters in these membranes in the lactating alveolar cell have not been clearly defined.

6.1.3.6 Nucleus

It seems unlikely that nuclear transport mechanisms are specific to the mammary gland, and so this important subject will not be covered here.

6.1.3.7 Mitochondria

Mitochondria have a large population of transporters, many encoded by SLC35 genes. A quick perusal of these genes indicates that they are expressed at meaningful levels in the lactating mammary epithelium, but that these levels are similar to expression during pregnancy. Because mitochondrial transport mechanisms are likely similar in most cells, they will not be covered in this treatise.

Much of the functional work on transport pathways has been carried out in dairy species, particularly cows and goats, and in rodents with functional studies of varying value for defining in vivo systems done on tissue culture cells that seldom mimic the in vivo process. More recently, some interesting studies have been done with the porcine species as well (Trott et al. 2009; Manjarin et al. 2014; Zhang et al. 2018), as piglet growth is of particular importance to human cultures where pork is a major source of protein. With the advent of extensive data on the human milk transcriptome (Maningat et al. 2007, 2009; Mohammad et al. 2012; Lemay et al. 2013a), the information gained from these studies can often be applied to human lactation, as well. In a comparative study of the transcriptomes of lactating cows, rats, and wallabies, few of the transporter genes described in this chapter showed up as differentially expressed (Farhadian et al. 2018), suggesting that the transport processes involved in milk synthesis are conserved across very different mammalian species.

6.1.4 Organization of Transporter Discussion

We start by reviewing the mechanisms by which divalent ions are transported from the extracellular space into the milk because these processes have been most completely characterized in the lactating cells, particularly transport of Ca^{2+}, Zn^{2+}, and Cu^{2+}. We then consider the synthesis of milk proteins, a process that takes place in the cytoplasm utilizing amino acids mostly brought into the cell by transporters in the basal plasma membranes. During the synthetic process, the proteins, destined for both the milk and the apical membrane of the alveolar cell, are transported into the endoplasmic reticulum for folding and some processing before being transferred to the Golgi system. In the Golgi system, both glucose and nucleotide sugars are utilized to synthesize the disaccharide lactose, which is unique to milk. Nucleotide sugars are utilized to glycosylate secreted and membrane proteins. The transport systems that bring glucose and these nucleotide sugars into the Golgi are outlined. Finally, we discuss the transport area that has actually received the most attention over the past six decades, the monovalent ion transport systems. Although the molecules brought into the cell by these systems, mainly, Na^+, K^+, and Cl^-, provide the essential milieu for all the processes we will be reviewing, the molecular identities of the transporters involved and, in particular, their cellular localization are not, with the exception of anoctamin, presently clear, and thus, no consistent model for these systems can currently be presented. Much work remains in this area.

One thing that must be made clear is the sources of energy for the transport of all the molecules discussed here. There are three: (1) For molecules like glucose, which are rapidly utilized within the cell, a simple facilitated transporter like GLUT1 through which glucose can pass down a concentration gradient is all that is necessary. (2) For molecules that must be highly concentrated ATPases like the Na^+/K^+ ATPase in the basal membrane and the Ca^{2+} ATPases in many membranes, the metabolic energy stored in ATP is used to transport ions against steep electrochemical gradients. (3) The gradients created by these processes can then be used to transport other molecules against their concentration gradients. As will be seen, a great many transport processes are driven by the electrochemical Na^+ gradient created by the Na^+/K^+ ATPase. Another interesting example in a very recently published article from the Reinhardt laboratory documents a Ca^{2+}/H^+ exchanger in the Golgi compartment that may maintain pH homeostasis in this compartment (Snyder et al. 2019).

6.1.5 Nomenclature

The nomenclature used to describe various transport functions has changed over the years and can be very confusing. In this article, we anchor each transport system to its mRNA expression as the nomenclature for almost all transporters is now quite clear and provides a reference framework for studies of function and localization based on the use of advanced techniques including electrophysiology, gene knockdown, immunohistochemistry, as well as more standard flux techniques. These techniques can be used to study the in situ gland and are therefore more reliable physiologically than many of the tissue culture models used for earlier studies. The rules for designating mRNA and proteins can be found at https://academic.oup.com/molehr/pages/Gene_And_Protein_Nomenclature. To summarize: All gene and mRNA symbols are in italics; human, bovine, and all other species that are not mice or rats are in upper case; and mouse and rat genes and mRNAs start with an upper case letter and then are in lower case. Protein symbols for all species are nonitalicized and in upper case. Protein names are in lower case. Thus, for the anoctamin proteins, the gene for anoctamin 1 is designated *ANO1* in humans and *Ano1* in mice; the protein symbol is ANO1.

6.2 Transport of Bivalent Ions

6.2.1 Calcium

6.2.1.1 Maintenance of Low Cytosolic Calcium

Calcium is present at very high concentrations in milk bound to casein and citrate and free in solution (Reinhardt and Horst 1999; Neville 2005). While the free calcium in milk is about 3.3 mM, its concentration in the cytoplasm is 3000 orders of magnitude lower, only around 100 nM. Because many cellular activities are critically sensitive to cytosolic calcium, the transfer of calcium across both the apical and basolateral cell membranes must be tightly regulated to maintain low cytoplasmic levels at the same as large quantities of calcium are transferred across the plasma membrane and into the Golgi to be secreted into milk. One mechanism of maintaining low cytosolic calcium is thought to be the presence of powerful P-type calcium ATPase transporters that sequester calcium in the ER, Golgi, and secretory compartments and move the ion across the apical plasma membrane (Reinhardt et al. 2014). The role of the endoplasmic reticulum calcium transporter SERCA in calcium transport into the ER is similar in most cells and will be described in the section on endoplasmic reticulum transporters. A different set of calcium ATPase transporters, SPCA1 and SPCA2, are thought to transfer calcium and magnesium into the secretory compartment where calcium complexes with casein to form the casein micelle as well as with alpha-lactalbumin and citrate (Wuytack et al. 2003). Transport of calcium across the apical membrane by specific plasma membrane calcium transporters, PMCA1 and PMCA2 (Reinhardt et al. 2004), may also help maintain low cytoplasmic calcium as may the presence of calcium binding proteins in the cytoplasm and Golgi. Each of these processes will be described in more detail below.

6.2.1.2 Calcium Binding in the Cytosol and Golgi

In many types of cells, calbindins (CALB1 and CALB2) are used to buffer cytosolic calcium levels (VanHouten and Wysolmerski 2007). While neither *CALB1* nor *CALB2* mRNA levels were above background in the Ramanathan dataset, consistent with the data for the mouse summarized by VanHouten and Wysolmerski (VanHouten and Wysolmerski 2007), *CALB2* was expressed in the Lemay dataset (Lemay et al. 2013a) at levels higher than any of the calcium ATPases (Fig. 6.4), and so calbindin 2 may serve to modulate cytosolic Ca^{2+} in the human gland. The only other calcium binding protein found to be expressed at higher levels in lactation than pregnancy was nucleobindin [NUCB2, an EF-hand protein also known as nefstatin (Kim et al. 2014)], found at high levels in lactation in both the Ramanathan and the Lemay datasets (Lemay et al. 2013a; Heinz et al. 2016). Nucleobindin was shown to be localized to the luminal surface of the Golgi compartment by the Farquhar laboratory (Lin et al. 1998) two decades ago. The detailed physiological role of these calcium binding proteins in calcium homeostasis in the lactating gland has yet to be elucidated.

6.2.1.3 Calcium Transfer into Milk

Calcium transfer into milk can be broken into four steps designated as CALTRANS by the Reinhardt laboratory (VanHouten and Wysolmerski 2007; Cross et al. 2014):

(a) Transfer of calcium across the basolateral membrane from the extracellular fluid.
(b) Intracellular sequestration of calcium in the endoplasmic reticulum to help maintain free cytosolic calcium in the micromolar range. This ER function is carried out in most cells (see section on transporters in the ER).
(c) Transfer of calcium into the Golgi and secretory compartments where it binds to proteins, phosphate, and citrate.
(d) Export of calcium into milk across the apical membrane.

6.2.1.4 Calcium Transfer into the Golgi Compartment

As stated above, calcium entry into the Golgi compartment is necessary for formation of the casein micelle, a structure responsible for up to 80% of the calcium in milk, depending on the species. SPCA1 is ubiquitously expressed in epithelia, whereas SPCA2 is expressed in tissues with high secretory activity such as the

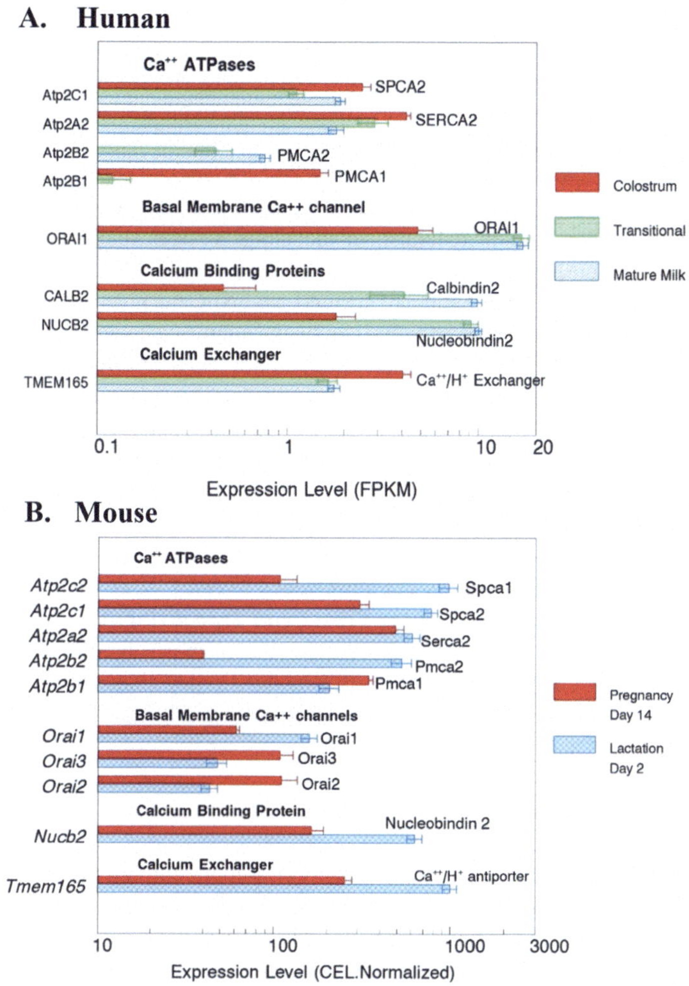

Fig. 6.4 Expression of mRNAs for CALTRANS molecules in the mammary glands of lactating humans and mice. Gene names are on the left axis, and protein names are in the body of the graphs. (**a**) Human mRNA levels are from Lemay and colleagues (2013a, b) analyzing mRNA from the milk fat droplets from colostrum (red; 3 subjects), transitional milk (green; 5 subjects), and mature milk (blue; 7 subjects). PMCA2 and ORAI1 are significantly elevated in mature milk, whereas PMCA1 is downregulated. (**b**) mRNA in isolated mouse cells from four 13.5 day pregnant (red) and four 2-day lactating mice (blue) was analyzed by RNASeq (Heinz et al. 2016). In the mouse, the mRNAs for both *Spca1* and *Spca2* as well as *Pmca2, Orai1, Nucb2*, and *Tmem165* are significantly upregulated in the lactating gland, whereas *Pmca1, Orai3*, and *Orai2* are downregulated. Expression of *SERCA2* was not significantly altered by reproductive state in either species. Note the expression of the calcium binding proteins calbindin in the human dataset and nucleobindin in both datasets; these are among the higher expressing genes

lactating mammary gland (Wuytack et al. 2003). The proteins coded by these genes transport calcium into the Golgi compartment where it is complexed with casein, alpha-lactalbumin, phosphate, and citrate. *Atp2c2* and *Atp2c1* mRNAs encoding SPCA1 and SPCA2, respectively, are the highest expressed calcium transport genes in the lactating mouse (Fig. 6.4), a finding consistent with the earlier results (Faddy et al. 2008; Cross et al. 2013), and *ATP2A1* for SPCA2 is expressed at a level slightly higher than *ATP2B2* for PMCA2 in the human gland. The formation of the casein micelle, dependent on calcium, in the Golgi vesicle suggests that a major portion of milk calcium is secreted by this pathway (Neville 2005). This hypothesis is supported first by the localization of SPCA1 and SPCA2 proteins specifically to the Golgi and secretory vesicle membranes of the mammary secretory cell (Faddy et al. 2008; Cross et al. 2013). Second, SPCA2 couples with ORAI1, a basolateral channel that may ferry calcium from the extracellular space into the cell bypassing the cytoplasmic compartment (Cross et al. 2013). Interestingly, SPCA1 has been shown to colocalize with a Ca^{2+}/H^{+} antiporter at the Golgi membrane, whereas SPCA2 has a broader distribution throughout the lactating mouse mammary gland (Reinhardt et al. 2004).

A type of plasma membrane calcium channel, the transient receptor potential (TRP or TRPV) channel, is highly calcium selective (Park et al. 2014a). However, expression of all TRP channels except *Trpm7* and *Trpm4* (Fig. 6.6b) in the mouse dataset is low in general (Heinz et al. 2016); the role of these channels is not clear at present. All TRPV channels are below 0.2 in the Lemay database (Lemay et al. 2013a), and so these channels may not be important in the formation of human milk.

6.2.1.5 Calcium Transfer Across the Apical Membrane

Definitive studies showed that PMCA2, an ATP-dependent calcium transporter, is localized exclusively in the apical plasma membrane of rodent epithelial cells (Reinhardt et al. 2004; VanHouten and Wysolmerski 2007). As in the Lemay and Ramanathan databases (Fig. 6.4), its expression increased markedly as lactation progressed in the rat (Reinhardt et al. 2000) and mouse (VanHouten et al. 2007). Knockdown of this protein decreased milk calcium markedly in mice, also decreasing milk secretion (Reinhardt et al. 2004; Mamillapalli et al. 2013). These experiments were interpreted by both Van Houten and Reinhardt as indicating that the major pathway for secretion of calcium into milk was across the apical plasma membrane mediated by PMCA2 (Reinhardt et al. 2004; Mamillapalli et al. 2013). However, this conclusion predated the findings that SPCA1 colocalized in the Golgi membranes with a Ca^{2+}/H^{+} antiporter and that SPCA2 interacts with a plasma membrane calcium channel, ORAI1, likely facilitating direct Ca^{2+} entry into the Golgi compartment. The plasma membrane transporter may be crucial for maintaining intracellular calcium homeostasis accounting in part for decreased calcium secretion when it is knocked down. However, there is a significant question as to whether it has access to sufficient Ca^{2+} in the cytoplasm to supply the large amount of Ca^{2+} present in milk.

6.2.1.6 Calcium Transfer Across the Basolateral Membrane

Work from the Rao and Reinhardt laboratories was the first elucidation of a mechanism by which calcium enters the lactating mammary epithelial cell from the interstitial space. The calcium channel ORAI1 (Feng et al. 2010; Cross et al. 2013) whose activity is linked to calcium stores in most tissues, e.g., store operated calcium entry or SOCE, was shown in mammary cells to produce what is now called store-independent calcium entry (SICE) from the extracellular space. More recent experiments have shown that the calcium content of milk of ORAI1 knockout mice was reduced by more than 50% (Davis et al. 2015), indicating that SICE plays a central role in calcium homeostasis in the lactating gland. As stated above, the ORAI1 channel in mammary epithelial cells may be coupled directly to SPCA2, potentially providing a direct path for entry of calcium from the extracellular space into the Golgi (Smaardijk et al. 2017). As an aside, ORAI1 was also found to be essential for the letdown reflex of the myoepithelial cells (Davis et al. 2015).

6.2.1.7 Summary

The current picture of calcium fluxes in the lactating mammary alveolar cell is that the calcium channel ORAI1 transfers calcium across the basal membrane into the cell, either into the cytoplasm or across the Golgi membrane via the calcium ATPase SPCA2. Cytosolic calcium is transported into the endoplasmic reticulum by the calcium ATPase SERCA2 helping to maintain low levels of cytosolic calcium. In mice, SPCA1 is localized only to the GM130-containing portion of the Golgi membranes near the apical membrane and may provide Ca^{2+} for the Ca^{2+}/H^{+} exchanger TMEM165 to help maintain the pH within the Golgi. Finally, calcium is transferred directly into milk via PMCA2 localized in the apical membrane. The proportion of milk calcium transported in this manner is controversial; the need for calcium in the Golgi for formation of the casein micelle may mean that apical membrane flux plays a regulatory role maintaining low cytoplasmic calcium levels. Calcium fluxes must be tightly regulated to maintain appropriate levels of free calcium in the cytosol and milk; to provide calcium for complexing with casein, alpha-lactalbumin, and citrate; and likely to provide an exit pathway for H^{+} ions generated during lactose synthesis in the Golgi compartment. It is becoming ever clearer how these pathways interact to produce the very large amount of calcium transferred into the milks of most mammals, but more remains to be done.

6.2.2 *Zinc*

Zinc is an important regulator of the molecular processes that control mammary proliferation, differentiation, and secretion (Lee and Kelleher 2016). In addition, during human lactation, 1–3 mg/day zinc is transferred from the plasma into milk (Kelleher and Lönnerda 2003), although the actual amount varies significantly because the concentration of zinc in human milk declines almost 20-fold during the first year of lactation. Very little zinc in milk is free (Zhang and Allen 1995b); the great proportion is bound to proteins such as casein, lactoferrin, and lysozyme as well as to small molecules like citrate. Measurement of the free zinc concentration in bovine milk suggests that zinc is at electrochemical equilibrium across the apical plasma membrane (Zhang and Allen 1995a), so that energy-dependent processes may not be necessary to transport this element across the apical membrane and, by extension, other cellular membranes.

Zinc is secreted through the Golgi-secretory vesicle pathway (Lee and Kelleher 2016) as well as directly across the apical plasma membrane. The supply of zinc in all the compartments of the mammary gland must be carefully managed; this is done by zinc transporting proteins, channels, and binding proteins. There are two major families of zinc transporters: the ZIP family (ZIP1–14 proteins, encoded by *SL39A1-SLC39A14*) imports zinc into the cytoplasm either across plasma membranes or out of vesicular compartments in the cell (Jeong and Eide 2013). Members of the ZNT family (ZNT1–10, encoded by *SLC30A1-SLC30A10*) export zinc from the cytosol into various cellular compartments including secretory vesicles, which then transfer zinc into milk (Lee et al. 2015). Elegant studies using bimolecular fluorescence complementation show that ZNT transporters undergo homodimerization (and occasionally heterodimerization) to form a zinc conducting channel in the membranes of various cellular compartments (Lasry et al. 2014). Balanced zinc concentrations in the various cellular compartments are thought to be maintained by the activity of at least two opposing transporters for each compartment (McCormick et al. 2014). mRNA expression levels for the zinc transporting molecules expressed in human and mouse mammary glands are shown for both humans and mice in Fig. 6.5. It is interesting that expression of the major zinc transporters differs markedly between these two species.

6.2.2.1 ZIP Transporters

Members of the ZIP family have 9 transmembrane domains with the N and C-termini localized extracellularly. ZIP2 and ZIP8 may engage in $Zn^{2+}/(HCO3^{-})_2$ cotransport (Jeong and Eide 2013). *SLC39A8* (ZIP8) is the most highly expressed Zn transporter in human milk and is highly expressed along with Slc39a14 (ZIP14) in mammary cells from the lactating mouse (Fig. 6.5). ZIP8 is expressed in lysosomal/endosomal membranes after transfection and may also transport manganese (Jeong and Eide 2013). The mRNA for ZIP6, which localizes to the plasma membrane and transports zinc into cells, is also upregulated between colostrum and mature milk in the human dataset and between pregnancy and lactation in mouse mammary cells. The mRNAs for ZIP7, ZIP10, and ZIP11 are upregulated between pregnancy and lactation in the mouse (Fig. 6.5). ZIP10 appears to be a plasma membrane transporter, which may dimerize with ZIP6 (Kong et al. 2014). Both have been shown to be regulated by

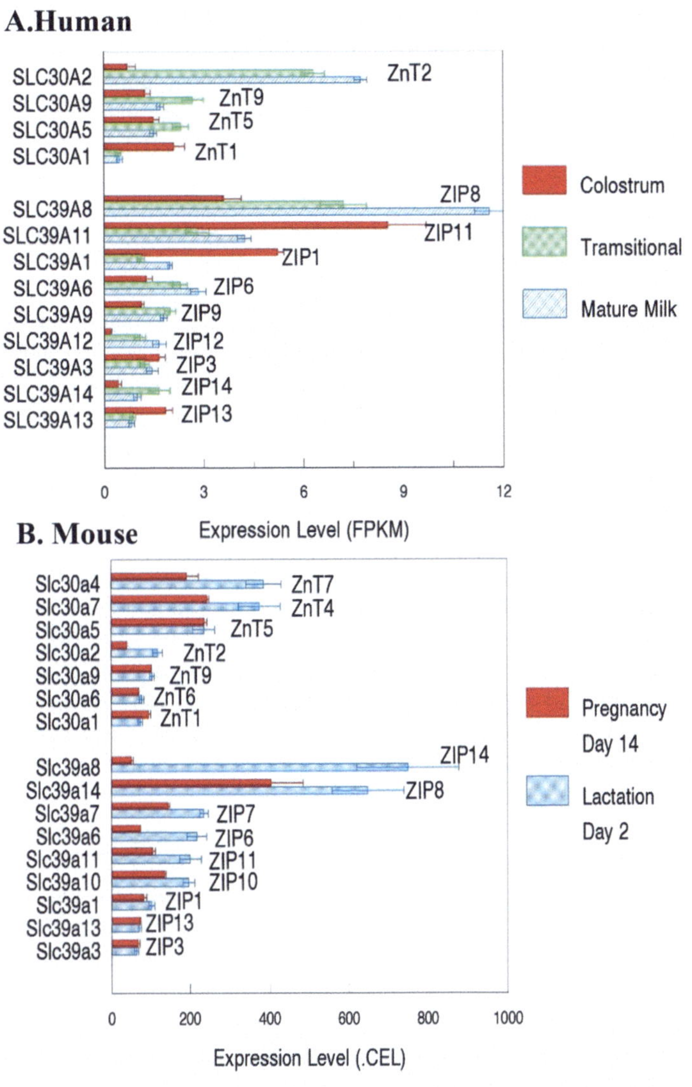

Fig. 6.5 Transporters for zinc in the lactating mammary epithelium. (**a**) Transcriptomic data for human transporters derived from the milk fat globule membrane of human colostrum and milk (Lemay et al. 2013a). (**b**) Transcriptome levels of zinc transporters in mammary cells from the mammary gland of pregnant and lactating mice (Heinz et al. 2016). MRNA names are on the left axis, and protein names are in the body of the graph

STAT3, often implicated in regulation of mammary development and involution (Sargeant et al. 2014). ZIP7 has been implicated in ER stress and therefore may be involved in zinc transport into the ER (Bin et al. 2017). ZIP11 has been mapped to the Golgi in mammary epithelial cells (Kelleher et al. 2012). In both species, ZIP3 is localized mainly to the apical membrane where it transports zinc from the luminal space to the cytoplasm and is thought to be important in the regulation of mammary function (Kelleher et al. 2009). ZIP14 has been implicated in manganese (Xin et al. 2017) and iron (Jenkitkasemwong et al. 2015) transport in other organs; its role in zinc transport in the mouse mammary epithelium is not well defined.

6.2.2.2 ZNT Transporters

Expression of the mRNA for ZNT2 is increased more than ten-fold between colostrum and mature milk in the human and more than three-fold between pregnancy and lactation in the mouse (Fig. 6.5a) consistent with its localization in secretory vesicles. ZNT2 is considered to be the major zinc exporter from the cytoplasm in both the secretory vesicles and apical membrane of the mouse and human mammary gland and appears to be essential for optimal development of lactation, for involution, and for the normal composition of mouse milk (McCormick and Kelleher 2012; Lee et al. 2015, 2017; Lee and Kelleher 2016; Rivera et al. 2018). Two forms of protein are translated from alternative splicing of *SLC39A2* mRNA; a 42 kD isoform is associated mainly with the endosomal secretory compartment and a 35 kD isoform is associated with the plasma membrane, possibly playing a role in zinc export (Lopez and Kelleher 2009). Extensive studies show that genetic defects in this transporter lead to changes in zinc secretion into milk. Genetic defects in ZNT2 (*SLC30A2*) lead to lactation defects in women (Alam et al. 2015; Itsumura et al. 2016), which are often sufficiently severe that exclusive breastfeeding is not possible.

The mRNAs for ZNT4, ZNT5, ZNT7, and ZNT9 are expressed at high levels along with ZNT2 in lactating mouse mammary cells, and ZNT5 and ZNT7 are increased between pregnancy and lactation although not as much as ZNT2 (Fig. 6.5b). ZNT2, ZNT4, and ZNT7 were the highest expressed ZNT proteins in the lactating mouse mammary gland in studies from the Kelleher laboratory (Kelleher et al. 2012). ZNT4 is localized to the trans-Golgi network and plasma membrane and is important in transporting zinc into mouse milk. Defects in this gene have been shown to lead to insufficient mammary development and premature mammary involution in the mouse (McCormick et al. 2016).

6.2.2.3 Summary

It is clear that there is extensive knowledge about zinc transport in the mammary epithelium, much of the work stemming from the Kelleher laboratory (McCormick et al. 2014, 2016; Hennigar et al. 2015; Lee et al. 2015, 2017; Lee and Kelleher 2016; Rivera et al. 2018). Similar to our understanding of calcium transport in the mammary gland, important advances will come if techniques for assessing the zinc concentration in the various compartments of the mammary gland can be developed.

6.2.3 Magnesium

The magnesium concentration in human milk is about 1 mM (Urzica et al. 2013). While the concentration is quite variable, it is not affected by a wide variety of maternal and environmental variables (Dorea 2000). It is, however, the second most abundant intracellular cation (Schäffers et al. 2018) with a cellular concentration up to 12 mM (Brandao et al. 2013) much of which is bound to nucleotides or proteins. The free intracellular concentration is thought to be around 0.5 to 1.0 mM, about the same as the extracellular fluid and is tightly regulated by influx, efflux, and exchange from organelles like mitochondria. The mechanism of its transport into milk has received little study although the two databases used in this chapter show that there are several magnesium transporters whose mRNA is increased at lactation in both mice and humans and several more that are expressed at lower levels in the mouse mammary gland (Fig. 6.6).

6.2.3.1 MAGT1, Magnesium Transporter 1

The highest expressed Mg^{2+} transporter in the mouse and human mammary gland mRNA datasets is the *Magtl* mRNA, which is increased nearly two-fold between pregnancy and lactation in the mouse (Fig. 6.6). This gene has been most studied in T-cells where it helps the cells import magnesium when they detect a foreign invader; Mg^{2+} is important for activation of the T-cell response (Brandao et al. 2013). It is also expressed at high levels in the human milk dataset although the

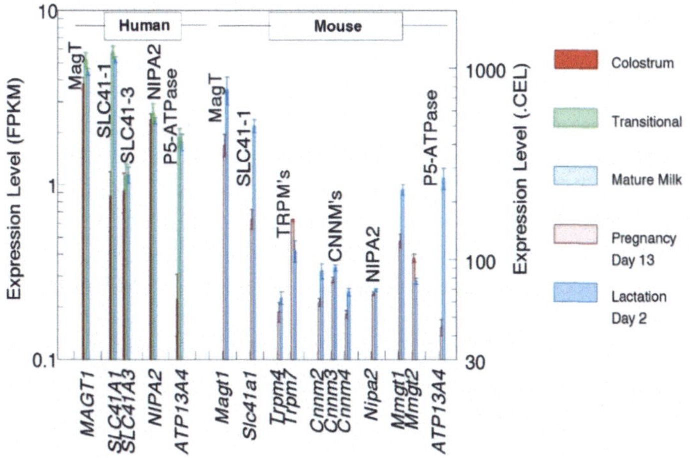

Fig. 6.6 Magnesium transporters in the lactating mammary epithelium. Transcriptomic data from human milk (Lemay et al. 2013a), left axis, and from mouse mammary alveolar cells (Heinz et al. 2016), right axis. Samples are described in the legend to Fig. 6.4. In general, the protein names (body of graph) reflect the gene names (bottom axis) in this series of transporters. Some of these transporters may transport other bivalent ions such as Zn^{2+}; Cd^{2+}, Co^{2+}, Cu^{2+}, and Fe^{2+}

increase from colostrum to mature milk is modest (Fig. 6.6). The protein is apparently an accessory to an N-oligosaccharide transferase complex in protein glycosylation; recently, Baaij and colleagues have questioned its role in Mg^{+2} transport (Schäffers et al. 2018).

6.2.3.2 The SLC41 Family

There are three members of the *SLC41* family, *SLC41A1*, *SLC41A2*, and *SLC41A3*. *SLC41A1* and *SLC41A3* are expressed at meaningful levels in the human milk database, and both are upregulated in mature milk compared to colostrum (Fig. 6.6a). Only *Slc41a3* is upregulated in lactating mouse mammary cells (Fig. 6.6b). Renal distal collecting duct *Slc41a3* has been most studied and responds to the magnesium content of the diet in mice (de Baaij et al. 2016). Evidence indicates that the protein encoded by *SLC41A1* is a Mg^{2+}/Na^{+} exchanger (NME) localized at the basolateral membrane (Kolisek et al. 2012; Fleig et al. 2013). Little is known about the activity of this family of magnesium transporters in the mammary gland.

6.2.3.3 TRPM Channels. Transient Receptor Potential Cation Channel Subfamily

Trpm7 encodes a Ca^{+2}/Mg^{+2} channel expressed in the mouse mammary gland at a level compatible with protein expression (Fig. 6.6). The protein is a channel but also contains a kinase domain and is considered to be important in regulating Mg fluxes in the kidney and intestine (Schäffers et al. 2018). Its biophysical properties have been extensively studied [reviewed in Park et al. (2014a)]. In the kidney, TRPM7 reabsorbs cations with a permeation profile order $Ba \geq Ni > Mg > Ca$. Although TRPM7 was expressed at low levels in the human breast in the Human Protein Atlas, it was associated most highly with cell types other than glandular cells (https://www.proteinatlas.org/ENSG00000092439-TRPM7/tissue/breast). TRPM4 is a calcium impermeable channel that has been extensively studied in relation to cardiovascular disease (Constantine et al. 2016) that may transport K^{+} and Na^{+}. There does not appear to be any specific literature on the role of either TRPM4 or TRPM7 in the mammary gland.

6.2.3.4 The CNNM Family

The CNNM proteins are integral membrane proteins implicated in Mg^{2+} homeostasis (Chen et al. 2018b). Despite the protein name cyclin M, these transport proteins are not related to the cyclins. Several cyclin M family members are expressed above background levels in the mouse mammary database, *Cnnm3, Cnnm3*, and *Cnnm4*, and all are increased in cells from the lactating mouse (Fig. 6.6). CNNM2 and

CNNM4 localize to the basal membrane of renal and intestinal epithelial cells where they are thought to mediate Mg^{2+} efflux. CNNMs have been found to bind PRL-2 (phosphatase of the regenerating liver-2), leading to an increase in intracellular Mg^{2+} (Chen et al. 2018b). The cytoplasmic domain of CNNM proteins has a region called the CNBH domain with significant homology to cyclic nucleotide binding domains. However, this domain has been convincingly shown not to bind cyclic nucleotides (Chen et al. 2018b) and is thought to be responsible for dimerization of the protein. The propensity of these proteins to dimerize appears to be related to Mg^{2+} efflux activity. No characterization of CNNM activity in the lactating mammary gland appears to be available.

6.2.3.5 Other Suggested Transporters

The products of several genes have been suggested to act as Mg^{2+} transporters including nonimprinted in Prader-Willi/Angelman syndrome (NIPA) proteins, membrane Mg^{2+} transporter 1–2 (MMGT1 and MMG2) proteins, Huntingtin-interacting protein 14 (HIP14) and HIP14-like protein (HIP14L), and ATP13A4, a potential Ca^{2+} or Mg^{2+} ATPase (Quamme 2010). Both *Mmgtl* and *Mmgt2* are expressed at levels clearly above baseline in the mouse dataset, and *Mmgtl* is increased between pregnancy and lactation (Fig. 6.6).

NIPA2 mRNA is expressed in the human milk and lactating mouse databases at levels comparable to other transporters (Fig. 6.6). Mg^{2+} transport by NIPA2 is electrogenic, voltage-dependent, and saturable with high specificity for Mg^{2+}. In oocytes where it has been most extensively studied, it was localized to the plasma membrane and to endosomes (Goytain et al. 2008).

Expression of the P5-ATPASE coded by *ATP13A4*, a potential Ca^{2+} or Mg^{2+} ATPASE (Quamme 2010), has generally been thought to be restricted to brain and stomach (Schultheis et al. 2004). It was found to be expressed in the endoplasmic reticulum in Cos-7 (monkey kidney) cells (Vallipuram et al. 2010). *ATP13A4* was expressed at moderately high levels in both the human and mouse datasets and was increased more than seven-fold between colostrum and mature milk and six-fold between pregnancy and lactation in the mouse mammary cells.

6.2.3.6 Summary

Evidence about mechanisms of Mg^{2+} transport into milk is in very short supply. Based on levels of expression in the mammary gland, the first choices for relevant transporters would be MAGT and *SLC41A1*; however, other Mg^{2+} transporters expressed at possibly relevant levels in the lactating mammary gland of both mouse and human are *NIPA2* and *APT13A4*. *Mmgtl* may also be a candidate in the rodent mammary gland.

6.2.4 Iron

Iron is present at a concentration in mouse milk three times the plasma level (Zhang et al. 2000). It is much lower in human milk as full-term human infants are born with large iron stores. Unless they and/or their mothers have suffered from iron deficiency in pregnancy, as is often the case in developing countries, human infants have little need for oral iron in the first 4–6 months postpartum (Lönnerdal 2017; Santos et al. 2018). In species where iron is secreted into milk iron is transported in the blood stream and arrives at the basal surface of the mammary epithelial cell bound to the protein transferrin; the complex is endocytosed into clathrin-coated vesicles; these vesicles fuse with acidic endosomes, and the freed iron is transported to the cytoplasm by the divalent metal on transporter, DMT1; the transferrin is recycled to the extracellular space, and the intracellular iron is bound to ferroportin on intracellular compartment membranes and may be secreted into milk by the secretory vesicle mechanism (Kelleher and Lönnerdal 2005). To support the large amount of iron needed in mouse milk, the mRNA for the transferrin receptor, *Tfrc*, is upregulated nearly three-fold between pregnancy and lactation in this species. The two iron transporters DMT1 (divalent metal ion transporter, *Slc11a2*) and ferroportin (FPN1, *Slc40a1*) are expressed at low levels in the mammary gland of pregnancy but are upregulated at lactation (Fig. 6.7b). The opposite takes place in the human mammary gland (Fig. 6.7a). Expression of the transferrin receptor is decreased almost ten-fold between colostrum and mature milk. DMT1 (*SLC11A2)* is expressed at a moderate level in the human database, decreasing about 40% between colostrum and mature milk. *SLC40A1* (FPN1) was found at low levels in a mRNA database derived from a longitudinal study of gene expression in the milk of women directly after birth until day 41 of lactation (Mohammad et al. 2012) falling dramatically between colostrum and mature milk. With the exception of identification of single nucleotide polymorphisms associated with iron transporters in the Hadsell laboratory (Hadsell et al. 2018), there is little recent work on iron transport into milk.

6.2.5 Copper

Copper is transported in the plasma by albumin and a macroglobulin; however, half of the copper is tightly bound to a protein oxidase, ceruloplasmin, which binds six Cu^{2+} ions. Mammary epithelial cells can take up ceruloplasmin bound copper (Ramos et al. 2016); however, uptake into the cell from ceruloplasmin presents a problem as ceruloplasmin binds only Cu^{2+} [also called Cu(II)], and the only copper plasma membrane transporter identified, CTR1, takes up Cu^{+} [also called Cu(I)]. Linder and colleagues postulated that the reductase STEAP2, found on their cultured mammary cells, could reduce Cu^{2+} to Cu^{+}, but they were unable to knock down this protein sufficiently to prove this hypothesis (Ramos et al. 2016). While *STEAP2* mRNA was found at fairly low levels in the mouse and human datasets, the

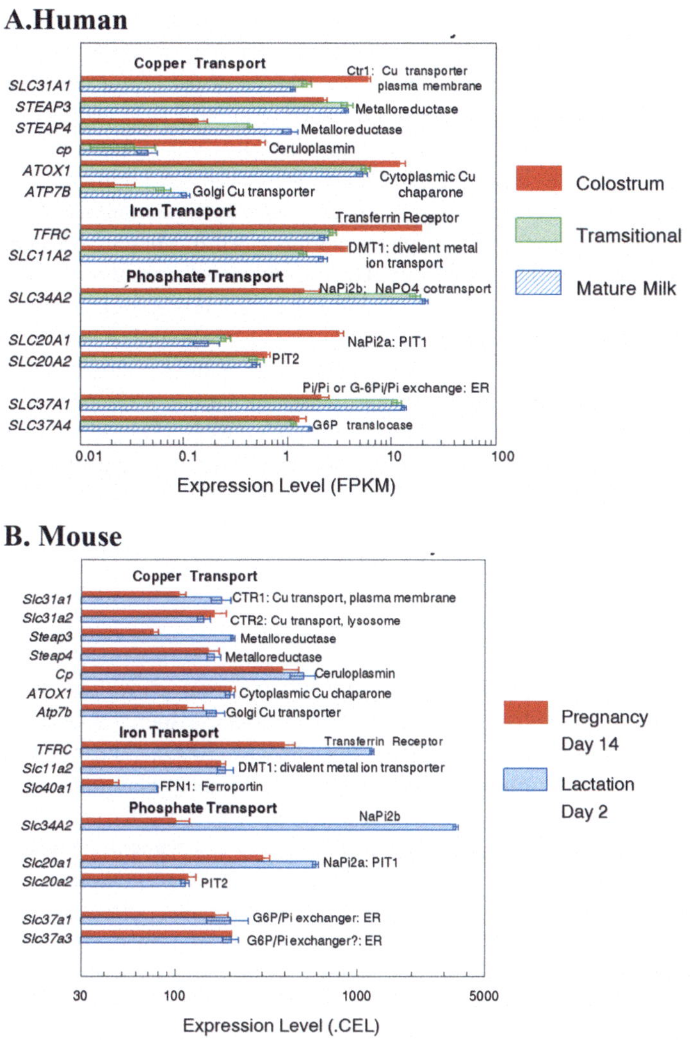

Fig. 6.7 Transcriptomic data for copper, iron, and phosphate transporters. (**a**) Human milk and (**b**) mouse mammary cells. Datasets are described in the legend to Fig. 6.4. PIT2 (*SLC11A2*), in addition to iron, is thought to transport Cd^{2+}, Co^{2+}, Cu^{2+}, and Mn^{2+}. Gene names are on the vertical axis; protein names are in the body of the graph

reductases *STEAP3* and *STEAP4* were found at substantial levels in both and were increased, in some cases substantially, in mature milk or lactating cells (Fig. 6.7). The picture of copper transport into the mammary epithelial cell that emerges from the data of Linder and the gene expression data reported here is that Cu-ceruloplasmin in the plasma binds to the basolateral surface of the lactating mammary cell, encounters STEAP3 or STEAP4, which reduces Cu^{2+} to Cu^{+}, and then enters the cell via CTR1.

However, free copper does not exist in biological systems; once in the cell, the ion is bound to ATOX1 or other cytoplasmic copper binding proteins (Bompiani et al. 2016). ATOX1 carries the copper to the membrane of the Golgi complex where it binds to an ATP-driven copper transporter ATP7a or ATP7b (Lutsenko et al. 2007; Polishchuk and Lutsenko 2013) and is reoxidized to Cu^{2+} and transported into the Golgi. In the case of the lactating mammary gland, *Atp7b* is highly expressed in the mouse dataset and expressed, albeit at lower levels, in the human database. Consistent with findings from other laboratories [summarized in Michalczyk et al. (2008)], *ATP7b* likely codes for the transporter. The protein structure and molecular function of this transporter have been well documented because it is associated with Wilson's disease (Pierson et al. 2018). Once in the Golgi, Cu^{2+} is bound to newly synthesized copper-dependent enzymes like ceruloplasmin and is then secreted into milk. The low expression of ceruloplasmin (*CP*) in the mature human milk dataset is puzzling as the protein has clearly been found in human milk (Puchkova et al. 2018). However, its concentration is higher in colostrum consistent with the higher level of ceruloplasmin mRNA in the colostral dataset (Fig. 6.7).

6.2.6 Phosphate

In normal goat milk, approximately 30% of the total phosphate is present as inorganic soluble phosphate; the remainder is either associated with proteins or covalently bound to casein (Shennan and Peaker 2000). These concentrations represent a 4 to 5 mmol/L concentration gradient over plasma phosphate (Muscher-Banse and Breves 2019). The question is how is this phosphate concentrated across the mammary epithelial cell. In early experiments, Shillingford and colleagues (Shillingford et al. 1996) found that phosphate was taken up largely via a Na^+-depending pathway in both tissue explants and the perfused rat mammary gland. The pathway was saturable with a K_m of about 1 mM. These researchers postulated that there was a Na^+-phosphate transporter in the basolateral membrane of the mammary epithelial cells.

The transporter coded by *SLC34A2* (NaPi2b) is thought to be the major phosphate transporter in the lactating mammary gland. The mRNA is upregulated at least ten-fold in lactation in both the mouse and human datasets (Fig. 6.7) and, in contrast to the postulate of Shillingford, was localized to the apical membrane in both mice and goats (Miyoshi et al. 2001; Huber et al. 2007), where it could transfer cellular phosphate into milk. The problem is that this transport mechanism does not account for the large amount of phosphate that must cross the basolateral membrane (Muscher-Banse and Breves 2019). There must be another sodium phosphate transporter in the basal membrane, or the localization of NaPi2b may change under various conditions as found in the kidney (Suyama et al. 2012). The identity of the basal transporter has been elusive.

Both the intestine and the kidney express another set of high affinity transporters for Na phosphate: the PIT1, NaPi2a and PIT2 systems encoded by the *SLC20A1* and

SLC20A2 genes, respectively (Collins et al. 2004). Both these systems are expressed in the mouse and human datasets, and the mRNA for PIT1 is upregulated in the lactating mouse cells (Fig. 6.7). While PIT1 is expressed at relatively high levels in colostrum, it is significantly downregulated in mature milk (Fig. 6.7). It is clear that phosphate transport into milk is currently poorly understood.

6.2.7 Other Bivalent Ions

We have not dealt with the transport of cadmium, cobalt, selenium, molybdenum, arsenic, and manganese although all are transported to some extent into milk, sometimes using transporters we have examined here (see Figs. 6.6 and 6.7). Selenium appears to be nutritionally important, but appears in milk almost entirely as selenoproteins (McConnell 1948; Dorea 2002); it does appear to share a transport pathway with sulfate (Shennan and McNeillie 1990), but its mammary metabolism and transport pathways have received little attention. The best information available about sulfate transport appears to be the role of the PAPS transporters that allow sulfation of glycoproteins in the Golgi compartment (see Sect. 6.5.3 on nucleotide transport).

6.3 Transport into the Endoplasmic Reticulum

6.3.1 Protein

The protein translocator located on the endoplasmic reticulum (ER) membrane, SEC61, transports ribosome-attached proteins as they are synthesized (cotranslational transport) as well as cytoplasmic proteins associated with molecular chaperones (Skach 2007). For cotranslational transport, newly synthesized peptides at the ribosome are targeted to SEC61 by a signal recognition peptide at the N-terminus followed by release of the completed protein into the phospholipid bilayer of the membrane or the lumen of the channel (Linxweiler et al. 2017). N-glycosylation of the protein can be accomplished during this transfer by an associated oligosaccharyltransferase (OST) complex (Braunger et al. 2018). For a complete listing of the many proteins thought to be associated with the SEC61 translocon, see Lang et al. (2017). In addition to SEC61 and its associated proteins, SEC63 and SEC62 as well as several OSTs, BIP (binding immunoglobulin protein, GRP-78, HSPA5), an ATP, and calcium-dependent heat-shock protein-like protein, and the multicomponent TRAP (translocon-associated protein) are essential components of the translocation machinery, binding translocated proteins and assisting with their folding as well as identification of misfolded proteins. Another protein thought to be associated with translocation is RAMP4 (gene name *SERC1* or stress-associated ER response protein), which is thought to stabilize membrane proteins during cellular stress and facilitate glycosylation. The mRNA for all of these proteins is expressed at high levels in both the human and mouse datasets (Fig. 6.8). All appear to be upregulated in the lactating alveolar cell of the mouse. In the human,

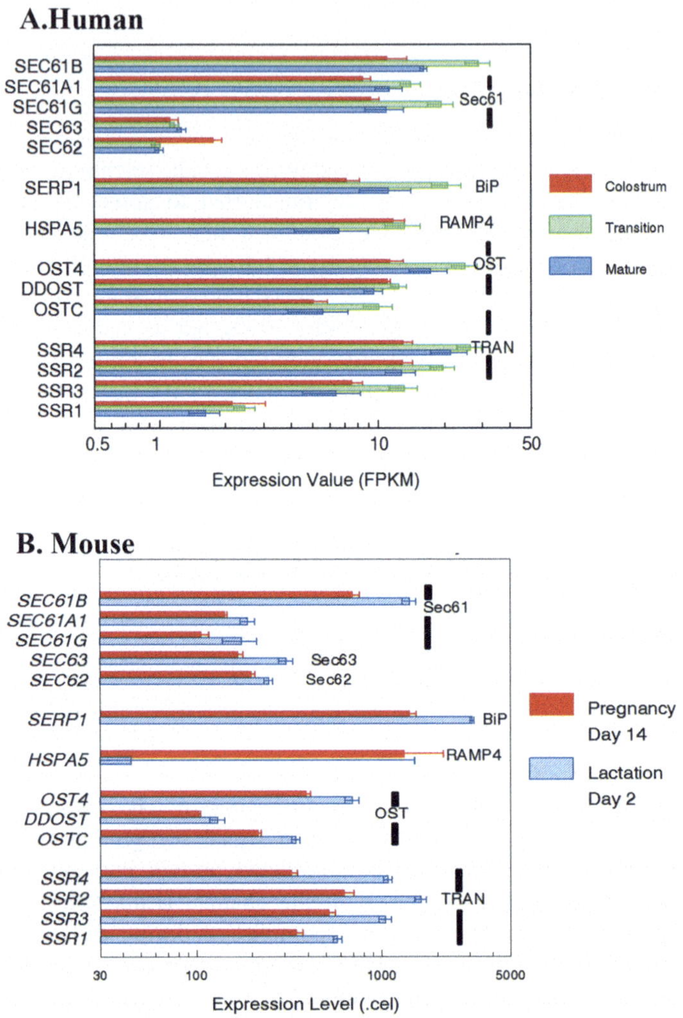

Fig. 6.8 The transcriptome of major proteins associated with translocation of membrane and secreted proteins in the endoplasmic reticulum. (**a**) SEC61 is the major protein translocating channel; its subunits are expressed at relatively high levels with auxiliary proteins Sec62 and Sec63 at considerably lower levels. *SERP1* (RAMP4) and *HSPA5* (BIP) are expressed at moderate level, whereas *OST4* and *SSR4* (TRANδ) are highly expressed. Note that expression of all these mRNAs is the highest in transitional milk in the human database. (**b**) In the mouse database, all the mRNAs are increased 1.7 to 3.3-fold between pregnancy day 14 and lactation day 2 (Heinz et al. 2016). Gene names are on the vertical axis; protein names are in the body of the graph. NOTE: TRAN is misplaced upward in A

almost all are expressed at highest levels in transitional milk, perhaps reflecting the high synthesis of proteins like lactoferrin at this time.

6.3.2 *Calcium*

The Ca^{2+} ATPase SERCA is responsible for calcium transport into the ER of most cells. Ca^{2+} levels in the ER signal to the calcium channel ORAI1 through the STIM protein, a transmembrane ER protein (Feng et al. 2010; Cross et al. 2013), leading to store-operated calcium entry or SOCE. The mRNA for STIM2 is increased in lactation in mice as well as humans (data not shown), suggesting that the ER calcium pool is increased in the lactating cell.

6.3.3 *Chloride Channel, CLIC-Like, CLCC1*

Expression of the gene *CLCC1* for a CLIC-like channel, also known as MCLC, is found is most tissues of the body. The protein was found to be localized to the endoplasmic reticulum, nucleus, and Golgi of Chinese hamster ovary cells and to have a high permeability to anions (Nagasawa et al. 2001). Knockdown of *CLCC1* in cultured cells led to increased ER stress, suggesting that this channel may promote the proper chloride concentrations necessary for protein folding (Jia et al. 2015). It is expressed in both the human and mouse datasets (Fig. 6.12).

6.3.4 *Glucose-6-Phosphate (G-6-P)*

The *SLC37* family consisting of four sugar-phosphate exchange proteins encoded by *SLC37A1–4*. *SLC37A1*, and *SLC37A4* is expressed in the human database, and *Slc37a1* and *Slc37a3* are expressed at equal levels in the mouse database (Fig. 6.7). *SLC37A1* (G6P/Pi exchange) encodes an ER-associated, P_i-linked antiporter, which can catalyze both homologous (Pi/Pi) and heterologous (G6P/Pi) exchange (Cappello et al. 2018). Because the mRNA is homologous to the bacterial glycerol 3-phosphate permeases, it is postulated to transport glycerol-3-phosphate (Cappello et al. 2018). The exchanger encoded by *SLC37A1* has received the most attention and is expressed at moderate levels in both the mouse and human dataset as well as being one of the most highly expressed genes in the lactating bovine mammary gland (Raven et al. 2016), where it is correlated with milk volume, milk phosphate, and the cheese-making properties of bovine milk (Sanchez et al. 2018). Its ER localization rules it out as the basolateral phosphate transporter. The function of *SLC37A3* is not well-understood at this time (Cappello et al. 2018). *Slc37a4* is the best characterized *SLC37* transporter, and its major function appears to be to transport glucose-6-phosphate into the ER lumen (Chou and Mansfield 2014). Why it is not expressed at meaningful levels in the mouse mammary gland is not known.

6.3.5 *Other Potential ER Transporters*

Zinc may be transported into the ER by ZIP7 (*Slc39a7*) as it has been implicated in ER stress (Bin et al. 2017), and its mRNA is increased during lactation in the mouse (Fig. 6.5).

6.4 Amino Acid Transport in the Lactating Mammary Gland

It has long been known that the uptake of amino acids by the mammary gland increases markedly during the transition from pregnancy to lactation (Verma and Kansal 1993). Many of the mechanisms involved have been elucidated in the intervening decades; the transporters are complex because they must move multiple amino acids with different structures across the cell membrane and often cotransport them in and out of cells (Manjarin et al. 2014). Na^+-dependent transporters use the energy generated by the Na^+/K^+ ATPase, exchanging amino acids transported into the cell with Na^+ as it flows down its electrochemical gradient. Non-Na^+-dependent transporters often exchange amino acids accumulated by Na^+-dependent transport for amino acids not accumulated by this process. However, some transporters act merely as conductive channels moving amino acids down their concentration gradients. The amino acid transporters have been categorized as Na^+-independent neutral, Na^+-dependent neutral, anionic branched chain, and Na^+-independent cationic (Shennan et al. 2002; Manjarin et al. 2014). Taurine transport is often treated separately. Each of these categories will be discussed in turn.

6.4.1 *Na^+-Independent Neutral Amino Acid Transporters*

This set of transporters has been known as System L; they are encoded by *SLC7A5* (LAT1) and *SLC7A8* (LAT2; Fig. 6.9). LAT transporters are known to transport Ala, Ser, Val, Thr, Leu, Ile, Met, Phe, Tyr, Trp, and His across the plasma membrane. *SLC3A2* encodes the heavy chain transport unit 4F2hc (Yan et al. 2019), which is linked with both LAT1 and LAT2 by disulfide linkages and appears to be essential for their activity. *Slc3a2* is upregulated in the mouse and expressed at levels above the corresponding mRNA for LAT1 and LAT2 in the human. *Slc3a2* is expressed at high levels in colostrum possibly because colostrum enhanced amino acid transport

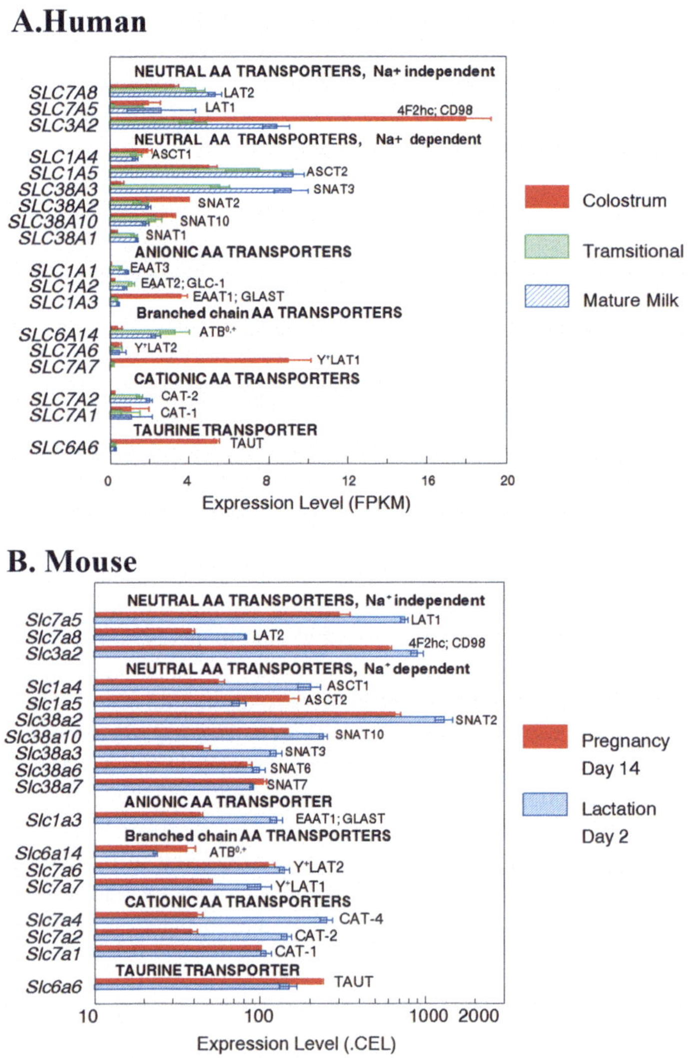

Fig. 6.9 Expression of mRNA for amino acid transporters. (**a**) Human colostrum, transitional, and mature milk. (**b**) Mouse, mid-pregnancy and day 2 lactation. Human values from Lemay et al. (2013a); values for cells from the mouse mammary gland from Heinz et al. (2016). See legend to Fig. 6.4 for details. Gene names are on the vertical axis; protein names are on the body of the graph

may be necessary for synthesis of lactoferrin in the human mammary gland in the colostral phase. System L transporters have been localized to the basolateral membrane in the mammary gland (Shennan et al. 2002); they may transport branched chain amino acids in addition to the listed neutral amino acids and may participate in an amino acid exchange mechanism (Baumrucker 1985), although the last suggestion requires verification. There is abundant evidence that the LATs are upregulated by prolactin (Shennan and Boyd 2014) and growth hormone treatment led to a threefold increase in the expression of *SLC3A2* in the mammary tissue of the lactating bovine (Sciascia et al. 2015). Shennan has suggested that apical LAT transporters are responsible for the free amino acids found in milk (Shennan and Boyd 2014); this suggestion remains to be verified.

6.4.2 Na^+-Dependent Neutral Amino Acid Transporters

The mRNA for the neutral amino acid transporter *Slc1a4* (ASCT1) was the highest expressed *Slc1a* transporter in the mouse dataset, whereas *SLC1A5* (ASCT2) was the highest expressed in the human dataset and was increased almost two-fold with lactation (Fig. 6.9). *Slc1a5* (ASCT2) was also found at a high level in the mouse dataset, but was decreased in lactation compared to pregnancy. In neither species, was it increased with lactation. These transporters are Na^+-dependent with the highest affinity for Ala, Ser, Gly, Val, Thr, and Cys. There is currently no evidence as to the basal or apical localization of ASCT1 or ASCT2 (Shennan and Boyd 2014). However, they would be expected to be largely localized to the basal membrane because protein synthesis relies on essential amino acids transported from the extracellular space to the cytoplasm.

Another set of Na^+-dependent amino acid transporters (the SNATs) is encoded by the *SLC38A* gene set of which *Slc38a2*, *Slc38a10*, *Slc38a3*, *Slc38a6*, and *Slc38a7* are expressed in the mouse lactation dataset; *SLC38A3, SLC38A2, SLC39A10*, and *SLC38A1* are expressed in the human dataset (Fig. 6.9). *Slc38a2* was studied in the rat and found to be upregulated at lactation (Lopez et al. 2006; Velazquez-Villegas et al. 2015) as it was in the sow (Chen et al. 2018a) and in the mouse database (Fig. 6.9b) but not in the human. SNAT2 expression was increased by prolactin in tissue culture, and its expression was altered by diet in the rat and the bovine (Dai et al. 2018).

6.4.3 Anionic Amino Acid Transporters

These transporters are known as excitatory amino acid transporters (EAATs) because they transport glutamate into neuronal cells, terminating its neurotransmitter action. Both anionic amino acid transporters for glutamate encoded by *SLC1A2* (EAAT2) and *SLC1A3* (EAAT1; GLAST) were expressed at low levels in the human milk dataset with only *Slc1a3* expressed at levels above background in the murine dataset (Fig. 6.9). Similar to the findings of Aleman (Aleman et al. 2009), neither of these genes showed substantial regulation with lactation state. *SLC1A1* (EAAC3), known to transport both glutamate and aspartate, was expressed at a low level in the human dataset during mature lactation.

6.4.4 Branched Chain Amino Acid (BCAA) Transporters

These transporters, ATB$^{0,+}$ (*SLC6A14)*, Y^{+}LAT1(*SLC7A7*), and Y^{+}LAT2 (*SLC7A6*), transport the cationic amino acids lys, arg, and his as well as the neutral amino acids leu, isoleu, met, ala, and ser and possibly thr and val across the plasma membrane (Manjarin et al. 2014). ATB$^{0,+}$ works as an exchanger, coupling transport of cationic amino acids to the inward movement of Na^+ and Cl^-. It was expressed in the human database but not in the mouse database, and its mRNA was found to decrease between pregnancy and lactation in the rat mammary gland (Shennan and Boyd 2014), suggesting that the transporter may not be functional in the rodent mammary gland. Like LAT1 and LAT2, Y^{+}LAT1 and Y^{+}LAT2 are coupled to the heavy chain transport unit 4F2hc (*SLC3A2*) by disulfide bonds (Torrents et al. 1998). These two transporters as well as 4F2hc are expressed in the human milk database although *SLC7A7 (*Y + LAT1) decreased to undetectable levels in mature milk. *Slc7a7* (Y^{+}LAT1) and *Slc7a6* (Y^{+}LAT2) along with *Slc3a2* (4F2hc) are increased with lactation in the mouse (Fig. 6.9b). Interestingly, we failed to detect significant levels of the mRNA for Slc6a14 (ATB$^{0,+}$) in the mouse database.

6.4.5 Na^{+}-Independent Cationic Amino Acid Transporters

Sodium-independent transporters for cationic amino acids CAT-1 (*SLC7A1*), CAT-2 (*SlC7A2*), and CAT-4 (*SLC1A4*) also known as y^{+} are specific for lys, arg, his, and orn. Transport is not dependent on Na^+ or protons but responds to changes in membrane potential as well as to the presence of intracellular amino acids (Manjarin et al. 2014). The mRNA for both CAT-1 (*SLC7A1)* and CAT-2 (*SLC7A2*) is present in the human and mouse datasets; *SLC7A2* was significantly increased in mature milk as well as in mammary cells from lactating mice. SLC7A1 was expressed at similar levels in colostrum and mature milk in humans and in cells from pregnant and lactating mouse mammary glands (Fig. 6.9). *Slc7a4* (CAT-4) was the highest expressed cation transporter in the mouse dataset.

6.4.6 Taurine Transport

Taurine, or 2-aminoethanesulfonic acid, is found at reasonable concentrations in the milk of many species including humans and mice (Rassin et al. 1978) and appears to be important in humans for retinal development (Heird 2004). It has been supplemented in most infant formulas since the early 1980s, although recommended levels have not been established. Taurine transport has been extensively studied by Shennan and colleagues (Shennan and McNeillie 1994; Shennan 1995) who found that the compound is concentrated in the mammary cell by a transporter dependent on both Na^+ and Cl^-. They suggested that milk taurine is derived from taurine crossing the apical membrane by a mechanism not clarified in this work, which was carried out before the molecular identity of the taurine transporter, *SLC6A6*, was established. The finding that the mRNA for this transporter is decreased in lactation in the mouse and transitional and mature milk in the human (Fig. 6.9) suggests that another transporter may be responsible for the passage of taurine from the cytosol into milk.

6.5 Transport of Sugars, Sugar Compounds, and Sulfate in the Lactating Mammary Cell

Lactose is the major carbohydrate in the milk of most species, and it is synthesized in the Golgi from glucose and UDP-galactose, increasing the osmolarity of the solutes within the Golgi and leading to an influx of water into this compartment. In addition to lactose, milk, particularly human milk, contains many oligosaccharides and glycoproteins, which are also synthesized in the Golgi vesicle system utilizing nucleotide sugars. Finally, the apical membrane of the mammary epithelial cell as well as the milk fat globule membrane contains many proteins like MUC-1, which are also glycosylated in the Golgi compartment (Liao et al. 2011; Becker et al. 2018). This means that transporters for glucose and nucleotide sugars must be present in the membranes of the Golgi compartment (Hirschberg 2018).

Glucose is the major sugar imported into the lactating mammary gland serving as both substrate for energy production and the source of carbohydrate for lactose, oligosaccharide, glycerol, glutamate, citrate, and fatty acid synthesis. The reader is referred to an excellent review by Zhao for an outline of all these pathways (Zhao 2014). In the lactating dairy cow, as much as 3 kg/day of glucose must be extracted from the blood stream to meet her needs and it has been shown in lactating sows and goats that the lactating gland extracts 9 mg/min/100 g of tissue (Linzell 1960; Linzell et al. 1969). Thus, glucose transport across the basal plasma membrane is powerful. It is thought to depend on three types of transporters, a facilitated diffusion process (Tanasova et al. 2018), primarily carried out by (a) the product of the *Slc2A1* gene, GLUT1, (b) a Na^+-dependent process carried out by the product of the *Slc5a1* gene, and (c) a fairly recently discovered glucose transporter SWEET (Chen et al. 2010), the product of the *Slc50a1* gene. This protein may be localized to the basolateral membrane of the lactating mammary cell.

6.5.1 Glucose

Glucose transport into the mammary epithelial cells of many species uses at least in part a facilitated diffusion process that is saturable, Na^+-independent, specific to D-glucose, D-mannose, D-xylose, D-galactose, and 2-deoxyglucose (2-DG), and inhibitable by cytochalasin-B or phloretin [summarized in Zhao (2014)]. However, it has been shown repeatedly that cytochalasin B only inhibits about 50% of this transport process. For example, Kuhn and colleagues (Threadgold et al. 1982) observed a second nonspecific transport process for 2-DG, which was also able to transport sorbitol, fructose, and sucrose and was not inhibited by cytochalasin B or even phloretin. A saturable facilitated diffusion process available to D-glucose and D-galactose but not sorbitol or fructose was present in the apical membrane as well (Faulkner and Peaker 1987). In the present study, *SLC2A1*, coding for GLUT1, was the dominant glucose transporting GLUT mRNA expressed in both the human and mouse mammary datasets (Fig. 6.10). This gene was expressed at high level in rat and bovine glands as well as in other studies of mouse and human gland (Zhao 2014). As in the mouse dataset, *Slc2a8* expression has been observed in the mammary gland of the lactating rat (Laporta et al. 2013) and bovine (Zhao et al. 2004). Expression of *Slc2a1* and *Slc2a8* in the mammary gland has been shown to make consistent and progressive increase from pregnancy to early lactation in mice (Nemeth et al. 2000), rats (Burnol et al. 1990; Camps et al. 1994), and cows (Zhao and Keating 2007; Mattmiller et al. 2011). *SLC2A8* was not observed in the human dataset.

GLUT1 protein has been localized by immunocytochemistry to the basal membrane of the lactating mammary cell in both the rat (Macheda et al. 2003) and the bovine (Finucane et al. 2008). It was localized to Golgi-associated vesicles by localization with 110-kD coatomer-associated protein beta-COP (Nemeth et al. 2000). Fairly strong GLUT8 staining was observed in both the apical and basolateral membrane of lactating bovine mammary epithelial cells (Zhao 2014). It was suggested by Chen and associates (2010) that SWEET (*SLC50A1*) is also located on the Golgi membrane of the lactating mammary cell, but Zhao pointed out that the selectivity of transport is not consistent with its presence in the Golgi membrane (Zhao 2014). GLUT1 has traditionally been thought of as the major glucose transporter in the Golgi membrane. It also transports glucose across the basal membrane and can be responsible for glucose transport across the apical membrane as well (Zhao 2014). GLUT8 has also been implicated in glucose transport in the lactating mammary gland (Zhao 2014) both in the basal membrane and the Golgi membrane. *Slc2a4* (GLUT4), the mRNA for an insulin-dependent glucose transporter, appears to be present in the mouse database, but may represent contaminating adipocytes, since it has not been found by others to be expressed in the lactating mammary cell. The role of the Na^+/glucose symporter represented by *SLC5A1*in the mammary gland is not at present clear.

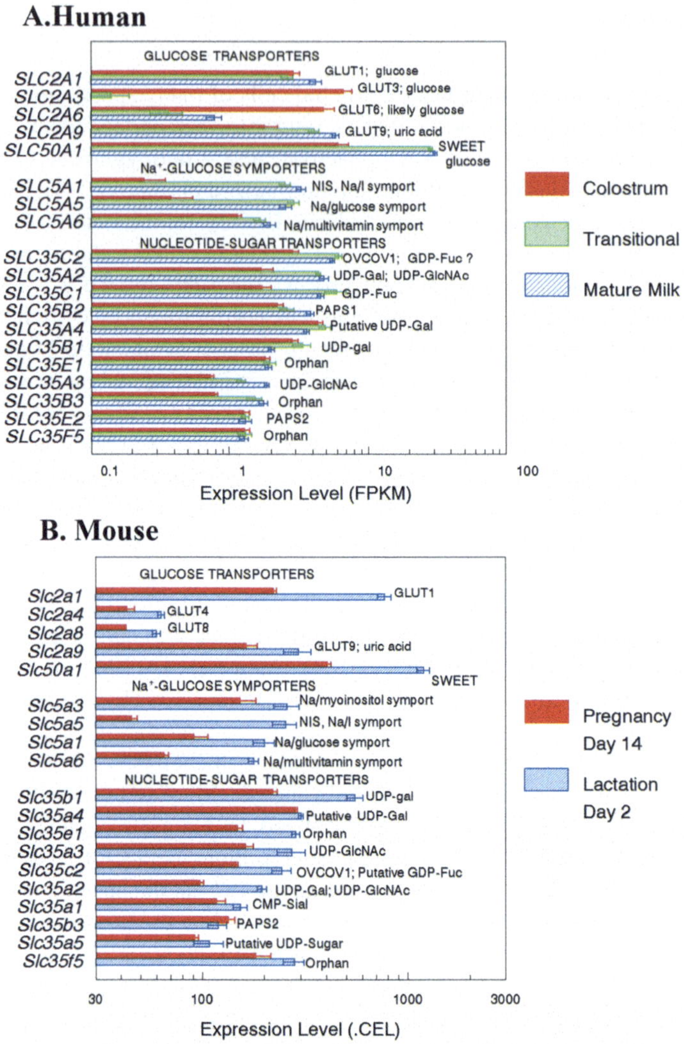

Fig. 6.10 mRNA expression of glucose and nucleotide transporters. (**a**) Human milk fat globule RNA (Lemay et al. 2013a) and (**b**) mouse mammary epithelial cells (Heinz et al. 2016). See the legend to Fig. 6.4 for details. Gene names are on the vertical axis; protein names are in the body of the graphs

6.5.2 *Sugar Nucleotides*

Free sugars enter the cytoplasm and may be converted to nucleotide sugars like UDP-galactose, UDP-N-acetylglucosamine, or GDP-fucose. Proteins to be glycosylated are synthesized on polysomes and then translocated into the endoplasmic reticulum by mechanisms described in Sect. 6.3 on the endoplasmic reticulum

(ER). They may be N-glycosylated in the ER by sugars linked to dolichol (Hirschberg 2018). There they enter vesicles to be translocated to the Golgi system from which they go either to the plasma membrane or are secreted (Hirschberg 2018). At the same time, nucleotide sugars made in the cytosol are transported into the endoplasmic reticulum or Golgi (Carey et al. 1980) via a series of specific transporters that are able to produce a 10- to 30-fold concentration gradient by coupling to the corresponding nucleoside monophosphate (Capasso et al. 1989). The nucleotide sugars in the Golgi are used by glycosyltransferases to synthesize lactose (in the presence of alpha-lactalbumin), glycosaminoglycans, or oligosaccharides. The coupled nucleoside is released to be transported back into the cytoplasm in exchange for new nucleotide sugar molecules.

Characteristics of transport by nucleotide sugar transporters are (Hadley et al. 2014):

- The entire nucleotide sugar is translocated
- The process is saturable and temperature and time dependent
- The nucleotide sugar can be concentrated within the ER or Golgi.
- The presence of ATP and ionophores does not alter translocation.
- The process is energized by coupled translocation of the corresponding nucleoside monophosphate.
- Translocation is not inhibited by the free sugar.
- The localization of a particular transporter can be the ER, the Golgi or both.

As expected, UDP-galactose (UDP-Gal), UDP-N-acetylglucosamine (UDP-GlcNAc), and GDP-fucose transporters are expressed in both the human and mouse datasets as is at least one of the transporters for PAPS (3′phosphoadenosine 5′phosphosulfate) (Fig. 6.10). Most are upregulated in mature milk or at lactation in the mouse.

6.5.2.1 Other Nucleotide Transporters

Certain *SLC35* genes whose function is poorly defined at present are expressed at fairly high levels in both datasets. These are *SLC35E1* and *SLC35F5* in the human dataset and *Slc35e1* and *Slc31f5* in the mouse (Fig. 6.10, designated as orphan). While they are considered to be transporters, no information appears to be available about their substrates, localization or function; all were expressed at similar levels in prelactating and lactating samples in both species in the Lemay and Ramanathan datasets (Lemay et al. 2013a; Heinz et al. 2016).

6.5.3 Sulfate

The mRNA for a sulfate transporter encoded by the *Slc26a2* gene and known as the diastrophic dysplasia sulfate transporter (DTDST) is found in the mouse dataset but not in the human (Fig. 6.14b). Mutations in this gene lead to chondrodysplasias as the protein is important for sulfation of cartilage proteins (Park et al. 2014b). DTDST has been found to downregulate the TRAIL receptors, DR4 and DR5 in breast cancer cell lines, suggesting that it plays a role in drug resistance in this disease (Dimberg et al. 2017). It seems likely that it plays an important role in sulfate transfer to milk or milk proteins, at least in some species, but this hypothesis has apparently not been tested.

6.5.3.1 Sulfate Donors Transferred into the Golgi

The PAPS transporters are universal sulfate donors for sulfation of proteins within the Golgi; the PAPS transporters are encoded by *SLC35B2* and *SLC35B3* mRNAs and provide sulfate for sulfation events within the Golgi. Both were expressed in the human database and are upregulated in mature milk (Fig. 6.10). Only *Slc35b3* was expressed in the mouse database.

6.6 Monovalent Ion Transport in the Lactating Mammary Epithelial Cell

Sodium (Na^+), potassium (K^+), and chloride (Cl^-) ions are not transformed in the cell or used in synthetic reactions. These ions, particularly Na^+, are important in the cotransport of many nutrients across cell membranes including amino acids (Mackenzie and Erickson 2004; Velazquez-Villegas et al. 2014), glucose (Wright and Turk 2004), phosphate (Collins et al. 2004; Murer et al. 2004; Huber et al. 2007), ascorbic acid (Takanaga et al. 2004), iodide, and even drugs (Gerk et al. 2002). Here, we are concerned with their transport across the basal and apical membranes and the forces that establish stable concentrations within the cytoplasm and in the milk. Concentrations of Na^+ and K^+ in milk, plasma, and mammary tissue for guinea pigs and mice are shown in Table 6.1. In all mammalian species, there is a large Na^+ gradient between the interstitial space (~150 mM) and the milk (5–20 mM depending on species). K^+ is high in the cytoplasm (~120 mM) and higher in milk (24 mM) than in plasma and interstitial space (4.5–5.1 mM). Cl^-, shown for guinea

Table 6.1 Monovalent ion concentrations in plasma and milk in various species compared to intracellular ion concentrations in the mouse and guinea pig (mM) (Shennan 1998)

	Human (Srisomboon et al. 2018)		Guinea Pig (Linzell and Peaker 1971a)			Mouse (Berga and Neville 1985)		
Ion	Plasma	Milk	Plasma	Milk	Intracellular	Plasma	Milk	Intracellular
Na^+	138	5–8	150	8	42	146	26	47
K^+	4.6	11–15	4.5	24	143	5.1	47	129
Cl^-			116	12	66			

pigs, is high in the interstitial space and the plasma (116 mM), moderate intracellularly (62 mM) and low in milk (12 mM).

On the basis of the ratios of Na^+ and K^+ between milk and their measured intracellular concentrations in the guinea pig, Linzell and Peaker nearly half a century ago proposed that the Na^+/K^+ ATPase in the basolateral membrane of the alveolar cell produced and maintained the Na^+ and K^+ gradients across this membrane with Cl^- entering through some sort of Na^+-K^+-Cl^- cotransport (Linzell and Peaker 1971b). Channels in the apical membrane were hypothesized to allow cellular Na^+ and K^+ to equilibrate with the milk; taking the membrane potential across the apical membrane into account by using the Hodgkin-Katz Goldman equation, the ratios across the apical membrane supported this hypothesis in the guinea pig. However, detailed measurements of Na^+ and K^+ in the mouse mammary gland led Berga and Neville to question the concept of Na^+ and K^+ equilibrating across the apical membrane (Berga and Neville 1985), and it is clear from data of many other species that this simple concept does not hold and more complex systems must determine the levels of Na^+ and K^+ in the milk. Cl^- is out of equilibrium across the basal membrane. Therefore some sort of Cl^- exchanger must allow passage of the correct amount of Cl^- into the cell with the membrane potential possibly driving Cl^- across the apical membrane. The molecular nature of these channels and transporters is only now becoming clear as we shall see (Schultz 2016). However, one aspect of the transport of these ions in the lactating mammary alveolar cells has not explicitly been taken into account in any studies we were able to find: The partitioning of monovalent ions between the cytoplasm and the many membrane-bound compartments within the cell. For an example of partitioning of an ion between these structures and the cytoplasm that has been studied extensively, the reader is referred to the section on calcium transport.

The most studied basic question is: What are the mechanisms by which the concentrations of the monovalent ions are maintained across the apical and basolateral membranes of the lactating mammary epithelial cell? Bruce Schultz defined the situation as an “unheavenly state” where a heavenly state means that only one transporter need be studied at a time; unfortunately, within the cell, changes in one system reverberate to all other systems and none of the transport systems can be studied unless the entire system is defined (Schultz 2011). Nonetheless, we will examine each ion in turn, hoping to proceed to a synthesis by the end of this section.

6.6.1 Na^+/K^+ ATPase

Perhaps the most important transporter in the basal membrane is the Na^+/K^+ ATPase, an enzyme that utilizes energy from ATP to pump sodium ions out and potassium ions into the cell. The ouabain-sensitive Na^+/K^+ ATPase present in the basolateral membrane of the mammary alveolar cells at all stages of development has been shown to be responsible for the low sodium and high potassium content of the mammary alveolar cell, pumping Na^+ out and K^+ into the cell using energy provided by ATP (Linzell and Peaker 1971a). This mechanism is supported by the results of many experiments both in the in vivo gland and in many cell culture models (Shennan and Peaker 2000). The Na^+/K^+ ATPase enzyme is composed of two subunits, a large catalytic α subunit (*ATP1A1*) and a smaller *β* glycoprotein subunit (*ATP1B3*, CD298). The beta subunit regulates, through assembly of alpha/beta heterodimers, the number of sodium pumps transported to the plasma membrane. The mRNA for the alpha subunit is expressed at reasonably high levels in both the human and mouse datasets (Fig. 6.11). Interestingly, the mRNA for the β-subunit of the enzyme is decreased in cells from the lactating gland compared to the pregnant gland in the mouse as well as in mRNA from colostrum compared to mature milk in the human. The question is whether the proteins for the two subunits have different stabilities so that less mRNA is needed to synthesize the appropriate amount of beta subunit.

6.6.2 Na^+ Transporters in the Apical Membrane

6.6.2.1 ENaC

The well-known amiloride-sensitive sodium channel, ENaC, is composed of three homologous subunits containing various combination of α, β, γ, and δ, subunits encoded by genes *SCNN1A*, *SCNN1B*, *SCNN1G*, and *SCNN1D* in the human and *Scnn1a*, *Scnn1b*, *Scnn1c*, and *Scnn1d* in the mouse. The combined protein channel is expressed in the apical membrane of renal and other epithelia (Klemens et al. 2017). Of the mRNA for these four components, *SCNN1B* is expressed at a very high level and is increased about ten-fold between colostrum and mature human milk; *SCNN1A* and *SCNN1G* are expressed at lower levels in mature milk, but are also substantially increased between colostrum and mature milk (Fig. 6.11a). Likely, ENaC is a physiologically relevant channel localized in the apical membrane of the human mammary alveolar cell; in sections of the nonlactating gland in the Human Protein Atlas (https://v17.proteinatlas.org/), the protein appears to be localized apically. In the mouse, the mRNA for *Scnn1b* is the highest expressed component of ENaC, consistent with previously published results (Boyd and Náray-Fejes-Tóth 2007), and it and the γ-component were upregulated in cells from the lactating gland although expression was relatively low. Both the β- and γ-subunits were upregulated by dexamethasone in an in vitro model of the bovine mammary gland (Quesnell et al. 2007); the upregulated transporter was sensitive to amiloride. ENaC has been studied in tissue culture models of primary human mammary epithelial cells where it is localized apically (Srisomboon et al. 2018) and appears to mediate apical sodium secretion (Blaug et al. 2001; Quesnell et al. 2007; Srisomboon et al. 2018). It was upregulated by prolactin in primary human mammary epithelial cells in culture (Lee et al. 2007) as well as by purinergic substrates like UTP, which

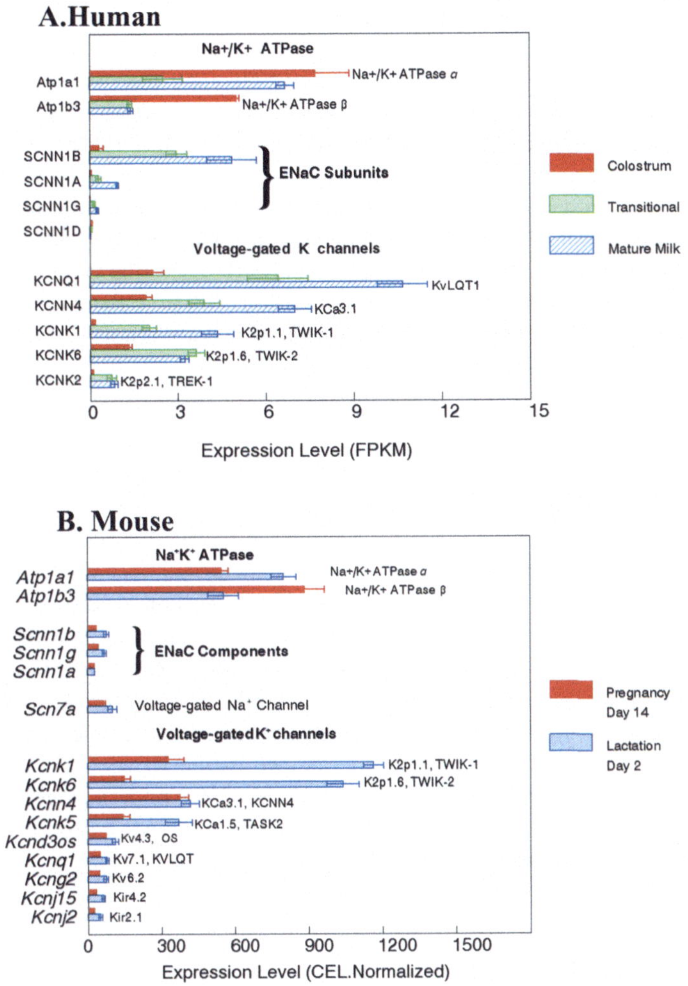

Fig. 6.11 Major cation transporters in the mammary gland of the human and mouse. (**a**) Expression values for the human mammary gland as represented by mRNA in the milk fat globule (Lemay et al. 2013a). (**b**) Expression values of mRNA for cation transporting proteins in the mammary gland of the day 14 pregnant mouse and the day 2 lactating mouse as contained in the Ramanathan database (Heinz et al. 2016). Gene names are on the vertical axis; corresponding protein names are in the body of the graphs

apparently acted by increasing cell calcium, which in turn upregulated ENaC (Lee et al. 2007). These findings provide support for the idea that ENaC is responsible for equilibration of Na^+ across the apical membrane of the lactating mammary epithelial cell.

6.6.2.2 Voltage-Gated Na Channel, Scn7a

Scn7a codes for a sodium-sensitive sodium channel that is present in the Ramanathan murine dataset at levels that are increased in lactation and higher than expression of the subunits of ENaC (Fig. 6.11b). While this mRNA is present at vanishingly low levels in the Lemay database (not shown), the protein can be seen in the nonpregnant human breast in the Human Protein Atlas where it appears to be localized to the apical surface of the epithelial cells. This channel has mainly been studied in the nervous system but has been localized to alveolar type II cells in the lungs, where it was postulated to absorb sodium through the alveolar surface (Xu et al. 2015). No additional information about its expression or localization in the mouse mammary gland appears to be available.

6.6.3 K^+ Channels

Potassium channels create a favorable electrochemical gradient for Na^+ entry and Cl^- exit from epithelial cells by altering the K^+ distribution across the plasma membrane. The graphs in Fig. 6.11 indicate that four voltage-sensitive K^+ channels are synthesized at relatively high levels in the lactating mammary glands of both human and mouse: These include the dual channel K^+ transporters $K_{2P}1.1$ and K_{2P} 1.6 encoded by the *Kcnk1* and *Kcnk6* genes in the mouse and the *KCNK1* and *KCNK6* genes in the human, respectively, as well as the KCa3.1 and the KvLQT channels encoded by the *KCNN4* or *Kcnn4* and *KCNQ1* or *Kcnq1* genes, respectively. *Kcnk5*, which encodes for the channel TASK2, is expressed at reasonably high levels in the lactating mouse gland.

6.6.3.1 Two Pore K^+ Channels

$K_{2P}1.1$ and K_{2P} 1.6 are two pore K^+ channels inhibited by bupivacaine and quinidine (Srisomboon et al. 2018). In humans, *KCNK1* codes for the protein TWIK1, a pH-sensitive dimeric potassium-selective channel, expressed in the apical membrane of the mammary epithelial cell. *Kcnk1*, encoding TWIK1, is the highest expressed channel in the lactating mouse mammary cells (Fig. 6.11b). TWIK1 and TWIK2 were localized diffusely at the apical membrane in the human bronchial epithelium (Zhao et al. 2012) and at the apical membrane of cultured human mammary epithelial cells (Srisomboon et al. 2018). These findings suggest that TWIK1 and TWIK2 transfer K^+ across the apical membrane; however, firm evidence would involve immunohistochemical studies on native lactating mammary epithelial cells, apparently not available. In the bronchial epithelium, the protein product of *KCNK5*, TASK2, appeared to be localized to a subapical compartment (Srisomboon et al. 2018) similar to findings of Davis and Colway on another type of airway epithelial cell (Davis and Cowley 2006). These observations raise the possibility that TASK2 could transfer potassium across the membrane of an intracellular compartment, possibly the Golgi.

6.6.3.2 KCa3.1

Another K^+ channel that is expressed at fairly high levels in mRNA from both human and mouse mammary glands is the KCa3.1 channel encoded by the *KCNN4* gene in the human and the corresponding *Kcnn4* gene in the mouse (Fig. 6.11). KCa3.1 gating depends on Ca^{2+} and is inhibited by the protein TRAM-34, The channel is generally localized to the basolateral membrane (Bertuccio et al. 2014; Devor et al. 2016) although it was recently identified in the apical membrane of mammary epithelial cells in culture (Srisomboon et al. 2018). Gating of this channel depends on calcium binding to calmodulin, one molecule of which is bound to each subunit of KCa3.1 (Bertuccio et al. 2014). Interestingly, several forms of calmodulin are highly expressed in both the Lemay and Ramanathan databases (data not shown). The basolateral transporter is important in maintaining transepithelial Cl-secretion across intestinal epithelia (Bertuccio et al. 2014), presumably by maintaining a favorable electrochemical gradient, and the channel was able to induce sufficient hyperpolarization of the human bronchial epithelium to drive Na^+ entry through ENaC in the apical membrane (Devor et al. 2000).

Significant studies on calcium-regulated K^+ fluxes were carried out in the Enomoto laboratory on cultured primary mouse mammary epithelial cells three decades ago. They found that the fluxes were hyperpolarizing and increased by EGF (Enomoto et al. 1986). Patch clamp studies showed that the channels had a high conductance, which was calcium activated and inhibited by barium (Furuya et al. 1989). Whole cell clamp recordings showed that these channels had slow conductance oscillations and were inhibited by internal Ba^+ and external tetraethyl ammonium (Enomoto et al. 1991). They were unable to relate these findings to a specific K^+ channel at that time.

KCa3.1 studies have also been carried out in a human mammary cell culture model. In the first of these studies UTP or ATP-γ-S, substrates that interact with purinergic P2Y and P2X receptors were applied to the basal surface of the cells. Activation of these receptors produced an increase in calcium that activated a K^+ current across the basal membrane of these cells (Lee et al. 2007). The increase was abolished by inhibitors of KCa3.1, charybdotoxin, and clotrimazole, providing strong evidence for the presence of these channels in the basal membrane. In the same laboratory, Palmer and colleagues found that UTP or ATP-γ-S, applied apically to cultures of human mammary epithelial cells, produced a decrease in short circuit current consistent with an increase in apical K^+ conductance (Palmer et al. 2011). This decreased current was not seen in the presence of the calcium antagonist BAPTA, or the inhibitors TRAM-34 or clotrimazole, providing evidence that purinergic activation may also regulate KCa3.1 channels in the apical membrane.

Interestingly, when UTP was applied apically, calcium was increased only in the apical part of the cell, whereas basal application resulted in a more diffuse increase in intracellular calcium. Is it possible that this apical effect of UTP is stimulating calcium transport into a trans-Golgi compartment, which is localized apically as shown in Fig. 6.3? It is not clear from the Palmer paper what channels are actually involved in the increased K^+ conductance although it could be that the increased calcium resulting from apical P2Y receptor stimulation is increasing KCa3.1 channel activity or number. Clearly, much work like the more complete studies of Kamikawa and associates below is necessary (Kamikawa and Ishikawa 2014).

Interested readers are referred to Devor et al. (2020) Chap. 22 in Volume 3 of this book series for additional information about KCa3.1.

6.6.3.3 K^+ Channel KVLQT1

KVLQT1 is encoded by *KCNQ1*, the highest expressed K^+ channel in the human mammary transcriptome; The expression of this gene, *Kcnq1*, is lower in the mouse mammary transcriptome. KVLQT1 is a basolateral calcium and cAMP-regulated channel widely distributed in epithelia throughout the body (Bleich and Warth 2000; Schultz and Devor 2016). It is inhibited by chromanol 293B. K^+ flux through these channels hyperpolarizes the cell providing an electrochemical gradient that could promote apical Cl^- efflux into the milk space.

6.6.3.4 Additional K^+ Channels in the Mouse

There are a few additional studies that bear on K^+ transport in the lactating mammary cell. Kamikawa and colleagues found a strong inwardly rectifying K^+ current that was sensitive to barium and cesium in membranes from freshly isolated cells from a lactating mouse mammary gland (Kamikawa and Ishikawa 2014). These are characteristics of the KIR2.1 channel, a product of the *Kcnj2* gene; its mRNA was increased in the Ramanathan database in cells from lactating compared to pregnant mice (Fig. 6.11b). Kamikawa and associates also found that the expression of KIR2.1 decreased with weaning (Kamikawa et al. 2015). The KIR2.1 channel forms a complex with the Kv6.2 channel (Zhu et al. 1999), the product of the *Kcng2* gene also expressed in the mouse database and increased in lactation. If this channel is present in the apical membrane, it would contribute to the electrochemical gradient that moves Na^+-dependent transported molecules into the cells (Kamikawa and Ishikawa 2014). However, Western blots and immunocytochemistry (Kamikawa and Ishikawa 2014) indicated a glycosylated form of the channel that appeared to be concentrated in cytoplasmic puncta in lactating cells. While the electrophysiological evidence indicates that the channel is present in the plasma membrane, the presence of the protein in cytoplasmic puncta raises the question of whether this channel may also ferry K^+ into the Golgi along with the water needed to maintain osmolality as lactose is synthesized in this compartment.

6.6.4 Anion Transporters

Cl^- makes its way from the interstitial space to the milk by mechanisms that are only recently becoming clear. The mRNAs for many types of chloride channels and transporters are expressed at meaningful levels, often increased in lactation in the mouse and increased in mature milk over colostrum in the human database (Fig. 6.12). These transporters include the anoctamins (TMEM16 proteins) encoded by the *ANO* genes and studied physiologically using electrophysiological techniques (Pedemonte and Galietta 2014), the CLIC channels and intracellular chloride channels associated with the endoplasmic reticulum (Nagasawa et al. 2001), and voltage-gated chloride channels encoded by *CLCN* genes in the human and *Clcn* genes in the mouse. Although Kamikawa and associates (Kamikawa et al. 2016) detected mRNA for BEST1 and BEST3, possible calcium-activated chloride channels in isolated mammary glands from late lactation mice, these transcripts were detected at very low levels in the Ramanathan and Lemay databases (data not shown). BEST channels will not be further discussed here. The extensively studied cystic fibrosis transmembrane conductance regulator, the protein CFTR, has been found to be present in a number of mammary cell culture models (Schultz 2016; Srisomboon et al. 2018); however, the level of expression of its gene, zero in the human database and baseline in the mouse mRNA database, rules it out as a functional Cl^- transporter under stable lactation conditions. This conclusion was also reached by Shennan (Shennan 1998). Additional volume-regulated anion transporters are encoded by the *LRRC8* genes in the human and *Lrrc8* and *Slco2a1* genes in the mouse.

6.6.4.1 Anoctamins

Because they have eight putative transmembrane segments thought to be involved in anion transport, the genes for TMEM16 proteins were named *anoctamins* of which there are 9 known genes (*ANO1–8* and *ANO10*; corresponding to TMEM16A-H and TMEM16K). Numerous studies (Pedemonte and Galietta 2014) make it abundantly clear that *ANO1* codes for a calcium-modulated chloride channel often called CaCC, and there is evidence that *ANO10* (TMEM16K) also acts as a channel (Pedemonte and Galietta 2014). *Ano1* (TMEM16A) is the most highly expressed chloride channel in the mouse database (Fig. 6.12b). In an elegant study from the Kamikawa laboratory (Kamikawa et al. 2016), it was identified as residing in or near the apical membrane of lactating mammary epithelial cells (Fig. 6.13). Electrophysiological studies allowed the following conclusions about TMEM16A or CaCC:

(a) It is activated by Ca^{2+}.
(b) Activation is voltage-dependent.
(c) Activation is slow and time-dependent.
(d) The channel carries an outwardly rectified steady state current.
(e) It has permeability ratios on the order of $I^- > NO3^- > Br^- > Cl^- >> \text{glutamate}^-$.

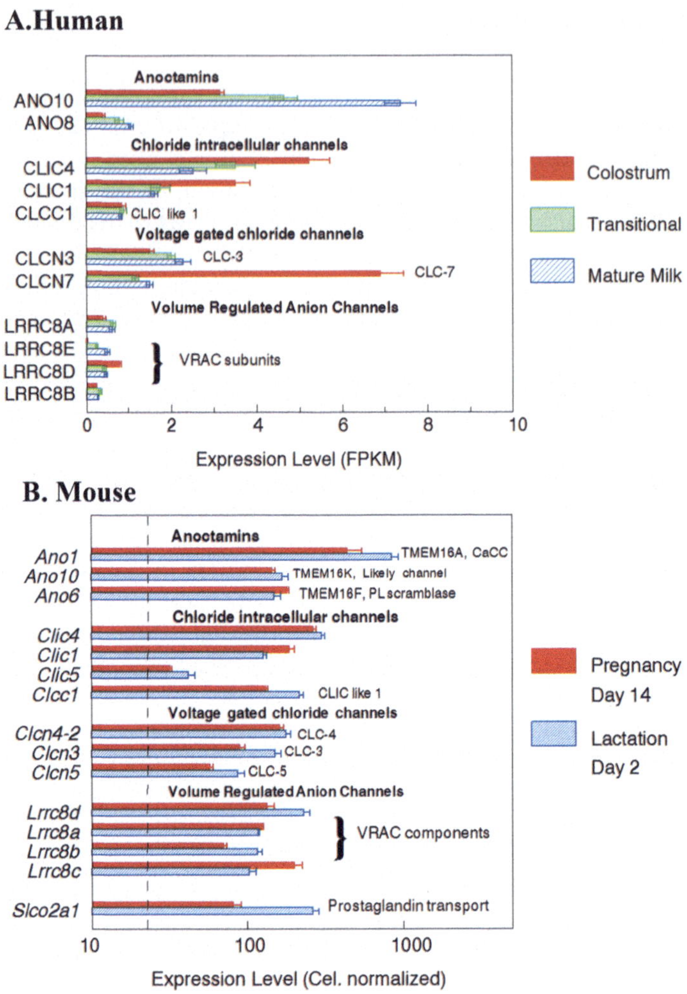

Fig. 6.12 Transcriptome of major anion transporters in the mammary gland. (**a**) Human and (**b**) mouse. See legend to Fig. 6.4 for the sources of expression data. Gene names are on the vertical axis; corresponding protein names are in the body of the graphs

(f) It is sensitive to the chloride channel blockers such as NFA (Niflumic Acid), DIDS (4′-diisothiocyanostilbene-2, 2′-disulfonic acid), and a newly found chloride channel blocker, CaCCinh-A01.

A number of CaCC channel variants, obtained by alternative splicing of the mRNA for *Ano1*, appeared to be expressed in the mouse mammary gland. The authors modestly conclude that the CaCC may influence the quality or quantity of milk.

In the Human Protein Atlas, *ANO1* was only expressed in myoepithelial cells in the normal human breast. *ANO1* was not expressed in the Lemay database derived from mature milk. It may be that *ANO10* codes for an apical chloride channel in the

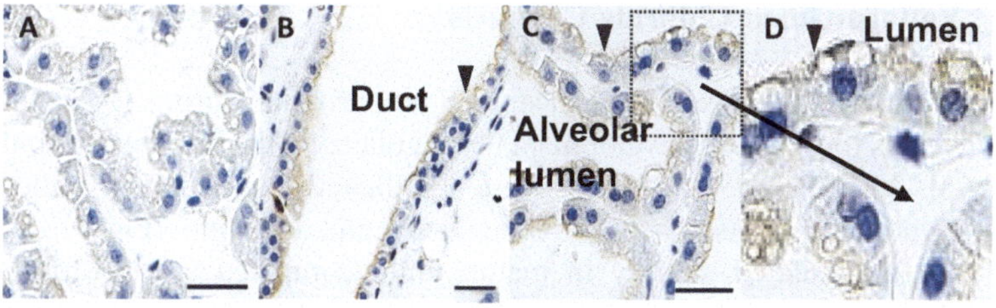

Fig. 6.13 Immunohistochemistry of anoctamin 1 (TMEM16A) in the epithelium of the lactating mammary gland of the mouse. The arrowheads point to apical stain, and the blow up image on the right, (**d**), shows distinct apical localization of protein, although not necessarily all associated with the apical membrane. (**a**) No antibody control. (**b**) Duct. (**c**) Multiple alveolar linings. (**d**) Enlargement of square in (**c**). Scale bar 25 μM. Image modified from Kamikawa et al. (2016)

lactating breast since it was the highest expressed chloride channel/transporter in the Lemay database (Lemay et al. 2013a) (Fig. 6.12a) and its expression was increased more than two-fold between colostrum and mature milk. Although anion currents have been found to be associated with other anoctamins, it is not clear that all TMEM16 proteins have an ion channel function (Pedemonte and Galietta 2014). The other anoctamins expressed in the mouse and human databases, *ANO8* and *Ano6* coding for TMEM16H and TMEM16F, may or may not function as channels. TMEM16F has clearly been identified as a calcium-dependent phospholipid scramblase (Veit et al. 2018; Watanabe et al. 2018). There is little information about anoctamin 8 (TMEM16H) other than it is thought to be associated with intracellular organelles (Yang and Jan 2016). For further reading about TMEM16, see Chap. 17 in Volume 3 of this series.

6.6.4.2 Chloride Intracellular Channels (CLICs)

CLICs are metamorphic proteins that localize to intracellular compartments, including endoplasmic reticulum (ER) (Ponnalagu et al. 2016b), mitochondria (Ponnalagu et al. 2016a), and possibly vesicles (Jiang et al. 2014). Mammalian CLIC proteins have no signal sequence and are synthesized in the cytoplasm where they may be associated with vesicle membranes. They appear to be highly permeable to anions, particularly chloride (Nagasawa et al. 2001) under certain conditions. The protein was localized to the ER and Golgi in cultured cells, and from electrophysiological measurements, the sequence of permeability ratios was $Br^{-} \geq Cl^{-} > SO4^{=}$. There appear to be no studies of CLICs specifically in the mammary gland, and interestingly, the expression of CLIC1 and CLIC4 decreased in the human dataset from colostrum to mature milk (Fig. 6.12a).

6.6.4.3 Voltage-Gated Chloride Channels

Voltage-gated Cl^- channels have been studied primarily in neurons (Qi et al. 2018), but *CLCN3* was found to be developmentally regulated in the pulmonary epithelium (Lamb et al. 2001). *Clcn3* and *Clcn5* were significantly upregulated in the mouse mammary epithelial dataset between pregnancy and lactation (Fig. 6.12b), and *CLCN3* was upregulated slightly in mature milk compared to colostrum in the human database (Fig. 6.12a). However, the role of this class of transporters in chloride secretion into milk has not been elucidated.

6.6.4.4 Volume-Regulated Anion Channels, LRRC8 or VRAC

Volume regulation is important in all cells. In epithelial cells, it has been long known that a swelling-activated Cl^- current was present with several distinctive activation characteristics and a permeability sequence $I^- > Br^- > Cl^- > F^-$. This anion current is activated by reduced intracellular ionic strength and is stimulated by the Rho-Rho kinase and PIP3 pathways (Larsen and Hoffmann 2015). Recently, the molecular identity of the gene for these volume regulated anion channels (VRACs) has been proposed to be heteromers of the *LRRC8*, leucine-rich repeat containing protein (Jentsch 2016). The mRNAs for monomers *d*, *a*, *b*, and *c* are all expressed at substantial levels in the mouse mammary transcriptome. Of these, *d* and *b* are significantly upregulated at lactation and *a* and *c* are downregulated (Fig. 6.12b). *LRRC8A, E, B,* and *D* are expressed at very low levels in the human transcriptome (Fig. 6.12a). However, the identity of VRAC is still not entirely clear as Kunzelmann and colleagues (Sirianant et al. 2016) have proposed anoctamin 6 as an essential component of the response to altered cell volume, by acting either as a Cl^- channel or as phospholipid scramblase. *Ano6* is expressed at reasonably high levels in the transcriptome from both the pregnant and the lactating mouse mammary gland but was not expressed at significant levels in the human transcriptome (Fig. 6.12b). Clearly, more targeted mammary research studies are needed here. The taurine transporter TAUT, described previously, has also been implicated in mammary cell volume regulation (Shennan et al. 1995).

The molecular identity of another channel, long termed the Maxi-Cl channel, which is also a prostaglandin transporter, has recently been confirmed as being coded by the *Slco2a1* gene (Sabirov et al. 2017). *Slco2a1* is found at reasonable levels in the mouse mammary transcriptome and is significantly upregulated at lactation (Fig. 6.12b). No form of *SLCO2* is expressed at significant levels in the human transcriptome.

6.6.4.5 Iodide Transport

It has been known for a long time that iodide is transported and even concentrated in milk (Grosvenor 1963). Expression of the Na^+/I^- cotransporter, NIS protein, and *Slc5a5* mRNA, was detected by RT-PCR in the mammary gland of lactating rats (Ajjan et al. 1998) and in the human breast (Heufelder et al. 1998). The mRNA is increased substantially at lactation in the human and mouse datasets (Fig. 6.10). The transporter is thought to be located in the basal membrane (Darrouzet et al. 2014) where its expression can be stimulated by prolactin (Rillema and Yu 1996). Expression of NIS in cultured mammalian cells induced sodium iodide symport (Perron et al. 2001). NIS expression was inhibited by iodine as well as IGF-1 and TGF-beta in explants from lactating mouse mammary cells (Yu et al. 2012). It is very clear that I^- easily enters the lactating mammary cell from the blood stream using the NIS symporter; the pathway by which it enters the milk space is not clear.

6.6.5 Monovalent Ion CoTransporters

In 1988, Blatchford and Peaker suggested the existence of a nonselective channel for Na^+ and K^+ in the apical membrane of the goat mammary gland (Blatchford and Peaker 1988). Such a channel could help explain fluxes of these ions across the apical membrane. A similar current was demonstrated in cultures of mammary epithelial cells by the Enomoto laboratory (Furuya et al. 1989). It is not clear whether any of the cotransporters currently described in mammary databases fit the criteria needed for apical cotransport of these two cations.

6.6.5.1 NKCC1

It has long been thought that Cl^- crosses the plasma membrane by cotransport with Na^+, K^+, or both. Shennan postulated that a Na^+-K^+-$2Cl^-$ symport system carried out such a function based on a Na^+- and Cl^--sensitive K^+ flux from gland explants that were also sensitive to furosemide (Shennan 1989). Such a transporter, NKCC1, encoded by the *Slc12a2* gene is widely distributed in the basal membranes of epithelial cells and could account for electroneutral movement of K^+ and Cl^- into the alveolar cell. It is highly expressed during lactation in both the human and mouse datasets (Fig. 6.14). The Hennighausen laboratory (Shillingford et al. 2002) reported that the mRNA for this transporter was upregulated in the mammary gland of the lactating mouse; by immunohistochemistry, they found that the protein localized to the basal surface of lactating alveolar cells. Interestingly, loss of the transporter in an *Slc12a2* negative mouse was associated with developmental defects in the mouse mammary gland. Whether the protein NKCC1 is responsible for the total flux of Cl^- into the mammary alveolar cell is not clear at present, but Schultz and Devor (2016)

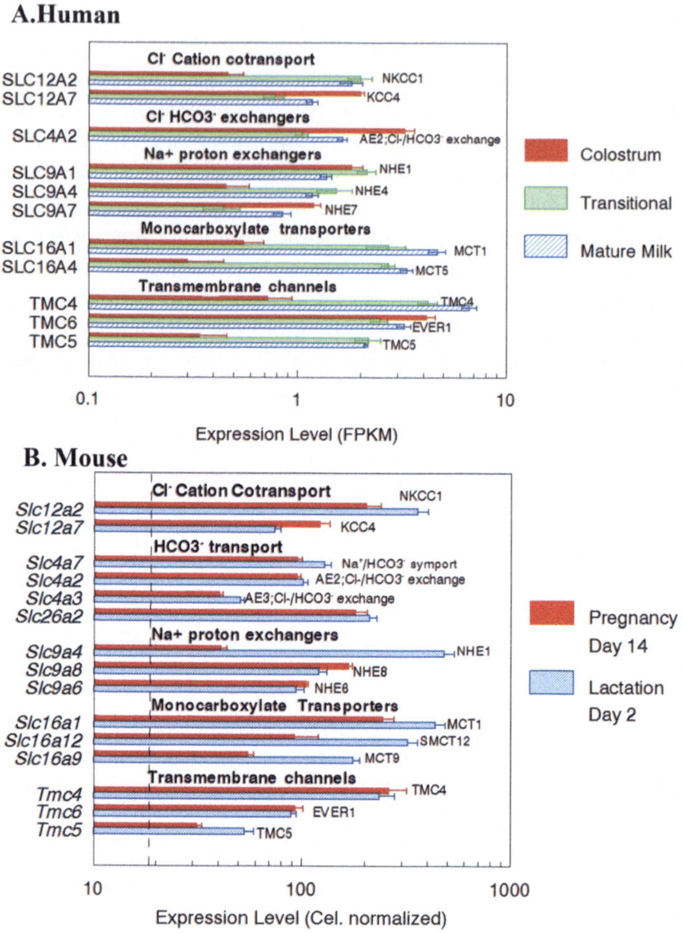

Fig. 6.14 Transcriptomes for heterogeneous transporters in the mammary gland. (**a**) Human and (**b**) mouse. See legend to Fig. 6.4 for the sources of expression data. Gene names are on the vertical axis; corresponding protein names are in the body of the graphs

postulate that the gradients of the three ions together provide a substantial driving force for Cl^- entry. For further information about NKCC1, readers are directed to Chap. 2 in Volume 3 of this series.

6.6.5.2 KCC4

KCC4 is a Na^+-independent K^+-coupled Cl^- cotransporter that is thought to transfer one K^+ and one Cl^- simultaneously across the plasma membrane (Marcoux et al. 2017). It is encoded by the *Slc12a7* gene and expressed in many epithelial tissues where it can be localized either apically or basolaterally. It appears to support other transport systems by maintaining intracellular K^+ at a level that is important for cellular operation. There do not appear to be any specific studies of KCC4 in the mammary gland, but the mRNA is somewhat downregulated in mature human milk compared to colostrum and downregulated between pregnancy and lactation in mouse mammary cells (Fig. 6.14), suggesting that other Cl^- transport pathways may be more important.

6.6.5.3 $Cl^-/HCO3^-$ Exchange

$Cl^-/HCO3^-$ exchange by proteins designated as AE1–3 is another means by which Cl^- can be transported in or out of the mammary epithelial cell. $Cl^-/HCO3^-$ exchangers are encoded by the *SLC4* gene family of which only *SLC4A2* encoding for AE2, a Na^+-independent Cl^-/bicarbonate exchanger, was found in the human database. *Slc4a2*, *Slc4a3*, and *Slc4a7* were found in the mouse database at meaningful levels (see Fig. 6.14). The exchanger AE2 is found in the basolateral membrane of many epithelia and has been extensively studied in the human airway epithelial cell line, Calu-3 (Huang et al. 2012). *Slc4a7* was the highest expressed *Slc4* gene, and it and *Slc4a3* were significantly increased in the gland of the lactating mouse. To what extent the proteins coded by these mRNA are responsible for Cl^- transport in the mammary gland as well as their cellular localizations is a question for future research.

6.7 Other Transporters

6.7.1 Water Transport

Milk is iso-osmolar with plasma and is assumed to be iso-osmolar with the interior of the alveolar cell. This means that water must permeate not only across the basal membrane as solutes enter the cell but also across the membranes of intracellular compartments and likely across the apical membrane as well. Water channel aquaporin 3 (AQP3) has been consistently observed in the lactating mammary alveolar cell in rats (Nazemi et al. 2014; Liu et al. 2015), humans (Mobasheri and Barrett-Jolley 2014), and cows (Mobasheri et al. 2011), and the levels of *AQP3* mRNA from the human milk fat globule and mouse mammary epithelial cells are consistent with these results (Fig. 6.15). AQP1 has been mainly associated with blood vessels in mammary glands from rats (Nazemi et al. 2014), humans (Mobasheri and Barrett-Jolley 2014), and cows (Mobasheri et al. 2011). Since the mRNA from the human milk fat globule is representative only of the alveolar cells, the lack of *AQP1* expression in the human study in Fig. 6.15 is to be expected. However, the finding of substantial mRNA for *Aqp1* in the mouse data in Fig. 6.15 suggests either that the isolated cells used for this analysis (Rudolph et al. 2009) contain vascular cells or that the mouse gland is different in its distribution of AQP1

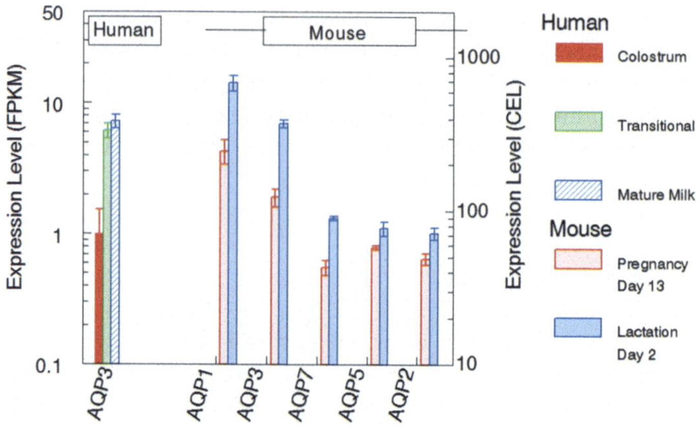

Fig. 6.15 mRNA expression of aquaporin in human and mouse mammary glands. Values for human lactation are from Lemay et al. (2013a); values for the mouse are from Heinz et al. (2016). Gene names are on the horizontal axis; corresponding protein names are in the body of the graph

from other species that have been studied. The expression of *Aqp7* may reflect the very high demand of the murine gland for glycerol for the synthesis of triglycerides. Interestingly, AQP7 facilitates the exit of glycerol from adipocytes, preventing the intracellular accumulation of glycerol and triglyceride (Verkman et al. 2014). In the mammary gland, it is more likely that it mediates entry of glycerol into the alveolar cell or into the endoplasmic reticulum where triglyceride synthesis takes place. AQP5 was observed in the rat mammary gland, mainly near the apical membrane, leading Knight and colleagues to postulate that it is involved in transfer of water from the alveolar cell directly into milk (Nazemi et al. 2014). However, in their study, AQP5 was actually decreased in the lactating gland; the increase in mRNA expression in mouse mammary epithelial cells (Fig. 6.15) is not significant. In this regard, however, it is worth noting that a major portion of water in milk must be derived from the secretory vesicles that contain lactose synthesized in the Golgi compartment. Lactose is the major osmotic constituent of the milk of all the species that have been studied for mammary aquaporin expression. It seems likely that AQP3 is found in the membranes of both the Golgi compartment and the secretory vesicles to allow osmotic equilibration of these compartments.

It may be that additional channels allow water permeation. For example, SGLT1, a sodium-coupled sugar transporter encoded by the *SLC5A1* gene (see Fig. 6.10), has been shown to be involved in passive water movement in the gut (Zeuthen et al. 2016). The intriguing idea that other solute channels may allow for water movements between compartments of the mammary cell needs exploration.

6.7.2 Na^+-Coupled Symporters and Antiporters

We have encountered many transporters that utilize the energy generated by the Na^+/K^+ ATPase to produce a sodium gradient to concentrate solutes within the cytoplasm. These include the phosphate transporters encoded by the *SLC34A2* (NaPi2b; Fig. 6.7) and the *SLC20A1* genes (NaPi2a; PIT1; Fig. 6.7), a number of amino acid transporters (Fig. 6.9) and symporters for glucose, iodide, myoinositol, and multivitamins encoded by the *SLC5A* family of transporters (Fig. 6.10). In addition, Na^+ promotes bicarbonate transport by the protein encoded by *Slc4a7* (Fig. 6.14) as well as proton exchange via Slc9a encoded transporters (Fig. 6.14).

Sodium/proton exchangers [NHE, encoded by *SLC9* genes (Donowitz et al. 2013)] are found in all cells and have somewhat diverse functions (Orlowski and Grinstein 2004). *SLC9A4* and *Slc9a4* are expressed in both the human and mouse datasets and increased in mature milk or lactation in both (Fig. 6.14). This Na^+/H^+ exchanger, NHE4, is found in the basolateral membrane of most epithelia where it is expressed (Orlowski and Grinstein 2004); it has been shown to play roles in pH regulation (Arena et al. 2012) and ammonia transport (Houillier and Bourgeois 2012) in other tissues. Its specific function in the mammary gland appears not to have been explored. The proteins NHE8 and NHE6 encoded by *Slc9a8* and *Slc9A6*, found in the mouse lactation database, are predominantly endosomal exchangers (Fuster and Alexander 2014) as is the human protein NHE7 encoded by *SLC9A7*. Unlike the gene for *SLC9A4*, expression of none of these is increased in lactation. NHE1 encoded by *SLC9A1*, found in the human database, is predominantly a plasma membrane transporter. The precise role of the NHE family in the mammary gland has not been elucidated.

6.7.3 Monocarboxylate Channels

Of the 14 members of the monocarboxylate (MCT) family, only MCT1–4 have been shown to transport compounds classified as monocarboxylates (e.g., lactate and pyruvate) coupled to proton movement. The gene for MCT1, *SLC16A1* and *Slc16a1*, is expressed in both the human and mouse datasets and is increased in mature milk in the human and in lactation in the mouse (Fig. 6.14). MCT1 stain has been shown to overlap with that of 4F2hc, implicated in amino acid transport (see Fig. 6.9) at the basolateral surface of various renal epithelial cells (Becker et al. 2010). The genes for other MCTs (MCT5, encoded by *SLC16A5,* in human milk and SMCT1, encoded by *Slc16a12*, and MCT9, encoded by *Slc16a9*, and in mouse mammary cells) are expressed in both the human and mouse datasets where they are upregulated in lactation. MCTs have been extensively studied in breast cancer where it has been shown that they interact with carbonic anhydrase (Noor et al. 2017); their function in the lactating mammary gland has not been elucidated. There may be some interesting work to be done here.

6.7.4 Transmembrane Channels, TMC

TMC genes are thought to be plasma membrane-localized proteins coding for calcium-regulated channel-like proteins or regulators of channel-like proteins (Kunzelmann et al. 2016). Their molecular structure bears a striking resemblance to the anoctamins (Hahn et al. 2009). Expression is relatively high in both the human and mouse transcriptomes (Fig. 6.14), and the isoforms that are expressed at higher levels are *TMC4*, *TMC6*, and *TMC5* in both species. The expression of the mRNA for *TMC5* is increased significantly in lactation in the mouse and in mature milk in the human; that for *TMC4* is increased between colostrum and mature milk in the human. These channels are thought to be associated with mechanotransduction in the inner ear (Kawashima et al. 2011), but nothing is known about their function in the mammary gland.

6.8 Where Are We Now?

From the summary presented here, it is clear that we have a good understanding of certain aspects of transport within the lactating mammary alveolar cell. We understand the nature of the ORAI channel that transfers Ca^{2+} into the lactating cells and the ATPases that transfer Ca^{2+} out of the cytoplasm and into the Golgi or across the plasma membrane. We have even recently come to understand a possible role of a Ca^{2+}/H^{+} exchanger in regulating secretory pathway pH. The molecules and pathways that transfer copper, amino acids, and glucose into the alveolar cell and across intracellular membranes have been well-characterized. The identity of many of the numerous monovalent ion channels, symporters, and exchangers that regulate the cytoplasmic milieu and the electrochemical gradients across the apical and basolateral membranes of the cell is known; how these molecules function together and their localization are important questions for which we now have tools—gene identities, gene knockouts, and antibodies to the relevant proteins as well as electrophysiological techniques—to begin to approach answers. For many ions, we can measure intracellular activities of the free species. All these technologies should foster rapid progress in a real characterization of the role of membrane transport molecules in the synthesis of an incredibly important nutrient—milk.

References

Ajjan R, Kamaruddin NA, Crisp M, Watson PF, Ludgate M, Weetman AP (1998) Regulation and tissue distribution of the human sodium iodide symporter gene. Clin Endocrinol 49:517–523

Alam S, Hennigar SR, Gallagher C, Soybel DI, Kelleher SL (2015) Exome sequencing of SLC30A2 identifies novel Loss- and gain-of-function variants associated with breast cell dysfunction. J Mammary Gland Biol Neoplasia 20:159–172

Aleman G, Lopez A, Ordaz G, Torres N, Tovar AR (2009) Changes in messenger RNA abundance of amino acid transporters in rat mammary gland during pregnancy, lactation, and weaning. Metabolism 58:594–601

Arena EA, Longo WE, Roberts KE, Geibel P, Nateqi J, Brandstetter M, Geibel JP (2012) Functional role of NHE4 as a pH regulator in rat and human colonic crypts. Am J Physiol Cell Physiol 302:C412–C418

Baumrucker CR (1985) Symposium: Nutrient uptake across the mammary gland. Amino acid transport systems in bovine mammary tissue. J Dairy Sci 68:2436–2451

Becker HM, Mohebbi N, Perna A, Ganapathy V, Capasso G, Wagner CA (2010) Localization of members of MCT monocarboxylate transporter family Slc16 in the kidney and regulation during metabolic acidosis. Am J Physiol Renal Physiol 299:F141–F154

Becker J, Tran D, Tabak L (2018) Members of the GalNAc-T family of enzymes utilize distinct Golgi localization mechanisms. Glycobiology 28:841–848

Benmoussa A, Gotti C, Bourassa S, Gilbert C, Provost P (2019) Identification of protein markers for extracellular vesicle (EV) subsets in cow's milk. J Proteome 192:78–88

Berga SE, Neville MC (1985) Sodium and potassium distribution in the lactating mouse mammary gland in vivo. J Physiol 361:219–230

Bertuccio CA, Lee S-L, Wu G, Butterworth MB, Hamilton KL, Devor DC (2014) Anterograde trafficking of KCa3.1 in polarized epithelia is Rab1- and Rab8-dependent and recycling endosome-independent. PLoS One 9:e92013–e92013

Bin BH, Bhin J, Seo J, Kim SY, Lee E, Park K, Choi DH, Takagishi T, Hara T, Hwang D, Koseki H, Asada Y, Shimoda S, Mishima K, Fukada T (2017) Requirement of zinc transporter SLC39A7/ZIP7 for dermal development to fine-tune endoplasmic reticulum function by regulating protein disulfide 1somerase. J Invest Dermatol 137:1682–1691

Blatchford DR, Peaker M (1988) Effect of ionic composition of milk on transepithelial potential in the goat mammary gland. J Physiol 402:533–541

Blaug S, Hybiske K, Cohn J, Firestone GL, Machen TE, Miller SS (2001) ENaC- and CFTR-dependent ion and fluid transport in mammary epithelia. Am J Physiol Cell Physiol 281:C633–C648

Bleich M, Warth R (2000) The very small-conductance K^+ channel KVLQT1 and epithelial function. Pflugers Arch 440:202–206

Bompiani KM, Tsai CY, Achatz FP, Liebig JK, Howell SB (2016) Copper transporters and chaperones CTR1, CTR2, ATOX1, and CCS as determinants of cisplatin sensitivity. Metallomics 8:951–962

Boyd C, Náray-Fejes-Tóth A (2007) Steroid-mediated regulation of the epithelial sodium channel subunits in mammary epithelial cells. Endocrinology 148:3958–3967

Brandao K, Deason-Towne F, Perraud A-L, Schmitz C (2013) The role of Mg^{2+} in immune cells. Immunol Res 55:261–269

Braunger K, Pfeffer S, Shrimal S, Gilmore R, Berninghause O, Mandon EC, Becker T, Förster F, Beckmann R (2018) Structural basis for coupling protein transport and N-glycosylation at the mammalian endoplasmic reticulum. Science 360:215

Brew K (1969) Secretion of alpha-lactalbumin into milk and its relevance to the organization and control of lactose synthetase. Nature 222:671–672

Burnol A-F, Leturque A, Loizeau M, Postic C, Girard J (1990) Glucose transporter expression in rat mammary gland. Biochem J 270:277–279

Camps M, Vilaro S, Testar X, Palacin M, Zorzano A (1994) High and polarized expression of GLUT1 glucose transporters in epithelial cells from mammary gland: acute down-regulation of GLUT1 carriers by weaning. Endocrinology 134:924–934

Capasso JM, Keenan TW, Abeijon C, Hirschberg CB (1989) Mechanism of phosphorylation in the lumen of the Golgi apparatus. Translocation of adenosine 5′-triphosphate into Golgi vesicles from rat liver and mammary gland. J Biol Chem 264:5233–5240

Cappello AR, Curcio R, Lappano R, Maggiolini M, Dolce V (2018) The physiopathological role of the exchangers belonging to the SLC37 family. Front Chem 6:122–122

Carey DJ, Sommers LW, Hirschberg CB (1980) CMP-N-Acetylneuraminic acid: isolation from and penetration into mouse liver microsomes. Cell 19:597–605

Chen LQ, Hou BH, Lalonde S, Takanag H, Hartung ML, Qu XQ, Guo WJ, Kim JG, Underwood W, Chaudhuri B, Chermak D, Antony G, White FF, Somerville SC, Mudgett MB, Frommer WB (2010) Sugar transporters for intercellular exchange and nutrition of pathogens. Nature 468:527–532

Chen Q, Wu Y, Zhang M, Xu W, Guo X, Yan X, Deng H, Jiang Q, Yang X, Lan G, Guo Y, Qin G, Jiang H (2016) Milk fat globule is an alternative to mammary epithelial cells for gene expression analysis in buffalo. J Dairy Res 83:202–208

Chen F, Zhang S, Deng Z, Zhou Q, Cheng L, Kim SW, Chen J, Guan W (2018a) Regulation of amino acid transporters in the mammary gland from late pregnancy to peak lactation in the sow. J Anim Sci Biotechnol 9:35

Chen YS, Kozlov G, Fakih R, Funato Y, Miki H, Gehring K (2018b) The cyclic nucleotide-binding homology domain of the integral membrane protein CNNM mediates dimerization and is required for Mg^{2+} efflux activity. J Biol Chem 293:19998–20007

Chou JY, Mansfield BC (2014) The SLC37 family of sugar-phosphate/phosphate exchangers. Curr Top Membr 73:357–382

Collins JF, Bai L, Ghishan FK (2004) The SLC20 family of proteins: dual functions as sodium-phosphate cotransporters and viral receptors. Pflugers Arch 447:647–652

Constantine M, Liew CK, Lo V, Macmillan A, Cranfield CG, Sunde M, Whan R, Graham RM, Martinac B (2016) Heterologously-expressed and liposome-reconstituted human transient receptor potential melastatin 4 channel (TRPM4) is a functional tetramer. Sci Rep 6:19352–19352

Cross BM, Hack A, Reinhardt TA, Rao R (2013) SPCA2 regulates Orai1 trafficking and store independent Ca^{2+} entry in a model of lactation. PLoS One 8:e67348–e67348

Cross BM, Breitwieser GE, Reinhardt TA, Rao R (2014) Cellular calcium dynamics in lactation and breast cancer: from physiology to pathology. Am J Phys Cell Physiol 306:C515–C526

Dai W, Wang Q, Zhao F, Liu J, Liu H (2018) Understanding the regulatory mechanisms of milk production using integrative transcriptomic and proteomic analyses: improving inefficient utilization of crop by-products as forage in dairy industry. BMC Genomics 19:403

Darrouzet E, Lindenthal S, Marcellin D, Pellequer J-L, Pourcher T (2014) The sodium/iodide symporter: state of the art of its molecular characterization. Biochim Biophys Acta Biomembr 1838:244–253

Davis KA, Cowley EA (2006) Two-pore-domain potassium channels support anion secretion from human airway Calu-3 epithelial cells. Pflugers Arch 451:631–641

Davis FM, Janoshazi A, Janardhan KS, Steinckwich N, D'Agostin DM, Petranka JG, Desai PN, Roberts-Thomson SJ, Bird GS, Tucker DK, Fenton SE, Feske S, Monteith GR, Putney JW Jr (2015) Essential role of Orai1 store-operated calcium channels in lactation. Proc Natl Acad Sci U S A 112:5827–5832

de Baaij JHF, Arjona F, van den Brand M, Lavrijsen M, Lameris ALL, Bindels RJM, Hoenderop JGJ (2016) Identification of SLC41A3 as a novel player in magnesium homeostasis. Sci Rep 6:28565–28565

Devor DC, Bridges RJ, Pilewski JM (2000) Pharmacological modulation of ion transport across wild-type and ΔF508 CFTR-expressing human bronchial epithelia. Am J Physiol Cell Physiol 279:C461–C479

Devor DC, Bertuccio CA, Hamilton KL (2016) KCa3.1 in epithelia. In: Hamilton KL, Devor DC (eds) Ion channels and transporters of epithelia in health and disease. American Physiological Society: Springer, New York, pp 659–705

Dimberg LY, Towers CG, Behbakht K, Hotz TJ, Kim J, Fosmire S, Porter CC, Tan A-C, Thorburn A, Ford HL (2017) A genome-wide loss-of-function screen identifies SLC26A2 as a novel mediator of TRAIL resistance. Mol Cancer Res 15:382–394

Donowitz M, Ming Tse C, Fuster D (2013) SLC9/NHE gene family, a plasma membrane and organellar family of Na^{+}/H^{+} exchangers. Mol Asp Med 34:236–251

Dorea JG (2000) Magnesium in human milk. J Am Coll Nutr 19:210–219
Dorea JG (2002) Selenium and breast-feeding. Br J Nutr 88:443–461
Enomoto K, Cossu MF, Maeno T, Edwards C, Oka T (1986) Involvement of the Ca^{2+}-dependent K^{+} channel activity in the hyperpolarizing response induced by epidermal growth factor in mammary epithelial cells. FEBS Lett 203:181–184
Enomoto K, Furuya K, Maeno T, Edwards C, Oka T (1991) Oscillating activity of a calcium-activated K^{+} channel in normal and cancerous mammary cells in culture. J Membr Biol 119:133–139
Faddy HM, Smart CE, Xu R, Lee GY, Kenny PA, Feng M, Rao R, Brown MA, Bissell MJ, Roberts-Thomson SJ, Monteith GR (2008) Localization of plasma membrane and secretory calcium pumps in the mammary gland. Biochem Biophys Res Commun 369:977–981
Farhadian M, Rafat SA, Hasanpur K, Ebrahimi M, Ebrahimie E (2018) Cross-species meta-analysis of transcriptomic data in combination with supervised machine learning models identifies the common gene signature of lactation process. Front Genet 9:235–235
Faulkner A, Peaker M (1987) Regulation of mammary glucose metabolism in lactation. In: Neville MC, Daniel CW (eds) The mammary gland: development, regulation and function. Plenum Press, New York, pp 536–562
Feng M, Grice DM, Faddy HM, Nguyen N, Leitch S, Wang Y, Muend S, Kenny PA, Sukumar S, Roberts-Thomson SJ, Monteith GR, Rao R (2010) Store-independent activation of Orai1 by SPCA2 in mammary tumors. Cell 143:84–98
Finucane KA, McFadden TB, Bond JP, Kennelly JJ, Zhao FQ (2008) Onset of lactation in the bovine mammary gland: gene expression profiling indicates a strong inhibition of gene expression in cell proliferation. Funct Integr Genomics 8:251–264
Fleig A, Schweigel-Röntgen M, Kolisek M (2013) Solute carrier family SLC41, what do we really know about it? Wiley interdiscip Rev Membr Transp Signal 2. https://doi.org/10.1002/wmts.95
Furuya K, Enomoto K, Furuya S, Yamagishi S, Edwards C, Oka T (1989) Single calcium-activated potassium channel in cultured mammary epithelial cells. Pflugers Arch 414:118–124
Fuster DG, Alexander RT (2014) Traditional and emerging roles for the SLC9 Na^{+}/H^{+} exchangers. Pflugers Arch 466:61–76
Gerk PM, Hanson L, Neville MC, McNamara PJ (2002) Sodium dependence of nitrofurantoin active transport across mammary epithelia and effects of dipyridamole, nucleosides, and nucleobases. Pharm Res 19:299–305
Goytain A, Hines RM, Quamme GA (2008) Functional characterization of NIPA2, a selective Mg^{2+} transporter. Am J Physiol Cell Physiol 295:C944–C953
Grosvenor CE (1963) I131 accumulation by the lactating rat mammary gland. Am J Physiol-Legacy Content 204:856–860
Hadley B, Maggioni A, Ashikov A, Day CJ, Haselhorst T, Tiralongo J (2014) Structure and function of nucleotide sugar transporters: current progress. Comput Struct Biotech J 10:23–32
Hadsell DL, Hadsell LA, Rijnkels M, Carcamo-Bahena Y, Wei J, Williamson P, Grusak MA (2018) In silico mapping of quantitative trait loci (QTL) regulating the milk ionome in mice identifies a milk iron locus on chromosome 1. Mamm Genome 29:632–655
Hahn Y, Kim DS, Pastan IH, Lee B (2009) Anoctamin and transmembrane channel-like proteins are evolutionarily related. Int J Mol Med 24:51–55
Heinz RE, Rudolph MC, Ramanathan P, Spoelstra NS, Butterfield KT, Webb PG, Babbs BL, Gao H, Chen S, Gordon MA, Anderson SM, Neville MC, Gu H, Richer JK (2016) Constitutive expression of microRNA-150 in mammary epithelium suppresses secretory activation and impairs de novo lipogenesis. Development 143:4236–4248
Heird WC (2004) Taurine in neonatal nutrition--revisited. Arch Dis Child Fetal Neonatal Ed 89: F473–F474
Hennigar SR, Seo YA, Sharma S, Soybel DI, Kelleher SL (2015) ZnT2 is a critical mediator of lysosomal-mediated cell death during early mammary gland involution. Sci Rep 5:8033
Heufelder AE, Spitzweg C, Joba W, Eisenmenger W (1998) Analysis of human sodium iodide symporter gene expression in extrathyroidal tissues and cloning of its complementary

deoxyribonucleic acids from salivary gland, mammary gland, and gastric mucosa. J Clin Endocrinol Metab 83:1746–1751
Hirschberg CB (2018) My journey in the discovery of nucleotide sugar transporters of the Golgi apparatus. J Biol Chem 293:12653–12662
Houillier P, Bourgeois S (2012) More actors in ammonia absorption by the thick ascending limb. Am J Physiol Renal Physiol 302:F293–F297
Huang J, Shan J, Kim D, Liao J, Evagelidis A, Alper SL, Hanrahan JW (2012) Basolateral chloride loading by the anion exchanger type 2: role in fluid secretion by the human airway epithelial cell line Calu-3. J Physiol 590:5299–5316
Huber K, Muscher A, Breves G (2007) Sodium-dependent phosphate transport across the apical membrane of alveolar epithelium in caprine mammary gland. Comp Biochem Physiol A Mol Integr Physiol 146:215–222
Huebner RJ, Lechler T, Ewald AJ (2014) Developmental stratification of the mammary epithelium occurs through symmetry-breaking vertical divisions of apically positioned luminal cells. Development 141:1085–1094
Itsumura N, Kibihara Y, Fukue K, Miyata A, Fukushima K, Tamagawa-Mineoka R, Katoh N, Nishito Y, Ishida R, Narita H, Kodama H, Kambe T (2016) Novel mutations in SLC30A2 involved in the pathogenesis of transient neonatal zinc deficiency. Pediatr Res 80:586
Jenkitkasemwong S, Wang CY, Coffey R, Zhang W, Chan A, Biel T, Kim JS, Hojyo S, Fukada T, Knutson MD (2015) SLC39A14 is required for the development of hepatocellular iron overload in murine models of hereditary hemochromatosis. Cell Metab 22:138–150
Jentsch TJ (2016) VRACs and other ion channels and transporters in the regulation of cell volume and beyond. Nat Rev Mol Cell Biol 17:293–307
Jeong J, Eide DJ (2013) The SLC39 family of zinc transporters. Mol Asp Med 34:612–619
Jia Y, Jucius TJ, Cook SA, Ackerman SL (2015) Loss of Clcc1 results in ER stress, misfolded protein accumulation, and neurodegeneration. J Neurosci 35:3001–3009
Jiang L, Phang JM, Yu J, Harrop SJ, Sokolova AV, Duff AP, Wilk KE, Alkhamici H, Breit SN, Valenzuela SM, Brown LJ, Curmi PMG (2014) CLIC proteins, ezrin, radixin, moesin and the coupling of membranes to the actin cytoskeleton: A smoking gun? Biochim Biophys Acta 1838:643–657
Kamikawa A, Ishikawa T (2014) Functional expression of a Kir2.1-like inwardly rectifying potassium channel in mouse mammary secretory cells. Am J Phys Cell Physiol 306:C230–C240
Kamikawa A, Sugimoto S, Ichii O, Kondoh D (2015) Decrease in an inwardly rectifying potassium conductance in mouse mammary secretory cells after forced weaning. PLoS One 10:e0141131
Kamikawa A, Ichii O, Sakazaki J, Ishikawa T (2016) Ca^{2+}-activated Cl^- channel currents in mammary secretory cells from lactating mouse. Am J Phys Cell Physiol 311:C808–C819
Kawashima Y, Géléoc GSG, Kurima K, Labay V, Lelli A, Asai Y, Makishima T, Wu DK, Della Santina CC, Holt JR, Griffith AJ (2011) Mechanotransduction in mouse inner ear hair cells requires transmembrane channel-like genes. J Clin Invest 121:4796–4809
Kelleher SL, Lönnerda B (2003) Zn transporter levels and localization change throughout lactation in rat mammary gland and are regulated by Zn in mammary cells. J Nutr 133:3378–3385
Kelleher SL, Lönnerdal B (2005) Molecular regulation of milk trace mineral homeostasis. Mol Asp Med 26:328–339
Kelleher SL, Lopez V, Lonnerdal B, Dufner-Beattie J, Andrews GK (2009) Zip3 (Slc39a3) functions in zinc reuptake from the alveolar lumen in lactating mammary gland. Am J Phys Regul Integr Comp Phys 297:R194–R201
Kelleher SL, Velasquez V, Croxford TP, McCormick NH, Lopez V, MacDavid J (2012) Mapping the zinc-transporting system in mammary cells: molecular analysis reveals a phenotype-dependent zinc-transporting network during lactation. J Cell Physiol 227:1761–1770
Kim J, Chung Y, Kim H, Im E, Lee H, Yang H (2014) The tissue distribution of nesfatin-1/NUCB2 in mouse. Dev Reprod 18:301–309
Klemens CA, Edinger RS, Kightlinger L, Liu X, Butterworth MB (2017) Ankyrin G expression regulates apical delivery of the epithelial sodium channel (ENaC). J Biol Chem 292:375–385

Kolisek M, Nestler A, Vormann J, Schweigel-Röntgen M (2012) Human gene SLC41A1 encodes for the Na^+/Mg^{2+} exchanger. Am J Physiol Cell Physiol 302:C318–C326

Kong BY, Duncan FE, Que EL, Kim AM, O'Halloran TV, Woodruff TK (2014) Maternally-derived zinc transporters ZIP6 and ZIP10 drive the mammalian oocyte-to-egg transition. Mol Hum Reprod 20:1077–1089

Kunzelmann K, Cabrita I, Wanitchakool P, Ousingsawat J, Sirianant L, Benedetto R, Schreiber R (2016) Modulating Ca^{2+} signals: a common theme for TMEM16, Ist2, and TMC. Pflugers Arch 468:475–490

Lamb FS, Graeff RW, Clayton GH, Smith RL, Schutte BC, McCray PB Jr (2001) Ontogeny of CLCN3 chloride channel gene expression in human pulmonary epithelium. Am J Respir Cell Mol Biol 24:376–381

Lang S, Pfeffer S, Lee P-H, Cavalié A, Helms V, Förster F, Zimmermann R (2017) An update on Sec61 channel functions, mechanisms, and related diseases. Front Physiol 8:887–887

Laporta J, Peters TL, Merriman KE, Vezina C, Hernandez LL (2013) Serotonin (5-HT) affects expression of liver metabolic enzymes and mammary gland glucose transporters during the transition from pregnancy to lactation. PLoS One 8:e57847

Larsen EH, Hoffmann EK (2015) Volume regulation in epithelia. In: Hamilton KL, Devor DC (eds) Ion channels and transporters in epithelia in health and disease. American Physiological Society: Springer, New York, pp 131–185

Lasry I, Golan Y, Berman B, Amram N, Glaser F, Assaraf YG (2014) In situ dimerization of multiple wild type and mutant zinc transporters in live cells using bimolecular fluorescence complementation. J Biol Chem 289:7275–7292

Lee S, Kelleher SL (2016) Molecular regulation of lactation: the complex and requisite roles for zinc. Arch Biochem Biophys 611:86–92

Lee SY, Palmer ML, Maniak PJ, Jang SH, Ryu PD, O'Grady SM (2007) P2Y receptor regulation of sodium transport in human mammary epithelial cells. Am J Phys Cell Physiol 293:C1472–C1480

Lee S, Hennigar SR, Alam S, Nishida K, Kelleher SL (2015) Essential role for Zinc transporter 2 (ZnT2)-mediated zinc transport in mammary gland development and function during lactation. J Biol Chem 290:13064–13078

Lee S, Rivera OC, Kelleher SL (2017) Zinc transporter 2 interacts with vacuolar ATPase and is required for polarization, vesicle acidification, and secretion in mammary epithelial cells. J Biol Chem 292:21598–21613

Lemay DG, Ballard OA, Hughes MA, Morrow AL, Horseman ND, Nommsen-Rivers LA (2013a) RNA sequencing of the human milk fat layer transcriptome reveals distinct gene expression profiles at three stages of lactation. PLoS One 8:e67531

Lemay DG, Hovey RC, Hartono SR, Hinde K, Smilowitz JT, Ventimiglia F, Schmidt KA, Lee JW, Islas-Trejo A, Silva PI, Korf I, Medrano JF, Barry PA, German JB (2013b) Sequencing the transcriptome of milk production: milk trumps mammary tissue. BMC Genomics 14:872

Liao Y, Alvarado R, Phinney B, Lönnerdal B (2011) Proteomic characterization of human milk fat globule membrane proteins during a 12 month lactation period. J Proteome Res 10:3530–3541

Lin P, Le-Niculescu H, Hofmeister R, McCaffery JM, Jin M, Hennemann H, McQuistan T, De Vries L, Farquhar MG (1998) The mammalian calcium-binding protein, nucleobindin (CALNUC), is a Golgi resident protein. J Cell Biol 141:1515–1527

Linxweiler M, Schick B, Zimmermann R (2017) Let's talk about Secs: Sec61, Sec62 and Sec63 in signal transduction, oncology and personalized medicine. Signal Transduct Target Ther 2:17002

Linzell JL (1960) Mammary-gland blood flow and oxygen, glucose and volatile fatty acid uptake in the conscious goat. J Physiol 153:492–509

Linzell JL (1974a) Mammary blood flow and methods of identifying and measuring precursors of milk. In: Larson BL, Smith VR (eds) Lactation, a comprehensive treatise. Academic, New York, pp 143–225

Linzell JL (1974b) The use of isotopes in the study of milk secretion. Proc Nutr Soc 33:17–23

Linzell JL, Peaker M (1971a) Intracellular concentrations of sodium, potassium and chloride in the lactating mammary gland and their relation to the secretory mechanism. J Physiol 216:683–700
Linzell JL, Peaker M (1971b) Mechanism of milk secretion. Physiol Rev 51:564–597
Linzell JL, Mepham TB, Annison EF, West CE (1969) Mammary metabolism in lactating sows: arteriovenous differences of milk precursors and the mammary metabolism of [14C]glucose and [14C]acetate. Br J Nutr 23:319–332
Lippolis JD, Powell EJ, Reinhardt TA, Thacker T, Casas E (2018) Symposium review: omics in dairy and animal science—promise, potential, and pitfalls. J Dairy Sci 102:4741–4754
Liu H, Hua Y, Luo H, Shen Z, Tao X, Zhu X (2015) An herbal galactagogue mixture increases milk production and aquaporin protein expression in the mammary glands of lactating rats. Evid Based Complement Alternat Med 2015:760585–760585
Lönnerdal B (2017) Development of iron homeostasis in infants and young children. Am J Clin Nutr 106:1575S–1580S
Lopez V, Kelleher SL (2009) Zinc transporter-2 (ZnT2) variants are localized to distinct subcellular compartments and functionally transport zinc. Biochem J 422:43–52
Lopez A, Torres N, Ortiz V, Aleman G, Hernandez-Pand R, Tovar AR (2006) Characterization and regulation of the gene expression of amino acid transport system A (SNAT2) in rat mammary gland. Am J Physiol Endocrinol Metab 291:11059–11066
Lutsenko S, Barnes NL, Bartee MY, Dmitriev OY (2007) Function and regulation of human copper-transporting ATPases. Physiol Rev 87:1011–1046
Macheda ML, Williams ED, Best JD, Wlodek ME, Rogers S (2003) Expression and localization of GLUT1 and GLUT12 glucose transporter in the pregnant and lactating rat mammary gland. Cell Tissue Res 311:91–97
Mackenzie B, Erickson JD (2004) Sodium-coupled neutral amino acid (System N/A) transporters of the SLC38 gene family. Pflugers Arch 447:784–795
Mamillapalli R, VanHouten J, Dann P, Bikle D, Chang W, Brown E, Wysolmerski J (2013) Mammary-specific ablation of the calcium-sensing receptor during lactation alters maternal calcium metabolism, milk calcium transport, and neonatal Calcium Accrual. Endocrinology 154:3031–3042
Maningat PD, Sen P, Sunehag AL, Hadsell DL, Haymond MW (2007) Regulation of gene expression in human mammary epithelium: effect of breast pumping. J Endocrinol 195:503–511
Maningat PD, Sen P, Rijnkels M, Sunehag AL, Hadsell DL, Bray M, Haymond MW (2009) Gene expression in the human mammary epithelium during lactation: the milk fat globule transcriptome. Physiol Genomics 37:12–22
Manjarin R, Bequette BJ, Wu G, Trottier NL (2014) Linking our understanding of mammary gland metabolism to amino acid nutrition. Amino Acids 46:2447–2462
Marcoux AA, Garneau AP, Frenette-Cotton R, Slimani S, Mac-Way F, Isenring P (2017) Molecular features and physiological roles of K^+-Cl^- cotransporter 4 (KCC4). Biochim Biophys Acta Gen Subj 1861:3154–3166
Mather IH, Masedunskas A, Chen Y, Weigert R (2019) Symposium review: intravital imaging of the lactating mammary gland in live mice reveals novel aspects of milk-lipid secretion. J Dairy Sci 102:1–23
Mattmiller SA, Corl CM, Gandy JC, Loor JJ, Sordillo LM (2011) Glucose transporter and hypoxia-associated gene expression in the mammary gland of transition dairy cattle. J Dairy Sci 94:2912–2922
McConnell KP (1948) Passage of selenium through the mammary glands of the white rat and the distribution of selenium in the milk proteins after subcutaneous injection of sodium selenate. J Biol Chem 173:653–657
McCormick NH, Kelleher SL (2012) ZnT4 provides zinc to zinc-dependent proteins in the trans-Golgi network critical for cell function and Zn export in mammary epithelial cells. Am J Phys Cell Physiol 303:C291–C917

McCormick NH, Hennigar SR, Kiselyov K, Kelleher SL (2014) The biology of zinc transport in mammary epithelial cells: implications for mammary gland development, lactation, and involution. J Mammary Gland Biol Neoplasia 19:59–71

McCormick NH, Lee S, Hennigar SR, Kelleher SL (2016) ZnT4 (SLC30A4)-null ("lethal milk") mice have defects in mammary gland secretion and hallmarks of precocious involution during lactation. Am J Phys Regul Integr Comp Phys 310:R33–R40

McManaman JL (2014) Lipid transport in the lactating mammary gland. J Mammary Gland Biol Neoplasia 19:35–42

Michalczyk A, Bastow E, Greenough M, Camakaris J, Freestone D, Taylor P, Linder M, Mercer J, Ackland ML (2008) ATP7B expression in human breast epithelial cells is mediated by lactational hormones. J Histochem Cytochem 56:389–399

Miyoshi K, Shillingford JM, Smith GH, Grimm SL, Wagner KU, Oka T, Rosen JM, Robinson GW, Hennighausen L (2001) Signal transducer and activator of transcription (Stat) 5 controls the proliferation and differentiation of mammary alveolar epithelium. J Cell Biol 155:531–542

Mobasheri A, Barrett-Jolley R (2014) Aquaporin water channels in the mammary gland: from physiology to pathophysiology and neoplasia. J Mammary Gland Biol Neoplasia 19:91–102

Mobasheri A, Kendall BH, Maxwell JE, Sawran AV, German AJ, Marples D, Luck MR, Royal MD (2011) Cellular localization of aquaporins along the secretory pathway of the lactating bovine mammary gland: an immunohistochemical study. Acta Histochem 113:137–149

Mohammad MA, Hadsell DL, Haymond MW (2012) Gene regulation of UDP-galactose synthesis and transport: potential rate-limiting processes in initiation of milk production in humans. Am J Physiol Endocrinol Metab 303:E365–E376

Monks J, Neville MC (2004) Albumin transcytosis across the epithelium of the lactating mouse mammary gland. J Physiol 560:267–280

Mostov KE, Kraehenbuhl J-P, Blobel G (1980) Receptor-mediated transcellular transport of immunoglobulin: synthesis of secretory component as multiple and larger transmembrane forms. Proc Natl Acad Sci U S A 77:7257–7261

Murer H, Forster I, Biber J (2004) The sodium phosphate cotransporter family SLC34. Pflugers Arch 447:763–767

Muscher-Banse AS, Breves G (2019) Mechanisms and regulation of epithelial phosphate transport in ruminants: approaches in comparative physiology. Pflugers Arch 471:185–191

Nagasawa M, Kanzaki M, Iino Y, Morishita Y, Kojima I (2001) Identification of a novel chloride channel expressed in the endoplasmic reticulum, golgi apparatus, and nucleus. J Biol Chem 276:20413–20418

Nazemi S, Rahbek M, Parhamifar L, Moghimi SM, Babamoradi H, Mehrdana F, Klaerke DA, Knight CH (2014) Reciprocity in the developmental regulation of aquaporins 1, 3 and 5 during pregnancy and lactation in the rat. PLoS One 9:e106809

Nemeth BA, Tsang SWY, Geske RS, Haney PM (2000) Golgi targeting of the GLUT1 glucose transporter in lactating mouse mammary gland. Pediatr Res 47:444–450

Neville MC (2005) Calcium secretion into milk. J Mammary Gland Biol Neoplasia 10:119–128

Nguyen DA, Neville MC (1998) Tight junction regulation in the mammary gland. J Mammary Gland Biol Neoplasia 3:233–246

Nguyen D-AD, Beema N, Neville MC (2001a) Regulation of tight junction permeability in the mammary gland. In: Cerejido M, Anderson JM (eds) Tight junctions, 2nd edn. CRC Press, New York, pp 395–414

Nguyen D-AD, Parlow AF, Neville MC (2001b) Hormonal regulation of tight junction closure in the mouse mammary epithelium during the transition from pregnancy to lactation. J Endocrinol 170:347–356

Noor SI, Pouyssegur J, Deitmer JW, Becker HM (2017) Integration of a 'proton antenna' facilitates transport activity of the monocarboxylate transporter MCT4. FEBS J 284:149–162

Orlowski J, Grinstein S (2004) Diversity of the mammalian sodium/proton exchanger SLC9 gene family. Pflugers Arch 447:549–565

Palmer ML, Peitzman ER, Mania PJ, Sieck GC, Prakash YS, O'Grady SM (2011) K(Ca)3.1 channels facilitate K^+ secretion or Na^+ absorption depending on apical or basolateral P2Y receptor stimulation. J Physiol 589:3483–3494

Park HS, Hong C, Kim BJ, So I (2014a) The pathophysiologic roles of TRPM7 hannel. Korean J Physiol Pharmacol 18:15–23

Park M, Ohana E, Choi SY, Lee MS, Park JH, Muallem S (2014b) Multiple roles of the $SO_4^{2-}/Cl^-/OH^-$ exchanger protein Slc26a2 in chondrocyte functions. J Biol Chem 289:1993–2001

Pedemonte N, Galietta LJV (2014) Structure and function of TMEM16 proteins (Anoctamins). Physiol Rev 94:419–459

Perron B, Rodriguez AM, Leblanc G, Pourcher T (2001) Cloning of the mouse sodium iodide symporter and its expression in the mammary gland and other tissues. J Endocrinol 170:185–196

Pierson H, Muchenditsi A, Kim B-E, Ralle M, Zachos N, Huster D, Lutsenko S (2018) The function of ATPase copper transporter ATP7B in intestine. Gastroenterology 154:168–180.e5

Polishchuk R, Lutsenko S (2013) Golgi in copper homeostasis: a view from the membrane trafficking field. Histochem Cell Biol 140:285–295

Ponnalagu D, Gururaja Rao S, Farber J, Xin W, Hussain AT, Shah K, Tanda S, Berryman M, Edwards JC, Singh H (2016a) Molecular identity of cardiac mitochondrial chloride intracellular channel proteins. Mitochondrion 27:6–14

Ponnalagu D, Rao SG, Farber J, Xin W, Hussain AT, Shah K, Tanda S, Berryman MA, Edwards JC, Singh H (2016b) Data supporting characterization of CLIC1, CLIC4, CLIC5 and DmCLIC antibodies and localization of CLICs in endoplasmic reticulum of cardiomyocytes. Data Brief 7:1038–1044

Puchkova LV, Babich PS, Zatulovskaia YA, Ilyechova EY, Di Sole F (2018) Copper metabolism of newborns is adapted to milk ceruloplasmin as a nutritive Source of copper: Overview of the current data. Nutrients 10:1591

Qi Y, Mair N, Kummer KK, Leitner MG, Camprubí-Robles M, Langeslag M, Kress M (2018) Identification of chloride channels CLCN3 and CLCN5 mediating the excitatory Cl^- currents activated by sphingosine-1-phosphate in sensory neurons. Front Mol Neurosci 11:33–33

Quamme GA (2010) Molecular identification of ancient and modern mammalian magnesium transporters. Am J Phys Cell Physiol 298:C407–C429

Quesnell RR, Han X, Schultz BD (2007) Glucocorticoids stimulate ENaC upregulation in bovine mammary epithelium. Am J Phys Cell Physiol 292:C1739–C1745

Ramos D, Mar D, Ishida M, Vargas R, Gaite M, Montgomery A, Linder MC (2016) Mechanism of copper uptake from blood plasma ceruloplasmin by mammalian cells. PLoS One 11:e0149516

Rassin DK, Sturman JA, Guall GE (1978) Taurine and other free amino acids in milk of man and other mammals. Early Hum Dev 2:1–13

Raven L-A, Cocks BG, Kemper KE, Chamberlain AJ, Vander Jagt CJ, Goddard ME, Hayes BJ (2016) Targeted imputation of sequence variants and gene expression profiling identifies twelve candidate genes associated with lactation volume, composition and calving interval in dairy cattle. Mamm Genome 27:81–97

Reinhardt TA, Horst RL (1999) Ca^{2+}-ATPases and their expression in the mammary gland of pregnant and lactating rats. Am J Phys Cell Physiol 276:C796–C802

Reinhardt TA, Filoteo AG, Penniston JT, Horst RL (2000) Ca^{2+}-ATPase protein expression in mammary tissue. Am J Phys Cell Physiol 279:C1595–C1602

Reinhardt TA, Lippolis JD, Shull GE, Horst RL (2004) Null mutation in the gene encoding plasma membrane Ca^{2+}-ATPase isoform 2 impairs calcium transport into milk. J Biol Chem 279:42369–42373

Reinhardt TA, Lippolis JD, Sacco RE (2014) The Ca^{2+}/H^+ antiporter TMEM165 expression, localization in the developing, lactating and involuting mammary gland parallels the secretory pathway Ca^{2+} ATPase (SPCA1). Biochem Biophys Res Commun 445:417–421

Rillema JA, Yu TX (1996) Prolactin stimulation of iodide uptake into mouse mammary gland explants. Am J Physiol Endocrinol Metab 271:E879–E882

Rivera OC, Hennigar SR, Kelleher SL (2018) ZnT2 is critical for lysosome acidification and biogenesis during mammary gland involution. Am J Physiol Reg Int Comp 315:R323–R335

Rudolph MC, Wellberg EA, Anderson SM (2009) Adipose-depleted mammary epithelial cells and organoids. J Mammary Gland Biol Neoplasia 14:381–386

Sabirov RZ, Merzlyak PG, Okada T, Islam MR, Uramoto H, Mori T, Makino Y, Matsuura H, Xie Y, Okada Y (2017) The organic anion transporter SLCO2A1 constitutes the core component of the Maxi-Cl channel. EMBO J 36:3309–3324

Sanchez MP, Wolf V, El Jabri M, Beuvier E, Rolet-Répécaud O, Gaüzère Y, Minéry S, Brochard M, Michenet A, Taussat S, Barbat-Leterrier A, Delacroix-Buchet A, Laithier C, Fritz S, Boichard D (2018) Short communication: confirmation of candidate causative variants on milk composition and cheesemaking properties in Montbéliarde cows. J Dairy Sci 101:10076–10081

Santos DCC, Angulo-Barroso RM, Li M, Bian Y, Sturza J, Richards B, Lozoff B (2018) Timing, duration, and severity of iron deficiency in early development and motor outcomes at 9 months. Eur J Clin Nutr 72:332–341

Sargeant TJ, Lloyd-Lewis B, Resemann HK, Ramos-Montoya A, Skepper J, Watson CJ (2014) Stat3 controls cell death during mammary gland involution by regulating uptake of milk fat globules and lysosomal membrane permeabilization. Nat Cell Biol 16:1057–1068

Schäffers OJM, Hoenderop JGJ, Bindels RJM, Baaij JHF (2018) The rise and fall of novel renal magnesium transporters. Am J Physiol Renal Physiol 314:F1027–F1033

Schultheis PJ, Hagen TT, O'Toole KK, Tachiban A, Burke CR, McGill DL, Okunade GW, Shull GE (2004) Characterization of the P5 subfamily of P-type transport ATPases in mice. Biochem Biophys Res Commun 323:731–738

Schultz BD (2011) Clarifying the role of (apical) K^+ channels in Na^+ and Cl^- transport. J Physiol 589:3689–3690

Schultz BD (2016) Making isotonic milk. Focus on "Ca^{2+}-activated Cl^- channel currents in mammary secretory cells from lactating mouse". Am J Phys Cell Physiol 311:C805–C807

Schultz B, Devor DC (2016) Fundamentals of epithelial chloride transport. In: Hamilton KL, Devor DC (eds) Ion channels and transporters of epithelia in health and disease. Springer, New York, pp 3–47

Sciascia QL, Pacheco D, McCoard SA (2015) Administration of exogenous growth hormone is associated with changes in plasma and intracellular mammary amino acid profiles and abundance of the mammary gland amino acid transporter SLC3A2 in midlactation dairy cows. PLoS One 10:e0134323

Shennan DB (1989) Evidence for furosemide-sensitive Na^+-K^+-Cl^--cotransport in lactating rat mammary tissue. Quart J Exp Physiol 74:927–938

Shennan DB (1995) Identification of a high affinity taurine transporter which is not dependent on chloride. Biosci Rep 15:231–239

Shennan DB (1998) Mammary gland membrane transport systems. J Mammary Gland Biol Neoplasia 3:247–258

Shennan DB, Boyd CA (2014) The functional and molecular entities underlying amino acid and peptide transport by the mammary gland under different physiological and pathological conditions. J Mammary Gland Biol Neoplasia 19:19–33

Shennan DB, McNeillie SA (1990) An interaction between selenate and a sulfate transporter in the lactating rat mammary gland. Biol Trace Elem Res 24:173–182

Shennan DB, McNeillie SA (1994) High affinity (Na^+ + Cl^-)-dependent taurine transport by lactating mammary tissue. J Dairy Res 61:335–343

Shennan DB, Peaker M (2000) Transport of milk constituents by the mammary gland. Physiol Rev 80:925–951

Shennan DB, McNeillie SA, Cliff MJ (1995) The relationship between volume-sensitive taurine efflux and the anion-exchanger in lactating mammary tissue. Biochem Soc Trans 23:392S

Shennan DB, Calvert DT, Travers MT, Kudo Y, Boyd CA (2002) A study of L-leucine, L-phenylalanine and L-alanine transport in the perfused rat mammary gland: possible involvement of LAT1 and LAT2. Biochim Biophys Acta 1564:133–139

Shillingford JM, Calvert DT, Shennan DB, Beechey RB (1996) Phosphate transport via Na^+-Pi cotransport and anion exchange in lactating rat mammary tissue. Exp Physiol 81:273–284

Shillingford JM, Miyoshi K, Flagella M, Shull GE, Hennighausen L (2002) Mouse mammary epithelial cells express the Na-K-Cl cotransporter, NKCC1: characterization, localization, and involvement in ductal development and morphogenesis. Mol Endocrinol 16:1309–1321

Sirianant L, Ousingsawat J, Wanitchakool P, Schreiber R, Kunzelmann K (2016) Cellular volume regulation by anoctamin 6: Ca^{2+}, phospholipase A2 and osmosensing. Pflugers Arch 468:335–349

Skach WR (2007) The expanding role of the ER translocon in membrane protein folding. J Cell Biol 179:1333–1335

Smaardijk S, Chen J, Wuytack F, Vangheluwe P (2017) SPCA2 couples Ca^{2+} influx via Orai1 to Ca^{2+} uptake into the Golgi/secretory pathway. Tissue Cell 49:141–149

Snyder NA, Palmer MV, Reinhardt TA, Cunningham KW (2019) Milk biosynthesis requires the Golgi cation exchanger TMEM165. J Biol Chem 294:3181–3191

Solari R, Kraehenbuhl J-P (1984) Biosynthesis of the IgA antibody receptor: a model for the transepithelial sorting of a membrane glycoprotein. Cell 36:61–71

Srisomboon Y, Zaidman NA, Maniak PJ, Deachapunya C, O'Grady SM (2018) P2Y receptor regulation of K2P channels that facilitate K^+ secretion by human mammary epithelial cells. Am J Phys Cell Physiol 314:C627–C639

Suyama T, Okada S, Ishijima T, Iida K, Abe K, Nakai Y (2012) High phosphorus diet-induced changes in NaPi-IIb phosphate transporter expression in the rat kidney: DNA microarray analysis. PLoS One 7:e29483–e29483

Takanaga H, Mackenzie B, Hediger MA (2004) Sodium-dependent ascorbic acid transporter family SLC23. Pflugers Arch 447:677–682

Tanasova M, Begoyan VV, Weselinski L (2018) Targeting sugar uptake and metabolism for cancer identification and therapy: An overview. Curr Top Med Chem 18:467–483

Threadgold LC, Coore HC, Kuhn NJ (1982) Monosaccharide transport into lactating rat mammary acini. Biochem J 204:493–501

Torrents D, Estevez R, Pineda M, Fernandez E, Lloberas J, Shi YB, Zorzano A, Palacin M (1998) Identification and characterization of a membrane protein (y+L amino acid transporter-1) that associates with 4F2hc to encode the amino acid transport activity y+L. A candidate gene for lysinuric protein intolerance. J Biol Chem 273:32437–32445

Trott JF, Horigan KC, Gloviczki JM, Costa KM, Freking BA, Farmer C, Hayashi K, Spencer T, Morabito JE, Hovey RC (2009) Tissue-specific regulation of porcine prolactin receptor expression by estrogen, progesterone, and prolactin. J Endocrinol 202:153–166

Urzica D, Gales C, Zamfir C, Nechifor M (2013) The influence of oral steroidal contraceptives on magnesium concentration in breast milk. Magnes Res 26:188–191

Vallipuram J, Grenville J, Crawford DA (2010) The E646D-ATP13A4 mutation associated with autism reveals a defect in calcium regulation. Cell Mol Neurobiol 30:233–246

VanHouten JN, Wysolmerski JJ (2007) Transcellular calcium transport in mammary epithelial cells. J Mammary Gland Biol Neoplasia 12:223–235

VanHouten JN, Neville MC, Wysolmerski JJ (2007) The calcium-sensing receptor regulates plasma membrane calcium adenosine triphosphatase isoform 2 activity in mammary epithelial cells: a mechanism for calcium-regulated calcium transport into milk. Endocrinology 148:5943–5954

Veit M, Koyro KI, Ahrens B, Bleibaum F, Munz M, Rövekamp H, Andrä J, Schreiber R, Kunzelmann K, Sommer A, Bhakdi S, Reiss K (2018) Anoctamin-6 regulates ADAM sheddase function. Biochim Biophys Acta 1865:1598–1610

Velazquez-Villegas LA, Ortiz V, Strom A, Torres N, Engler DA, Matsunami R, Ordaz-Rosado D, Garcia-Becerra R, Lopez-Barradas AM, Larrea F, Gustafsson JA, Tovar AR (2014) Transcriptional regulation of the sodium-coupled neutral amino acid transporter (SNAT2) by 17beta-estradiol. Proc Natl Acad Sci U S A 111:11443–11448

Velazquez-Villegas LA, Lopez-Barradas AM, Torres N, Hernandez-Pando R, Leon-Contreras JC, Granados O, Ortiz V, Tovar AR (2015) Prolactin and the dietary protein/carbohydrate ratio regulate the expression of SNAT2 amino acid transporter in the mammary gland during lactation. Biochim Biophys Acta 1848:1157–1164

Verkman AS, Anderson MO, Papadopoulos MC (2014) Aquaporins: important but elusive drug targets. Nat Rev Drug Discov 13:259–277

Verma N, Kansal VK (1993) Characterisation of the routes of methionine transport in mouse mammary glands. Indian J Med Res 98:297–304

Watanabe R, Sakuragi T, Noji H, Nagata S (2018) Single-molecule analysis of phospholipid scrambling by TMEM16F. Proc Natl Acad Sci U S A 115:3066–3079

West CE, Bickerstaffe R, Annison EF, Linzell JL (1972) Studies on the mode of uptake of blood triglycerides by the mammary gland of the lactating goat. The uptake and incorporation into milk fat and mammary lymph of labelled glycerol, fatty acids and triglycerides. Biochem J 126:477–490

Wright EM, Turk E (2004) The sodium/glucose cotransport family SLC5. Pflugers Arch 447:510–518

Wuytack F, Raeymaekers L, Missiaen L (2003) PMR1/SPCA Ca^{2+} pumps and the role of the Golgi apparatus as a Ca^{2+} store. Pflugers Arch 446:148–153

Xin Y, Gao H, Wang J, Qiang Y, Imam MU, Li Y, Wang J, Zhang R, Zhang H, Yu Y, Wang H, Luo H, Shi C, Xu Y, Hojyo S, Fukada T, Min J, Wang F (2017) Manganese transporter Slc39a14 deficiency revealed its key role in maintaining manganese homeostasis in mice. Cell Discov 3:17025

Xu W, Hong SJ, Zhong A, Xie P, Jia S, Xie Z, Zeitchek M, Niknam-Bienia S, Zhao J, Porterfield DM, Surmeier DJ, Leung KP, Galiano RD, Mustoe TA (2015) Sodium channel Nax is a regulator in epithelial sodium homeostasis. Sci Transl Med 7:312ra177

Yan R, Zhao X, Lei J, Zhou Q (2019) Structure of the human LAT1–4F2hc heteromeric amino acid transporter complex. Nature 568:127–130

Yang H, Jan LY (2016) Chapter 7 – TMEM16 membrane proteins in health and disease. In: Pitt GS (ed) Ion channels in health and disease. Academic, Boston, pp 165–197

Yu X, Shen H, Liu L, Lin L, Gao M, Wang S (2012) Changes of sodium iodide symporter regulated by IGF-I and TGF-β1 in mammary gland cells from lactating mice at different iodine levels. Biol Trace Elem Res 146:73–78

Zeuthen T, Gorraitz E, Her K, Wright EM, Loo DDF (2016) Structural and functional significance of water permeation through cotransporters. Proc Natl Acad Sci U S A 113:E6887–E6894

Zhang P, Allen JC (1995a) A novel dialysis procedure measuring free Zn^{2+} in bovine milk and plasma. J Nutr 125:1904–1910

Zhang PF, Allen JC (1995b) Free zinc concentration in bovine milk measured by analytical affinity chromatography with immobilized metallothionein. Biol Trace Elem Res 50:135–148

Zhang P, Sawicki V, Lewis A, Hanson L, Monks J, Neville MC (2000) The effect of serum iron concentration on iron secretion into mouse milk. J Physiol 522:479–491

Zhang S, Chen F, Zhang Y, Lv Y, Heng J, Min T, Li L, Guan W (2018) Recent progress of porcine milk components and mammary gland function. J Anim Sci Biotechnol 9:77–77

Zhao FQ (2014) Biology of glucose transport in the mammary gland. J Mammary Gland Biol Neoplasia 19:3–17

Zhao FQ, Keating AF (2007) Expression and regulation of glucose transporters in the bovine mammary gland. J Dairy Sci 90(Suppl 1):E76–E86

Zhao FQ, Miller PJ, Wall EH, Zheng YC, Dong B, Neville MC, McFadden TB (2004) Bovine glucose transporter GLUT8: cloning, expression, and developmental regulation in mammary gland. Biochim Biophys Acta 1680:103–113

Zhao K-Q, Xiong G, Wilber M, Cohen NA, Kreindler JL (2012) A role for two-pore K^+ channels in modulating Na^+ absorption and Cl^- secretion in normal human bronchial epithelial cells. Am J Physiol Lung Cell Mol Physiol 302:L4–L12

Zhu XR, Netzer R, Bohlke K, Liu Q, Pongs O (1999) Structural and functional characterization of Kv6.2 a new gamma-subunit of voltage-gated potassium channel. Recept Channels 6:337–350

Chapter 7
Lipid Transport Across the Mammary Gland

James L. McManaman

Abstract Mammary secretory cells have evolved robust processes for transporting and secreting nutrients, including lipids, proteins, and sugars, required for neonatal growth and development. The processes responsible for secreting lipids are particularly unique to mammary secretory cells and represent novel adaptations of apocrine secretion. All eukaryotic cells depend on phospholipid (PL) and fatty acid (FA) transport to maintain membrane structure and organization, and to fuel and regulate cellular functions. The copious milk secretion, including large quantities of lipid in some species, requires adaptation and integration of PL and FA synthesis and transport processes to meet secretion demands. At present, few details exist about how these processes are regulated within the mammary gland. However, recent advances in our understanding of the structural and molecular biology of membrane systems and cellular lipid trafficking provide insights into the mechanisms underlying the regulation and integration of PL and FA transport processes in the secretion of milk lipids. This review discusses the PL and FA transport processes required to maintain the structural integrity and organization of the mammary gland and support its secretory functions within the context of current molecular and cellular models of their regulation.

Keywords Biosynthesis, Lactation · Lipids · Mammary gland · Membranes · Transport mechanisms

---

J. L. McManaman (✉)
Division of Reproductive Sciences, University of Colorado Anschutz Medical Campus, Aurora, CO, USA
e-mail: jim.mcmanaman@cuanschutz.edu

K. L. Hamilton, D. C. Devor (eds.), *Ion Transport Across Epithelial Tissues and Disease*, Physiology in Health and Disease,
https://doi.org/10.1007/978-3-030-55310-4_7

7.1 The Mammary Secretory Cell

The mammary gland is a mammalian-specific exocrine organ comprised of a network of ducts and acini lined by single layers of cuboidal or columnar epithelial cells that serve barrier, structural, and/or secretory functions. Epithelial cells lining acini (hereafter referred to as mammary secretory cells) undergo functional differentiation during pregnancy and lactation to synthesize and secrete a complex fluid, which contains the diverse array of nutritive and bioactive substances that comprise milk. Epithelial cells lining ducts serve structural and barrier functions forming the conduits that supply milk to nursing young. Unlike ductal epithelial cells of other exocrine organs, such as the salivary gland, ductal epithelial cells of the mammary gland lack recognized synthetic or secretory functions (McManaman et al. 2006), and little is known about their specific transport functions. In contrast, mammary secretory cells require efficient and integrated cellular transport mechanisms to produce molecularly complex milk and meet the robust secretory functions of the mammary gland during lactation (Shennan and Peaker 2000). This chapter focuses on the molecular, cellular, and physiological processes that mediate lipid transport in mammary secretory cells emphasizing milk lipid production during lactation.

7.1.1 Lipid Secretion

Lipids are major constituents of the milk of most mammals and a singularly important source of the calories required for neonatal growth and signaling molecules that promote postnatal development (Oftedal 1984). The principal lipid constituents of milk are neutral lipids, which include fatty acid esters of glycerol (diacyl- and triacylglycerol esters) and cholesterol (cholesteryl esters), and polar phospholipids that include glycerophospholipids and sphingolipids. Triglycerides, cholesterol, and phospholipids comprise 98–99%, 0.3–1.3%, and 0.6–1.3%, respectively, of total milk lipids (Jensen et al. 1990; Jensen 1999).

Milk lipids are secreted uniquely by an apocrine mechanism. Estimates of neutral lipid secretion rates during lactation from multiple species indicate that, depending on the lipid content of milk, rates of apocrine lipid secretion by mammary secretory cells exceed those of vesicle-mediated secretion of serum lipids by hepatocytes and other lipid secreting cell types (Table 7.1). Lactating women, who produce milk with an average fat content of 4.2%, secrete approximately 0.5 g of fat/day/kg body weight (Kent et al. 2018) into milk, whereas mice that produce milk containing an average of 28% fat (Sadowska et al. 2015) secrete approximately 135 g/day/kg body weight into milk, and rats with 14% milk fat secrete 13 g/day/kg body weight.

The relatively large amount of lipid secreted during lactation by most mammals, and the apocrine mechanism of lipid secretion, which removes a portion of the apical plasma membrane with each secretory event, increases demands on neutral and

Table 7.1 Milk and serum lipid secretion rates

	Milk lipid content (vol %)	Milk lipid secretion rate (g/d/kg)	Serum lipid (VLDL) secretion rate (g/d/kg)	Ratio milk lipid to serum lipid secretion rates
Human	4.7%[a]	0.6[a]	0.2[b]	3.0
Mouse	28%[c]	135[c]	1.6[d]	85
Rat	15%[e]	13[e]	4.3[f]	3

[a]Based on an average milk production of 860 ml/day and a milk fat content of 4.7% (Kent et al. 2018)
[b]Based on VLDL secretion rates of non-lactating women (Mittendorfer et al. 2016)
[c]Based on average milk production of 14.5 g/day and milk fat content of 28% (Sadowska et al. 2015)
[d]Based on average VLDL secretion rates for male and non-pregnant female mice (Tsukamoto et al. 2000)
[e]Based on average milk production of 14 g/day and milk fat content of 15% (Morag 1970; Keen et al. 1981)
[f]Based on VLDL secretion rates for mid-lactating rats (Agius et al. 1981)

phospholipid synthesis and transport processes in mammary secretory cells above those of other cell types (McManaman 2012).

7.1.2 Lipid Transport Processes

Mammary secretory cells depend on distinct transport activities to secrete lipids (Fig. 7.1): Fatty acids and cholesterol transported into the cell from the circulation are primary substrates in the synthesis of secreted lipids; Intracellular transport mechanisms transfer fatty acids to the endoplasmic reticulum (ER), Golgi, and mitochondria for neutral and phospholipid synthesis; Phospholipids are transported within the cell by intra-organellar or carrier-mediated mechanisms for membrane synthesis and maintenance; Neutral lipids are synthesized from glycerol and fatty acids and packaged into cytoplasmic lipid droplets (CLD); CLD are transported on microtubules to the apical plasma membrane for secretion, where they dock in preparation for secretion; Neutral lipids are transferred between CLD to adjust their dimensions prior to secretion; and CLD docked at the apical plasma membrane are secreted as membrane enveloped structures, referred to as milk fat globules (MFG), into the acinar lumen by an apocrine mechanism (Mather and Keenan 1998). Although specific details about the molecular mediators and mechanisms of many of these transport activities have yet to be established rigorously, the general features, and in some cases the molecular details, are known.

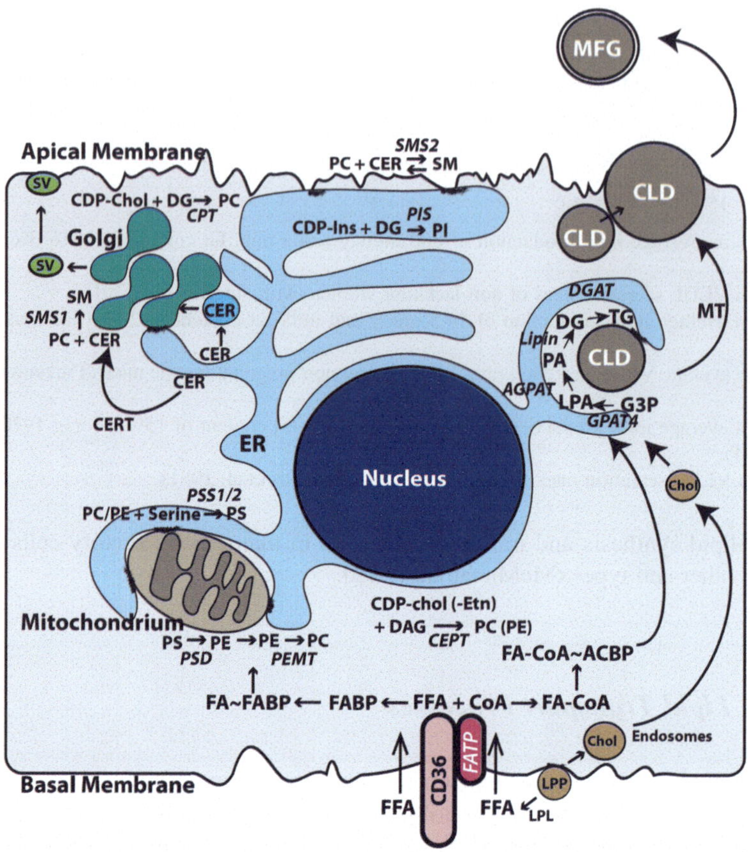

Fig. 7.1 The figure depicts the primary pathways responsible for uptake, intracellular transport and secretion of fatty acids, phospholipids, neutral lipids, and cholesterol in mammary secretory cells. Free fatty acids bound to serum albumin (FFA) are transported into the cell by the actions of transporters on the basal plasma membrane, exemplified by CD36 and fatty acid transport proteins (FATP), which also transport FFA derived by lipoprotein lipase hydrolysis of neutral lipids in lipoprotein particles (LPP) (Sect. 7.2). Cholesterol and phospholipids in LPP are transported into the cell by a receptor-mediated processes and packaged into endosomes, which are transported within endosomal pathways to become part of endoplasmic reticulum (ER) and Golgi membranes (Sect. 7.3). Inside the cell, FFA bind to carrier proteins, such as fatty acid binding proteins (FABP), or are converted to thioesters of coenzyme A (FA-CoA), which bind to acyl-CoA binding proteins (ACBP) that serve as specific intracellular carries of FA-CoAs (Sect. 7.3). Carrier bound fatty acids are transported to the nucleus where they serve as transcriptional regulators, to the mitochondria where they serve as substrates for energy production or to the ER, mitochondria, and Golgi where they serve as substrates for the synthesis of neutral lipids and/or ceramide and phospholipids. The ER is primarily responsible for cellular phospholipid synthesis, which involves multiple substrate and product transport steps (Sect. 7.3) Phosphatidylcholine (PC) is synthesized by two ER enzymes. Choline-ethanolamine phosphotransferase (Cept) is located throughout the ER and catalyzes the synthesis of both PC and PE. Phosphatidylethanolamine N-methyltransferase (Pemt) localizes to specific ER domains that contact the mitochondria and selectively synthesizes PC by sequential methylation of phosphatidylethanolamine (PE), which is synthesized in the mitochondria by decarboxylation of phosphatidylserine (PS). PC is also synthesized in the Golgi by a Golgi-specific form of CPT. Phosphatidylserine is synthesized within the specialized ER membrane domains that contact mitochondria by phosphatidylserine synthases-1 and -2 (PSS1/2). PC, PE,

7.2 Fatty Acid Transport Across the Plasma Membrane

Mammary secretory cells use fatty acids derived from the diet or synthesized de novo for lipid synthesis. In humans and other non-ruminate species, dietary fatty acids used in lipid synthesis by mammary secretory cells are predominantly long-chain (C16 and greater), whereas fatty acids synthesized de novo are medium chain (C8–C14), owing to the expression of a thioesterase that terminates elongation after completion of 8–14 cycles (Thompson and Smith 1985; Neville and Picciano 1997). Dietary long-chain fatty acids (LCFA) exist in the circulation in the form of triglycerides within lipoprotein particles (e.g., chylomicrons and VLDL) or as free fatty acids (FFA) complexed to albumin. Evidence from isotope tracer studies indicates that the majority of dietary LCFA traffic to adipose stores before being released into the circulation as FFA (Fidler et al. 2000).

Transport of LCFA into cells is a multistep process involving: (1) the transfer FFA from serum carrier proteins to binding sites on the external plasma membrane leaflet; (2) transport across the bilayer to the inner membrane leaflet; and (3) desorption from the inner membrane leaflet by transfer to an intracellular carrier protein or enzyme-mediated CoA thioacylation (3) (Glatz et al. 2010). LCFA readily diffuse across membrane bilayers by a flip-flop mechanism (Hamilton 2007; Jay and Hamilton 2018). However, at physiological concentrations, the uptake of serum FFA into cells is mediated (or facilitated) by specific membrane proteins (Glatz et al.

Fig. 7.1 (continued) and PS are transferred between cellular membrane compartments by vesicular and non-vesicular contact or carrier-mediated pathways. Phosphatidylinositol (PI) is synthesized by phosphatidylinositol synthase (PIS), which localizes to ER-specific domains that contact the plasma membrane and Golgi. Phosphatidylinositol transport between these membranes is mediated by vesicular and carrier-mediated mechanisms. Sphingomyelin (SM) is synthesized by the Golgi-specific enzyme sphingomyelin synthase-1 (SMS1), which catalyzes transfer of choline or ethanolamine from PC and PE, respectively, to ceramide, a fatty acid amide. Ceramide is synthesized by enzymes on the cytosolic leaflet ER membranes and transported to the Golgi by vesicular trafficking (blue oval labeled CER) or by the ceramide transport protein (CERT). Interconversion of PC and SM is also catalyzed by SMS2 on the plasma membrane. Neutral lipids are synthesized by actions of ER membrane enzymes glycerol-3-phosphate O-acyltransferase (GPAT) and 1-acylglycerol-3-phosphate O-acyltransferase (AGPAT), which catalyze sequential esterification of glycerol-3-phosphate (G3P) to form lysophosphatidic acid (LPA) and phosphatidic acid (PA), respectively, (Sect. 7.4). Phosphatidic acid is dephosphorylated by the phosphatidate phosphatase, lipin, to form diacylglycerol (DG), which is converted to triacylglycerol (TG) by acyl-CoA: diacylglycerol O-acyltransferase (DGAT) activity (Sect. 7.4). DG and TG are packaged into cytoplasmic lipid droplets (CLD), which are specialized neutral lipid storage structures that originate from specific ER membrane domains (Sect. 7.4). CLD are transported by microtubules (MT) from the ER to the apical plasma membrane (Sect. 7.5). During transport, CLD undergo fusion, transferring neutral lipids from smaller to larger structures (Sect. 7.5), which is indicated by arrow between CLD structures. At the apical plasma membrane, CLD dock and become progressively enveloped by elements of the apical plasma membrane (Sect. 7.5), before being secreted by a contraction-dependent apocrine mechanism (Sect. 7.5) as membrane coated milk fat globules (MFG). Enzymes are italicized. Membrane contact sites are indicated by fuzzy lines

2010). Similar processes and membrane proteins mediate membrane transport of fatty acids released from circulating lipoprotein particles by the action of lipoprotein lipase (LPL) (Goldberg et al. 2009).

In mammalian cells, membrane-associated proteins including CD36/SR-B2/Fatty acid translocase (CD36/FAT); membrane and cytoplasmic fatty acid binding proteins (FABPs); caveolins; and members of the solute carrier 27A/fatty acid transport protein (SLC27A/FATP) and acyl-coA synthetase (ACSL) families are implicated directly or indirectly in plasma membrane transport of LCFA (Glatz et al. 2010). CD36/FAT (Coburn et al. 2000; Nassir et al. 2007), FABPs (Luiken et al. 2003; Nickerson et al. 2009), and certain members of the SLC27A/FATP (Wu et al. 2006; Nickerson et al. 2009) have been identified as physiologically important transporters of LCFA into adipose, heart, skeletal muscle, intestine, and/or liver in mice using transgenic approaches (Richards et al. 2006; Goldberg et al. 2009; Nickerson et al. 2009). Nevertheless, the precise mechanisms regulating LCFA plasma membrane transport into mammalian cells remain incompletely understood, and the mechanisms may involve distinct cell-specific contributions from these proteins acting individually or in combination with each other.

7.2.1 CD36/FAT

CD36/FAT is a ubiquitous multifunctional double-pass transmembrane glycoprotein with high affinity (nM range) for LCFA. CD36/FAT has a binding site for one or more fatty acids on its external surface (Jay and Hamilton 2018). Homology modeling based on the structure of LIMP2, a structurally similar family member, indicates that the ectodomain of CD36/FAT possesses a hydrophobic fatty acid binding pocket, which connects to a large cavity that runs the length of the protein and provides a tunnel for transferring fatty acids and other lipids to the outer leaflet of the plasma membrane (Neculai et al. 2013; Pepino et al. 2014). Evidence of the physiological importance of this tunnel in lipid transport comes from mutation studies of SNMP1, a structurally similar *Drosophila* CD36 homologue (Gomez-Diaz et al. 2016). Targeted mutations of amino acids within the tunnel of SNMP1 that sterically constrain its size interfere with membrane transport of cis-vaccenyl acetate (a vaccenic acid-related pheromone) in *Drosophila* cilia.

The ability of CD36/FAT to enhance fatty acid membrane transport in muscle is increased by plasma membrane and/or cytosolic forms of FABP (FABPpm and FABPcyt) (Luiken et al. 1999; Glatz and Luiken 2017), and in their absence, the effects of CD36/FAT on fatty acid transport are significantly attenuated or absent. These data indicate that CD36/FAT may act in concert with other FA binding proteins to enrich LCFA at the external leaflet of the plasma membrane and facilitate their diffusional membrane transport and desorption to cytosolic acceptors (Glatz et al. 2010). At the plasma membrane, CD36/FAT localizes to lipid raft domains or is found within caveolae, in a process dependent on cavolin-1 (Ring et al. 2006). Translocation of CD36/FAT to the plasma membrane is increased by insulin and other metabolic stimulators and is associated with increased plasma membrane fatty acid transport (Glatz et al. 2010). While precise details are limited, the available evidence indicates that CD36/FAT facilitates LCFA membrane transport.

CD36/FAT is identical to PAS IV, a glycoprotein that is highly enriched on the apical membrane of mammary secretory cells in the bovine mammary gland (Greenwalt and Mather 1985), and on membranes surrounding secreted MFG (Greenwalt et al. 1990). CD36/FAT transcripts are expressed in mammary secretory cells of multiple mammalian species including humans, and their expression is increased several fold in response to lactation (Abumrad et al. 1993; Aoki et al. 1997; Coburn et al. 2000; Bionaz and Loor 2008b; Han et al. 2010; Mohammad and Haymond 2013). Immunolocalization studies demonstrated that CD36/FAT is enriched on the basal lateral membranes of goat mammary glands, which is consistent with a role in facilitating LCFA transport into mammary secretory cells (Wooding and Sargeant 2015). However, as indicated, it is also found on the apical plasma membrane (Wooding and Sargeant 2015) and is an abundant membrane protein of MFG. Significantly, CD36/FAT in bovine MFGs has was shown to bind to FABPcyt (Spitsberg et al. 1995). This study provides the primary data implicating direct interactions between CD36/FAT and FABPcyt as an underlying mechanism of LCFA desorption from the inner plasma membrane leaflet. The presence of CD36/FAT-FABPcyt complexes within MFG suggests possible functions in LCFA transport across the MFG membrane. The specific contribution of CD36/FAT to plasma membrane transport of LCFA in mammary secretory cells, however, remains uncertain. The observation that CD36/FAT deletion in mice significantly reduces LCFA content of milk by over 50% without altering overall milk fat content (O'Byrne et al. 2010), implicates CD36/FAT in the regulation of LCFA uptake by mammary secretory cells. However, in light of the established role of CD36/FAT in the transport of LCFA into adipose tissue (Coburn et al. 2000), and evidence that adipose-derived LCFA are primary contributors to the LCFA composition of milk (Fidler et al. 2000), the observed effects of CD36/FAT on milk lipid composition may be an indirect effect of altered LCFA transport into adipose rather than direct action on mammary secretory cells.

7.2.2 *Slc27A/FATP and ACSL Families*

The six members of the SLC27A/FATP family and the five members of the ACSL family are enzymes that catalyze formation of thioesters between LCFA and Coenzyme A. Members of each family are hypothesized to function as LCFA membrane transporters due to their ability to facilitate cellular LCFA uptake. However, specific fatty acid transporter functions of these proteins remain uncertain, since thioesterification of LCFA within the cell would indirectly facilitate their membrane transport by acting as a metabolic trap (Watkins 2008; Anderson and Stahl 2013). The genetic, cellular, and functional multiplicities of SLC27A/FATP and ACSL family members have created difficulties in identifying precise physiological functions of individual family members in regulating uptake of LCFA. Nevertheless, transgenic studies have documented that altered levels of SLC27A/FATP-1,4, and 5 influence fatty acid transport into skeletal muscle, heart, and adipose (Anderson and Stahl 2013), skin (Jia et al. 2007), and hepatocytes (Doege et al. 2006), respectively.

SLC27A and ACSL family member are expressed in mammary glands of human, murine, or bovine species. Human mammary secretory cells express SLC27A-1, -3, and -5 and ACSL-1 and -4 (Mohammad and Haymond 2013). Transcript levels of SLC27A-5, ACSL-1, and ACSL-4 increased 3-, 2-, and 6-times, respectively, in human mammary secretory cells in conjunction with the activation of milk secretion and milk lipid synthesis. Conversely, transcripts for SLC27A/FATP-1 in human mammary secretory cells decrease over this period. Transcripts for SLC27A-1, -2, -3, -5, and -6 and ACSL-1, -3, -4, -5, and -6 have been identified in mammary glands of cattle (Bionaz and Loor 2008a). Of these, SLC27A-6 and ACSL-1 are the most abundantly expressed members of their respective families, accounting for greater than 80% and 45–84% of the total number of transcripts of each family, respectively. Activation of milk secretion significantly increased relative transcript levels of SLC27A-2, -5, and -6 and ACLS-1 in the bovine mammary gland. Interestingly, although SLC27A3 transcript abundance is low in bovine mammary tissue and did not change in response to lactation initiation, SNP analyses demonstrated its localization to a QTL linked to increased milk fat production in cattle and sheep (Calvo et al. 2006a, b). In the mouse mammary gland, the expression of SLC27A-3 increases approximately 30-times during lactation. In contrast, transcripts for SLC27A-1, -2, -4, and -6 are weakly expressed and their levels either decrease or do not significantly change in response to lactation, and transcripts for SLC27A5 are not detected (Han et al. 2010).

Differences in substrate specificities, reaction kinetics, tissue, and cellular distributions patterns of individual members the SLC27A and ACSL families have led to suggestions that they may possess physiologically distinct FA transport functions (Soupene and Kuypers 2008; Watkins 2008). Although the intracellular localizations of SLC27A and ACSL family members in mammary secretory cells are likely to be similar to those of other cell types, their relative contributions to fatty acid transport or acyl-CoA synthetase activity in mammary secretory cells have not been documented, and defects in mammary gland function related to their loss or increased expression have not been identified in existing genetic models. Thus, their physiological in lipid transport or secretion by mammary secretory cells remains uncertain.

7.2.3 Caveolins

Caveolins are a three-member family of integral membrane proteins and the principal components of caveolae, which are specialized plasma membrane lipid raft-microdomain invaginations implicated in endocytosis, transcytosis, and cell signaling processes (Parton and Simons 2007). Functional screens for high-affinity fatty acid binding proteins in plasma membranes identified caveolins as potential LCFA membrane transporters (Trigatti et al. 1999). The precise role of caveolins in fatty acid transport remains uncertain, and conflicting, due to both positive, negative, and neutral effects of deleting individual caveolin family members on LCFA uptake in specific tissues (Glatz et al. 2010). Moreover, structural and functional properties suggest that putative effects of caveolins on LCFA transport are related to plasma membrane organization of CD36/FAT or SLC27A family members or membrane signaling complexes (Pohl et al. 2005).

Caveolin-1 is expressed in human and mouse mammary epithelial cells (Park et al. 2001). In the mouse, caveolin transcripts decrease during pregnancy and lactation, and the actions of caveolin-1 have been shown to negatively regulate functional differentiation of the mammary gland by inhibiting prolactin signaling through the Jak/Stat5 pathway (Park et al. 2002). In addition, genetic deletion studies demonstrated that caveolin-1 loss promotes milk lipid formation (Sotgia et al. 2006). While effects on LCFA transport in mammary secretory cells have not been documented, the existing data are not consistent with caveolin-1 functions contributing to this process.

7.3 Intracellular Lipid Transport

Lipids are transported within eukaryotic cells by protein carriers, vesicular pathways, and/or direct membrane contacts (van Meer et al. 2008). The substantial quantities of lipid secreted during lactation, and the expansion of endomembrane systems responsible for synthesis and packaging of lipids and other milk cargo during lactation (Hollmann 1974; Wooding 1977; Clermont et al. 1993; Ron and Hampton 2004), increases demands on cellular lipid synthesis and transport mechanisms in mammary secretory cells. Interference with these mechanisms impairs organellar biogenesis and milk production (Hasegawa et al. 2015; Davis et al. 2016).

7.3.1 Fatty Acids

As noted earlier, LCFA exist in cells as unesterified free fatty acids (LC-FFA) or are converted by fatty acyl transferases (Mashek et al. 2007) to metabolically active esters of coenzyme A (Acyl-CoAs). LC-FFA and Acyl-CoAs differ in their cellular and metabolic functions and their mechanisms and patterns of intracellular transport. LC-FFA are substrates for energy production, synthesis of eicosanoid signaling lipids, and ligands for nuclear transcription factors regulating cellular lipid metabolism. Acyl-CoAs are transcription factor ligands and serve as substrates for neutral lipid, phospholipid, and ceramide synthesis; protein and fatty acid modification; and oxidative metabolism (Grevengoed et al. 2014). Due to their extremely limited aqueous solubility (Weisiger 2002), LC-FFA and Acyl-CoAs are transferred between cellular membrane compartments bound to carrier proteins, which increases their cytosolic concentrations and diffusional flux (Weisiger 2002). Carriers for LC-FFA include the 10 members of the fatty binding protein (FABP) family and sterol carrier protein-2 (SCP-2), which bind LC-FFA with nM affinities (Richieri et al. 1994; Gallegos et al. 2001). FABPs also bind Acyl-CoAs and other lipophilic ligands including eicosanoids, bile acids, and endocannabinoids, depending on the specific isoform (Haunerland and Spener 2004; Smathers and Petersen 2011). In addition to LC-FFA, SCP-2 binds Acyl-CoAs, sterols, and phospholipids with nM affinities (Gallegos et al. 2001). Acyl-CoAs are also specific ligands for acyl-CoA binding protein (ACBP), which is the primary intracellular Acyl-CoA carrier (Grevengoed et al. 2014). Each type of carrier physically interacts with membrane systems to promote fatty acid transfer (Chao et al. 2002; Storch et al. 2002; Smathers and Petersen 2011).

7.3.1.1 Fatty Acid Binding Proteins

Cellular levels of specific FABPs differ between cell types and correspond to an individual cell's lipid metabolizing activity and/or exposure to LC-FFA (Haunerland and Spener 2004; Grevengoed et al. 2014). In cells that specialize in lipid metabolism, such as hepatocytes or intestinal epithelial cells, specific FABP isoforms can comprise as much as 5% of the total cellular protein (Grevengoed et al. 2014). Mammary secretory cells express transcripts for multiple FABP isoforms (Rudolph et al. 2007; Bionaz and Loor 2008a). However, FABP-3 and -4 are the most abundantly expressed isoforms in species where data exist (Nielsen et al. 1994; Haunerland and Spener 2004; Rudolph et al. 2007; Bionaz and Loor 2008a). Consistent with the importance of FABP to cellular lipid metabolic functions in mammary secretory cells, the expression of these isoforms increases several fold in response to the stimulation of lipid synthesis during their functional differentiation (Bionaz and Loor 2008b; Mohammad and Haymond 2013). In rat mammary secretory cells, which secrete milk containing as much as 15% lipid (Paul et al. 2015), FABP-3 levels increase over 60-fold between non-lactating and lactating states, and during lactation FABP-3 accounts for ~6% of total cellular protein (Nielsen et al. 1994). The precise physiological role(s) of FABP-3 and-4 isoforms in mammary secretory cell lipid transport remain uncertain. In cultured bovine mammary secretory cells, FABP-3 levels directly influence activation of lipogenic pathways and cellular lipid accumulation (Liang et al. 2014). FABP-3 deletion reduces the amount of unsaturated fatty acids in the milk of mice by 10% (Clark et al. 2000), and bovine FABP-4 SNPs are associated with significant differences in milk fatty acid compositions (Nafikov et al. 2013). However, the cellular processes and molecular mechanisms underlying these actions have not been identified and specific effects of FABP deletion on mammary secretory cell development or function remain uncertain.

7.3.1.2 Sterol Carrier Protein-2

Sterol carrier protein-2 is found in the cytoplasm and peroxisomes. In peroxisomes, SCP-2 mediates the import of branched chain lipids for oxidation (Schroeder et al. 2007). Transport functions of SCP-2 in the cytoplasm are less well established. In cultured cells, it transfers multiple lipid species, including LC-FFA, Acyl-CoAs, cholesterol, and phospholipids, and can remodel the lipid composition of lipid rafts (Schroeder et al. 2007), but its precise physiological roles have not been identified (Vance 2015). Sterol carrier protein-2 is expressed in lactating mammary glands of mice (http://biogps.org) and dairy cattle (Dai et al. 2018) as well as in a mammary epithelial cell line (Doria et al. 2014). In dairy cattle, mammary gland levels of SCP-2 are influenced by diet and increase in conjunction with upregulation of genes involved in fatty acid β-oxidation and reduced milk production in cattle are fed nutritionally poor diets (Dai et al. 2018). The data are consistent with SCP-2 participating in trafficking lipids for energy production in mammary secretory cells and the down-regulation of this pathway during active milk secretion, but details about how SCP-2 specifically influences mammary secretory cell function are lacking.

7.3.1.3 Acyl-coA Binding Protein

In the mouse, transcript levels of acyl-coA binding proteins in the mammary secretory cells increase in conjunction with lactation and are in the range of those found in tissues with high lipid metabolic activity, such as brown and white adipose, and liver in lactating animals (http://biogps.org). Acyl-coA binding domain containing 5 protein is expressed in a mouse mammary epithelial cell line, and its expression levels are modestly elevated by hormones that stimulate differentiation to a secretory phenotype (Doria et al. 2014). However, ACBP functions of mammary secretory cells of intact animals have not been formally investigated and remain uncharacterized.

7.3.2 Phospholipids

Mammalian membranes are comprised of two major phospholipid classes: glycerophospholipids and sphingolipids. Glycerophospholipids are composed of a glycerol backbone esterified in the sn1 and sn2 positions with saturated or cis-unsaturated fatty acids. The sn3 position of glycerol is esterified with phosphate or with phosphoesters of choline, ethanolamine, serine, or inositol, which form, respectively, phosphatidic acid (PA), phosphatidylcholine (PC), phosphatidylethanolamine (PE), phosphatidylserine (PS), and phosphatidylinositol (PI), respectively. The predominant sphingolipid in mammalian membranes is sphingomyelin (SM), which is a phosphodiester of ceramide and either choline or ethanolamine as the hydrophilic head groups.

In cells with high secretory activities, such as the mammary epithelial cell during the period of lactation, phospholipid synthesis and transport mechanisms are stimulated as organellar systems responsible for cargo biosynthesis expand, and vesicular trafficking mechanisms that mediate cargo secretion are activated (Ron and Hampton 2004). The physiological importance of phospholipid synthesis and/or transport to mammary secretory cells is exemplified by data showing that interference with the stimulation of these processes inhibits organelle biogenesis and impairs mammary secretory functions (Sriburi et al. 2007; Hasegawa et al. 2015; Davis et al. 2016).

7.3.2.1 Phospholipid Composition of Cellular Membrane Compartments

Organelles have defined phospholipid abundances (Drin 2014), which are important determinants of their biochemical and biophysical properties and cellular functions (van Meer et al. 2008). The phospholipid compositions of the membrane compartments involved in lipid synthesis, transport, and secretion by mammary secretory cells are shown in Table 7.2. With the exception of the plasma membrane of mammary secretory cells, which is enriched in SM and reduced in PC compared

Table 7.2 Phospholipid contents of cellular membranes

	PC	PE	PI	PS	SM/SL	Cardiolipin
Plasma membrane[a,b]	29%[a] (40%)[b]	25%[a] (24%)[b]	11%[a] (8%)[b]	9%[a] (9%)[b]	22%[a] (17%)[b]	0.8%[a] NR
Endoplasmic reticulum[b,c]	61%[c] (60%)[b]	20%[c] (23%)[b]	9%[c] (10%)[b]	3%[c] (2%)[b]	4%[c] (3%)[b]	0.8%[c] (1%)[b]
Golgi[b,c]	50%[c] (51%)[b]	25%[c] (24%)[b]	8%[c] (8%)[b]	4%[c] (9%)[b]	11%[c] (17%)[b]	NR (1%)[b]
Mitochondria[b,d]	43%[d] (44%)[b]	25%[d] (34%)[b]	5%[d] (5%)[b]	0.4%[d] (1%)[b]	3%[d] (1%)[b]	18%[d] (14%)[b]
Cytoplasmic lipid droplets[e]	NR (47%)	NR (20%)	NR (8%)	NR (1%)	NR (2%)	NR NR
Milk fat globule membrane[f]	31%	29%	9%	9%	23%	1%

Percentages of major phospholipid species in cell membrane compartments of mammary secretory cells. The phospholipid composition of membranes from representative mammalian cells are shown in parentheses. *PC* phosphatidylcholine, *PE* phosphatidylethanolamine, *PI* Phosphatidylinositol, *PS* phosphatidylserine, *SM/SL* sphingomyelin/sphingolipids, *NR* not reported

[a]Compiled from Keenan et al. (1970) and Sharma and Dahiya (1988)

[b]Compiled from Van Meer et al. (2008) and Horvath and Daum (2013)

[c]Compiled from Keenan et al. (1972) and Keenan and Huang (1972)

[d]Huang and Keenan (1971)

[e]Bartz et al. (2007)

[f]Compiled from Keenan et al. (1970), Contarini and Povolo (2013) and Keenan et al. (1972)

to values typically reported for mammalian cells (van Meer et al. 2008) (Table 7.2), the phospholipid compositions of mammary secretory cell membrane compartments are comparable to those of other mammalian cell types.

The phospholipid composition of cellular membrane compartments is regulated by the localization of enzymes that catalyze their synthesis and modification, and by intermembrane transport mechanisms (van Meer et al. 2008; Gatta and Levine 2017). In addition, there is significant membrane bilayer asymmetry of phospholipids within the Golgi, plasma membrane, and vesicular compartments, which is controlled by the location of biosynthetic enzyme active sites and transmembrane transporters. Details about how mammalian cells regulate the composition and distribution of phospholipids among cellular membrane compartments have emerged recently (Vance 2015). However, significant gaps remain in our understanding of the mechanisms and regulatory processes controlling phospholipid distribution in mammalian cells, and few details exist about specific processes in mammary secretory cells.

7.3.2.2 Transport of De Novo Synthesized Phospholipids

Enzymes in the ER are responsible for synthesis of the majority of cellular lipids, including phospholipids (Jacquemyn et al. 2017). Newly synthesized phospholipids are inserted initially into the cytosolic leaflet of the ER membrane and undergo enzyme-mediated transfer to the luminal leaflet (Sanyal and Menon 2009) to become uniformly distributed between the two leaflets (van Meer et al. 2008). ER-synthesized phospholipids are transferred to other cellular membrane compartments by both vesicular and non-vesicular pathways (Wong et al. 2019). The transport of ER-synthesized phospholipids to other membrane compartments occurs at variable rates. For example, the half-time for transfer of PC or PE from the ER to the plasma membrane is approximately 1 min (Kaplan and Simoni 1985), whereas PC is transported from the ER to the mitochondria with a half-time of approximately 5 min and ER-synthesized PE is transported to the mitochondria with a half-time of 2 h (Voelker 1991). Higher transport rates are consistent with a non-vesicular mechanism, whereas transport requiring hours fit a vesicular mechanism (Vance 2015).

Non-vesicular phospholipid transport is facilitated by membrane contact (Vance 2015). ER membranes form contacts with other cellular membrane systems, including the Golgi, plasma membrane, mitochondria, endosomes, and cytoplasmic lipid droplets, and serve as sites for non-vesicular phospholipid transfer (Prinz et al. 2020). Membrane contacts between different ER cisterna, and between ER and mitochondria, Golgi, and cytoplasmic lipid droplets have been detected in mammary secretory cells of lactating and pregnant rats (Ladinsky et al. 2019; Monks et al. 2020). While the structural organization of these contacts is consistent with facilitated lipid transfer between these membrane compartments (Ladinsky et al. 2019), their role in lipid transfer has not been established.

7.3.2.2.1 Phosphatidylcholine

There are two enzymatic pathways for PC synthesis in the ER of mammalian cells. The primary pathway is the transfer of choline from CDP-choline to diacylglycerol (DAG), which is catalyzed by choline-ethanolamine phosphotransferase (CEPT), a bifunctional ER membrane enzyme that catalyzes the synthesis of both PC and PE. PC is synthesized to a lesser extent by sequential methylation of PE by phosphatidylethanolamine N-methyltransferase (PEMT) (Vance 2013), which localizes to specialized ER membrane domains that form contacts with the mitochondria, referred to as mitochondrial-associated membranes (MAM) (Vance 1990). PEMT expression is highest in hepatocytes, where it is physiologically important determinant of PC synthesis and VLDL secretion (Vance 2013), which occurs by vesicular transport. Mammary secretory cells also express PEMT, and PEMT activity contributes to their PC synthesis (Yang et al. 1988). However, compared to hepatocytes, PEMT activity is low in mammary tissue (Vance and de Kruijff 1980) and loss of PEMT in mice is not known to affect mammary gland function (Vance 2013). Consequently, it is likely that the CDP-choline pathway is responsible for the majority of ER-synthesized PC in mammary secretory cells. Nevertheless, PEMT activity influences the triglyceride accumulation and cytoplasmic lipid droplet size in cultured bovine mammary secretory cells (Cohen et al. 2017) and may thus modulate their lipid utilization.

The physiological processes mediating transfer of ER-synthesized PC to other organelles in mammalian cells are uncertain. Multiple classes of lipid transfer proteins, including members of the steroidogenic acute regulatory protein (StART), SCP2, and tubular lipid binding protein (TULIP) families, potentially mediate non-vesicular PC transport (Wong et al. 2019). However, physiological roles of these proteins in cellular PC transport have not been established. Transcripts for many lipid transfer protein family members are detected in mammary glands of mice (http://biogps.org) and are present in mammary epithelial cell lines (Doria et al. 2014), although in most cases, their abundances are several fold less than those found in other lipogenic tissues, such as liver and adipose (http://biogps.org).

ER-synthesized PC can also be transferred to organelles in the secretory pathway by vesicular trafficking (Holthuis and Levine 2005). However, several lines of evidence, including the disruption of vesicular trafficking pathway, indicate that this mechanism is not the primary pathway for transporting ER-synthesized PC to the plasma membrane (Vance 2015). The importance of the vesicular pathway for transport of ER-synthesized PC may depend on the type and functional properties of the cell. Although PC trafficking has not been formally investigated in mammary secretory cells, secretory vesicle trafficking to the apical plasma membrane, which adds membrane, is closely related to apocrine lipid secretion, which removes membrane during secretion (Kralj and Pipan 1992). Consequently, during milk secretion, vesicular trafficking is likely to be an important pathway for transferring ER-derived PC, and other phospholipids, to the plasma membrane in the mammary secretory cell.

7.3.2.2.2 Phosphatidylethanolamine

In addition to the CEPT pathway described above, PE is synthesized by decarboxylation of ER-synthesized phosphatidylserine (PS) by phosphatidylserine decarboxylase (PSD), and inner-mitochondrial membrane enzyme in mammalian cells (Zborowski et al. 1983). PE synthesized by the CEPT pathway does not undergo significant transport to the mitochondria (Vance 2015), whereas mitochondrial-synthesized PE is transported to the ER for PC synthesis by PEMT (Vance 1991). Indeed, there is evidence that the majority of PE in mammalian cell membranes originates from the mitochondrial synthetic pathway (Voelker 1984). Transport of ER-synthesized PS to the mitochondria, and mitochondrial-synthesized PE to the ER, is facilitated by processes found at MAM domains of the ER (Vance 1991) and include metabolic conversion of the transferred substrate (Prinz et al. 2020).

Like PC, ER-synthesized PE appears to be transported to the plasma membrane primarily by non-vesicular pathways, which may include membrane contact sites and/or lipid transport proteins, based on data from common mammalian cell models (Vance 2015). The extent to which non-vesicular transport processes contribute to intermembrane transfer of PE in mammary secretory cells has not been investigated. However, as outlined above for PC, there is reason to suspect that during lactation vesicular transport may play a role in transferring PE from the ER to the plasma membrane in mammary secretory cells.

7.3.2.2.3 Phosphatidylserine

Phosphatidylserine synthesis occurs in the ER by base-exchange reactions catalyzed by phosphatidylserine synthase-1 and -2 (PSS1, PSS2), which, respectively, exchange the choline and ethanolamine head groups of PC and PE with serine (Vance 2008). Pss1 and Pss2 localize to MAM domains of the ER (Stone and Vance 2000). PS is transported from its site of synthesis in MAM domains to mitochondrial outer membranes and then to the inner-mitochondrial membranes, where it serves as a substrate for PE synthesis. The precise mechanism of these transfers is uncertain, but current studies indicate that they are driven by the metabolic conversion of PS to PE on the inner-mitochondrial membrane catalyzed by phosphatidylserine decarboxylase (Vance 2014).

Phosphatidylserine is enriched on plasma membrane and Golgi and milk fat globule membranes compared to ER, CLD, and mitochondrial membranes (van Meer et al. 2008) (Table 7.2). In a mammalian cell culture model (Fairn et al. 2011), PS localized to the cytosolic leaflets of the plasma membrane, trans-Golgi complex, and endocytic vesicles, and on inner membranes of the ER, Golgi, and mitochondria. How cells achieve membrane-specific distributions of PS is uncertain, although vesicular and non-vesicular mechanisms are proposed (Fairn et al. 2011; Balla et al. 2019). In cultured mammalian cells, non-vesicular PS transfer from the ER to the plasma membrane occurs by counter exchange with phosphatidylinositol 4-phosphate (PI4P) on the plasma membrane. This exchange occurs at ER-plasma membrane contact sites and is mediated by the integral ER membrane proteins—oxysterol-binding protein related proteins-5 and -8 (ORP-5 and -8) (Chung et al. 2015). Transcripts for ORP-5 and -8 are present in low abundances in mammary glands of lactating mice (http://biogps.org), and contacts between the ER and the plasma membrane have not been observed in mammary secretory cells. Consequently, the role of PS-PI4P counter exchange mechanism in PS transport to the plasma membrane in mammary secretory cells is uncertain but cannot be ruled out.

7.3.2.2.4 Phosphatidylinositol

Phosphatidylinositol biosynthesis is mediated by phosphatidylinositol synthase (PIS), which catalyzes the transesterification of CDP-DAG with myo-inositol forming phosphatidylinositol and CMP. Phosphatidylinositol synthase is an integral membrane enzyme that is found specifically on ER membranes and localizes to a highly mobile ER membrane domain that forms contacts with membranes of other organelles (Pemberton et al. 2020). Transfer of PI from the ER to other organelles in mammalian cells occurs by both vesicular and non-vesicular pathways (Pemberton et al. 2020), although the individual contributions of these pathways to intermembrane PI transfer have not been defined.

The localization of PI synthesis to ER-membrane contact sites is proposed to facilitate rapid PI distribution among cellular membranes by non-vesicular mechanisms (Kim et al. 2011). Non-vesicular intermembrane transfer of PI is proposed to be facilitated by phosphatidylinositol-transfer proteins belonging to the StARkin and Sec14 super families (Pemberton et al. 2020). In addition, PI transfer between the ER and the plasma membrane can occur by PI4P counter exchange mediated by ORP-5 and -8 as described for PS (Balla et al. 2019). However, the physiological functions and mechanisms of action of these transporters remain uncertain (Pemberton et al. 2020), and their role in PI transfer in mammary secretory cells has not been investigated.

7.3.2.2.5 Sphingomyelin

Sphingomyelin is synthesized by transfer of the choline or ethanolamine headgroups from PC and PE, respectively, to ceramide, a fatty acid amide that is synthesized by enzymes on the cytosolic leaflet ER membranes. Ceramide is transported from ER to Golgi membranes by vesicular trafficking or by the ceramide transport protein (CERT) (Hanada et al. 2007). In the Golgi, ceramide undergoes rapid transfer between Golgi membrane leaflets and is converted to sphingomyelin on the luminal leaflet of Golgi membrane by sphingomyelin synthase-1 (SMS1) (van Meer et al. 2008). Other enzymes within the Golgi lumen catalyze transfer of sugar or carbohydrate groups to ceramide to produce cerebroside or ganglioside glycosphingolipids.

In addition to synthesis by Golgi membranes, SM is also synthesized by sphingomyelin synthase-2 (SMS2), a ubiquitously expressed enzyme that localizes exoplasmic leaflet of the plasma membrane and catalyzes bidirectional transfer of choline between PC and SM (Milhas et al. 2010). SMS2 activity has been shown to influences membrane levels of SM in HeLa cells (Ding et al. 2008) and may contribute to the unique PS and PC composition noted for the plasma membrane of mammary secretory cells (Table 7.2).

7.3.2.2.6 Cardiolipin

Cardiolipin is synthesized from phosphatidylglycerol and CDP-DAG in a reaction catalyzed by cardiolipin synthase (CS) on the inner-mitochondrial membrane. Phosphatidylglycerol and CD-DAG can originate from either ER and inner-mitochondrial synthetic processes. However, CS is only expressed on inner-mitochondrial membranes and cardiolipin is considered to be a mitochondrial restricted phospholipid (Vance 2015).

7.3.2.2.7 Cholesterol

Cholesterol is synthesized primarily by ER membrane enzymes and is rapidly transferred to other cellular membranes by both vesicular and non-vesicular mechanisms (Litvinov et al. 2018). Multiple cholesterol binding proteins, including SBP2 (Seedorf et al. 2000), and members of the StARTkin and OSBP families (Balla et al. 2019) have been implicated in non-vesicular transport of cholesterol in mammalian cells. In model systems, OSBP and oxysterol-related binding protein-2 (OSRBP2) mediate cholesterol transport between the ER and the trans-Golgi and plasma membranes, respectively, by a counter exchange mechanism with PI4P (Wang et al. 2019). It remains uncertain how mammalian cells transport cholesterol from the ER to the mitochondria. However, homologs of OSBP mediate this transfer in yeast (Tian et al. 2018). Members of the StARTkin family mediate transmembrane cholesterol transport within the mitochondria (Vance 2014). Members of these cholesterol transport protein families are detected in mammary glands of mice (http://biogps.org) and in mammary epithelial cell lines (Doria et al. 2014) but their physiological functions are uncertain.

7.3.2.3 Transport of Exogenous Lipids

Mammalian cells, including mammary secretory cells, also take up cholesterol from serum lipoproteins by receptor-mediated processes (Ontsouka and Albrecht 2014). Serum-derived cholesterol is transported within the cell by vesicular trafficking mechanisms of the endocytic pathway (Litvinov et al. 2018). Cholesterol uptake and trafficking mechanisms are present in mammary secretory cells, and their activities are increased during lactation (Ontsouka and Albrecht 2014). In the lactating mammary gland, there is considerable apical membrane trafficking of endocytotic vesicles associated with transcytosis of serum-derived hormones, proteins, and minerals into milk (Hunziker and Kraehenbuhl 1998; Monks and Neville 2004; Monks and McManaman 2013). The mechanisms of transcytosis involve a series of complex sorting events in the basolateral early endosome and the common endosome recycling compartments that intersect with Golgi and secretory vesicle pathways (Welsch et al. 1984; Grant and Donaldson 2009; Golachowska et al. 2010). These sorting events provide mechanisms for transferring cholesterol from external sources and basolateral membrane phospholipids to multiple internal membrane compartments as well as to the apical plasma membrane (Ikonen and Simons 1998). Because there appears to be minimal apical membrane recycling, or trafficking of vesicles from the apical membrane to the basolateral membrane, during lactation (Monks and Neville 2004), it seems likely that PL transport by vesicular pathways involves net basolateral to apical membrane flow during active milk secretion. Few details exist about basal-apical vesicular trafficking during lactation, and the contributions of this pathway to PL transfer or exchange mechanisms between membrane compartments of mammary secretory cells are unknown. Nevertheless, tracer experiments demonstrated that transcytosis mediates the transport of serum lipoprotein particles into milk, and have provided evidence that this process contributes to phospholipid secretion into the whey fraction of milk (Monks et al. 2001).

7.3.2.4 Transport Mechanisms Mediating Membrane Lipid Asymmetry

The distribution of individual phospholipids among membrane bilayers is determined by inherent differences in their ability to spontaneously cross membrane bilayers, the presence specific transmembrane transporters and to processes that regulate their retention within a membrane leaflet. Phospholipids are asymmetrically distributed between leaflets of the Golgi, vesicular, and plasma membranes (van Meer et al. 2008). PE, PS, and phosphoinositides are enriched on the cytosolic leaflets of these compartments, whereas PC, SM, and glycosphingolipids are enriched on their outward facing leaflets (Daleke 2007). In the ER, newly synthesized phospholipids undergo transbilayer equilibration with half-times of approximately 20 min (Voelker 1991) through the actions of ER membrane-associated proteins with phospholipid transbilayer transport activities (Bishop and Bell 1985; Holthuis and Levine 2005). Sphingomyelin and other sphingolipids synthesized on luminal membranes of the Golgi do not undergo significant exchange between membrane leaflets. As a consequence, they are exclusively transported within the cell by the vesicular pathway and retain an asymmetric distribution on exoplasmic leaflets of the plasma membrane and vesicular membranes (van Meer et al. 2008).

Compared to typical values for other cell types (van Meer et al. 2008), the plasma membrane of mammary secretory cells appears to have relatively less PC and greater SM (Table 7.2). The reason for these differences and their physiological significance are uncertain. As noted in Sect. 7.3.2.2.5, sphingomyelin synthase-2 activity located on the external leaflet of the plasma membrane may influence the plasma membrane compositions and asymmetric distributions of PC and SM and contribute to their unique distributions on plasma membranes of mammary secretory cells.

7.3.2.4.1 Transmembrane Lipid Transfer

Transmembrane transport of lipids is an energy-dependent process mediated by distinct membrane transporters (Muthusamy et al. 2009). Plasma membranes, trans-Golgi membranes, and membranes of secretory vesicles possess aminophospholipid translocase activities that utilize ATP hydrolysis to transport PE and PS from the outer to the inner membrane leaflet, a process referred to as flipping (Seigneuret and Devaux 1984; Devaux and Morris 2004). Flippase activity is encoded by a large family of Type IV P-type ATPases (P4-ATPases) (Muthusamy et al. 2009). Conversely, members of the ATP-binding cassette (ABC) transporter family catalyze inner to outer membrane transport of PC, PS, PE, SM and glycosphingolipids, and cholesterol (Pomorski et al. 2004; van Meer et al. 2008), a process referred to as flopping.

Cultured mammary secretory cells express ABCA1 and ABCG1, which are present on the apical plasma membrane and on the milk fat globule membrane and facilitate phospholipid and cholesterol transport to exoplasmic membrane leaflet (Mani et al. 2010; Ontsouka et al. 2017). ABC-transporter-dependent transport of cholesterol has been detected on both the apical and basolateral sides of polarized mammary secretory cells cultured on transwell plates and is consistent with a general involvement of these proteins in the exoplasmic membrane transport of lipids (Ontsouka et al. 2013). However, physiological roles of ABC transporters in regulating membrane transport of PL in mammary secretory cells in vivo have not been identified, and their contributions to the control of PL transbilayer asymmetry in the mammary gland remain to be established.

7.4 Neutral Lipid Synthesis and Transport

In eukaryotic cells, neutral lipids (triglycerides and cholesteryl esters) are synthesized by ER membrane resident enzymes and assembled into cytoplasmic lipid droplets (CLD), which are unique ER-derived organellar structures functionally specialized for neutral lipid storage, intracellular trafficking, and metabolism (Salo and Ikonen 2019). In mammary secretory cells, CLD are uniquely transported to the apical plasma membrane where they are secreted to form milk lipids. Neutral lipid synthesis, and CLD formation and trafficking require multiple lipid transport processes, several of which involve novel mechanisms and membrane interactions.

7.4.1 Neutral Lipid Synthesis

Neutral lipids are synthesized by sequential esterification of glycerol-3-phosphate by fatty acyl-CoAs to form di- and tri-glycerides (Takeuchi and Reue 2009), or by fatty acyl-CoA-dependent esterification of cholesterol to form cholesteryl esters (Chang et al. 2001). These biosynthetic reactions are catalyzed by integral ER membrane enzymes (Joyce et al. 2000; Takeuchi and Reue 2009) and require transport of fatty acyl-CoAs to sites of synthesis and the transfer of lipid intermediates between catalytic enzymes (Fig. 7.1).

7.4.1.1 Fatty Acyl-CoA Substrates

Fatty acyl-CoAs are not membrane permeable but readily diffuse within the plane of the membrane (Boylan and Hamilton 1992). Consequently, neutral lipid synthesis requires transfer of fatty acyl-CoAs to membrane binding sites or catalytic sites of neutral lipid-synthesizing enzymes on the cytoplasmic membrane leaflet of the ER. This transfer potentially is carried out by carrier proteins (Sect. 7.3) or by membrane diffusion of fatty acyl-CoAs generated by fatty acyl-CoA synthetases located on ER membranes (Digel et al. 2009). Of the fatty acyl-CoA synthetases expressed in mammary secretory cells (Sect. 7.2.2), ACSL-1, -3, and -5 have been detected ER membranes in other cell types (Digel et al. 2009; Poppelreuther et al. 2012).

7.4.1.2 Lipid Intermediates

Glycerol-3-phosphate O-acyltransferase (GPAT), 1-acylglycerol-3-phosphate O-acyltransferase (AGPAT), phosphatidic acid phosphatase, and acyl-CoA: diacylglycerol O-acyltransferase (DGAT) enzymes catalyze sequential reactions in the synthesis of triglycerides. Respectively, these enzymes form lysophosphatidic acid, phosphatidic acid, diacylglycerol, and triacylglycerol (Fig. 7.1). Lysophosphatidic acid, phosphatidic acid, and diacylglycerol are able to diffuse within membrane planes and can be transferred between membranes by physical contact or carrier proteins for use as signaling molecules or as substrates in other lipid synthesis reactions. The ability to specifically channel these intermediates to the synthesis and storage of triglycerides in CLD is facilitated by formation of spatial distinct complexes of triglyceride-synthesizing enzymes on ER membranes and the CLD surface (Wilfling et al. 2013; Lee and Ridgway 2020).

Mammals express four GPAT isoforms. GPAT-1 and -2 localize to the mitochondria whereas GPAT-3 and -4 are ER membrane enzymes (Takeuchi and Reue 2009). Although lysophosphatidic acid synthesized by GPAT1 in the mitochondria can be transferred to the ER for triglyceride synthesis (Grevengoed et al. 2014), lysophosphatidic acid used in the synthesis of triglycerides secreted into milk is synthesized by GPAT-4, which is found on ER membranes and the CLD surface (Lee and Ridgway 2020). In both mice and cattle, transcript levels of GPAT 4 increase in response to lactation (Beigneux et al. 2006; Bionaz and Loor 2008a), and in mice, loss of GPAT4 is associated with the nearly complete absence of lipids in milk and the absence of CLD in mammary secretory cells (Beigneux et al. 2006; Chen et al. 2008; Takeuchi and Reue 2009). The identity and localization of functionally important AGPATs in mammary secretory cells have not been established. Of the three members of the Lipin family of phosphatidic acid phosphatases (Takeuchi and Reue 2009), Lipin1 has the highest expression in mammary secretory cells from mice (http://biogps.org) and cattle (Bionaz and Loor 2008a). In cattle, Lipin-1 is one of the most abundant transcripts in milk-secreting cells during the period of lactation (Bionaz and Loor 2008a), and polymorphisms in the Lipin-1 gene are associated with significant effects on the content and production of milk fat (Han et al. 2019). Two genetically different DGAT isoforms (DGAT1, DGAT2) are expressed by mammary secretory cells (Cases et al. 2001; Cases et al. 1998). DGAT1 appears to be specifically responsible for synthesis of triglycerides secreted into milk. Mice lacking DGAT1 in mammary secretory cells fail to produce CLD and have impaired functional differentiation (Cases et al. 2004), and in cattle DGAT1 polymorphisms are associated with alter contents of milk lipids (Grisart et al. 2004).

7.4.2 CLD Biogenesis

CLD are composed of a neutral lipid core encased by a monolayer of phospholipids containing associated proteins (Murphy 2001). The overall structural organization of CLD is similar to that of plasma lipoprotein particles, which are responsible for vascular transport of neutral lipids (Ohsaki et al. 2009). CLD and lipoproteins originate in the ER (Tiwari and Siddiqi 2012) and utilize common enzymatic pathways for synthesis of their neutral lipid cores (Weiss et al. 1960; Suckling and Stange 1985; Yen et al. 2008). However, they originate from the ER by different means, have different protein compositions, and are secreted by different mechanisms. Lipoprotein particles are synthesized within the lumen of smooth ER (sER) using members of the apolipoprotein-B family as essential structural proteins for their assembly. Nascent lipoprotein particles emerge from the ER within specialized vesicles and are transported to the cis-Golgi where they undergo maturation and additional repackaging into vesicles for transport to the plasma membrane for exocytotic secretion. The molecular and cellular processes responsible for plasma lipoprotein particle synthesis, trafficking, and secretion are known in some detail (Mansbach and Siddiqi 2010; Tiwari and Siddiqi 2012). Conversely, CLD are thought to form cytoplasmic protrusions of neutral lipids that accumulate between ER membrane leaflets (Walther et al. 2017). The molecular and cellular processes mediating interleaflet neutral lipid deposition, and their packaging and cytoplasmically directed export as CLD, are poorly understood, and multiple mechanisms have been proposed (Robenek et al. 2006; Ploegh 2007; Choudhary et al. 2011).

7.4.2.1 Mammary Secretory Cell Mechanisms

A prevailing view is that CLD originate from specific tubular microdomains of smooth ER based on genetic studies showing that disruption of these microdomains interferes with interleaflet neutral lipid accumulation (Kassan et al. 2013; Walther et al. 2017). However, mammary secretory cells are highly enriched in rough ER (Jarasch et al. 1977; Wooding 1977), which possesses the enzymes required for neutral lipid synthesis (Bauman and Davis 1974). Evidence that in mammary secretory cells, neutral lipids are synthesized by enzymes on the rough ER and subsequently packaged into CLD and secreted into milk by a non-vesicular mechanism was obtained over 50 years ago by Stein and Stein (1967). These seminal studies showed that within 1–3 min after injecting radioactive palmitic or oleic acid into tail veins of lactating mice, labeled neutral lipids were detected over rough ER cisternae in mammary secretory cells. Subsequently, labeled neutral lipids were detected in CLD that localized to rough ER, and were later found in milk. Labeled lipids were not observed in the Golgi or any secretory granules, which demonstrated that unlike neutral lipids in lipoprotein particles secreted by hepatocytes or enterocytes, neutral lipids secreted into milk are not packaged into vesicles and secreted by an exocytotic mechanism.

Rough ER in mammary secretory cells forms extensive connections with CLD that are comprised of concentrically layered stacks of flattened cisterna, which have ribosomes located on their surfaces at sites that selectively face the cytosol (Ladinsky et al. 2019). Cisterna that contact CLD also form contacts with elements of the trans-Golgi and mitochondria (Ladinsky et al. 2019; Monks et al. 2020). Concentrically organized stacks of ER surrounding CLD have also been observed in hepatic cells, and it is proposed that this type of structural organization facilitates direct intermembrane transfer of lipids and proteins needed for increased neutral lipid storage in highly lipogenic cell types (Ladinsky et al. 2019).

7.5 Apocrine Lipid Secretion Transport Processes

Apocrine secretion of CLD as intact structures is the primary lipid secretion mechanism of mammary secretory cells (McManaman 2012). This mechanism is distinct from holocrine mechanism used by epithelial cells of sebaceous glands to secrete lipids and waxes (Wooding 1980) or the exocrine mechanisms used hepatocytes and intestinal epithelial cells to secrete lipoprotein particles (McManaman 2012). Apocrine lipid secretion by mammary secretory cells involves at least four distinct lipid transport steps: (1) directional transport of neutral lipids packaged into CLD from their site of synthesis on the ER to the apical pole of the cell; (2) inter-CLD lipid transfer; (3) docking and envelopment of CLD by elements of the apical plasma membrane; and (4) release of membrane enveloped CLD into the lumen as MFG in response to hormone-regulated contraction.

7.5.1 Apical Membrane Directed CLD Transport

Directional intracellular CLD movement has been observed in multiple cell types (Welte 2009), and several lines of evidence demonstrate that this movement largely occurs through microtubule interactions (Spandl et al. 2009; Welte 2009; Orlicky et al. 2013). Consistent with a microtubule-based mechanism of apical plasma transport of CLD in mammary secretory cells, their microtubules are oriented perpendicular to the apical plasma membrane with minus ends directed toward the apically located centriole (Dylewski and Keenan 1984), their CLD express the minus-end microtubule motor, dynein, on their surface (Wu et al. 2000), and live cell imaging has demonstrated apical CLD movement with a superdiffusive behavior, which is similar to that of microtubule-dependent secretory vesicle transport (Masedunskas et al. 2017).

7.5.2 Inter-CLD Lipid Transfer

Lipid transfer between CLD can occur prior to their secretion. CLD increase in size as they are transported to the apical plasma membrane (Stemberger et al. 1984). This process, which has been observed in real-time by intravital microscopy (Mather et al. 2019), involves fusion-mediated transfer of lipid between CLD. CLD fusion is mediated by interaction between molecules of the cell death-inducing DFFA-like effector (CIDE-A) (Barneda et al. 2015), which is found on the surface of CLD in mammary secretory cells (Monks et al. 2016). CIDE-A functions are important physiological regulators of milk lipid secretion and milk production in mice (Wang et al. 2012), although precise mechanisms have not been delineated (Monks et al. 2020).

7.5.3 CLD: Apical Plasma Membrane Docking and Envelopment

It is generally accepted that CLD dock with and become progressive enveloped by elements of the apical plasma membrane before they are released into milk as MFG (Mather and Keenan 1998; Heid and Keenan 2005). Apical membrane docking of

CLD is not an intrinsic property of mammary secretory cells. Rather, it occurs in response to, or is strengthened by, cellular events associated with lactation initiation (McManaman et al. 2002). Although many details remain to be elucidated, there is substantial evidence that CLD docking at the apical plasma membrane is mediated by specific protein interactions.

CLD are connected to the cytoplasmic face of the apical membrane by a 10–20 nm wide hexagonally ordered electron dense layer (Mather and Keenan 1998), which is composed primarily of the cytoplasmic tail of the transmembrane protein butyrophilin (BTN), the cytosolic protein xanthine oxidoreductase (XOR), and the CLD coat protein perilipin-2 (PLIN2). Information about how CLD interact with the apical plasma membrane is still incomplete. However, BTN, XOR, and PLIN2 are linked covalently by disulfide bonding in the apical membrane that coats secreted milk fat globules (McManaman et al. 2002), and in mice, BTN and XOR are recruited to and become co-localized and enriched along CLD-apical plasma membrane contact sites (McManaman et al. 2002; Monks et al. 2016). XOR recruitment to the plasma membrane is mediated by binding to a specific cytoplasmic domain of BTN (Jeong et al. 2009), and in the absence of XOR, CLD are unable to form stable docking interactions with the apical plasma membrane (Monks et al. 2016). The importance of Plin2 to CLD-apical membrane docking has not been formally established. However, Plin2 possesses a phospholipid binding domain and there is evidence that its loss negatively influences lipid secretion from mammary secretory cells (Chong et al. 2011). CLD-apical membrane contact also recruits and concentrates the CIDE-A along sites of contact (Monks et al. 2016). Like PLIN2, CIDE-A possesses a phospholipid binding domain (Barneda et al. 2015) and appears to influence lipid secretion by mammary secretory cells (Wang et al. 2012). As noted in Sect. 7.5.2, CIDE-A mediates contact and lipid transfer between CLD. It is speculated that CIDE-A, or combinations of CIDE-A and PLIN2 are CLD-associated determinants of docking and apical membrane-envelopment of CLD (Monks et al. 2016).

Direct contact between CLD and apical membrane elements is proposed to represent an evolutionary adaptation of the apocrine secretion mechanism to facilitate secretion of large amounts of lipid into milk with minimal loss of cytoplasm and the attendant stress on cellular biosynthetic systems (Oftedal 2012). However, CLD-apical membrane contact is not required for membrane envelopment of CLD or for their release from the cell as MFG. In mice harboring mammary secretory cell-specific deletion of XOR, CLD fail to dock and/or form stable contacts with the apical membrane, but they still undergo apical membrane envelopment and apocrine secretion (Monks et al. 2016). Nevertheless, data showing that XOR deletion delays lipid secretion and retards the initial growth rate of litters (Monks et al. 2016) support the concept that CLD-apical membrane docking interactions facilitate lipid secretion in mammary secretory cells.

7.5.4 Contraction-Coupled Secretion of CLD as Milk Fat Globules

Release of apical membrane docked CLD into the alveolar lumen is contingent on oxytocin-stimulated myoepithelial cell contractions induced in response to nursing signals. In mice, the absence of nursing signals, brought about by removing lactating dams from their litters, cause CLD to remain in contact with and become progressively enveloped by the apical membrane, forming extensive membrane protrusions into the alveolar lumen without separation from the cytoplasm (Monks et al. 2016; Masedunskas et al. 2017). In response to oxytocin injection and myoepithelial contraction, apical membrane-docked CLD are rapidly released into the luminal space as MFG (Masedunskas et al. 2017). The mechanism(s) that mediate this release are currently unknown. Oxytocin-stimulated myoepithelial cell contraction produces mechanical forces on mammary alveoli that deform mammary secretory cells and eject secreted milk (Davis et al. 2015). Such mechanically induced deformation of mammary secretory cells may also cause release of apical membrane-enveloped CLD. In addition to mechanical forces, oxytocin stimulation of myoepithelial cells may activate release of membrane-enveloped CLD into milk by increasing intracellular Ca^{2+} levels and stimulating contraction of the apical membrane-associated actin-myosin network in mammary secretory cells (Masedunskas et al. 2017; Mather et al. 2019). Alternatively, evidence that rabbit mammary secretory cells express oxytocin receptors and that oxytocin signaling enhances exocytic secretory processes within these cells (Lollivier et al. 2006), indicates that direct oxytocin activation of signaling processes in mammary secretory cells may contribute to CLD secretion. Whether external mechanical forces, internal processes, or combinations thereof mediate CLD release into the luminal space, it is now evident that lipid secretion from mammary secretory cells is a stimulated process regulated by intermittent oxytocin release in response to nursing and is not a continuous process as previously thought (Mather et al. 2019).

7.6 Other Lipid Secretion Mechanisms

While the apocrine process is the primary mechanism for secreting neutral lipids by mammary secretory cells, exocytic vesicles may contribute to this process (Kralj and Pipan 1992; Wooding and Sargeant 2015; Monks et al. 2016). However, the involvement of exocytic processes in neutral lipid secretion has not been fully validated and its contribution to overall lipid secretion by mammary secretory cells is not well-understood. In addition, to the apocrine mechanism, mammary secretory cells release small amounts of phospholipids into milk by an exocytic process in conjunction with the secretion of exosome and lactosome nanovesicle structures (Argov-Argaman et al. 2010; Ortega-Anaya and Jimenez-Flores 2019).

7.7 Summary

The quantities of milk secreted during the course of a normal lactation, the concentration of lipid found in the milk of most species, and the unique mechanism by which milk lipids are secreted, combine to place unique demands on lipid transport processes in mammary secretory cells. Despite significant progress in our knowledge of the genes regulated by lactation (Lemay et al. 2007), our understanding of the pathway machinery for lipid transport and secretion, and how diverse types of lipid transport processes are integrated to meet the demands of lactation remains very limited. Nevertheless, numerous mechanistic advances in how lipids are synthesized, trafficked, and metabolized in other cell types provide experimental frameworks for understanding the molecular and physiological details of lipid transport in mammary secretory cells.

Acknowledgments JLM is supported by NIH grant R01HD093729.

References

Abumrad NA, el-Maghrabi MR, Amri EZ, Lopez E, Grimaldi PA (1993) Cloning of a rat adipocyte membrane protein implicated in binding or transport of long-chain fatty acids that is induced during preadipocyte differentiation. Homology with human CD36. J Biol Chem 268 (24):17665–17668

Agius L, Blackshear PJ, Williamson DH (1981) Rates of triacylglycerol entry into the circulation in the lactating rat. Biochem J 196(2):637–640

Anderson CM, Stahl A (2013) SLC27 fatty acid transport proteins. Mol Asp Med 34(2–3):516–528. https://doi.org/10.1016/j.mam.2012.07.010

Aoki N, Ishii T, Ohira S, Yamaguchi Y, Negi M, Adachi T, Nakamura R, Matsuda T (1997) Stage specific expression of milk fat globule membrane glycoproteins in mouse mammary gland: comparison of MFG-E8, butyrophilin, and CD36 with a major milk protein, á-casein. Biochim Biophys Acta, Gen Subj 1334(2–3):182–190

Argov-Argaman N, Smilowitz JT, Bricarello DA, Barboza M, Lerno L, Froehlich JW, Lee H, Zivkovic AM, Lemay DG, Freeman S, Lebrilla CB, Parikh AN, German JB (2010) Lactosomes: structural and compositional classification of unique nanometer-sized protein lipid particles of human milk. J Agricultural Food Chem 58:11234–11242

Balla T, Kim YJ, Alvarez-Prats A, Pemberton J (2019) Lipid dynamics at contact sites between the endoplasmic reticulum and other organelles. Annu Rev Cell Dev Biol 35:85–109. https://doi.org/10.1146/annurev-cellbio-100818-125251

Barneda D, Planas-Iglesias J, Gaspar ML, Mohammadyani D, Prasannan S, Dormann D, Han GS, Jesch SA, Carman GM, Kagan V, Parker MG, Ktistakis NT, Klein-Seetharaman J, Dixon AM, Henry SA, Christian M (2015) The brown adipocyte protein CIDEA promotes lipid droplet fusion via a phosphatidic acid-binding amphipathic helix. eLife 4:e07485. https://doi.org/10.7554/eLife.07485

Bartz R, Li WH, Venables B, Zehmer JK, Roth MR, Welti R, Anderson RG, Liu P, Chapman KD (2007) Lipidomics reveals that adiposomes store ether lipids and mediate phospholipid traffic. J Lipid Res 48(4):837–847. https://doi.org/10.1194/jlr.M600413-JLR200

Bauman DE, Davis CL (1974) Biosynthesis of milk fat. In: Larson BL, Smith VR (eds) Lactation. Academic, New York, pp 31–75

Beigneux AP, Vergnes L, Qiao X, Quatela S, Davis R, Watkins SM, Coleman RA, Walzem RL, Philips M, Reue K, Young SG (2006) Agpat6--a novel lipid biosynthetic gene required for triacylglycerol production in mammary epithelium. J Lipid Res 47(4):734–744

Bionaz M, Loor JJ (2008a) ACSL1, AGPAT6, FABP3, LPIN1, and SLC27A6 are the most abundant isoforms in bovine mammary tissue and their expression is affected by stage of lactation. J Nutr 138(6):1019–1024

Bionaz M, Loor JJ (2008b) Gene networks driving bovine milk fat synthesis during the lactation cycle. BMC Genomics 9:366. https://doi.org/10.1186/1471-2164-9-366

Bishop WR, Bell RM (1985) Assembly of the endoplasmic reticulum phospholipid bilayer: the phosphatidylcholine transporter. Cell 42(1):51–60

Boylan JG, Hamilton JA (1992) Interactions of acyl-coenzyme A with phosphatidylcholine bilayers and serum albumin. Biochemistry 31(2):557–567. https://doi.org/10.1021/bi00117a037

Calvo JH, Martinez-Royo A, Beattie AE, Dodds KG, Marcos-Carcavilla A, Serrano M (2006a) Fine mapping of genes on sheep chromosome 1 and their association with milk traits. Anim Genet 37 (3):205–210. https://doi.org/10.1111/j.1365-2052.2006.01412.x

Calvo JH, Martinez-Royo A, Silveri L, Floriot S, Eggen A, Marcos-Carcavilla A, Serrano M (2006b) Isolation, mapping and identification of SNPs for four genes (ACP6, CGN, ANXA9, SLC27A3) from a bovine QTL region on BTA3. Cytogenet Genome Res 114(1):39–43. https://doi.org/10.1159/000091926

Cases S, Smith SJ, Zheng YW, Myers HM, Lear SR, Sande E, Novak S, Collins C, Welch CB, Lusis AJ, Erickson SK, Farese RV Jr (1998) Identification of a gene encoding an acyl CoA: diacylglycerol acyltransferase, a key enzyme in triacylglycerol synthesis. Proc Natl Acad Sci U S A 95(22):13018–13023

Cases S, Stone SJ, Zhou P, Yen E, Tow B, Lardizabal KD, Voelker T, Farese RV Jr (2001) Cloning of DGAT2, a second mammalian diacylglycerol acyltransferase, and related family members. J Biol Chem 276(42):38870–38876

Cases S, Zhou P, Shillingford JM, Wiseman BS, Fish JD, Angle CS, Hennighausen L, Werb Z, Farese RV Jr (2004) Development of the mammary gland requires DGAT1 expression in stromal and epithelial tissues. Development 131(13):3047–3055

Chang TY, Chang CC, Lu X, Lin S (2001) Catalysis of ACAT may be completed within the plane of the membrane: a working hypothesis. J Lipid Res 42(12):1933–1938

Chao H, Martin GG, Russell WK, Waghela SD, Russell DH, Schroeder F, Kier AB (2002) Membrane charge and curvature determine interaction with acyl-CoA binding protein (ACBP) and fatty acyl-CoA targeting. Biochemistry 41(33):10540–10553. https://doi.org/10.1021/bi0259498

Chen YQ, Kuo MS, Li S, Bui HH, Peake DA, Sanders PE, Thibodeaux SJ, Chu S, Qian YW, Zhao Y, Bredt DS, Moller DE, Konrad RJ, Beigneux AP, Young SG, Cao G (2008) AGPAT6 is a novel microsomal glycerol-3-phosphate acyltransferase. J Biol Chem 283(15):10048–10057. https://doi.org/10.1074/jbc.M708151200

Chong BM, Reigan P, Mayle-Combs KD, Orlicky DJ, McManaman JL (2011) Determinants of adipophilin function in milk lipid formation and secretion. Trends Endocrinol Metab 22 (6):211–217. https://doi.org/10.1016/j.tem.2011.04.003

Choudhary V, Jacquier N, Schneiter R (2011) The topology of the triacylglycerol synthesizing enzyme Lro1 indicates that neutral lipids can be produced within the luminal compartment of the endoplasmatic reticulum: implications for the biogenesis of lipid droplets. Commun Integr Biol 4(6):781–784. https://doi.org/10.4161/cib.17830

Chung J, Torta F, Masai K, Lucast L, Czapla H, Tanner LB, Narayanaswamy P, Wenk MR, Nakatsu F, De Camilli P (2015) INTRACELLULAR TRANSPORT. PI4P/phosphatidylserine countertransport at ORP5- and ORP8-mediated ER-plasma membrane contacts. Science 349 (6246):428–432. https://doi.org/10.1126/science.aab1370

Clark AJ, Neil C, Gusterson B, McWhir J, Binas B (2000) Deletion of the gene encoding H-FABP/MDGI has no overt effects in the mammary gland. Transgenic Res 9(6):439–444

Clermont Y, Xia L, Rambourg A, Turner JD, Hermo L (1993) Structure of the Golgi apparatus in stimulated and nonstimulated acinar cells of mammary glands of the rat. Anat Rec 237:308–317

Coburn CT, Knapp FF Jr, Febbraio M, Beets AL, Silverstein RL, Abumrad NA (2000) Defective uptake and utilization of long chain fatty acids in muscle and adipose tissues of CD36 knockout mice. J Biol Chem 275(42):32523–32529. https://doi.org/10.1074/jbc.M003826200

Cohen BC, Raz C, Shamay A, Argov-Argaman N (2017) Lipid droplet fusion in mammary epithelial cells is regulated by phosphatidylethanolamine metabolism. J Mammary Gland Biol Neoplasia 22(4):235–249. https://doi.org/10.1007/s10911-017-9386-7

Contarini G, Povolo M (2013) Phospholipids in milk fat: composition, biological and technological significance, and analytical strategies. Int J Mol Sci 14(2):2808–2831. https://doi.org/10.3390/ijms14022808

Dai W, Wang Q, Zhao F, Liu J, Liu H (2018) Understanding the regulatory mechanisms of milk production using integrative transcriptomic and proteomic analyses: improving inefficient utilization of crop by-products as forage in dairy industry. BMC Genomics 19(1):403. https://doi.org/10.1186/s12864-018-4808-5

Daleke DL (2007) Phospholipid flippases. J Biol Chem 282(2):821–825. https://doi.org/10.1074/jbc.R600035200

Davis FM, Janoshazi A, Janardhan KS, Steinckwich N, D'Agostin DM, Petranka JG, Desai PN, Roberts-Thomson SJ, Bird GS, Tucker DK, Fenton SE, Feske S, Monteith GR, Putney JW Jr (2015) Essential role of Orai1 store-operated calcium channels in lactation. Proc Natl Acad Sci U S A 112(18):5827–5832. https://doi.org/10.1073/pnas.1502264112

Davis KR, Giesy SL, Long Q, Krumm CS, Harvatine KJ, Boisclair YR (2016) XBP1 regulates the biosynthetic capacity of the mammary gland during lactation by controlling epithelial expansion and endoplasmic reticulum formation. Endocrinology 157(1):417–428. https://doi.org/10.1210/en.2015-1676

Devaux PF, Morris R (2004) Transmembrane asymmetry and lateral domains in biological membranes. Traffic 5(4):241–246. https://doi.org/10.1111/j.1600-0854.2004.0170.x

Digel M, Ehehalt R, Stremmel W, Fullekrug J (2009) Acyl-CoA synthetases: fatty acid uptake and metabolic channeling. Mol Cell Biochem 326(1–2):23–28. https://doi.org/10.1007/s11010-008-0003-3

Ding T, Li Z, Hailemariam T, Mukherjee S, Maxfield FR, Wu MP, Jiang XC (2008) SMS overexpression and knockdown: impact on cellular sphingomyelin and diacylglycerol metabolism, and cell apoptosis. J Lipid Res 49(2):376–385. https://doi.org/10.1194/jlr.M700401-JLR200

Doege H, Baillie RA, Ortegon AM, Tsang B, Wu Q, Punreddy S, Hirsch D, Watson N, Gimeno RE, Stahl A (2006) Targeted deletion of FATP5 reveals multiple functions in liver metabolism: alterations in hepatic lipid homeostasis. Gastroenterology 130(4):1245–1258. https://doi.org/10.1053/j.gastro.2006.02.006

Doria ML, Ribeiro AS, Wang J, Cotrim CZ, Domingues P, Williams C, Domingues MR, Helguero LA (2014) Fatty acid and phospholipid biosynthetic pathways are regulated throughout mammary epithelial cell differentiation and correlate to breast cancer survival. FASEB J 28 (10):4247–4264. https://doi.org/10.1096/fj.14-249672

Drin G (2014) Topological regulation of lipid balance in cells. Annu Rev Biochem 83:51–77. https://doi.org/10.1146/annurev-biochem-060713-035307

Dylewski DP, Keenan TW (1984) Centrioles in the mammary epithelium of the rat. J Cell Sci 72:185–193

Fairn GD, Schieber NL, Ariotti N, Murphy S, Kuerschner L, Webb RI, Grinstein S, Parton RG (2011) High-resolution mapping reveals topologically distinct cellular pools of phosphatidylserine. J Cell Biol 194(2):257–275. https://doi.org/10.1083/jcb.201012028

Fidler N, Sauerwald T, Pohl A, Demmelmair H, Koletzko B (2000) Docosahexaenoic acid transfer into human milk after dietary supplementation: a randomized clinical trial. J Lipid Res 41 (9):1376–1383

Gallegos AM, Atshaves BP, Storey SM, Starodub O, Petrescu AD, Huang H, McIntosh AL, Martin GG, Chao H, Kier AB, Schroeder F (2001) Gene structure, intracellular localization, and functional roles of sterol carrier protein-2. Prog Lipid Res 40(6):498–563. https://doi.org/10.1016/s0163-7827(01)00015-7

Gatta AT, Levine TP (2017) Piecing Together the Patchwork of Contact Sites. Trends Cell Biol 27 (3):214–229. https://doi.org/10.1016/j.tcb.2016.08.010

Glatz JF, Luiken JJ (2017) From fat to FAT (CD36/SR-B2): understanding the regulation of cellular fatty acid uptake. Biochimie 136:21–26. https://doi.org/10.1016/j.biochi.2016.12.007

Glatz JF, Luiken JJ, Bonen A (2010) Membrane fatty acid transporters as regulators of lipid metabolism: implications for metabolic disease. Physiol Rev 90(1):367–417. https://doi.org/10.1152/physrev.00003.2009

Golachowska MR, Hoekstra D, van IJzendoorn SC (2010) Recycling endosomes in apical plasma membrane domain formation and epithelial cell polarity. Trends Cell Biol 20(10):618–626. https://doi.org/10.1016/j.tcb.2010.08.004

Goldberg IJ, Eckel RH, Abumrad NA (2009) Regulation of fatty acid uptake into tissues: lipoprotein lipase- and CD36-mediated pathways. J Lipid Res 50(Suppl):S86–S90. https://doi.org/10.1194/jlr.R800085-JLR200

Gomez-Diaz C, Bargeton B, Abuin L, Bukar N, Reina JH, Bartoi T, Graf M, Ong H, Ulbrich MH, Masson JF, Benton R (2016) A CD36 ectodomain mediates insect pheromone detection via a putative tunnelling mechanism. Nat Commun 7:11866. https://doi.org/10.1038/ncomms11866

Grant BD, Donaldson JG (2009) Pathways and mechanisms of endocytic recycling. Nat Rev Mol Cell Biol 10(9):597–608. https://doi.org/10.1038/nrm2755

Greenwalt DE, Mather IH (1985) Characterization of an apically derived epithelial membrane glycoprotein from bovine milk, which is expressed in capillary endothelia in diverse tissues. J Cell Biol 100:397–408

Greenwalt DE, Watt KWK, So OY, Jiwani N (1990) PAS IV, an integral membrane protein of mammary epithelial cells, is related to platelet and endothelial cell DC36 (GPIV). Biochemistry 29:7054–7059

Grevengoed TJ, Klett EL, Coleman RA (2014) Acyl-CoA metabolism and partitioning. Annu Rev Nutr 34:1–30. https://doi.org/10.1146/annurev-nutr-071813-105541

Grisart B, Farnir F, Karim L, Cambisano N, Kim JJ, Kvasz A, Mni M, Simon P, Frere JM, Coppieters W, Georges M (2004) Genetic and functional confirmation of the causality of the DGAT1 K232A quantitative trait nucleotide in affecting milk yield and composition. Proc Natl Acad Sci U S A 101(8):2398–2403

Hamilton JA (2007) New insights into the roles of proteins and lipids in membrane transport of fatty acids. Prostaglandins Leukot Essent Fatty Acids 77(5–6):355–361. https://doi.org/10.1016/j.plefa.2007.10.020

Han LQ, Li HJ, Wang YY, Zhu HS, Wang LF, Guo YJ, Lu WF, Wang YL, Yang GY (2010) mRNA abundance and expression of SLC27A, ACC, SCD, FADS, LPIN, INSIG, and PPARGC1 gene isoforms in mouse mammary glands during the lactation cycle. Genet Mol Res 9(2):1250–1257. https://doi.org/10.4238/vol9-2gmr814

Han B, Yuan Y, Liang R, Li Y, Liu L, Sun D (2019) Genetic effects of LPIN1 polymorphisms on milk production traits in dairy cattle. Genes (Basel) 10(4):265. https://doi.org/10.3390/genes10040265

Hanada K, Kumagai K, Tomishige N, Kawano M (2007) CERT and intracellular trafficking of ceramide. Biochim Biophys Acta 1771(6):644–653. https://doi.org/10.1016/j.bbalip.2007.01.009

Hasegawa D, Calvo V, Avivar-Valderas A, Lade A, Chou HI, Lee YA, Farias EF, Aguirre-Ghiso JA, Friedman SL (2015) Epithelial Xbp1 is required for cellular proliferation and differentiation during mammary gland development. Mol Cell Biol 35(9):1543–1556. https://doi.org/10.1128/MCB.00136-15

Haunerland NH, Spener F (2004) Fatty acid-binding proteins--insights from genetic manipulations. Prog Lipid Res 43(4):328–349. https://doi.org/10.1016/j.plipres.2004.05.001

Heid HW, Keenan TW (2005) Intracellular origin and secretion of milk fat globules. Eur J Cell Biol 84(2–3):245–258. https://doi.org/10.1016/j.ejcb.2004.12.002

Hollmann KH (1974) Cytology and fine structure of the mammary gland. In: Larson BL, Smith VR (eds) Lactation, vol I. Academic, New York, pp 3–95

Holthuis JC, Levine TP (2005) Lipid traffic: floppy drives and a superhighway. Nat Rev Mol Cell Biol 6(3):209–220. https://doi.org/10.1038/nrm1591

Horvath SE, Daum G (2013) Lipids of mitochondria. Prog Lipid Res 52(4):590–614. https://doi.org/10.1016/j.plipres.2013.07.002

Huang CM, Keenan TW (1971) Membranes of mammary gland. I. Bovine mammary mitochondria. J Dairy Sci 54(10):1395–1405. https://doi.org/10.3168/jds.S0022-0302(71)86038-1

Hunziker W, Kraehenbuhl JP (1998) Epithelial transcytosis of immunoglobulins. J Mammary Gland Biol Neoplasia 3(3):287–302

Ikonen E, Simons K (1998) Protein and lipid sorting from the trans-Golgi network to the plasma membrane in polarized cells. Semin Cell Dev Biol 9(5):503–509. https://doi.org/10.1006/scdb.1998.0258

Jacquemyn J, Cascalho A, Goodchild RE (2017) The ins and outs of endoplasmic reticulum-controlled lipid biosynthesis. EMBO Rep 18(11):1905–1921. https://doi.org/10.15252/embr.201643426

Jarasch ED, Bruder G, Keenan TW, Franke WW (1977) Redox constituents in milk fat globule membranes and rough endoplasmic reticulum from lactating mammary gland. J Cell Biol 73 (1):223–241

Jay AG, Hamilton JA (2018) The enigmatic membrane fatty acid transporter CD36: new insights into fatty acid binding and their effects on uptake of oxidized LDL. Prostaglandins Leukot Essent Fatty Acids 138:64–70. https://doi.org/10.1016/j.plefa.2016.05.005

Jensen RG (1999) Lipids in human milk. Lipids 34(12):1243–1271

Jensen RG, Ferris AM, Lammi-Keefe CJ, Henderson RA (1990) Lipids of bovine and human milks: a comparison. J Dairy Sci 73:223–240

Jeong J, Rao AU, Xu J, Ogg SL, Hathout Y, Fenselau C, Mather IH (2009) The PRY/SPRY/B30.2 domain of butyrophilin 1A1 (BTN1A1) binds to xanthine oxidoreductase: implications for the function of BTN1A1 in the mammary gland and other tissues. J Biol Chem 284 (33):22444–22456

Jia Z, Moulson CL, Pei Z, Miner JH, Watkins PA (2007) Fatty acid transport protein 4 is the principal very long chain fatty acyl-CoA synthetase in skin fibroblasts. J Biol Chem 282 (28):20573–20583. https://doi.org/10.1074/jbc.M700568200

Joyce CW, Shelness GS, Davis MA, Lee RG, Skinner K, Anderson RA, Rudel LL (2000) ACAT1 and ACAT2 membrane topology segregates a serine residue essential for activity to opposite sides of the endoplasmic reticulum membrane. Mol Biol Cell 11(11):3675–3687. https://doi.org/10.1091/mbc.11.11.3675

Kaplan MR, Simoni RD (1985) Intracellular transport of phosphatidylcholine to the plasma membrane. J Cell Biol 101(2):441–445

Kassan A, Herms A, Fernandez-Vidal A, Bosch M, Schieber NL, Reddy BJ, Fajardo A, Gelabert-Baldrich M, Tebar F, Enrich C, Gross SP, Parton RG, Pol A (2013) Acyl-CoA synthetase 3 promotes lipid droplet biogenesis in ER microdomains. J Cell Biol 203(6):985–1001. https://doi.org/10.1083/jcb.201305142

Keen CL, Lonnerdal B, Clegg M, Hurley LS (1981) Developmental changes in composition of rat milk: trace elements, minerals, protein, carbohydrate and fat. J Nutr 111:226–230

Keenan TW, Huang CM (1972) Membranes of mammary gland. VI. Lipid and protein composition of Golgi apparatus and rough endoplasmic reticulum from bovine mammary gland. J Dairy Sci 55(11):1586–1596. https://doi.org/10.3168/jds.S0022-0302(72)85725-4

Keenan TW, Morre DJ, Olson DE, Yunghans WN, Patton S (1970) Biochemical and morphological comparison of plasma membrane and milk fat globule membrane from bovine mammary gland. J Cell Biol 44(1):80–93. https://doi.org/10.1083/jcb.44.1.80

Keenan TW, Huang CM, Morre DJ (1972) Membranes of mammary gland. 3. Lipid composition of Golgi apparatus from rat mammary gland. J Dairy Sci 55(1):51–57. https://doi.org/10.3168/jds.s0022-0302(72)85431-6

Kent JC, Gardner H, Lai CT, Hartmann PE, Murray K, Rea A, Geddes DT (2018) Hourly breast expression to estimate the rate of synthesis of milk and fat. Nutrients 10(9):1144. https://doi.org/10.3390/nu10091144

Kim YJ, Guzman-Hernandez ML, Balla T (2011) A highly dynamic ER-derived phosphatidylinositol-synthesizing organelle supplies phosphoinositides to cellular membranes. Dev Cell 21(5):813–824. https://doi.org/10.1016/j.devcel.2011.09.005

Kralj M, Pipan N (1992) The role of exocytosis in the apocrine secretion of milk lipid globules in mouse mammary gland during lactogenesis. Biol Cell 75(3):211–216

Ladinsky MS, Mardones GA, Orlicky DJ, Howell KE, McManaman JL (2019) Electron tomography revels that milk lipids originate from endoplasmic reticulum domains with novel structural features. J Mammary Gland Biol Neoplasia 24:293–304

Lee J, Ridgway ND (2020) Substrate channeling in the glycerol-3-phosphate pathway regulates the synthesis, storage and secretion of glycerolipids. Biochim Biophys Acta Mol Cell Biol Lipids 1865(1):158438. https://doi.org/10.1016/j.bbalip.2019.03.010

Lemay DG, Neville MC, Rudolph MC, Pollard KS, German JB (2007) Gene regulatory networks in lactation: identification of global principles using bioinformatics. BMC Syst Biol 1:56. https://doi.org/10.1186/1752-0509-1-56

Liang MY, Hou XM, Qu B, Zhang N, Li N, Cui YJ, Li QZ, Gao XJ (2014) Functional analysis of FABP3 in the milk fat synthesis signaling pathway of dairy cow mammary epithelial cells. In Vitro Cell Dev Biol Anim 50(9):865–873. https://doi.org/10.1007/s11626-014-9780-z

Litvinov DY, Savushkin EV, Dergunov AD (2018) Intracellular and plasma membrane events in cholesterol transport and homeostasis. J Lipids 2018:3965054. https://doi.org/10.1155/2018/3965054

Lollivier V, Marnet PG, Delpal S, Rainteau D, Achard C, Rabot A, Ollivier-Bousquet M (2006) Oxytocin stimulates secretory processes in lactating rabbit mammary epithelial cells. J Physiol 570(Pt 1):125–140. https://doi.org/10.1113/jphysiol.2005.097816

Luiken JJ, Turcotte LP, Bonen A (1999) Protein-mediated palmitate uptake and expression of fatty acid transport proteins in heart giant vesicles. J Lipid Res 40(6):1007–1016

Luiken JJ, Koonen DP, Coumans WA, Pelsers MM, Binas B, Bonen A, Glatz JF (2003) Long-chain fatty acid uptake by skeletal muscle is impaired in homozygous, but not heterozygous, heart-type-FABP null mice. Lipids 38(4):491–496

Mani O, Korner M, Sorensen MT, Sejrsen K, Wotzkow C, Ontsouka CE, Friis RR, Bruckmaier RM, Albrecht C (2010) Expression, localization, and functional model of cholesterol transporters in lactating and nonlactating mammary tissues of murine, bovine, and human origin. Am J Physiol Regul Integr Comp Physiol 299(2):R642–R654. https://doi.org/10.1152/ajpregu.00723.2009

Mansbach CM, Siddiqi SA (2010) The biogenesis of chylomicrons. Annu Rev Physiol 72:315–333. https://doi.org/10.1146/annurev-physiol-021909-135801

Masedunskas A, Chen Y, Stussman R, Weigert R, Mather IH (2017) Kinetics of milk lipid droplet transport, growth, and secretion revealed by intravital imaging: lipid droplet release is intermittently stimulated by oxytocin. Mol Biol Cell 28(7):935–946. https://doi.org/10.1091/mbc.E16-11-0776

Mashek DG, Li LO, Coleman RA (2007) Long-chain acyl-CoA synthetases and fatty acid channeling. Futur Lipidol 2(4):465–476

Mather IH, Keenan TW (1998) Origin and secretion of milk lipids. J Mammary Gland Biol Neoplasia 3(3):259–273

Mather IH, Masedunskas A, Chen Y, Weigert R (2019) Symposium review: Intravital imaging of the lactating mammary gland in live mice reveals novel aspects of milk-lipid secretion. J Dairy Sci 102(3):2760–2782. https://doi.org/10.3168/jds.2018-15459

McManaman JL (2012) Milk lipid secretion: recent biomolecular aspects. Biomol Concepts 3:581–591. https://doi.org/10.1515/bmc-2012-0025
McManaman JL, Palmer CA, Wright RM, Neville MC (2002) Functional regulation of xanthine oxidoreductase expression and localization in the mouse mammary gland: evidence of a role in lipid secretion. J Physiol 545(Pt 2):567–579
McManaman JL, Reyland ME, Thrower EC (2006) Secretion and fluid transport mechanisms in the mammary gland: comparisons with the exocrine pancreas and the salivary gland. J Mammary Gland Biol Neoplasia 11(3–4):249–268
Milhas D, Clarke CJ, Hannun YA (2010) Sphingomyelin metabolism at the plasma membrane: implications for bioactive sphingolipids. FEBS Lett 584(9):1887–1894. https://doi.org/10.1016/j.febslet.2009.10.058
Mittendorfer B, Yoshino M, Patterson BW, Klein S (2016) VLDL triglyceride kinetics in lean, overweight, and obese men and women. J Clin Endocrinol Metab 101(11):4151–4160. https://doi.org/10.1210/jc.2016-1500
Mohammad MA, Haymond MW (2013) Regulation of lipid synthesis genes and milk fat production in human mammary epithelial cells during secretory activation. Am J Physiol Endocrinol Metab 305(6):E700–E716. https://doi.org/10.1152/ajpendo.00052.2013
Monks J, McManaman JL (2013) Secretion and fluid transport mechanisms in the mammary gland. In: Zibadi S, Watson RR, Preedy VR (eds) Handbook of dietary and nutritional aspects of human breast milk, vol 5. Wageningen Academic, Wageningen, pp 35–56
Monks J, Neville MC (2004) Albumin transcytosis across the epithelium of the lactating mouse mammary gland. J Physiol 560(Pt 1):267–280
Monks J, Huey PU, Hanson L, Eckel RH, Neville MC, Gavigan S (2001) A lipoprotein-containing particle is transferred from the serum across the mammary epithelium into the milk of lactating mice. J Lipid Res 42(5):686–696
Monks J, Dzieciatkowska M, Bales ES, Orlicky DJ, Wright RM, McManaman JL (2016) Xanthine oxidoreductase mediates membrane docking of milk-fat droplets but is not essential for apocrine lipid secretion. J Physiol 594:5899–5921. https://doi.org/10.1113/JP272390
Monks J, Ladinsky MS, McManaman JL (2020) Organellar contacts of milk lipid droplets. Contact 3:1–10
Morag M (1970) Estimation of milk yield in the rat. Lab Anim 4(2):259–272. https://doi.org/10.1258/002367770781071671
Murphy DJ (2001) The biogenesis and functions of lipid bodies in animals, plants and microorganisms. Prog Lipid Res 40(5):325–438
Muthusamy BP, Natarajan P, Zhou X, Graham TR (2009) Linking phospholipid flippases to vesicle-mediated protein transport. Biochim Biophys Acta 1791(7):612–619. https://doi.org/10.1016/j.bbalip.2009.03.004
Nafikov RA, Schoonmaker JP, Korn KT, Noack K, Garrick DJ, Koehler KJ, Minick-Bormann J, Reecy JM, Spurlock DE, Beitz DC (2013) Association of polymorphisms in solute carrier family 27, isoform A6 (SLC27A6) and fatty acid-binding protein-3 and fatty acid-binding protein-4 (FABP3 and FABP4) with fatty acid composition of bovine milk. J Dairy Sci 96(9):6007–6021. https://doi.org/10.3168/jds.2013-6703
Nassir F, Wilson B, Han X, Gross RW, Abumrad NA (2007) CD36 is important for fatty acid and cholesterol uptake by the proximal but not distal intestine. J Biol Chem 282(27):19493–19501. https://doi.org/10.1074/jbc.M703330200
Neculai D, Schwake M, Ravichandran M, Zunke F, Collins RF, Peters J, Neculai M, Plumb J, Loppnau P, Pizarro JC, Seitova A, Trimble WS, Saftig P, Grinstein S, Dhe-Paganon S (2013) Structure of LIMP-2 provides functional insights with implications for SR-BI and CD36. Nature 504(7478):172–176. https://doi.org/10.1038/nature12684
Neville MC, Picciano MF (1997) Regulation of milk lipid secretion and composition. Annu Rev Nutr 17:159–183. https://doi.org/10.1146/annurev.nutr.17.1.159
Nickerson JG, Alkhateeb H, Benton CR, Lally J, Nickerson J, Han XX, Wilson MH, Jain SS, Snook LA, Glatz JF, Chabowski A, Luiken JJ, Bonen A (2009) Greater transport efficiencies of the

membrane fatty acid transporters FAT/CD36 and FATP4 compared with FABPpm and FATP1 and differential effects on fatty acid esterification and oxidation in rat skeletal muscle. J Biol Chem 284(24):16522–16530. https://doi.org/10.1074/jbc.M109.004788
Nielsen SU, Rump R, Hojrup P, Roepstorff P, Spener F (1994) Differentiational regulation and phosphorylation of the fatty acid-binding protein from rat mammary epithelial cells. Biochim Biophys Acta 1211(2):189–197. https://doi.org/10.1016/0005-2760(94)90268-2
O'Byrne SM, Kako Y, Deckelbaum RJ, Hansen IH, Palczewski K, Goldberg IJ, Blaner WS (2010) Multiple pathways ensure retinoid delivery to milk: studies in genetically modified mice. Am J Physiol Endocrinol Metab 298(4):E862–E870. https://doi.org/10.1152/ajpendo.00491.2009
Oftedal OT (1984) Milk composition, milk yield and energy output at peak lactation: a comparative review. Symp Zool Soc Lond 51:33–85
Oftedal OT (2012) The evolution of milk secretion and its ancient origins. Animal 6(3):355–368. https://doi.org/10.1017/S1751731111001935
Ohsaki Y, Cheng J, Suzuki M, Shinohara Y, Fujita A, Fujimoto T (2009) Biogenesis of cytoplasmic lipid droplets: from the lipid ester globule in the membrane to the visible structure. Biochim Biophys Acta 1791(6):399–407. https://doi.org/10.1016/j.bbalip.2008.10.002
Ontsouka EC, Albrecht C (2014) Cholesterol transport and regulation in the mammary gland. J Mammary Gland Biol Neoplasia 19(1):43–58. https://doi.org/10.1007/s10911-014-9316-x
Ontsouka EC, Huang X, Stieger B, Albrecht C (2013) Characteristics and functional relevance of apolipoprotein-A1 and cholesterol binding in mammary gland tissues and epithelial cells. PLoS One 8(7):e70407. https://doi.org/10.1371/journal.pone.0070407
Ontsouka CE, Huang X, Aliyev E, Albrecht C (2017) In vitro characterization and endocrine regulation of cholesterol and phospholipid transport in the mammary gland. Mol Cell Endocrinol 439:35–45. https://doi.org/10.1016/j.mce.2016.10.016
Orlicky DJ, Monks J, Stefanski AL, McManaman JL (2013) Dynamics and molecular determinants of cytoplasmic lipid droplet clustering and dispersion. PLoS One 8(6):e66837. https://doi.org/10.1371/journal.pone.0066837
Ortega-Anaya J, Jimenez-Flores R (2019) Symposium review: the relevance of bovine milk phospholipids in human nutrition-evidence of the effect on infant gut and brain development. J Dairy Sci 102(3):2738–2748. https://doi.org/10.3168/jds.2018-15342
Park DS, Lee H, Riedel C, Hulit J, Scherer PE, Pestell RG, Lisanti MP (2001) Prolactin negatively regulates caveolin-1 gene expression in the mammary gland during lactation, via a Ras-dependent mechanism. J Biol Chem 276(51):48389–48397. https://doi.org/10.1074/jbc.M108210200
Park DS, Lee H, Frank PG, Razani B, Nguyen AV, Parlow AF, Russell RG, Hulit J, Pestell RG, Lisanti MP (2002) Caveolin-1-deficient mice show accelerated mammary gland development during pregnancy, premature lactation, and hyperactivation of the Jak-2/STAT5a signaling cascade. Mol Biol Cell 13(10):3416–3430. https://doi.org/10.1091/mbc.02-05-0071
Parton RG, Simons K (2007) The multiple faces of caveolae. Nat Rev Mol Cell Biol 8(3):185–194. https://doi.org/10.1038/nrm2122
Paul HA, Hallam MC, Reimer RA (2015) Milk collection in the rat using capillary tubes and estimation of milk fat content by creamatocrit. J Vis Exp 106:e53476. https://doi.org/10.3791/53476
Pemberton JG, Kim YJ, Balla T (2020) Integrated regulation of the phosphatidylinositol cycle and phosphoinositide-driven lipid transport at ER-PM contact sites. Traffic 21(2):200–219. https://doi.org/10.1111/tra.12709
Pepino MY, Kuda O, Samovski D, Abumrad NA (2014) Structure-function of CD36 and importance of fatty acid signal transduction in fat metabolism. Annu Rev Nutr 34:281–303. https://doi.org/10.1146/annurev-nutr-071812-161220
Ploegh HL (2007) A lipid-based model for the creation of an escape hatch from the endoplasmic reticulum. Nature 448(7152):435–438. https://doi.org/10.1038/nature06004

Pohl J, Ring A, Korkmaz U, Ehehalt R, Stremmel W (2005) FAT/CD36-mediated long-chain fatty acid uptake in adipocytes requires plasma membrane rafts. Mol Biol Cell 16(1):24–31. https://doi.org/10.1091/mbc.e04-07-0616

Pomorski T, Holthuis JC, Herrmann A, van Meer G (2004) Tracking down lipid flippases and their biological functions. J Cell Sci 117(Pt 6):805–813. https://doi.org/10.1242/jcs.01055

Poppelreuther M, Rudolph B, Du C, Grossmann R, Becker M, Thiele C, Ehehalt R, Fullekrug J (2012) The N-terminal region of acyl-CoA synthetase 3 is essential for both the localization on lipid droplets and the function in fatty acid uptake. J Lipid Res 53(5):888–900. https://doi.org/10.1194/jlr.M024562

Prinz WA, Toulmay A, Balla T (2020) The functional universe of membrane contact sites. Nat Rev Mol Cell Biol 21(1):7–24. https://doi.org/10.1038/s41580-019-0180-9

Richards MR, Harp JD, Ory DS, Schaffer JE (2006) Fatty acid transport protein 1 and long-chain acyl coenzyme A synthetase 1 interact in adipocytes. J Lipid Res 47(3):665–672. https://doi.org/10.1194/jlr.M500514-JLR200

Richieri GV, Ogata RT, Kleinfeld AM (1994) Equilibrium constants for the binding of fatty acids with fatty acid-binding proteins from adipocyte, intestine, heart, and liver measured with the fluorescent probe ADIFAB. J Biol Chem 269(39):23918–23930

Ring A, Le Lay S, Pohl J, Verkade P, Stremmel W (2006) Caveolin-1 is required for fatty acid translocase (FAT/CD36) localization and function at the plasma membrane of mouse embryonic fibroblasts. Biochim Biophys Acta 1761(4):416–423. https://doi.org/10.1016/j.bbalip.2006.03.016

Robenek H, Hofnagel O, Buers I, Robenek MJ, Troyer D, Severs NJ (2006) Adipophilin-enriched domains in the ER membrane are sites of lipid droplet biogenesis. J Cell Sci 119 (Pt 20):4215–4224

Ron D, Hampton RY (2004) Membrane biogenesis and the unfolded protein response. J Cell Biol 167(1):23–25. https://doi.org/10.1083/jcb.200408117

Rudolph MC, McManaman JL, Phang T, Russell T, Kominsky DJ, Serkova NJ, Stein T, Anderson SM, Neville MC (2007) Metabolic regulation in the lactating mammary gland: a lipid synthesizing machine. Physiol Genomics 28(3):323–336

Sadowska J, Gebczynski AK, Paszko K, Konarzewski M (2015) Milk output and composition in mice divergently selected for basal metabolic rate. J Exp Biol 218(Pt 2):249–254. https://doi.org/10.1242/jeb.111245

Salo VT, Ikonen E (2019) Moving out but keeping in touch: contacts between endoplasmic reticulum and lipid droplets. Curr Opin Cell Biol 57:64–70. https://doi.org/10.1016/j.ceb.2018.11.002

Sanyal S, Menon AK (2009) Flipping lipids: why an' what's the reason for? ACS Chem Biol 4 (11):895–909. https://doi.org/10.1021/cb900163d

Schroeder F, Atshaves BP, McIntosh AL, Gallegos AM, Storey SM, Parr RD, Jefferson JR, Ball JM, Kier AB (2007) Sterol carrier protein-2: new roles in regulating lipid rafts and signaling. Biochim Biophys Acta 1771(6):700–718. https://doi.org/10.1016/j.bbalip.2007.04.005

Seedorf U, Ellinghaus P, Roch Nofer J (2000) Sterol carrier protein-2. Biochim Biophys Acta 1486 (1):45–54. https://doi.org/10.1016/s1388-1981(00)00047-0

Seigneuret M, Devaux PF (1984) ATP-dependent asymmetric distribution of spin-labeled phospholipids in the erythrocyte membrane: relation to shape changes. Proc Natl Acad Sci U S A 81 (12):3751–3755

Sharma A, Dahiya R (1988) The lipid composition of plasma membrane from goat mammary gland. Biochem Cell Biol 66(12):1355–1359. https://doi.org/10.1139/o88-157

Shennan DB, Peaker M (2000) Transport of milk constituents by the mammary gland. Physiol Rev 80(3):925–951

Smathers RL, Petersen DR (2011) The human fatty acid-binding protein family: evolutionary divergences and functions. Hum Genomics 5(3):170–191. https://doi.org/10.1186/1479-7364-5-3-170

Sotgia F, Schubert W, Pestell RG, Lisanti MP (2006) Genetic ablation of caveolin-1 in mammary epithelial cells increases milk production and hyper-activates STAT5a signaling. Cancer Biol Ther 5(3):292–297

Soupene E, Kuypers FA (2008) Mammalian long-chain acyl-CoA synthetases. Exp Biol Med (Maywood) 233(5):507–521. https://doi.org/10.3181/0710-MR-287

Spandl J, White DJ, Peychl J, Thiele C (2009) Live cell multicolor imaging of lipid droplets with a new dye, LD540. Traffic 10(11):1579–1584. https://doi.org/10.1111/j.1600-0854.2009.00980.x

Spitsberg VL, Matitashvili E, Gorewit RC (1995) Association and coexpression of fatty-acid-binding protein and glycoprotein CD36 in the bovine mammary gland. Eur J Biochem/FEBS 230(3):872–878

Sriburi R, Bommiasamy H, Buldak GL, Robbins GR, Frank M, Jackowski S, Brewer JW (2007) Coordinate regulation of phospholipid biosynthesis and secretory pathway gene expression in XBP-1(S)-induced endoplasmic reticulum biogenesis. J Biol Chem 282(10):7024–7034. https://doi.org/10.1074/jbc.M609490200

Stein O, Stein Y (1967) Lipid synthesis, intracellular transport, and secretion. II. Electron microscopic radioautographic study of the mouse lactating mammary gland. J Cell Biol 34 (1):251–263

Stemberger BH, Walsh RM, Patton S (1984) Morphometric evaluation of lipid droplet associations with secretory vesicles, mitochondria and other components in the lactating cell. Cell Tissue Res 236(2):471–475

Stone SJ, Vance JE (2000) Phosphatidylserine synthase-1 and -2 are localized to mitochondria-associated membranes. J Biol Chem 275(44):34534–34540. https://doi.org/10.1074/jbc.M002865200

Storch J, Veerkamp JH, Hsu KT (2002) Similar mechanisms of fatty acid transfer from human anal rodent fatty acid-binding proteins to membranes: liver, intestine, heart muscle, and adipose tissue FABPs. Mol Cell Biochem 239(1–2):25–33

Suckling KE, Stange EF (1985) Role of acyl-CoA: cholesterol acyltransferase in cellular cholesterol metabolism. J Lipid Res 26(6):647–671

Takeuchi K, Reue K (2009) Biochemistry, physiology, and genetics of GPAT, AGPAT, and lipin enzymes in triglyceride synthesis. Am J Physiol Endocrinol Metab 296(6):E1195–E1209. https://doi.org/10.1152/ajpendo.90958.2008

Thompson BJ, Smith S (1985) Biosynthesis of fatty acids by lactating human breast epithelial cells: an evaluation of the contribution to the overall composition of human milk fat. Pediatr Res 19 (1):139–143. https://doi.org/10.1203/00006450-198501000-00036

Tian S, Ohta A, Horiuchi H, Fukuda R (2018) Oxysterol-binding protein homologs mediate sterol transport from the endoplasmic reticulum to mitochondria in yeast. J Biol Chem 293 (15):5636–5648. https://doi.org/10.1074/jbc.RA117.000596

Tiwari S, Siddiqi SA (2012) Intracellular trafficking and secretion of VLDL. Arterioscler Thromb Vasc Biol 32(5):1079–1086. https://doi.org/10.1161/ATVBAHA.111.241471

Trigatti BL, Anderson RG, Gerber GE (1999) Identification of caveolin-1 as a fatty acid binding protein. Biochem Biophys Res Commun 255(1):34–39. https://doi.org/10.1006/bbrc.1998.0123

Tsukamoto K, Maugeais C, Glick JM, Rader DJ (2000) Markedly increased secretion of VLDL triglycerides induced by gene transfer of apolipoprotein E isoforms in apoE-deficient mice. J Lipid Res 41(2):253–259

van Meer G, Voelker DR, Feigenson GW (2008) Membrane lipids: where they are and how they behave. Nat Rev Mol Cell Biol 9(2):112–124. https://doi.org/10.1038/nrm2330

Vance JE (1990) Phospholipid synthesis in a membrane fraction associated with mitochondria. J Biol Chem 265(13):7248–7256

Vance JE (1991) Newly made phosphatidylserine and phosphatidylethanolamine are preferentially translocated between rat liver mitochondria and endoplasmic reticulum. J Biol Chem 266 (1):89–97

Vance JE (2008) Phosphatidylserine and phosphatidylethanolamine in mammalian cells: two metabolically related aminophospholipids. J Lipid Res 49(7):1377–1387. https://doi.org/10.1194/jlr.R700020-JLR200
Vance DE (2013) Physiological roles of phosphatidylethanolamine N-methyltransferase. Biochim Biophys Acta 1831(3):626–632. https://doi.org/10.1016/j.bbalip.2012.07.017
Vance JE (2014) MAM (mitochondria-associated membranes) in mammalian cells: lipids and beyond. Biochim Biophys Acta 1841(4):595–609. https://doi.org/10.1016/j.bbalip.2013.11.014
Vance JE (2015) Phospholipid synthesis and transport in mammalian cells. Traffic 16(1):1–18. https://doi.org/10.1111/tra.12230
Vance DE, de Kruijff B (1980) The possible functional significance of phosphatidylethanolamine methylation. Nature 288(5788):277–279
Voelker DR (1984) Phosphatidylserine functions as the major precursor of phosphatidylethanolamine in cultured BHK-21 cells. Proc Natl Acad Sci U S A 81(9):2669–2673. https://doi.org/10.1073/pnas.81.9.2669
Voelker DR (1991) Organelle biogenesis and intracellular lipid transport in eukaryotes. Microbiol Rev 55(4):543–560
Walther TC, Chung J, Farese RV Jr (2017) Lipid droplet biogenesis. Annu Rev Cell Dev Biol 33:491–510. https://doi.org/10.1146/annurev-cellbio-100616-060608
Wang W, Lv N, Zhang S, Shui G, Qian H, Zhang J, Chen Y, Ye J, Xie Y, Shen Y, Wenk MR, Li P (2012) Cidea is an essential transcriptional coactivator regulating mammary gland secretion of milk lipids. Nat Med 18(2):235–243. https://doi.org/10.1038/nm.2614
Wang H, Ma Q, Qi Y, Dong J, Du X, Rae J, Wang J, Wu WF, Brown AJ, Parton RG, Wu JW, Yang H (2019) ORP2 delivers cholesterol to the plasma membrane in exchange for phosphatidylinositol 4, 5-bisphosphate (PI(4,5)P2). Mol Cell 73(3):458–473 e457. https://doi.org/10.1016/j.molcel.2018.11.014
Watkins PA (2008) Very-long-chain acyl-CoA synthetases. J Biol Chem 283(4):1773–1777. https://doi.org/10.1074/jbc.R700037200
Weisiger RA (2002) Cytosolic fatty acid binding proteins catalyze two distinct steps in intracellular transport of their ligands. Mol Cell Biochem 239(1–2):35–43
Weiss SB, Kennedy EP, Kiyasu JY (1960) The enzymatic synthesis of triglycerides. J Biol Chem 235:40–44
Welsch U, Singh S, Buchheim W, Patton S (1984) Internalization of ferritin-concanavalin A by the lactating mammary cell in vivo. Cell Tissue Res 235(2):433–438
Welte MA (2009) Fat on the move: intracellular motion of lipid droplets. Biochem Soc Trans 37 (Pt 5):991–996
Wilfling F, Wang H, Haas JT, Krahmer N, Gould TJ, Uchida A, Cheng JX, Graham M, Christiano R, Frohlich F, Liu X, Buhman KK, Coleman RA, Bewersdorf J, Farese RV Jr, Walther TC (2013) Triacylglycerol synthesis enzymes mediate lipid droplet growth by relocalizing from the ER to lipid droplets. Dev Cell 24(4):384–399. https://doi.org/10.1016/j.devcel.2013.01.013
Wong LH, Gatta AT, Levine TP (2019) Lipid transfer proteins: the lipid commute via shuttles, bridges and tubes. Nat Rev Mol Cell Biol 20(2):85–101. https://doi.org/10.1038/s41580-018-0071-5
Wooding FBP (1977) Comparative mammary fine structure. In: Peaker M (ed) Comparative aspects of lactation. Academic, London, pp 1–41
Wooding FB (1980) Lipid droplet secretion by the rabbit harderian gland. J Ultrastruct Res 71 (1):68–78
Wooding FB, Sargeant TJ (2015) Immunocytochemical evidence for golgi vesicle involvement in milk fat globule secretion. J Histochem Cytochem 63(12):943–951. https://doi.org/10.1369/0022155415608918
Wu CC, Howell KE, Neville MC, Yates JR 3rd, McManaman JL (2000) Proteomics reveal a link between the endoplasmic reticulum and lipid secretory mechanisms in mammary epithelial cells. Electrophoresis 21(16):3470–3482

Wu Q, Ortegon AM, Tsang B, Doege H, Feingold KR, Stahl A (2006) FATP1 is an insulin-sensitive fatty acid transporter involved in diet-induced obesity. Mol Cell Biol 26 (9):3455–3467. https://doi.org/10.1128/MCB.26.9.3455-3467.2006

Yang EK, Blusztajn JK, Pomfret EA, Zeisel SH (1988) Rat and human mammary tissue can synthesize choline moiety via the methylation of phosphatidylethanolamine. Biochem J 256 (3):821–828

Yen CL, Stone SJ, Koliwad S, Harris C, Farese RV Jr (2008) Thematic review series: glycerolipids. DGAT enzymes and triacylglycerol biosynthesis. J Lipid Res 49(11):2283–2301

Zborowski J, Dygas A, Wojtczak L (1983) Phosphatidylserine decarboxylase is located on the external side of the inner mitochondrial membrane. FEBS Lett 157(1):179–182

Chapter 8
Ion Transport Across Inner Ear Epithelia

Daniel C. Marcus

Abstract The sensory functions of the auditory and vestibular peripheral organs (inner ear) depend on vigorous transepithelial ion transport processes under the control of hormones and other signal pathways. The most salient of these are secretion of K^+ and absorption of Na^+ and Ca^{2+} by the concerted action of multiple epithelial cell types. Much has been learned of the molecular bases and development of these processes, as well as the genetic basis of common cochlear and vestibular diseases. The salient aspects are concisely reviewed in this chapter.

Keywords Inner ear · K^+ secretion · Na^+ absorption · Ca^{2+} absorption · Regulation of transport

8.1 Introduction

The inner ear is the site of transduction of mechanical stimuli originating in sounds and in body accelerations, both linear and rotational, into coded neural transmissions that produce the sensations of hearing and balance. The cells that perform the transductions are neuroepithelia, that is, they have characteristics of epithelia (embedded as an integral part of the epithelial sheet, with apical and basolateral membrane domains) as well as of excitable cells (apical sensory structures that produce receptor potential responses to stimuli and synaptic machinery at basolateral domains that synapse with afferent and efferent nerve endings). The lumen of the inner ear epithelia is continuous among the several sensory regions, with connections among them that are either patent or blocked at different times during development, as described in Sect. 8.2.2.

Although much research has focused on the function of the sensory cells and their neural connections, most of the epithelial lumen is bounded by epithelial cells that

D. C. Marcus (✉)
Department of Anatomy and Physiology, Kansas State University, Manhattan, KS, USA
e-mail: marcus@ksu.edu

K. L. Hamilton, D. C. Devor (eds.), *Ion Transport Across Epithelial Tissues and Disease*, Physiology in Health and Disease,
https://doi.org/10.1007/978-3-030-55310-4_8

form a closed space and that exhibit many transport properties and regulatory pathways that will be described in this chapter. Due to its closed nature, the inner ear does not produce a major bulk flow of fluid, in contrast to most other epithelial organs (such as kidney nephrons, gastrointestinal tract, glands). Nonetheless, luminal volume is subject to perturbation and regulation (Sect. 8.4.4).

The net transepithelial transport of solutes by the specialized epithelia of the inner ear provides the required ionic environment for the generation and maintenance of the unusual composition of the luminal fluid (Sect. 8.3), which is essential for normal hearing and balance. Genetic aberrations and pharmacologic treatments can lead to disturbances of the sensory function via disruption of transport and its regulation in the extrasensory epithelial cells.

The primary transport functions of the inner ear epithelia result in unusually high luminal [K^+] and astonishingly-low [Na^+] and [Ca^{2+}], especially in the cochlea. Our current understanding of inner ear transport is based on canonical transport molecules and regulatory pathways that are also found in other epithelial tissues, but also in less usual constellations. Salient transport systems are reviewed in this chapter. Additional material can be found in other recent reviews (Marcus and Wangemann 2009, 2010; Wangemann and Marcus 2017).

8.2 Structure

The anatomical organization of the inner ear epithelia consists of the hearing organ (the cochlea), five vestibular sensory organs that sense linear (utricle and saccule) and rotational (three semicircular canals including their respective sensory ampullae) acceleration. In addition, there is a nonsensory but essential epithelial organ, the endolymphatic sac, which is connected to the rest of the system through the endolymphatic duct (Fig. 8.1).

8.2.1 Cochlea

The cochlea in mammals is spirally coiled to efficiently package a long, highly frequency-selective sensory structure (organ of Corti). The luminal cross-section is triangular in shape (Fig. 8.2) and bounded by about 12 cell types that each make their own contributions to the maintenance of the luminal fluid composition (Sect. 8.3). The epithelium is a monolayer except for the structure on the lateral wall, stria vascularis, which is multicellular and highly complex in structure and function (Sect. 8.2.1.1). The major cell types are labeled in Fig. 8.2 and salient aspects are described in the following subsections.

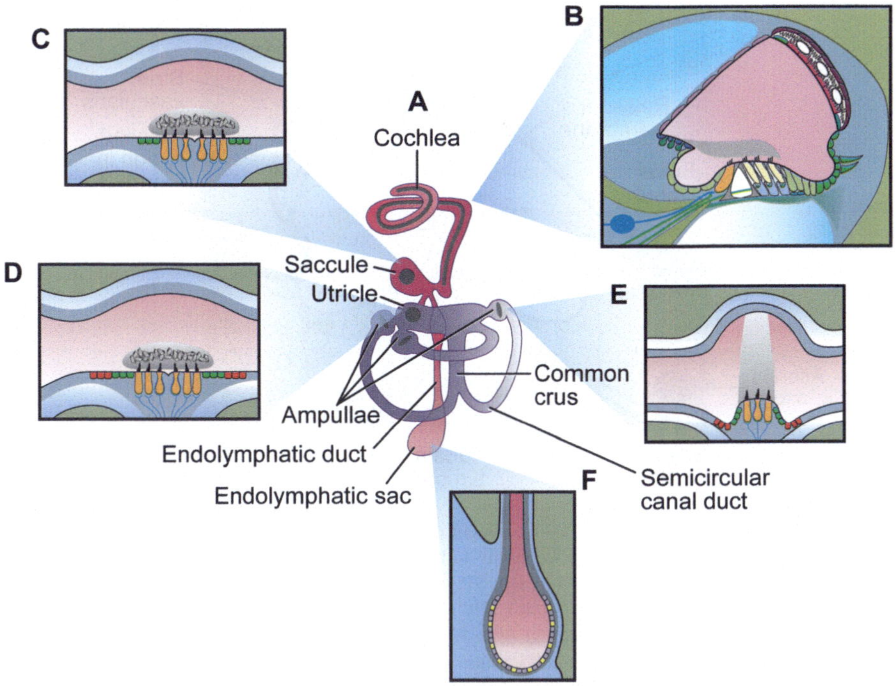

Fig. 8.1 Structure of the inner ear epithelium. (**a**) Diagram of inner ear, comprising the organ of hearing (cochlea), the five organs of balance (saccule, utricle, and 3 ampullae of the semicircular canals) and the endolymphatic duct and sac. Two semicircular canal ducts combine to form the common crus before joining the utricle. (**b**) Cross-section of one cochlear turn (see Fig. 8.2 for more detail and nomenclature). (**c**) Cross-section of the saccule. Sensory hair cells (orange) with their stereocilia inserted in the macula; transitional cells (green) surround the sensory cells. (**d**) Cross-section of the utricle. In common with the saccule are the sensory hair cells (orange) with their stereocilia inserted in the macula; transitional cells (green) surround the sensory cells. In addition, the K^+-secretory dark cells (red) occur outside of the macula in a pattern depicted in Fig. 8.4. (**e**) Cross-section of one of the three ampullae. In common with the utricle are sensory hair cells (orange), transitional cells (green), and dark cells (red; Fig. 8.4). The hair cells are in a ridge of tissue (the crista) and insert their long stereocilia in the gelatinous cupula, which extends to the roof. (**f**) Cross-section of the endolymphatic duct and endolymphatic sac. The sac is characterized by the absence of sensory cells, the presence of two types of epithelial cell (mitochondrial-rich cells and ribosome-rich cells (yellow) and in contrast to parts **a–e**, the luminal fluid is low in [K^+] and high in [Na^+]). All regions are lined by a continuous epithelium; cells with unknown transport properties and function are not shown. Redrawn and adapted from Wangemann and Marcus (2017) with permission

8.2.1.1 Stria Vascularis

This tissue is so named for its dense capillary network, which provides blood gas exchange for the mitochondria-packed marginal cell layer that borders the cochlear lumen (Figs. 8.2 and 8.3). An unusual feature of this juxtaposition of capillaries and epithelial cells is that there is no basement membrane between them. The high

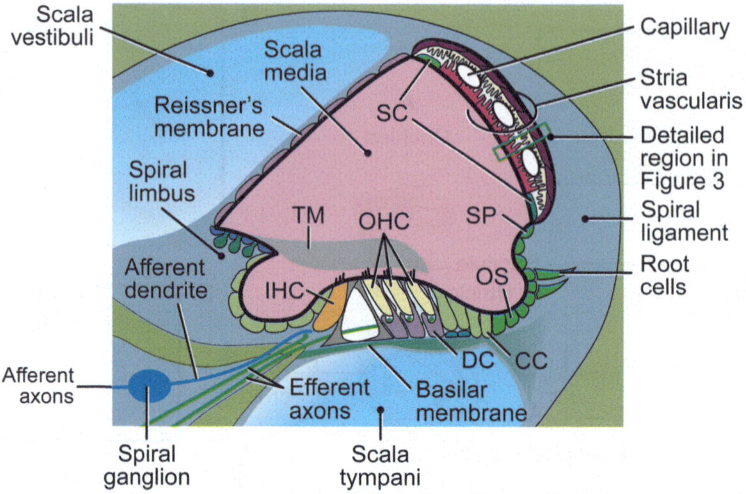

Fig. 8.2 Diagram of cross-section of the cochlear duct. The extracellular fluids (composition given in Table 8.1) are perilymph in Scala vestibuli and Scala tympani (blue), and endolymph in Scala media (pink). SC, strial spindle cells (green); SP, spiral prominence cells (green); OS, outer sulcus cells (green); OHC, outer hair cells (yellow); IHC, inner hair cells (orange); TM, tectorial membrane (gray); DC, Deiters' cells; CC, Claudius cells (light green). Heavy black line denotes the apical epithelial barrier. Details of Claudius cell – Outer sulcus cell juxtaposition is not shown (Spicer and Schulte 1996). Stria vascularis comprises multiple cell types and is depicted in Fig. 8.3. Redrawn and adapted from Wangemann and Marcus (2017) with permission

density of mitochondria in the narrow basolateral fingers of the marginal cells that interdigitate with the intermediate cells is consistent with the enormous energy use by the marginal cell ion transport processes, described later. Surprisingly, the strial capillaries do not appear to supply ion substrates or nutrients (other than oxygen) to the marginal cells, but the source is rather the vessels that exchange these substances via the perilymph (Wada et al. 1979; Kambayashi et al. 1982) (Sect. 8.4.1.1). The marginal cells form a continuous layer joined by tight junctions but exhibit no lateral communication with each other via gap junctions. The usual basement membrane that underlies the basolateral membrane of epithelial cells is notably absent, in contrast to the homologous dark cells of the vestibular labyrinth (Sects. 8.2.2.2 and 8.4.1.1).

In addition to contact of the marginal cells with the capillaries, they are also in intimate juxtaposition to a discontinuous layer (no tight junctions) of intermediate cells (Fig. 8.3). Lateral to these two layers is a third layer, consisting of the basal cells, that "seals" the outer border of the stria via dense tight junctions between the basal cells. This juxtaposition of the three cell layers results in a very small but essential intrastrial fluid space. The ionic composition of this fluid acts to communicate between the cell layers, which contribute in a coordinated way to K^+ secretion and to generation of a high luminal electrical potential difference (~+80 to 100 mV, lumen positive) across the entire strial structure and thereby also across all of the

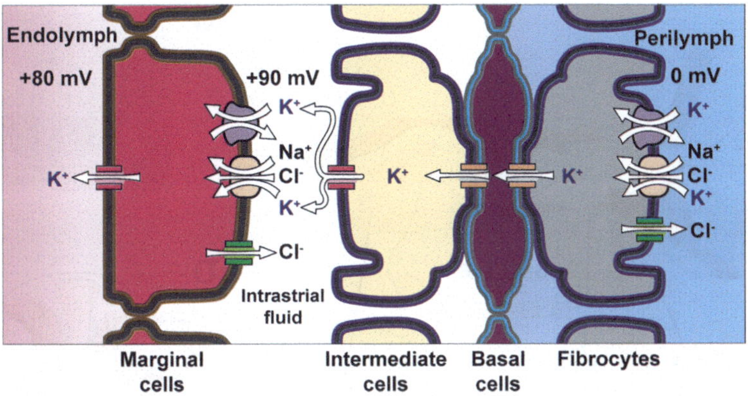

Fig. 8.3 Structural diagram of stria vascularis. The apical border to the endolymph (pink) is formed by the strial marginal cell epithelium, with the minor population of spindle cells at the edges of the stria (not shown; see Fig. 8.2). The basolateral membrane of the marginal cells is in contact with the intrastrial fluid (white), as are the intermediate cells and inner surface of the basal cells. The outer surface of the basal cells is in contact with perilymph (blue), which pervades the spiral ligament (extracellular space of the fibrocytes). Tight junctions among the basal cells and among the marginal/spindle cell layers form barriers to fluid and ion movements among the three fluids. The primary functions of the stria vascularis are (a) secretion of K^+ and (b) generation of the endocochlear potential (described in body of text). K^+ from the fibrocytes passes by diffusion through gap junctions (orange channels) to the basal cell cytosol and further through additional gap junctions to the intermediate cells. The endocochlear potential is generated by the large $[K^+]$ difference across the KCNJ10 K^+ channels in the intermediate cell membrane. K^+ is taken up (Fig. 8.5) by the Na^+,K^+-ATPase and Na^+, $2Cl^-$, K^+-cotransporter (SLC12A2) from the intrastrial fluid across the basolateral membrane of the marginal cells to the remarkably low level of about 1.2 mM. Cl^- recirculates through the ClC-K chloride channels. K^+ is secreted into endolymph through KCNQ1/KCNE1 K^+ channels in the apical membrane. Redrawn and adapted from Wangemann and Marcus (2017) with permission

other epithelial cell types of the cochlear lumen. The transport processes are described later (Sect. 8.4.1). The upper and lower edges of the stria facing the lumen are populated by spindle cells with different transport properties (Sect. 8.2.1.3) than the adjacent marginal cells (Fig. 8.2).

The vestibular system has a monolayer of epithelial cells that are homologous in function to the strial marginal cell layer, but does not have the intermediate cell and basal cell sublayers. These vestibular epithelial cells, dark cells, secrete K^+ with the same constellation of transporters as the cochlear marginal cells, but by contrast the luminal voltage is not significantly polarized (Marcus et al. 1994). The term "dark cells" refers to the electron-dense images of the cytoplasm in uranyl-stained tissue sections in contrast to their clear appearance in isolated native tissue. This terminology can be confusing, since melanocytes grow in a layer exclusively under the dark cells and thereby actually appear "dark" in isolated native tissue, simplifying orientation during dissections of inner ear tissues from pigmented experimental animals (Fig. 8.4) (Kimura 1969). The strial intermediate cells also express melanin.

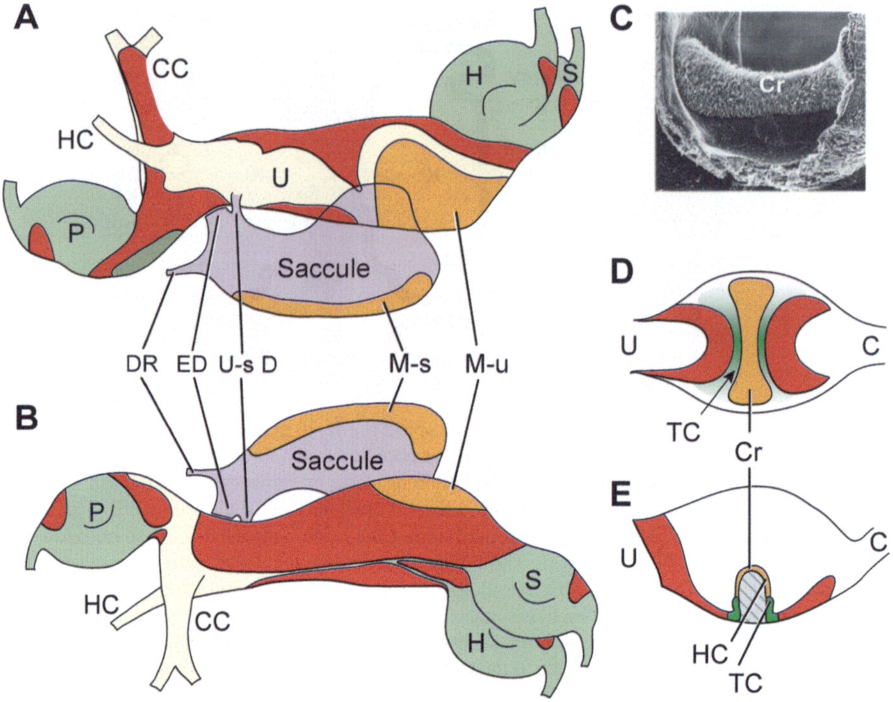

Fig. 8.4 Diagram of distribution of vestibular dark cells. The distribution of the K^+-secreting vestibular dark cell epithelium is shown in red (based on observations of isolated tissues (Kimura 1969); semicircular canal ducts, endolymphatic duct, and endolymphatic sac are not completely shown, but contain no dark cells and are represented in Fig. 8.1). The "dark cells" appear dark in osmium-stained histological sections but are clear in native tissue (An underlying discontinuous layer of melanocytes, however, gives the "dark cell" regions a dark appearance in native tissue of pigmented animals.). (**a**) Major structures of the membranous vestibular labyrinth. P, H, S: Ampullae of the posterior, horizontal, and superior semicircular canals; the arc adjacent to each label indicates the approximate location of the sensory structure, crista ampullaris (Cr in **c**, **d**, **e**). CC, common crus; U, utricle; M-u, utricular macula; M-s, saccular macula; ED, endolymphatic duct; U-s D, utriculo-saccular duct; HC, horizontal semicircular canal duct. The epithelia indicated with light yellow in the utricle, with light green in the ampullae and with light purple in the saccule may be functionally similar to each other, but the transport functions and homogeneity has not been established, although at least part of the extramacular saccular epithelium absorbs Na^+ via apical ENaC. (**b**) The vestibular labyrinth of the upper panel is rotated about a horizontal axis. (**c**) Scanning electron microscopic image of an ampulla (Hunter-Duvar 1983) illustrating the 3-D saddle appearance of the crista ampullaris (Cr), (**d**) Diagram of cell distributions in an ampulla, viewed from above the crista. TC, transitional cells (green); Cr, crista covered with sensory hair cells (orange); C, connection to semicircular canal duct; U, connection to utricle. (**e**) Vestibular labyrinth of Fig. 8.4(**d**) rotated about a horizontal axis. Gray hatched fill of the crista represents the connective tissue that contains sensory nerves and blood vessels. Redrawn and adapted from Kimura (1969) and Hunter-Duvar (1983) with permission

8.2.1.2 Reissner's Membrane

Reissner's membrane (Fig. 8.2) is a monolayer of extremely thin epithelial cells (similar in this regard to epithelial cells of the cortical thick ascending limb in the kidney and to alveolar Type I cells in the lung) about 2–3 μm thick at the nucleus, but only about 1 μm thick peripheral from the nucleus. The thinness of the cells, paucity of intracellular organelles, and lack of any remarkable membrane folding suggested from its histology alone that they were relatively inactive and served mostly as a barrier between the luminal and basolateral fluids, and in the case of the inner ear served as a flaccid acoustically-transparent membrane. However, Reissner's membrane exhibits an active regulated absorption of Na^+ (Sect. 8.4.2).

8.2.1.3 Spiral Prominence, Outer Sulcus, and Claudius Cells

A monolayer of epithelial cells extends between the stria vascularis and organ of Corti in the order stria vascularis, spiral prominence, outer sulcus, and Claudius cells (Fig. 8.2). The spiral prominence cells are much thinner than the outer sulcus cells, although some of the transport properties are in common (Cl/HCO-exchange). There is some growth of the Claudius cells over the outer sulcus cells and the extent of overlap varies by cochlear turn and species (Spicer and Schulte 1996). The reduced overlap in the upper turns provides the experimental advantage that the apical surface of the outer sulcus cells are best exposed for measurements of transepithelial currents by vibrating probe and of membrane currents by patch clamp (Spicer and Schulte 1996; Marcus and Chiba 1999; Chiba and Marcus 2000).

The spiral prominence, outer sulcus, as well as the spindle cells of the stria vascularis all express the anion exchanger pendrin (SLC26A4) in the apical membrane (Wangemann et al. 2004). Transport by Claudius cells has not been extensively characterized, but Na^+ absorption has been observed (Sect. 8.4.2).

8.2.1.4 Organ of Corti

The organ of Corti (Fig. 8.2) is a complex multicellular structure, including the hair cells and Deiters' cells, that serves as the site of auditory transduction. Briefly, mammals have two types of sensory cell; the outer and the inner hair cells, which have apical cilia that are deflected by sound wave input to the cochlea. The deflections result from acoustically driven fluid movements that in turn deflect the basilar membrane on which the organ of Corti sits and the tectorial membrane, which juxtaposes the apical stereocilia. Stereocilia of the outer hair cells are directly stimulated by being embedded in the tectorial membrane, while the inner hair cell stereocilia are stimulated by fluid pumped along the thin channel between the tectorial membrane and hair cells. Stimulation of the stereocilia leads to modulated cationic currents from the lumen into the cell, resulting in receptor potentials that modulate synaptic activity at their basolateral membranes (Schnee and Ricci 2017). The sensory hair cells are individually surrounded by supporting cells that participate in the homeostasis of the cations flowing out the basolateral membrane (Sect. 8.4.1.3) and of the neurotransmitters.

8.2.2 Vestibular Labyrinth

The epithelial lumen at the base of the cochlea continues to the vestibular labyrinth (Fig. 8.1c, d, e) via the narrow lumen of the ductus reuniens, opening into the saccule, one of the two organs that sense linear acceleration. The lumen then bifurcates after the saccule. One leg, the endolymphatic duct, leads to the endolymphatic sac (Fig. 8.1f), a blind sac with transepithelial ion transport activity but no sensory cells. The transport function of the endolymphatic sac is critical for the development of both hearing and balance during embryonic maturation. Even though the SLC26A4 Cl^-/HCO_3^- exchanger is expressed in the sense organs as well as the endolymphatic sac, the expression in the endolymphatic sac is critically essential and sufficient during embryonic development for postnatal hearing and balance, while expression in the sense organs is needed for stable mature function (Sect. 8.4.3).

The short leg of the bifurcation, the utriculo-saccular duct, connects from the saccule to the system of the utricle (which in addition to the saccule, also senses linear acceleration) and the three semicircular canals (which sense rotational acceleration in the three orthogonal planes). The utriculo-saccular duct is open only during embryonic development (until about 16.5 days after conception in mice), becoming closed to paint tracer (but possibly remaining minimally open) thereafter in contrast to the ductus reuniens, which although constricted relative to the cochlear duct and saccule is patent to paint tracer (Cantos et al. 2000). See, however, Sect. 8.2.2.1.

8.2.2.1 Utricle and Saccule

The sensory cell tissues (maculae) of the utricle and saccule consist of spatially organized arrangements of hair cells that are in contact at their stereocilia with a fine extracellular aggregate of crystals and organic matrix (Lundberg et al. 2015) that transduce linear acceleration in the plane of the crystal sheet into nerve impulses (Eatock and Songer 2011). The sensory planes of the utricle and saccule are at right angles to each other, providing sensory information in all three orthogonal planes. Vestibular dark cells and their K^+-secretory activity are abundantly present in the utricle but are absent in the saccule (Fig. 8.4) (Kimura 1969). Thus, the saccule relies on maintenance of luminal K^+ by diffusion from the cochlea through the ductus reuniens (Sellick and Johnstone 1972), while the utricle (and the three ampullae and the common crus) are self-sufficient in maintaining the K^+ composition of the lumen.

This difference in transport properties of the epithelia in the saccule from that in the other four vestibular organs is consistent with the open ductus reuniens and the closed (restricted) utriculo-saccular duct. A structure in the utriculo-saccular duct ("valve of Bast") has been judged by its morphology to be capable of controlling patency, although direct functional demonstration and characterization have not been reported (Hofman et al. 2008).

8.2.2.2 Semicircular Canals

Each of the three canals (anterior, posterior, and horizontal) consists of a semicircular canal duct and the sensory structure of the ampulla. The canal duct is lined by a monolayer of very thin epithelial cells, as described for the cochlear Reissner's membrane. One end of the duct opens into the associated ampulla and the other is joined to the utricle. The canals from the anterior (superior) canal and the posterior canal join together at the common crus before opening into the utricle. Although the canal ducts have no vestibular dark cells, the common crus has both canal duct cells and dark cells (Fig. 8.4). The sensory cells of the ampullae are organized in cristae and the long stereocilia are embedded in a gelatinous structure (cupula) that extends to the roof of each ampulla (Flock and Goldstein Jr 1978). Rotational motion of the head results in fluid flows in the canal ducts that are oriented along vector components of the direction of rotation. These fluid flows press against the cupula, which transmits the motion to the stereocilia, which leads to transduction by the neuroepithelial hair cells in the stimulated ampullae.

8.3 Fluid Composition

The composition of the luminal fluid, endolymph, varies markedly among the different organs. The cochlear endolymph has a highly unusual composition for an extracellular fluid, with slightly less extreme luminal vs abluminal ion concentration differences in the vestibular labyrinth (Table 8.1). The most salient features are the high [K^+], low [Na^+], and low [Ca^{2+}].

The extremely low Na^+ and Ca^{2+} concentrations correlate with the high positive luminal voltage in the cochlea (endocochlear potential; EP), which provides a driving force for cation efflux and is absent in the vestibular labyrinth (Marcus et al. 1994). In spite of the absence of the positive EP in the vestibular labyrinth, the endolymphatic Na^+ and Ca^{2+} levels are still markedly lower than in perilymph, pointing to active absorptive processes. It has been noted that the [K^+] of cochlear perilymph is higher in scala vestibuli than in scala tympani (Salt and Ohyama 1993) (not shown in Table 8.1; Sect. 8.4.1.2). By contrast to the endolymph in the sensory organs, the endolymph of the endolymphatic sac has the reverse monovalent cation profile (Table 8.1).

Table 8.1 Fluid compositions of inner ear and adjoining fluids

	Cochlear perilymph	Cochlear endolymph	Utricular endolymph	Endolymphatic sac endolymph	Cerebrospinal fluid	Plasma
Na^+ (mM)	148	1.3	9	129	149	145
K^+ (mM)	4.2	157	149	10	3.1	5.0
Cl^- (mM)	119	132	–	124	129	106
HCO_3^- (mM)	21	31	–	–	19	18
Ca^{2+} (mM)	1.3	0.023	0.28	–	–	2.6
Protein (mg/dl)	178	38	–	–	24	4238
pH	7.3	7.5	7.5	6.9	7.3	7.3
Potential (mV)	0	+80 to +100	0 to +4	+6 to +15	–	0

Reproduced with permission Marcus and Wangemann (2010)

This basic ion profile in the sensory organs is essential for transduction of stimuli into nerve impulses and mutation of key transporters in production of endolymph and/or the endocochlear potential is associated with a number of genetic losses of hearing, including Jervell Lange-Nielsen syndrome due to mutation of KCNQ1 and/or its regulatory subunit KCNE1 (Tranebjaerg et al. 1999; Faridi et al. 2019), KCNJ10 mutations that abolish the endocochlear potential (Chen and Zhao 2014) and mutations of SLC26A4 that cause Pendred syndrome through disruption of HCO_3^- secretion and the endocochlear potential (Wangemann 2013). In addition to the effects of this ion profile on transport, the mechanical properties of the tectorial membrane are highly sensitive to their ionic environment (Kronester-Frei 1979; Shah et al. 1995) and can thereby affect changes to transduction via direct and indirect alterations to the input stimulus to the stereocilia.

Energy must be expended to maintain the K^+/Na^+ levels so that K^+ is the primary ion that carries the transduction current from endolymph into the hair cells. The transduction ion current ion species requires little energy for removal at the basolateral membrane when it is K^+, but is energetically expensive if it were Na^+. The delicate and highly sensitive cochlear sensory cells, therefore, do not require a "noisy" local vasculature nor a high density of intracellular organelles to provide the large energy needs of a Na^+-mediated transduction process. Rather, the energy for K^+ secretion and EP generation is produced and consumed remotely by the stria vascularis. This high aerobic energy consumption by the stria vascularis leaves the organ highly susceptible to damage by oxidative stress, which mediates several pathological conditions, including fluctuating hearing loss (Pendred syndrome) (Singh and Wangemann 2008; Ito et al. 2014). The K^+ secretion by vestibular dark cells at a site remote from the vestibular sensory cells is an earlier evolutionary specialization, but the lack of a large transepithelial voltage suggests these organs do not require for adequate sensitivity the elevated driving force associated with producing the cochlear EP.

The abluminal fluid, perilymph, is in contact with the basolateral membranes of the epithelial cells, with the exception of marginal cells of the stria vascularis. In this instance, the marginal cells are in contact with the intrastrial space fluid, while perilymph bathes the outer surface of the strial basal cells that border and connect to the fibrocytes of the spiral ligament.

8.4 Transport

8.4.1 K^+-Secretion, Reabsorption, and Recycling

8.4.1.1 K^+-Secretion

K^+ is secreted by strial marginal cells in the cochlea and vestibular dark cells in the vestibular labyrinth (Figs. 8.3 and 8.5) employing the same constellation of transporters. The transporters in these cells that support K^+ secretion are an apical K^+

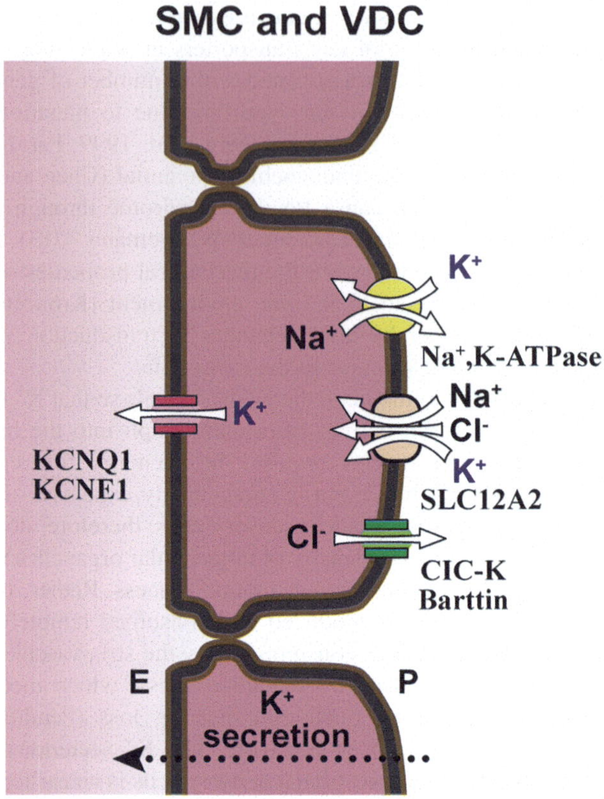

Fig. 8.5 Cell model of K^+ secretion by strial marginal cells and vestibular dark cells. The strial marginal cells (SMC) and vestibular dark cells (VCD) secrete K^+ into the endolymph (E) by expending energy from ATP via the Na^+,K^+-ATPase to take up K^+ from the basolateral fluid (P) into the cytosol in exchange for the removal of cytosolic Na^+. The basolateral Na^+,K^+-ATPase (Na^+-pump) consists of the α1 & β1 subunits in the cochlea and the α1 and β2 subunits in the vestibular dark cells. The Na^+ gradient thus created across the basolateral membrane drives the further uptake of K^+ via the SLC12A2 Na^+, $2Cl^-$, K^+-cotransporter. Cl^- taken up by the cotransporter then recycles via the ClC-K/Barttin Cl^- channels. The K^+ taken up at the basolateral membrane then is secreted through the KCNQ1/KCNE1 K^+ channels in the apical membrane. Redrawn and adapted from Wangemann and Marcus (2017) with permission

channel and basolateral Na^+,K^+-ATPase, Na^+-K^+-2 Cl^- cotransporter, and Cl^- channel. As with most mammalian epithelial ion transport, the energy for transport is provided by ATP, which drives cellular Na^+ extrusion and K^+ cellular uptake via the "sodium pump" (Na^+,K^+-ATPase; α1, β1 isoform subunits in the cochlea and α1, β2 in the vestibular dark cells (Schulte and Steel 1994; Ding et al. 2018).

K^+ is taken up across the basolateral membrane into the marginal cell/vestibular dark cell cytoplasm by the Na^+ pump and the concurrent extrusion of Na^+ produces a strong inward Na^+ concentration gradient that drives additional uptake of K^+ (accompanied by Na^+ and Cl^- through the Na^+,K^+,2 Cl^- cotransporter (NKCC1;

SLC12A2)). The uptake of K^+ is thereby highly efficient: one cycle of the pump utilizes the energy from one molecule of ATP to take up two molecules of K^+ and removes three molecules of Na^+. The energy now transferred into the ion gradients across the basolateral membrane then drives the three molecules of Na^+ back into the cell, pulling three additional molecules of K^+ into the cell, for a net win of five molecules K^+ uptake for an expenditure of only one molecule of ATP. Although there are additional obligatory Na^+-coupled transporters in these cells (Wangemann et al. 1996a), the overriding importance of the coupling between the Na^+ pump and NKCC is observed by the near-complete collapse of transepithelial electrogenic transport of K^+ by individual inhibition of either the Na^+ pump or NKCC (SLC12A2) (Marcus et al. 1994). The Cl^- taken up by the NKCC recycles out across the basolateral membrane via a high density of channels (ClC-K/Barttin) (Wangemann and Marcus 1992; Marcus et al. 1993; Sage and Marcus 2001; Rickheit et al. 2008).

Finally, the accumulated K^+ diffuses through the cell cytosol and exits the cell across the apical membrane via K^+ channels (KCNQ1/KCNE1) that are unusual in several respects (Marcus and Shen 1994; Shen et al. 1997). They activate very slowly (over seconds) by depolarization of the membrane and have a single-channel conductance of only circa 1.6 pS under usual conditions (high-K^+ inside, low-K^+ outside) (Shen et al. 1997). With the high-K^+ concentration outside that occurs in the cochlea and vestibular endolymph, the conductance is circa 5–18 pS (Sunose et al. 1994) (Fig. 8.5). The channels are mostly closed at common epithelial membrane polarizations (–80 to –40 mV), but both the marginal cells and vestibular dark cells are by comparison constitutively highly depolarized by a dominant membrane Cl^- conductance in the basolateral membrane in concert with an apparently high intracellular $[Cl^-]$ (Wangemann and Marcus 1992; Marcus et al. 1993). The voltage activation is strongly dependent on divalent cation levels in the cytosol (Shen and Marcus 1998). Hereditary defects in any of these transport components result in hereditary deafness and/or vertigo (Sect. 8.4.4).

The K^+ uptake at the basolateral membrane is from the pervasive perilymph in the vestibular labyrinth, but in the cochlear stria vascularis the uptake is from the extremely limited intrastrial fluid volume. This powerful uptake by the strial marginal cells (described above) draws the intrastrial fluid $[K^+]$ down to about 1.2 mM (Takeuchi et al. 2000). As a result, the intermediate cell membrane, also in contact with the intrastrial fluid, becomes highly polarized via the dominant membrane K^+ permeability (Takeuchi et al. 2000). This polarization, however, can initially seem counterintuitive since the intrastrial extracellular fluid is *positive* with respect to the perilymph (described later).

That the extracellular $[K^+]$ would be pulled so low is also counterintuitive since the $[K^+]$ of the serum flowing through the strial vasculature is presumably normal at about 3.6 mM and one would thereby assume that capillary K^+ would provide the major source for secretion by the marginal cells. Interestingly, this is not the case; removal of K^+ from the strial capillaries by vascular perfusion leads to a highly delayed (mean: 25 min, range: 20–42 min) drop in endocochlear potential (Wada et al. 1979), whereas removal of K^+ from perilymph (which pervades the organ of

Corti, spiral ligament, and adjoining strial basal cell surface) reduces endocochlear potential with a delay of less than 2 min (Marcus et al. 1981). Further support for the concept that K^+ secreted by the stria originates from perilymph rather than the intimate strial capillaries comes from radiotracer and chemical analog tracer experiments (Marcus et al. 1981; Sterkers et al. 1982). This body of evidence suggests that the strial capillaries are actually relatively tight to K^+.

The effect on the endocochlear potential of changes in marginal cell ion transport rate is greatly amplified via the change in $[K^+]$ of the intrastrial fluid. For example, rapid inhibition of marginal cell K^+ uptake by sudden-onset delivery of furosemide to the basolateral marginal cell membrane via vascular perfusion results in an immediate fall of the endocochlear potential that is far more rapid than furosemide delivered by perilymphatic perfusion (Kobayashi et al. 1984), suggesting a high permeability of these capillaries to furosemide. The immediate blockage of K^+ uptake from the intrastrial fluid via inhibition of the marginal cell SLC12A2 leads to a sharp rise in intrastrial $[K^+]$ that depolarizes the K^+-permeable intermediate cell membrane, which generates the endocochlear potential. This is reminiscent of the nearly instantaneous effect of furosemide on cortical thick ascending limb of kidney, mediated through changes in $[Cl^-]$ of the small volume of cytosol in these thin cells (Greger et al. 1983).

8.4.1.2 K^+-Exit from Endolymph

K^+ exits the cochlear endolymph by transcellular pathways primarily in the sensory cells, but also via a parasensory pathway in the outer sulcus cells (Lee et al. 2001; Kim and Marcus 2011). Both of these transport pathways result from uptake through nonselective cation channels in the apical membrane. The sensory cell apical channels (currently of incompletely understood molecular identity (Fettiplace and Kim 2014)) are modulated by the mechanical stimulus and those in the outer sulcus cells (also of unknown molecular identity) are constitutively active, although purine-activated nonselective cation channels (P2X2R) are also present (Kim and Marcus 2011). These apical pathways have been functionally and pharmacologically well characterized (Kim and Marcus 2011; Fettiplace and Kim 2014), and the sensory cell transduction channel identity has been vigorously pursued (Fettiplace and Kim 2014). In addition, an elevated $[K^+]$ in the perilymph of scala vestibuli of about twice that in scala tympani (Salt and Ohyama 1993) suggests possible K^+ absorption by transcellular and/or paracellular pathways across Reissner's membrane.

8.4.1.3 K^+ Recycling Pathway

With perilymph established as the source of K^+ for K^+-secretion by the stria, the notion of K^+ secretion as part of a local K^+ recycling pathway is a natural consequence. K^+ in endolymph passes through the hair cell apical transduction channels into the cytosol, out of the cell at the basolateral membrane via basolateral K^+

channels and into the perilymph pervading the organ of Corti. This concept has been borne out by measurements of elevated [K^+] in the extracellular space of the organ of Corti in response to acoustic stimulation (Johnstone et al. 1989). From there, the K^+ can diffuse into the perilymph of scala tympani and/or can be taken up by supporting cells of the organ of Corti and travel laterally by diffusion to the cochlear lateral wall through a gap-junction network between the epithelial cells of this region (Spicer and Schulte 1996). There is a second gap-junction network among the fibrocytes of the lateral wall that further connects by gap junctions to the strial basal and intermediate cells. Thus, there are two nonexclusive views that K^+ can (1) travel along the lateral wall to the strial basal cells by extracellular diffusion among the connective tissue fibers to the strial basal cells and/or (2) by intracellular diffusion through the gap-junction connected fibrocytes and then into the stria, as described earlier. Consistent with the latter hypothesis, the fibrocytes are known to express the same transport constellation used by marginal cells and vestibular dark cells for K^+ uptake (Crouch et al. 1997), even though the fibrocytes do not have discrete membrane domains.

8.4.1.4 Electrical Potential Profile in the Stria

As alluded, the positive potential of the intrastrial fluid is measured with respect to the perilymph, which is taken as the reference ($V = 0$) for voltage measurements in the cochlea. The perilymph bathes the basolateral side of all of the cochlear epithelial monolayer, as well as the "outside" of the strial basal cell layer, and is about the same potential as other extracellular fluids in the body (e.g., blood, CSF, etc.). The positive intrastrial potential results from (1) the postulated low potential difference of basal cell cytosol with respect to perilymph, proposed to be a result of nonselective cation channels (Salt et al. 1987), (2) which then holds the intracellular potential of the intermediate cells low via the extensive gap junction connection between the basal and intermediate cells (Kikuchi et al. 2000). This electrical syncytium between these cells virtually "grounds" the cytosol of the intermediate cells, so that the highly polarized intermediate cell membrane generates a positive potential in the intrastrial fluid with respect to perilymph. The difference to the "canonical" view of a negative intracellular potential stems from the fact that the outside is taken as ground for most cells vs the case of the intermediate cells of the cochlea, where the intracellular potential is clamped near ground.

With the intermediate cell cytoplasm held to a low voltage with respect to perilymph, the large voltage generated across the intermediate cells by the dense K^+ channels (KCNJ10) and the low [K^+] in the intrastrial fluid result in a strong *positive* polarization (with respect to perilymph) of the intrastrial fluid. This polarization of the intrastrial fluid is maintained by the very tight electrical resistance of the basal cell tight junctions (whose principal defining component is Claudin-11), which seal the lateral border of the stria (Gow et al. 2004) and the tight junctions of the marginal cells at the luminal border. The importance of the basal cell tight-junction seal has been demonstrated by the collapse of the EP in mice with defective junctional claudin-11, which is expressed exclusively at basal cell junctions (Gow et al. 2004). The marginal cells produce very little transepithelial potential difference (similar to the vestibular dark cell epithelium (Sect. 8.2.1.1)), in spite of their vigorous K^+ transport. In essence, one can view the stria as two batteries in series, the first battery produces about +90 mV at the intrastrial fluid and the second battery (marginal cell layer) adds little to that, but provides a necessary seal of the luminal barrier; the cochlear endolymphatic fluid then sits at about +80 mV with respect to perilymph.

8.4.2 Na^+ Absorption and Leak

The [Na^+] of cochlear endolymph is extremely low compared to all other bodily fluids, including cytosol. Electrochemical equilibrium for Na^+ across the cochlear tight junctions would be approximately 7 mM in endolymph, and values at this level or higher were historically often reported. However, using extremely stringent techniques to avoid contamination of samples by perilymph, Bosher and Warren (1968) showed that the true level was at 1 mM or less, suggesting that net transepithelial transport of Na^+ was dominated by active absorption and influx was likely a passive leak. Consistent with this view is the finding that the resting absorptive flux of Na^+ from the lumen is about 1% of the K^+ secretory flux (Konishi et al. 1978).

Transepithelial absorption of Na^+ by Reissner's membrane in the cochlea and by semicircular canal ducts and extramacular saccule epithelial cells is mediated by apical epithelial Na^+ channels (ENaC) and by the combined action of basolateral Na^+/K^+-ATPase and K^+ channels (Lee and Marcus 2003; Kim and Marcus 2011) (Fig. 8.6). This transport is regulated by glucocorticoid receptors (Kim and Marcus 2011). Additional transcellular absorption of Na^+ (as well as K^+) occurs via the same basolateral transporters, but via apical nonselective cation (NSC) channels in outer sulcus cells (Kim and Marcus 2011) in the cochlea and by transitional cells of the vestibular labyrinth (Kim and Marcus 2011) (Fig. 8.7). The cochlear and vestibular sensory cell transduction channels in the apical membrane at the tips of the stereocilia are of another molecular identity, but are also nonselective cation channels that pass Na^+, K^+ , and Ca^{2+} (Fettiplace and Kim 2014). Additional nonselective cation channel-mediated cation efflux occurs under purinergic control via P2X2 apical receptors in several cochlear and vestibular epithelial cell types (Kim and Marcus 2011; Morton-Jones et al. 2015). Na^+ absorption via ENaC was observed in Claudius cells and this Na^+ absorption and possibly a Cl^- secretory pathway is regulated by P2Y purinergic signaling (Yoo et al. 2012).

8.4.3 Ca^{2+} Absorption and Secretion

The low [Ca^{2+}] of endolymph is maintained within a critical range by the balance of a "push-pull" system of both active absorption and secretion in both the cochlea and

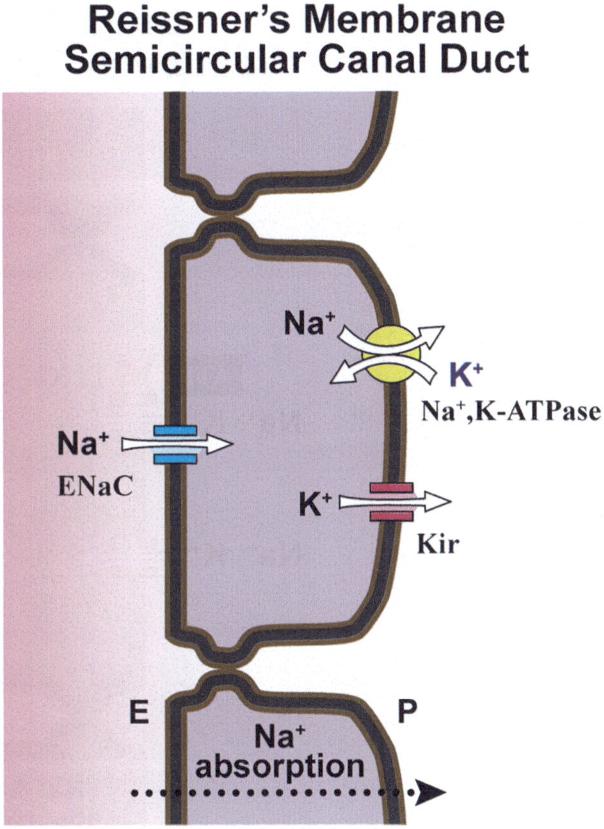

Fig. 8.6 Cell model of Na^+ absorption by Reissner's membrane and Semicircular canal duct epithelia. Apical Na^+ entry is via epithelial Na^+ channels (ENaC). Na^+ is removed at the basolateral membrane from the cytosol by the Na^+-pump and the K^+ taken up by the pump is recycled through basolateral inward rectifier (Kir) K^+ channels. Redrawn and adapted from Wangemann and Marcus (2017) with permission

the vestibular labyrinth. Ca^{2+} is secreted by apical plasma membrane Ca^{2+}-ATPase (PMCA2), which has been localized to sensory hair cell stereocilia and Reissner's membrane (Wood et al. 2004). Deletion or mutation of PMCA2 leads to loss of hearing and vestibular function, and to a reduced endolymphatic [Ca^{2+}] (Kozel et al. 1998; Wood et al. 2004; Chen et al. 2011).

All salient components of a Ca^{2+} absorptive process found in several other epithelia (Boros et al. 2009) (Fig. 8.8) were found to be expressed at the transcript and protein level in several epithelial cell types of the inner ear (Yamauchi et al. 2010).

The Ca^{2+} absorption process in other epithelia was found to be mediated by apical entry of Ca^{2+} via Ca^{2+}-selective channels of the isoforms TRPV5 (e.g., kidney) and/or TRPV6 (e.g., intestine) (Boros et al. 2009). Ca^{2+} entering the cytosol was buffered by soluble calbindin proteins, which diffused to the basolateral membrane, where the Ca^{2+} dissociated from the calbindin and was extruded by Na^+/Ca^{2+}-exchangers and a Ca^{2+}-ATPase (Fig. 8.8). This transport is regulated both by the extracellular pH at the apical membrane TRPV5/6 channel and by vitamin D, which increases TRPV5/6 expression (Nijenhuis et al. 2005).

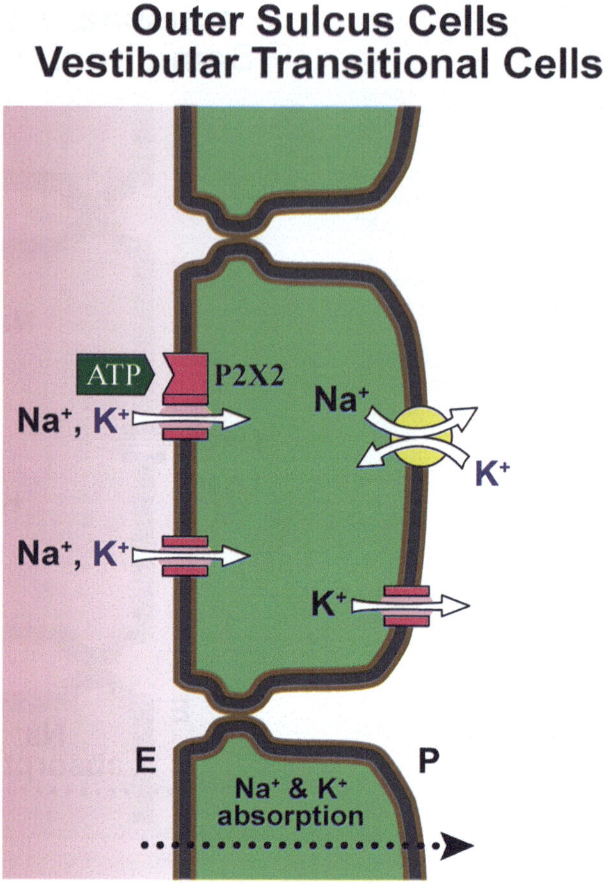

Fig. 8.7 Cell model of Na^+ and parasensory K^+ absorption by cochlear outer sulcus cells and vestibular transitional cells. Apical Na^+ and K^+ entry is via nonselective cation channels (both P2X2R ionotropic purinergic receptors and nonpurinergic channels). Na^+ is removed at the basolateral membrane from the cytosol by the Na^+-pump and the K^+ taken up through the apical channels and basolateral pump is removed through basolateral K^+ channels. Redrawn and adapted from Kim and Marcus (2011) with permission

The semicircular canal duct epithelial cells were observed to express mRNA for all components of this Ca^{2+} absorptive system and function was demonstrated via radiolabeled Ca^{2+} fluxes across primary cultures on permeable supports, as was the sensitivity of the flux to apical pH (inhibition by acidification), to vitamin D and to known TRPV5/6 channel inhibitors (Nakaya et al. 2007; Yamauchi et al. 2010). Both TRPV5 and TRPV6 proteins were demonstrated by immunofluorescence laser confocal microscopy near the apical membranes of native semicircular canal in the vestibular labyrinth. In the cochlea, TRPV5 was found near the apical membrane of strial marginal cells and both TRPV5 and TRPV6 were found in outer and inner sulcus cells (Yamauchi et al. 2010). The other components of the Ca^{2+} absorption system were also expressed in the cochlea (Yamauchi et al. 2010).

The pH sensitivity of TRPV5/6 suggests that endolymphatic pH may be an important regulator of endolymphatic [Ca^{2+}]. In support of this notion, deletion of the Cl^-/HCO_3^- exchanger SLC26A4 (pendrin; mutation of which results in Pendred syndrome) leads to acidification of endolymph in *Slc26a4* knockout mice (Nakaya

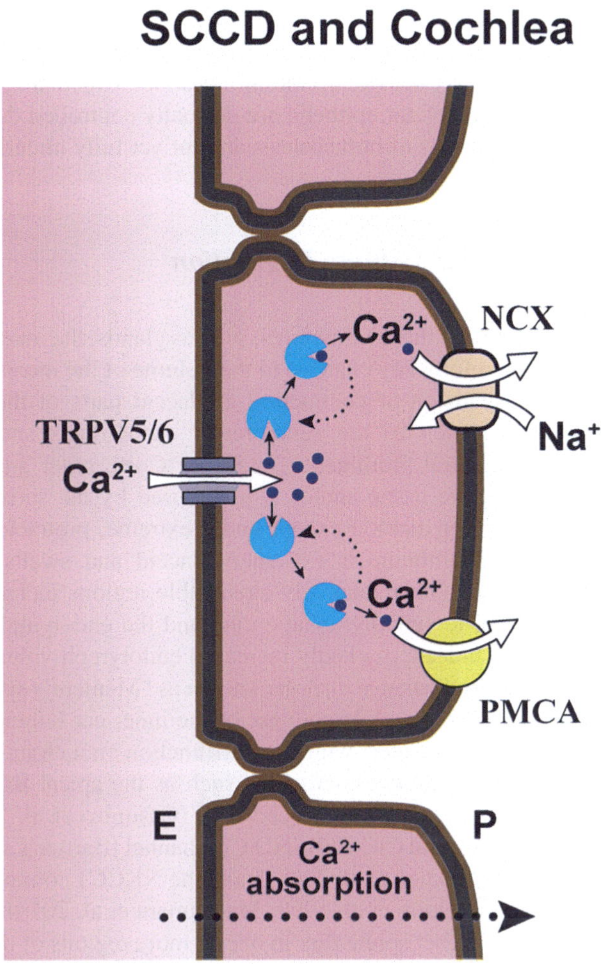

Fig. 8.8 Cell model of Ca^{2+} absorption by semicircular canal duct cells and cochlear cells. Ca^{2+} is taken up from endolymph via apical TRPV5 and TRPV6 Ca^{2+}-selective channels. The Ca^{2+} is subsequently buffered by soluble calbindin (blue), transported by diffusion to the basolateral membrane where the Ca^{2+} dissociates from the calbindin and is removed from the cytosol by basolateral Na^{+}/Ca^{2+}-exchanger (NCX) and plasma membrane Ca^{2+}-ATPase (PMCA). There is evidence that the stria vascularis, outer sulcus, and inner sulcus epithelial cells may be the sites for this transport system. Redrawn and adapted from Boros et al. (2009) and Wangemann and Marcus (2017) with permission

et al. 2007; Wangemann et al. 2007). Indeed, these knockout mice also exhibited a toxic elevation of luminal $[Ca^{2+}]$ associated with loss of hearing and balance.

It was subsequently found, however, that conditional SLC26A4 knockout mice in which SLC26A4 was expressed during critical embryonic development and subsequently removed in postnatal mice had normal levels of endolymphatic pH, suggesting that additional pH regulatory mechanisms were present in the conditional knockout mice that could compensate for the loss of SLC26A4 (Choi et al. 2011). The conditional knockout of postnatal SLC26A4, however, led to a fluctuating hearing loss that was related to oxidative stress in the stria vascularis, providing an alternate explanation for the postnatal requirement for SLC26A4 expression (Li et al. 2013; Nishio et al. 2016; Honda et al. 2017).

Observations in humans and rodents were consistent with the involvement of systemic vitamin D in maintenance of normal hearing, although whether the etiology of these results were related to direct local effects of vitamin-D on TRPV5/6 transport rates in the inner ear was not addressed (Ikeda et al. 1987; Ikeda et al. 1989). The physiological conditions under which the TRPV5/6 Ca^{2+} absorption pathway in inner ear epithelia are normally controlled during normal and pathological perturbations of homeostasis are not yet fully understood.

8.4.4 *Fluid Volume Regulation*

Unlike other epithelial structures, such as glands, the inner ear lumen is a closed fluid space. Therefore, any changes to the volume of the incompressible endolymph must lead to expansion or contraction of flaccid parts of the epithelium. Much of the epithelium (such as the semicircular canal ducts and ampullae, cochlear stria vascularis, spiral prominence, organ of Corti, outer and inner sulcus) has a very stiff connective tissue and/or is juxtaposed by the surrounding bone, giving little opportunity for dilation except under extreme, protracted conditions. By contrast, Reissner's membrane is extremely flaccid and swells or collapses under small perturbations. Other relatively stretchable regions include the extramacular utricle and saccule, the endolymphatic duct and the endolymphatic sac, with its complex infoldings. Indeed, markedly increased endolymph volume is often associated with the hearing and balance disorder known as "Meniere's syndrome" in humans (Foster and Breeze 2013), while collapse of the inner ear lumen ("Scheibe's deformity" in humans) is associated with loss-of-function mutations of genes involved in K^+-secretion by strial marginal cells, such as the apical KCNQ1/KCNE1 K^+ channel [Jervell and Lange-Nielsen syndrome (Casimiro et al. 2001; Warth and Barhanin 2002)], basolateral ClC-K/BSND Cl^- channel [Bartter's syndrome III and IV (Simon et al. 1997; Rickheit et al. 2008)], and the NKCC1 cotransporter SLC12A2 (Kilquist syndrome) (Delpire et al. 1999; Macnamara et al. 2019).

Changes in net solute flux in one or more regions of the inner ear epithelium will lead to osmotically driven volume changes in the lumen, limited by the water permeability of the epithelium. Water-permeable pathways across epithelia are transcellular and/or paracellular. The transcellular pathways are mediated by integral membrane proteins in the apical and basolateral cell membranes such as isoforms of aquaporin (AQP) (Verkman et al. 2014; Rosenthal et al. 2017b) and some other integral membrane proteins such as the sodium glucose cotransporter SGLT1(Beitz et al. 1999; Erokhova et al. 2016).

Paracellular pathways are gated by some isoforms of claudin in tight junctions (Rosenthal et al. 2017a). For example, circa 23–30% of transepithelial water flux across proximal kidney tubules is paracellular through claudin-2 tight junctions, which also passes cations via the same pores (Rosenthal et al. 2017a, b). It is to be expected that most ion and water flow in the sensory organs of the inner ear is transcellular rather than paracellular due to the high transepithelial Na^+ and K^+ concentration gradients sustained by these epithelia. The transepithelial ion gradients in the endolymphatic duct (Table 8.1) and sac however are small, suggesting a comparatively leaky epithelium and potentially a more permeable paracellular pathway.

Two possible volume regulatory pathways have been identified and recently reviewed (Wangemann and Marcus 2017). First, muscarinic receptor activation leads to translocation of AQP-5 from the cytosol to the apical membrane in outer sulcus cells of the upper cochlear turns and AQP-4 is constitutively expressed in the basolateral membrane (Eckhard et al. 2015). Second, a regulated water pathway mediated by AQP-2 may be active in the endolymphatic sac under the control of vasopressin, although the veracity of this mechanism has been a subject of discussion (Wangemann and Marcus 2017). Deletion of expression of AQP-1, -3, or -5 in mice had no effect on auditory brainstem response to clicks, while deletion of AQP-4 led to significant hearing loss (Li and Verkman 2001).

8.5 Regulation of Transport

Several regulatory pathways that control the rates of ion transport in inner ear epithelia have been mentioned in Sect. 8.4. These and other known pathways are presented in Table 8.2, along with representative references to the literature that describe these effects.

Table 8.2 Regulation of cochlear and vestibular ion transport

Ion transport	Signal pathway	Stimulate	Inhibit	References
K^+	β1-adrenergic agonist, basolateral	X		Wangemann et al. (1999, 2000)
	cAMP	X		Sunose et al. (1997a, b)
	Purinergic agonist, apical (P2Y4)	Transient	X	Lee and Marcus (2008); Liu et al. (1995); Lee and Marcus (2008)
	Purinergic agonist, basolateral		X	Liu et al. (1995)
	Muscarinic agonist (M3, M4)	Transient	X	Wangemann et al. (2001)
	Cell swelling, hypo-osmotic challenge	X		Wangemann et al. (1995)
	Elevated basolateral $[K^+]$	X		Wangemann et al. (1996b)
Na^+	Glucocorticoid	Intracellular		Kim and Marcus (2011)
	Purinergic agonist, apical (P2X)	X		Kim and Marcus (2011), Lee and Marcus (2008)
Ca^{2+}	Acidification, apical absorption		X	Nakaya et al. (2007)
Cl^-	β2-adrenergic agonist, basolateral	X		Milhaud et al. (2002)
	cAMP	X		Pondugula et al. (2013)
	Glucocorticoid + cAMP	X		Pondugula et al. (2013)

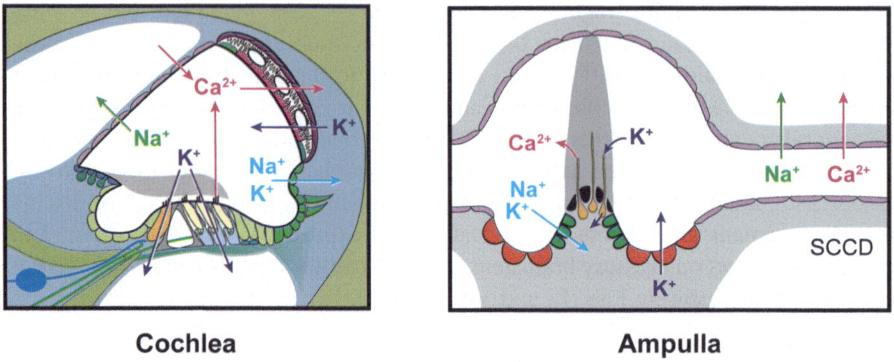

Fig. 8.9 Summary diagram of cochlear and vestibular transepithelial cation transport. K^+ is secreted by stria vascularis and vestibular dark cells (purple arrows; Figs. 8.3 and 8.5), and exits the endolymph via transduction channels in the stereocilia of hair cells (purple arrows) and via parasensory nonselective cation channels (ionotropic purinergic receptors and purine-insensitive channels) in the outer sulcus and transitional cells (blue arrows; Fig. 8.7). Na^+ is thought to enter the endolymph via a leak from perilymph and is absorbed both via Na^+-selective ENaC (green arrows; Fig. 8.6) and cation nonselective channels (blue arrows; Fig. 8.7). Ca^{2+} enters endolymph primarily via an ATP-driven pump (PMCA) at the tips of stereocilia and Reissner's membrane (red arrows; Fig. 8.8) and is absorbed by a TRPV5/6-mediated transport system at several sites in the cochlea (stria vascularis, outer sulcus, and inner sulcus epithelial cells) and by semicircular canal duct (SCCD) epithelial cells (red arrows; Fig. 8.8). The SCCD also actively secretes Cl^- under β2-adrenergic control (not shown) (Milhaud et al. 2002). Redrawn and adapted from Wangemann (1995) with permission

8.6 Transport Summary

The salient transepithelial transport processes described earlier and their locations are summarized in Fig. 8.9. Several preliminary reports and hypotheses point to emerging fields of thought on novel mechanisms of regulation of inner ear fluid volume and their effects on hearing and balance. These include volume regulation via a pathway in the cochlea that would utilize the stretch-sensitive nonselective cation channel TRPV4 (Heard and Jagger 2019) and via the volume sensor protein SPLUNC1 (Gaillard et al. 2010). SPLUNC1 is a soluble protein found to control airway surface liquid fluid volume via inhibition of ENaC activity (Gaillard et al. 2010) and which could conceivably be active in controlling endolymph volume through the well-characterized ENaC-mediated Na^+ absorption in the cochlea and vestibular labyrinth (Kim and Marcus 2011). There is also an exciting preliminary report of regulated contractions by inner ear epithelia, which could conceivably be active in the mechanical longitudinal movement of endolymph (Wangemann 2018).

Acknowledgments I thank Dr. Philine Wangemann for her long-term meaningful collaboration over the years, and her creation and publication of truly outstanding illustrations of inner ear structure and function. I thank Mallory Hoover for her assistance in adapting, redrawing, and developing the figures in this chapter. I appreciate the support provided by the College of Veterinary Medicine at Kansas State University.

References

Beitz E, Kumagami H, Krippeit-Drews P, Ruppersberg JP, Schultz JE (1999) Expression pattern of aquaporin water channels in the inner ear of the rat—The molecular basis for a water regulation system in the endolymphatic sac. Hear Res 132:76–84

Boros S, Bindels RJ, Hoenderop JG (2009) Active Ca^{2+} reabsorption in the connecting tubule. Pflügers Arch 458:99–109

Bosher SK, Warren RL (1968) Observations on the electrochemistry of the cochlear endolymph of the rat: a quantitative study of its electrical potential and ionic composition as determined by means of flame spectrophotometry. Proc R Soc Lond B Biol Sci 171:227–247

Cantos R, Cole LK, Acampora D, Simeone A, Wu DK (2000) Patterning of the mammalian cochlea. Proc Natl Acad Sci USA 97:11707–11713

Casimiro MC, Knollmann BC, Ebert SN, Vary JC Jr, Greene AE, Franz MR, Grinberg A, Huang SP, Pfeifer K (2001) Targeted disruption of the Kcnq1 gene produces a mouse model of Jervell and Lange-Nielsen syndrome. Proc Natl Acad Sci USA 98:2526–2531

Chen J, Zhao HB (2014) The role of an inwardly rectifying K^+ channel (Kir4.1) in the inner ear and hearing loss. Neuroscience 265:137–146

Chen Q, Chu H, Wu X, Cui Y, Chen J, Li J, Zhou L, Xiong H, Wang Y, Li Z (2011) The expression of plasma membrane Ca^{2+}-ATPase isoform 2 and its splice variants at sites A and C in the neonatal rat cochlea. Int J Ped Otorhinolaryngolgy 75:196–201

Chiba T, Marcus DC (2000) Nonselective cation and BK channels in apical membrane of outer sulcus epithelial cells. J Mem Biol 174:167–179

Choi BY, Kim HM, Ito T, Lee KY, Li X, Monahan K, Wen Y, Wilson E, Kurima K, Saunders TL, Petralia RS, Wangemann P, Friedman TB, Griffith AJ (2011) Mouse model of enlarged vestibular aqueducts defines temporal requirement of Slc26a4 expression for hearing acquisition. J Clin Invest 121:4516–4525

Crouch JJ, Sakaguchi N, Lytle C, Schulte BA (1997) Immunohistochemical localization of the Na-K-Cl co-transporter (NKCC1) in the gerbil inner ear. J Histochem Cytochem 45:773–778

Delpire E, Lu J, England R, Dull C, Thorne T (1999) Deafness and imbalance associated with inactivation of the secretory Na-K-2Cl co-transporter. Nat Genet 22:192–195

Ding B, Walton JP, Zhu X, Frisina RD (2018) Age-related changes in Na, K-ATPase expression, subunit isoform selection and assembly in the stria vascularis lateral wall of mouse cochlea. Hear Res 367:59–73

Eatock RA, Songer JE (2011) Vestibular hair cells and afferents: two channels for head motion signals. Annu Rev Neurosci 34:501–534

Eckhard A, Dos SA, Liu W, Bassiouni M, Arnold H, Gleiser C, Hirt B, Harteneck C, Muller M, Rask-Andersen H, Löwenheim H (2015) Regulation of the perilymphatic-endolymphatic water shunt in the cochlea by membrane translocation of aquaporin-5. Pflügers Arch 467:2571–2588

Erokhova L, Horner A, Ollinger N, Siligan C, Pohl P (2016) The sodium glucose cotransporter SGLT1 is an extremely efficient facilitator of passive water transport. J Biol Chem 291:9712–9720

Faridi R, Tona R, Brofferio A, Hoa M, Olszewski R, Schrauwen I, Assir MZK, Bandesha AA, Khan AA, Rehman AU, Brewer C, Ahmed W, Leal SM, Riazuddin S, Boyden SE, Friedman TB (2019) Mutational and phenotypic spectra of KCNE1 deficiency in Jervell and Lange-Nielsen syndrome and Romano-Ward syndrome. Hum Mutat 40:162–176

Fettiplace R, Kim KX (2014) The physiology of mechanoelectrical transduction channels in hearing. Phys Rev 94:951–986

Flock A, Goldstein MH Jr (1978) Cupular movement and nerve impulse response in the isolated semicircular canal. Brain Res 157:11–19

Foster CA, Breeze RE (2013) Endolymphatic hydrops in Meniere's disease: cause, consequence, or epiphenomenon? Otol Neurotol 34:1210–1214

Gaillard EA, Kota P, Gentzsch M, Dokholyan NV, Stutts MJ, Tarran R (2010) Regulation of the epithelial Na^+ channel and airway surface liquid volume by serine proteases. Pflügers Arch 460:1–17

Gow A, Davies C, Southwood CM, Frolenkov G, Chrustowski M, Ng L, Yamauchi D, Marcus DC, Kachar B (2004) Deafness in Claudin 11-null mice reveals the critical contribution of basal cell tight junctions to stria vascularis function. J Neurosci 24:7051–7062

Greger R, Oberleithner H, Schlatter E, Cassola AC, Weidtke C (1983) Chloride activity in cells of isolated perfused cortical thick ascending limbs of rabbit kidney. Pflügers Arch 399:29–34

Heard G, Jagger D (2019) TRPV4 receptors as multimodal sensors of inner ear fluids. ARO Abstracts 42:90

Hofman R, Segenhout JM, Buytaert JA, Dirckx JJ, Wit HP (2008) Morphology and function of Bast's valve: additional insight in its functioning using 3D-reconstruction. Eur Arch Otorhinolaryngol 265:153–157

Honda K, Kim SH, Kelly MC, Burns JC, Constance L, Li X, Zhou F, Hoa M, Kelley MW, Wangemann P, Morell RJ, Griffith AJ (2017) Molecular architecture underlying fluid absorption by the developing inner ear. Elif 6. https://doi.org/10.7554/eLife.26851

Hunter-Duvar IM (1983) An electron microscopic study of the vestibular sensory epithelium. Acta Otolaryngol 95:494–507

Ikeda K, Kusakari J, Kobayashi T, Saito Y (1987) The effect of vitamin D deficiency on the cochlear potentials and the perilymphatic ionized calcium concentration of rats. Acta Otolaryngol Suppl 435:64–72

Ikeda K, Kobayashi T, Itoh Z, Kusakari J, Takasaka T (1989) Evaluation of vitamin D metabolism in patients with bilateral sensorineural hearing loss. Am J Otol 10:11–13

Ito T, Li X, Kurima K, Choi BY, Wangemann P, Griffith AJ (2014) Slc26a4-insufficiency causes fluctuating hearing loss and stria vascularis dysfunction. Neurobiol Dis 66:53–65

Johnstone BM, Patuzzi R, Syka J, Sykova E (1989) Stimulus-related potassium changes in the organ of Corti of guinea-pig. J Physiol 408:77–92

Kambayashi J, Kobayashi T, DeMott JE, Marcus NY, Thalmann I, Thalmann R (1982) Effect of substrate-free vascular perfusion upon cochlear potentials and glycogen of the stria vascularis. Hear Res 6:223–240

Kikuchi T, Kimura RS, Paul DL, Takasaka T, Adams JC (2000) Gap junction systems in the mammalian cochlea. Brain Res Brain Res Rev 32:163–166

Kim SH, Marcus DC (2011) Regulation of sodium transport in the inner ear. Hear Res 280:21–29

Kimura RS (1969) Distribution, structure, and function of dark cells in the vestibular labyrinth. Ann Otol Rhinol Laryngol 78:542–561

Kobayashi T, Rokugo M, Marcus DC, Comegys TH, Thalmann R (1984) Prolonged maintenance of endocochlear potential by vascular perfusion with media devoid of oxygen carriers. Arch Otorhinolaryngol 239:243–247

Konishi T, Hamrick PE, Walsh PJ (1978) Ion transport in guinea pig cochlea. I. Potassium and sodium transport. Acta Otolaryngol 86:22–34

Kozel PJ, Friedman RA, Erway LC, Yamoah EN, Liu LH, Riddle T, Duffy JJ, Doetschman T, Miller ML, Cardell EL, Shull GE (1998) Balance and hearing deficits in mice with a null mutation in the gene encoding plasma membrane Ca^{2+}-ATPase isoform 2. J Biol Chem 273:18693–18696

Kronester-Frei A (1979) The effect of changes in endolymphatic ion concentrations on the tectorial membrane. Hear Res 1:81–94

Lee JH, Marcus DC (2003) Endolymphatic sodium homeostasis by Reissner's membrane. Neurosci 119:3–8

Lee JH, Marcus DC (2008) Purinergic signaling in the inner ear. Hear Res 235:1–7

Lee JH, Chiba T, Marcus DC (2001) P2X2 receptor mediates stimulation of parasensory cation absorption by cochlear outer sulcus cells and vestibular transitional cells. J Neurosci 21:9168–9174

Li J, Verkman AS (2001) Impaired hearing in mice lacking aquaporin-4 water channels. J Biol Chem 276:31233–31237

Li X, Sanneman JD, Harbidge DG, Zhou F, Ito T, Nelson R, Picard N, Chambrey R, Eladari D, Miesner T, Griffith AJ, Marcus DC, Wangemann P (2013) SLC26A4 targeted to the endolymphatic sac rescues hearing and balance in Slc26a4 mutant mice. PLoS Genetics 9:e1003641

Liu J, Kozakura K, Marcus DC (1995) Evidence for purinergic receptors in vestibular dark cell and strial marginal cell epithelia of the gerbil. Audit Neurosci 1:331–340

Lundberg YW, Xu Y, Thiessen KD, Kramer KL (2015) Mechanisms of otoconia and otolith development. Dev Dyn 244:239–253

Macnamara EF, Koehler AE, D'Souza P, Estwick T, Lee P, Vezina G, Fauni H, Braddock SR, Torti E, Holt JM, Sharma P, Malicdan MCV, Tifft CJ (2019) Kilquist syndrome: a novel syndromic hearing loss disorder caused by homozygous deletion of SLC12A2. Hum Mutat 40:532–538

Marcus DC, Chiba T (1999) K^+ and Na^+ absorption by outer sulcus epithelial cells. Hear Res 134:48–56

Marcus DC, Shen Z (1994) Slowly activating, voltage-dependent K^+ conductance is apical pathway for K^+ secretion in vestibular dark cells. Am J Physiol Cell Physiol 267:C857–C864

Marcus DC, Wangemann P (2009) Cochlear and vestibular function and dysfunction. In: Alvarez-Leefmans FJ, Delpire E (eds) Physiology and pathology of chloride transporters and channels in the nervous system—from molecules to diseases. Elsevier, New York, pp 425–437

Marcus DC, Wangemann P (2010) Inner ear fluid homeostasis. In: Fuchs PA (ed) The Oxford handbook of auditory science: the ear. Oxford University Press, Oxford, pp 213–230

Marcus DC, Marcus NY, Thalmann R (1981) Changes in cation contents of stria vascularis with ouabain and potassium-free perfusion. Hear Res 4:149–160

Marcus DC, Takeuchi S, Wangemann P (1993) Two types of chloride channel in the basolateral membrane of vestibular dark cells. Hear Res 69:124–132

Marcus DC, Liu J, Wangemann P (1994) Transepithelial voltage and resistance of vestibular dark cell epithelium from the gerbil ampulla. Hear Res 73:101–108

Milhaud PG, Pondugula SR, Lee JH, Herzog M, Lehouelleur J, Wangemann P, Sans A, Marcus DC (2002) Chloride secretion by semicircular canal duct epithelium is stimulated via b2-adrenergic receptors. Am J Physiol Cell Physiol 283:C1752–C1760

Morton-Jones RT, Vlajkovic SM, Thorne PR, Cockayne DA, Ryan AF, Housley GD (2015) Properties of ATP-gated ion channels assembled from P2X2 subunits in mouse cochlear Reissner's membrane epithelial cells. Purinergic Signal 11:551–560

Nakaya K, Harbidge DG, Wangemann P, Schultz BD, Green E, Wall SM, Marcus DC (2007) Lack of pendrin HCO_3^- transport elevates vestibular endolymphatic $[Ca^{2+}]$ by inhibition of acid-sensitive TRPV5 and TRPV6 channels. Am J Physiol Renal Physiol 292:F1314–F1321

Nijenhuis T, Hoenderop JG, Bindels RJ (2005) TRPV5 and TRPV6 in Ca^{2+} (re)absorption: regulating Ca^{2+} entry at the gate. Pflugers Arch 451:181–192

Nishio A, Ito T, Cheng H, Fitzgerald TS, Wangemann P, Griffith AJ (2016) Slc26a4 expression prevents fluctuation of hearing in a mouse model of large vestibular aqueduct syndrome. Neurosci 329:74–82

Pondugula SR, Kampalli SB, Wu T, De Lisle RC, Raveendran NN, Harbidge DG, Marcus DC (2013) cAMP-stimulated Cl- secretion is increased by glucocorticoids and inhibited by bumetanide in semicircular canal duct epithelium. BMC Physiol 136. https://doi.org/10.1186/1472-6793-13-6

Rickheit G, Maier H, Strenzke N, Andreescu CE, De Zeeuw CI, Muenscher A, Zdebik AA, Jentsch TJ (2008) Endocochlear potential depends on Cl^- channels: mechanism underlying deafness in Bartter syndrome IV. EMBO J 27:2907–2917

Rosenthal R, Gunzel D, Krug SM, Schulzke JD, Fromm M, Yu AS (2017a) Claudin-2-mediated cation and water transport share a common pore. Acta Physiol 219:521–536

Rosenthal R, Gunzel D, Theune D, Czichos C, Schulzke JD, Fromm M (2017b) Water channels and barriers formed by claudins. Ann N Y Acad Sci 1397:100–109

Sage CL, Marcus DC (2001) Immunolocalization of ClC-K chloride channel in strial marginal cells and vestibular dark cells. Hear Res 160:1–9

Salt AN, Ohyama K (1993) Accumulation of potassium in scala vestibuli perilymph of the mammalian cochlea. Ann Otol Rhinol Laryngol 102:64–70

Salt AN, Melichar I, Thalmann R (1987) Mechanisms of endocochlear potential generation by stria vascularis. Laryngoscope 97:984–991

Schnee ME, Ricci A (2017) Hair cells and their synapses. In Manley GA et al (eds) Understanding the cochlea. Springer, Cham, pp 183–213

Schulte BA, Steel KP (1994) Expression of alpha and beta subunit isoforms of Na,K-ATPase in the mouse inner ear and changes with mutations at the Wv or Sld loci. Hear Res 78:65–76

Sellick PM, Johnstone BM (1972) The electrophysiology of the saccule. Pflügers Arch 336:28–34

Shah DM, Freeman DM, Weiss TF (1995) The osmotic response of the isolated, unfixed mouse tectorial membrane to isosmotic solutions: effect of Na^+, K^+, and Ca^{2+} concentration. Hear Res 87:187–207

Shen Z, Marcus DC (1998) Divalent cations inhibit IsK/KvLQT1 channels in excised membrane patches of strial marginal cells. Hear Res 123:157–167

Shen Z, Marcus DC, Sunose H, Chiba T, Wangemann P (1997) I_{sK} channel in strial marginal cells: voltage-dependence, ion-selectivity, inhibition by 293B and sensitivity to clofilium. Aud Neurosci 3:215–230

Simon DB, Bindra RS, Mansfield TA, Nelson-Williams C, Mendonca E, Stone R, Schurman S, Nayir A, Alpay H, Bakkaloglu A, Rodriguez-Soriano J, Morales JM, Sanjad SA, Taylor CM, Pilz D, Brem A, Trachtman H, Griswold W, Richard GA, John E, Lifton RP (1997) Mutations in the chloride channel gene, CLCNKB, cause Bartter's syndrome type III. Nat Genet 17:171–178

Singh R, Wangemann P (2008) Free radical stress mediated loss of Kcnj10 protein expression in stria vascularis contributes to deafness in Pendred syndrome mouse model. Am J Physiol Renal Physiol 294:F139–F148

Spicer SS, Schulte BA (1996) The fine structure of spiral ligament cells relates to ion return to the stria and varies with place-frequency. Hear Res 100:80–100

Sterkers O, Saumon G, Tran Ba HP, Amiel C (1982) K, Cl, and H2O entry in endolymph, perilymph, and cerebrospinal fluid of the rat. Am J Physiol 243:F173–F180

Sunose H, Ikeda K, Suzuki M, Takasaka T (1994) Voltage-activated K channel in luminal membrane of marginal cells of stria vascularis dissected from guinea pig. Hear Res 80:86–92

Sunose H, Liu J, Shen Z, Marcus DC (1997a) cAMP increases apical I_{sK} channel current and K^+ secretion in vestibular dark cells. J Mem Biol 156:25–35

Sunose H, Liu J, Shen Z, Marcus DC (1997b) cAMP increases K^+ secretion via activation of apical I_{sK}/KvLQT1 channels in strial marginal cells. Hear Res 114:107–116

Takeuchi S, Ando M, Kakigi A (2000) Mechanism generating endocochlear potential: role played by intermediate cells in stria vascularis. Biophys J 79:2572–2582

Tranebjaerg L, Bathen J, Tyson J, Bitner-Glindzicz M (1999) Jervell and Lange-Nielsen syndrome: a Norwegian perspective. Am J Med Gen 89:137–146

Verkman AS, Anderson MO, Papadopoulos MC (2014) Aquaporins: important but elusive drug targets. Nat Rev Drug Discov 13:259–277

Wada J, Kambayashi J, Marcus DC, Thalmann R (1979) Vascular perfusion of the cochlea: effect of potassium-free and rubidium-substituted media. Arch Otorhinolaryngol 225:79–81

Wangemann P (1995) Comparison of ion transport mechanisms between vestibular dark cells and strial marginal cells. Hear Res 90:149–157

Wangemann P (2013) Mouse models for pendrin-associated loss of cochlear and vestibular function. Cell Physiol Biochem 32:157–165

Wangemann P (2018) Videomicroscopy of the developing inner ear: contractions of the endolymphatic sac expand scala media of the cochlea. In: Keynote presentation. 55th inner ear biology workshop, Berlin

Wangemann P, Marcus DC (1992) The membrane potential of vestibular dark cells is controlled by a large Cl^- conductance. Hear Res 62:149–156

Wangemann P, Marcus DC (2017) Ion and fluid homeostasis in the cochlea. In: Manley GA et al (eds) Understanding the cochlea. Springer, Cham, pp 253–286

Wangemann P, Liu J, Shen Z, Shipley A, Marcus DC (1995) Hypo-osmotic challenge stimulates transepithelial K^+ secretion and activates apical IsK channel in vestibular dark cells. J Mem Biol 147:263–273

Wangemann P, Liu J, Shiga N (1996a) Vestibular dark cells contain the Na^+/H^+ exchanger NHE-1 in the basolateral membrane. Hear Res 94:94–106

Wangemann P, Shen Z, Liu J (1996b) K^+-induced stimulation of K^+ secretion involves activation of the IsK channel in vestibular dark cells. Hear Res 100:201–210

Wangemann P, Liu J, Shimozono M, Scofield MA (1999) $beta_1$-adrenergic receptors but not $beta_2$-adrenergic or vasopressin receptors regulate K^+ secretion in vestibular dark cells of the inner ear. J Mem Biol 170:67–77

Wangemann P, Liu J, Shimozono M, Schimanski S, Scofield MA (2000) K^+ secretion in strial marginal cells is stimulated via beta 1-adrenergic receptors but not via beta 2-adrenergic or vasopressin receptors. J Mem Biol 175:191–202

Wangemann P, Liu J, Scherer EQ, Herzog M, Shimozono M, Scofield MA (2001) Muscarinic receptors control K^+ secretion in inner ear strial marginal cells. J Mem Biol 182:171–181

Wangemann P, Itza EM, Albrecht B, Wu T, Jabba SV, Maganti RJ, Lee JH, Everett LA, Wall SM, Royaux IE, Green ED, Marcus DC (2004) Loss of KCNJ10 protein expression abolishes endocochlear potential and causes deafness in Pendred syndrome mouse model. BMC Med 2:30. https://doi.org/10.1186/1741-7015-2-30

Wangemann P, Nakaya K, Wu T, Maganti RJ, Itza EM, Sanneman JD, Harbidge DG, Billings S, Marcus DC (2007) Loss of cochlear HCO_{3-} secretion causes deafness via endolymphatic acidification and inhibition of Ca^{2+} reabsorption in a Pendred syndrome mouse model. Am J Physiol Renal Physiol 292:F1345–F1353

Warth R, Barhanin J (2002) The multifaceted phenotype of the knockout mouse for the KCNE1 potassium channel gene. Am J Physiol Regul Integr Comp Physiol 282:R639–R648

Wood JD, Muchinsky SJ, Filoteo AG, Penniston JT, Tempel BL (2004) Low endolymph calcium concentrations in deafwaddler2J mice suggest that PMCA2 contributes to endolymph calcium maintenance. J Assoc Res Otolaryngol 5:99–110

Yamauchi D, Nakaya K, Raveendran NN, Harbidge DG, Singh R, Wangemann P, Marcus DC (2010) Expression of epithelial calcium transport system in rat cochlea and vestibular labyrinth. BMC Physiol 101. https://doi.org/10.1186/1472-6793-10-1

Yoo JC, Kim HY, Han KH, Oh SH, Chang SO, Marcus DC, Lee JH (2012) Na^+ absorption by Claudius' cells is regulated by purinergic signaling in the cochlea. Acta Otolaryngol 132(Suppl 1):S103–S108

Chapter 9
Regulation of Ion Transport Through Retinal Pigment Epithelium: Impact in Retinal Degeneration

Nadine Reichhart and Olaf Strauß

Abstract The retinal pigment epithelium (RPE) is a monolayer of pigmented cells that faces with its apical membrane the light-sensitive outer segments of photoreceptors and with its basolateral membrane the retinal blood supply by the choriocapillaris. The RPE is a close interaction partner of the photoreceptors and fulfills several tasks that are essential for photoreceptor function and, thus, for vision. As a tight epithelium, the RPE represents the outer blood/retina barrier and is essential to maintain the immune privilege of the eye. As a transporting epithelium, the RPE removes water from the retina, a transport driven by a transepithelial transport of Cl^- from the retinal to the blood side. The transport activity increases with increasing light intensity and generates transmembrane and transepithelial potentials that can be recorded by electroretinogram and electro-oculogram. The transepithelial transport has further roles in retinal pathology. Macula edema can be treated by increasing the transport rates by carbonic anhydrase inhibitors. A disturbance of epithelial transport might represent a major pathomechanism in an inherited macular dystrophy, Best's disease.

Keywords Retina · Pigment epithelium · Cl^- transport · Water transport · Blood/retina barrier · Light-dependent transport · K^+ homeostasis

9.1 Introduction

The retinal pigment epithelium (RPE) is unique among epithelia. The RPE is part of a sandwich-like structure composed of photoreceptor layer, RPE and choroid in the outer retina (Strauss 2005; Steinberg 1985; Bok 1993; Sparrow et al. 2010). It is not

N. Reichhart · O. Strauß (✉)
Experimental Ophthalmology, Department of Ophthalmology, Charité - Universitätsmedizin Berlin, a corporate member of Freie Universität, Humboldt-University, the Berlin Institute of Health, Berlin, Germany
e-mail: Nadine.reichhart@charite.de; olaf.strauss@charite.de

K. L. Hamilton, D. C. Devor (eds.), *Ion Transport Across Epithelial Tissues and Disease*, Physiology in Health and Disease,
https://doi.org/10.1007/978-3-030-55310-4_9

only a transporting epithelium, but fulfills at the same time many additional tasks, including phagocytosis, secretion, glia, immune barrier, and metabolic maintenance of photoreceptors. In this way, the RPE is a close interaction partner of the photoreceptors and participates in the visual process. The fact that failure or dysregulation of one of the RPE's functions ultimately leads to photoreceptor degeneration and blindness highlights the importance of the RPE (Sparrow et al. 2010; Strauss 2005). The following chapter describes the basic mechanisms of epithelial transport by the RPE, its illumination-dependent adaptation, and the pathologic changes in retinal degeneration.

9.2 Basic Properties of the RPE

The evolution of light-sensitive organs generated structures that are composed of a light-sensitive compound (organelle or cell) and a pigmented compound (organelle or cell) (Martinez-Morales et al. 2004). Thus, the interaction between a light-sensitive cell and a pigmented cell represents the basis of vision. The RPE fulfills the part of the pigmented cell in the vertebrate eye. First, the RPE absorbs the light energy that is focused onto the retina/photoreceptor layer by the lens (Bok 1993; Strauss 2005; Beatty et al. 1999; Boulton 1998). The RPE helps to maintain the structural integrity of the photoreceptors by secretion of neurotrophic factors such as PEDF (pigment epithelium–derived factor) (Polato and Becerra 2016; Wimmers et al. 2007) or by phagocytosing shed photoreceptor outer segments as part of the photoreceptors' daily renewal process (Bok and Hall 1971; LaVail 1976, 1980; Mazzoni et al. 2014). Furthermore, the RPE becomes part of the vision process by reisomerizing all-trans retinal back into 11-cis retinal, the chromophore of rhodopsin that ignites the vision process (Xue et al. 2004; Daruwalla et al. 2018). Since the reisomerization needs to be adapted to different light conditions between dark-adapted and light-adapted vision, the RPE plays an essential role in the visual process (Lamb and Pugh 2004). The RPE forms a part of the blood/retina barrier and helps to maintain the immune privilege of the inner eye by secretion of immune regulating factors in response to local changes in the immune activity (Ishida et al. 2003).

During embryonic development, the RPE forms from the neuroectoderm and differentiates into its final functional phenotype in concert with the developing neuronal retina (Bell 1906; Braekevelt and Hollenberg 1970; Gonzalez-Fernandez and Healy 1990; Jeffery 1998). In fact, proper differentiation requires the exchange of signaling molecules such as retinal between both developing tissues. During this process, the RPE changes the protein composition of its tight junctions finally resulting in a tight epithelium (Ban and Rizzolo 2000; Rizzolo 1997). Indeed, Miller and Steinberg (1977a, b) reported that the transcellular resistance of the RPE is ten times higher than that of the paracellular resistance, which functionally classifies the RPE as a tight epithelium. The RPE transepithelial resistance was reported from 350 to more 400 Ω cm^2 (Tsuboi and Pederson 1988; Nabi et al. 1993; Hu et al. 1996;

Hernandez et al. 1995; Frambach et al. 1988, 1989); in some reports more than 1000 Ω cm^2 (Ablonczy et al. 2011). Usually, cultured cells show higher transepithelial resistance. The reason is that native RPE might have deeper infoldings of the basolateral membrane than cultured cells. Thus, the true surface area in Ussing chamber preparations might be larger in preparations of native RPE than in cultured cells. With its tight junctions the RPE also represents a physical barrier for immune cells to maintain the immune privilege of the eye (Streilein 2003). Furthermore, it allows a directed secretion of different factors either to the apical/retinal or the basolateral/blood side. Finally, a tight epithelium ensures a very selective supply of nutrients for photoreceptor metabolism. The RPE faces with its apical side the photoreceptors. The RPE engulfs the light-sensitive outer segments by long apical microvilli. A specialized interphotoreceptor matrix ensures close exchange of molecules between the RPE and photoreceptors (Hollyfield 1999; Petit 2001; Carter-Dawson and Burroughs 1992; Gonzalez-Fernandez et al. 1993). At its basolateral side, the RPE faces the blood stream through choriocapillaris. Bruch's membrane, a multilayered basal membrane, and the endothelium of the choriocapillaris separate the RPE from the blood stream (Das et al. 1990; Guymer et al. 1999; Sumita 1961). Since the choroidal endothelium is a fenestrated endothelium, Bruch's membrane forms the exchange matrix for the RPE to the blood stream.

9.3 The Epithelial Transport of Water and Cl^- Across the RPE

The spatial proximity between the RPE and the photoreceptors is essential for the photoreceptor function. That results from a constant removal of water from the subretinal space to the blood side (Marmor 1990). This active transport establishes an adhesion force between the photoreceptor layer and the RPE that is lost by inhibition of the Na^+/K^+-ATPase (Kita and Marmor 1992; Marmor 1993). The transportation rate for water by the RPE is quite high with 1.4–11 μl/cm^2/h depending on the species (Hughes et al. 1998). The high transportation rate is required because water constantly accumulates at the outer retina and cannot passively leave the eye through the tight junctions of the RPE. The increase of extracellular volume increase is due to production of metabolic water from the highly active neurons and water from the anterior parts of the eye driven toward the retina by the intraocular pressure (Marmor 1999; Hamann 2002). Furthermore, the extracellular volume increases with the transition from dark to light. Thus, water transport needs to be increased beyond the baseline activity (Hughes et al. 1998; Li et al. 1994a, b).

Intracellular accumulation of Cl^- in the RPE is the base of the transportation pathway through the RPE (Hughes et al. 1998; Bialek and Miller 1994; Joseph and Miller 1991; Miller and Edelman 1990; Miller and Steinberg 1977b; Quinn and

Miller 1992; la Cour 1992). In fact, the RPE shows a quite high intracellular Cl^- concentration between 40 and 60 mM, depending on the species (Yamashita and Yamamoto 1991; Wiederholt and Zadunaisky 1984). The intracellular Cl^- accumulation occurs via active transport mechanisms across the RPE apical membrane. An apical bumetanide-sensitive $Na^+/2Cl^-/K^+$-cotransporter takes up Na^+, K^+ , and Cl^- from the subretinal space through the apical membrane into the RPE (Bialek and Miller 1994; Joseph and Miller 1991; Miller and Edelman 1990; Miller and Steinberg 1977b; Quinn and Miller 1992; Bialek et al. 1995; Kennedy 1990). The transport of the $Na^+/2Cl^-/K^+$-cotransporter is fueled by the primary active transport of the apically localized Na^+/K^+-ATPase that removes Na^+ from the RPE intracellular space and pumps it into the subretinal space. The apical localization of the Na^+/K^+-ATPase is another unique feature of the RPE (Marmorstein 2001; Defoe et al. 1994; Hu et al. 1994; Korte and Wanderman 1993; McGrail and Sweadner 1986; Rizzolo 1999). K^+ that is taken up by the $Na^+/2Cl^-/K^+$-cotransporter recycles back to the subretinal space across the apical membrane through Kir7.1 inward rectifier K^+ channels (Kusaka et al. 2001; Shimura et al. 2001; Yang et al. 2003; Yuan et al. 2003). The same pathway additionally supports the pump activity of the Na^+/K^+-ATPase because the recycling of K^+ from RPE intracellular to subretinal space keeps the K^+ gradient across the apical membrane small. The Na^+/K^+-ATPase itself transports Na^+ that enters the RPE via the $Na^+/2Cl^-/K^+$-cotransporter back to the subretinal space. In summary, the combined transport activities of $Na^+/2Cl^-/K^+$-cotransporter, Na^+/K^+-ATPase, and Kir7.1 K^+ channel leads to an accumulation of Cl^- inside the RPE cells because Na^+ and K^+ are transported back into the subretinal space. In contrast to the apical membrane, the basolateral membrane of the RPE shows a high Cl^- conductance (Fujii et al. 1992; Gallemore et al. 1993; Joseph and Miller 1991; Miller and Edelman 1990). Thus, following its electrochemical gradient, Cl^- leaves the RPE cell across the basolateral membrane through a variety of Cl^- channels (Wimmers et al. 2007). Analysis of the molecular identity of these channels revealed the presence of ClC2 (Bosl et al. 2001), CFTR (Blaug et al. 2003), bestrophin-1 (Xiao et al. 2010; Johnson et al. 2017), and anoctamin-2 (Keckeis et al. 2017b). Recently, we demonstrated also the expression of anoctamin-4 in the RPE (Reichhart et al. 2019a). The combined apical and basolateral transport mechanisms lead to transport of Cl^- across the RPE, which results in basolateral negative transepithelial potential of about –5 to –10 mV, depending on the species (la Cour 1992; Hughes et al. 1998). A transepithelial transport of K^+ accompanies the transport of Cl^- (Hughes et al. 1998; Arrindell et al. 1992; la Cour 1985; la Cour et al. 1986). That K^+ transport is not strong enough to counterbalance the Cl^- transport, as the transepithelial potential at the basolateral side is negative. Basically, the K^+ transport depends on the K^+ uptake across the apical membrane through the activity of the Na^+/K^+-ATPase and the $Na^+/2Cl^-/K^+$-cotransporter (Hughes and Takahira 1996; Hughes et al. 1995b; Segawa and Hughes 1994; Shimura et al. 2001). The recently discovered K^+ channels KCNQ4 and KCNQ5 provide an efflux pathway for K^+ out of RPE cells across the basolateral membrane (Pattnaik and Hughes 2012; Zhang and Hughes 2013; Zhang et al. 2011). The overall electrolyte transport across the RPE osmotically pulls water from the retinal to the blood side of

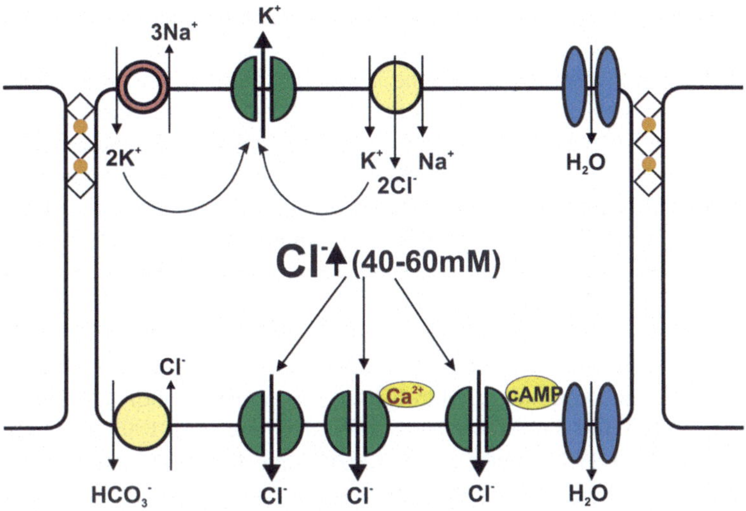

Fig. 9.1 Summary of transepithelial water and Cl^- -transport: the apical Na^+/K^+-ATPase and inward rectifier K^+ channel Kir7.1 cause a driving force for Na^+ into the cell. The $Na^+/K^+/2Cl^-$ cotransporter activity leads to an intracellular Cl^- concentration of 40–60 mM. Cl^- leaves the cell across the basolateral membrane via different types of Cl^- channels, water exits by osmotic forces via aquaporins. Apical membrane: Na^+/K^+ ATPase, Kir7.1, inward-rectifier K^+ channel 7.1, the $Na^+/K^+/2Cl^-$ cotransporter, aquaporine (water transport) cotransporter. Basolateral membrane: Cl^-/HCO_3^- exchanger, ClC-2, voltage-dependent Cl^- channels of the ClC family, Bestrophin-1, Ca^{2+}-dependent Cl^- channel, CFTR, cystic fibrosis transmembrane conductance regulator, cAMP-dependent, aquaporine (water transport)

the RPE (Hamann 2002; Hamann et al. 1998; Stamer et al. 2003). Aquaporins of the subtype AQP1 provide a transportation pathway for water across the RPE via the transcellular route (Fig. 9.1).

9.4 Coupling of Transepithelial Transport with pH Regulation

The photoreceptor metabolism leads to the production of lactic acid that reaches concentrations of about 25 mM in the subretinal space (Adler and Southwick 1992; Kenyon et al. 1994; Padnick-Silver and Linsenmeier 2002). The RPE removes lactic acid from the subretinal space by transepithelial transport of Lac^- and the corresponding control of extracellular and intracellular pH. Lac^- is taken up from subretinal space into the RPE by the transport activity of the apically localized MCT1 (monocaboxylate-transporter-1) and leaves the RPE through the basolateral localized MCT3 (Hamann et al. 2000; Kenyon et al. 1994; la Cour et al. 1994; Yoon et al. 1997; Zeuthen et al. 1996; Philp et al. 1998). Since Lac^- itself generates protons and the MCT-transporter work as carboxylate/H^+-cotransporter, the Lac^- handling requires efficient extra- and intracellular pH buffering. Simultaneous with the uptake of Lac^- into the RPE's intracellular space, protons invade the cell as well. These protons will be transported back to the subretinal space by the secondary active transportation activity of the Na^+/H^+-exchanger (la Cour et al. 1994; Keller et al. 1986, 1988). Furthermore, the secondary active Na^+/HCO_3^--cotransporter increases the intracellular buffer capacity of the RPE cell. Both transporters efficiently maintain the intracellular pH (la Cour 1991a, b). In addition, the Na^+/H^+-cotransporter maintains a H^+ gradient across the apical membrane that promotes the activity of MCT1 to remove Lac^- from the subretinal space. Intracellular protons and Lac^- leave the cell across the basolateral membrane by the activity of MCT3 that also transports Lac^- in cotransport with H^+. At the basolateral membrane, a HCO_3^-/Cl^--exchanger that reduces intracellular HCO_3^- driven by a Cl^- influx into the RPE cell across the basolateral membrane, controls intracellular pH (Tsuboi et al. 1986; Miller and Edelman 1990; Lin and Miller 1991, 1994; Kenyon et al. 1997; DiMattio et al. 1983; Marmor 1990, 1999). Since a decrease of intracellular pH inhibits the HCO_3^-/Cl^--exchanger (Lin and Miller 1994), the resulting increase in intracellular HCO_3^- would increase intracellular buffer capacity against acidification. This way of pH regulation affects also the transepithelial transport of Cl^- and water across the RPE. At low Lac^- transportation rates the HCO_3^-/Cl^--exchanger recycles back Cl^- that has left the RPE cell through the variety of basolateral Cl^- channels (Wimmers et al. 2007). This decreases the efficiency of transepithelial Cl^- transport because the Cl^- net flux across the basolateral membrane is reduced. In case of an increased Lac^- transport, the MCT3 transporter activity would result in an extracellular acidification that in turn activates the ClC2 Cl channel in the basolateral membrane (Bosl et al. 2001). Thus, the transport of Lac^- across the RPE enhances in addition the transepithelial transport of water and Cl^- (Fig. 9.2).

9.5 Light Dependence of the Transepithelial Transport

The transepithelial transport activity is modulated by changes in the illumination of the retina and is crucial for the maintenance of subretinal K^+ homeostasis and extracellular volume (Li et al. 1994b; Reichenbach et al. 1992; Linsenmeier and Steinberg 1984; Yuan et al. 2003; Shimura et al. 2001; Hughes et al. 1995a, b, 1998; Gallemore and Steinberg 1989a; la Cour et al. 1986; la Cour 1985). The transepithelial transport of ions through the RPE includes transport of K^+. As discussed, the amount of transported K^+ is smaller than that of Cl^- because the overall ion transport from the apical to basolateral side of the RPE results in a basolateral negative transepithelial potential. Furthermore, the K^+ transport has dynamic properties that differ from that of the Cl^- transport. The K^+ transport is coupled to the maintenance of the subretinal K^+ concentration. Changes in the illumination of photoreceptors lead to an increase/decrease in the subretinal K^+ concentration (Steinberg 1985; Oakley 1977; Linsenmeier and Steinberg 1983; la Cour 1985; Hughes et al. 1995a; Griff 1990, 1991; Edelman et al. 1994b;

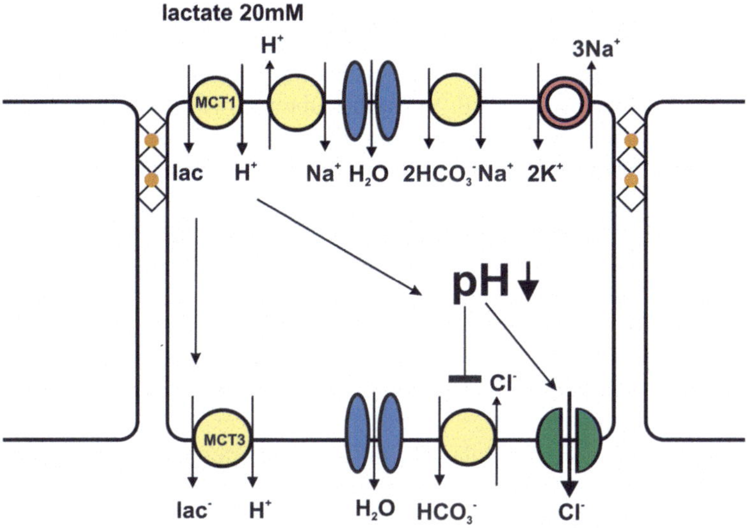

Fig. 9.2 pH regulation and water transport across the RPE. Lactate is transported across the RPE via monocarboxylate transporters, MCT1 and MCT3. pH is regulated by the activity of the apically localized Na^+/HCO_3^- cotransporter, the Na^+/H^+ exchanger, and the basolaterally localized Cl^-/HCO_3^- exchanger. apical membrane: MCT1, monocarboxylate transporter 1, Na^+/H^+ exchanger, aquaporine (water transport) cotransporter, Na^+/HCO_3^- cotransporter, Na^+/K^+ ATPase. LP, light peak. Basolateral membrane: MCT3, monocarboxylate transporter 3, aquaporine (water transport), Cl^-/HCO_3^- exchanger, ClC-2, voltage-dependent Cl^- channels of the ClC family

Dornonville de la Cour 1993). In the transition from dark to light, the dark current of the photoreceptors is reduced going along with a reduced K^+ outflow resulting in a decrease in the subretinal K^+ concentration (Baylor 1996). The reduced K^+ concentration would change the excitability of the photoreceptors. Therefore, apical membrane transport mechanisms of the RPE compensate this (la Cour et al. 1986; la Cour 1985; Oakley 1977). The apical membrane has a high K^+ conductance that hyperpolarizes the apical membrane when extracellular (subretinal) K^+ concentration decreases. The hyperpolarization opens inwardly rectifying K^+ channels of the Kir7.1 subtype that in turn leads to an efflux of K^+ out of the RPE cell into the subretinal space across the apical membrane (Yuan et al. 2003; Yang et al. 2003; Shimura et al. 2001; Kusaka et al. 2001). The c-wave in the electroretinogram corresponds with the apical membrane hyperpolarization (Linsenmeier and Steinberg 1983, 1986, 1987; Griff et al. 1985). The electrophysiological properties of Kir7.1 are ideal for that compensatory mechanism. Kir7.1 is active within a wide range of potentials, shows only mild rectification, and increases its conductance with decreasing extracellular K^+ concentrations. Furthermore, the fact that Kir7.1 is activated by acidification would support the pH-dependent regulation of transepithelial transport activity as described earlier (Yuan et al. 2003). In the transition from dark to light, the transport reverses to decrease subretinal K^+ concentration has increased due to ignition of the dark current. Using Ussing chamber experiments and K^+-sensitive electrodes, la Cour suggested a transepithelial model for the K^+ buffering mechanism (la Cour et al. 1986; la Cour 1985; Dornonville de la Cour 1993). That model is based on a constant transport of K^+ from the subretinal space to the blood side of the RPE. When the subretinal K^+ concentration decreases (changing from dark to light), an increase in the apical K^+ conductance occurs to recycle K^+ back from the RPE intracellular space to the subretinal space across the apical membrane (through Kir7.1 channels). This reduction in the net K^+ uptake reduces the net transepithelial transport of K^+ across the RPE. In contrast, transitioning from light to dark the subretinal K^+ concentration is increased causing depolarization of the apical membrane, closure of Kir7.1 channels, and an increase of the K^+ net flux from the subretinal space into the RPE across the apical membrane by the activity of the Na^+/K^+-ATPase and the $Na^+/2Cl^-/K^+$-cotransporter. The increased K^+ uptake decreases the subretinal K^+ concentration by increasing the net K^+ transport across the RPE.

Experiments using extracellular electrolyte electrodes showed that during the transition from dark to light it comes to a transient increase in the extracellular volume of the retinal tissue (Li et al. 1994a; Huang and Karwoski 1992; Dmitriev et al. 1999; Hughes et al. 1998). Together with the pump activity by Muller cells (Reichenbach et al. 1992), the volume increase is compensated by an adaptation of the transepithelial transport of water across the RPE (Hughes et al. 1998). Two principal mechanisms account for that reaction. One is the formation of extracellular agonists that bind to receptors at the apical membrane of the RPE and modulate the transport activity by second messenger pathways. These mechanisms will be described in more detail in the next section. The other mechanism is based on a direct modulation of the transporters and ion channels in a voltage-dependent manner. In the dark, the net Cl^- transport rate is rather low, and the apical membrane is depolarized. Under this condition, the Na^+/HCO_3^--cotransporter shows a high transport activity, because, due to the stoichiometry of transported ions, the transporter is electrogenic (la Cour 1989, 1991a). In consequence, the intracellular HCO_3^- increases that in turn reduces the net flux of Cl^- across the basolateral membrane by the activity of the basolateral Cl^-/HCO_3^- exchanger (Lin and Miller 1991; Li et al. 1994a; Joseph and Miller 1991). A change from dark to light would decrease the Na^+/HCO_3^--cotransporter due to hyperpolarization and increase transepithelial transport via acidification resulting in increased basolateral Cl^- conductance (Gallemore et al. 1993; Edelman et al. 1994a). The resulting decrease in intracellular HCO_3^- reduces the activity of the Cl^-/HCO_3^- exchanger that further increases the net flux of Cl^- across the basolateral membrane (Hughes et al. 1998; Dawis et al. 1985; DiMattio et al. 1983; Lin and Miller 1991; Maruiwa et al. 1999; Tsuboi 1987; Tsuboi et al. 1986). However, it is to mention that this process might result in an only short-term increase in the transepithelial water transport, whereas the regulation of Cl^- absorption from the subretinal space by second messenger regulation might be more efficient (Fig. 9.3).

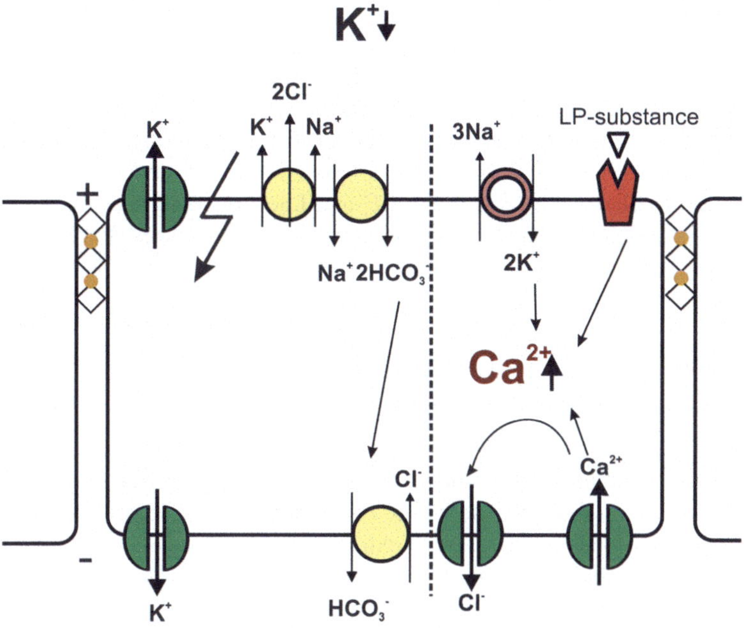

Fig. 9.3 Illumination-dependent changes in ion transport activity. Dark-to-light transition decreases the [K^+] in the space between the photoreceptors, modulating transepithelial K^+ transport (left side). Right side: illumination triggers the release of a light-peak substance that ignites a Ca^{2+}-dependent second messenger cascade that increases the basolateral Cl^- conductance. Apical membrane: Kir7.1, inward-rectifier K^+ channel 7.1, the $Na^+/K^+/2Cl^-$ cotransporter, Na^+/HCO_3^- cotransporter, Na^+/K^+ ATPase. LP, light peak. Basolateral membrane: KCN, potassium channel; Cl^-/HCO_3^- exchanger, Bestrophin-1, Ca^{2+} dependent Cl^- channel, L-Type channel: voltage-dependent Calcium channel $Ca_V1.3$

9.6 Regulation of Ion Transport Across the RPE

Since the voltage-dependent and ionic mechanisms of Cl^- transport modulation certainly have rather short-term effects, other regulatory mechanisms are required to permit a long-term transport regulation. In fact, measurement of light-dependent extracellular volume changes and their compensation by the RPE indicates that corresponding changes in transport activity are within minutes, whereas the light-dependent changes in the membrane potential of photoreceptors and the RPE happen within seconds (Hughes et al. 1998; Steinberg 1985; Steinberg et al. 1983). Given the transportation rates of cotransporters in comparison to that of ion channels, the regulation of the basolateral Cl^- channels provides the most efficient way to regulate transepithelial net flux of Cl^- and water across the RPE (Gallemore et al. 1993; Wimmers et al. 2007). As mentioned earlier, the basolateral membrane displays a high Cl^- conductance that results from the activity of a variety of different Cl^- channels (Gallemore et al. 1993; Wimmers et al. 2007; Bialek et al. 1995; Fujii et al. 1992; Gallemore and Steinberg 1993; Quinn and Miller 1992; Wills et al. 2000).

These ion channels show differential regulation that permits a modulation of the transepithelial Cl^- transport according to the metabolic needs of the photoreceptors RPE. Mice deficient for ClC2 show retinal degeneration along with the absence of a transepithelial potential across the RPE (Bosl et al. 2001). Furthermore, ClC2 channels might contribute to the pH-dependent increase in transepithelial transport as described in the section earlier. It is discussed that the knockout of the ClC2 channel leads to disruption of both the Cl^- and Lac^- transport causing a metabolic and pH misbalance and therefore degeneration of photoreceptors (Bosl et al. 2001).

9.7 Second-Messenger-Dependent Regulation of Transepithelial Transport

The long-term regulation of Cl^- and water transport represents the reaction to tonic changes in the illumination of the retina. As mentioned earlier, an increase in light intensity leads to a transient increase in extracellular fluid that is compensated by an increase in transepithelial transport of water and Cl^- across the RPE. Indeed, the increased Cl^- transport results from an increase in the basolateral Cl^- conductance by activation of Cl^- channels (Bialek et al. 1995; Fujii et al. 1992; Gallemore and Steinberg 1993; Quinn and Miller 1992). This leads to depolarization of the basolateral RPE membrane and to a more negative transepithelial potential across the RPE due to the corresponding efflux of Cl^- out of the RPE across the basolateral membrane (Fujii et al. 1992; Gallemore et al. 1988; Gallemore and Steinberg 1989b). That reaction is the basis of the so-called light-peak of the electro-oculogram (EOG) (Arden and Constable 2006; Gallemore et al. 1988; Gallemore and Steinberg 1989b, 1993). The EOG measures the potential difference between the anterior and posterior pole of the bulbus. The posterior part facing the RPE is negatively charged against the anterior or corneal part. The light peak results from the fact that anterior to posterior potential becomes more negative. Basolateral application of Cl^- channel blockers evidenced that the more negative potential directly results from increased Cl^- conductance of the RPE basolateral membrane (Gallemore et al. 1988; Gallemore and Steinberg 1989b, 1993). The exact mechanism of the light-peak generation is unknown. It is likely that the increased illumination of photoreceptors leads to the release of a molecule, the so-called light-peak substance, that diffuses to the RPE and ignites an intracellular second-messenger pathway that in turn activates basolateral Cl^- channels (Gallemore et al. 1988). The analysis of transepithelial transport of Cl^- through the RPE in Ussing chamber experiments revealed agonists that bind to plasma membrane receptors (Edelman and Miller 1991; Peterson et al. 1997; Quinn et al. 2001; Gallemore and Steinberg 1990) and subsequently activated second messenger pathways that contribute to stimulation of Cl^- transport (Miller and Farber 1984; Gallemore et al. 1994). Apical application of ATP increased basolateral negative potential and net Cl^- transport across the RPE (Peterson et al. 1997; Quinn and Miller 1992). The mechanism most likely occurs via an increased intracellular free Ca^{2+} after release of Ca^{2+} from cytosolic Ca^{2+} stores in response to activation of P2Y (ATP) receptors (Gallemore et al. 1994; Peterson et al. 1997; Quinn and Miller 1992; Neussert et al. 2010; Lu et al. 2007; Mitchell and Reigada 2008). Intracellular Ca^{2+} in turn activates basolateral Ca^{2+}-dependent Cl^- channels (CaCC) to increase the net flux of Cl^- across the basolateral membrane and thus Cl^- transport (Gallemore et al. 1994; Hartzell and Qu 2003; Joseph and Miller 1992; Maminishkis et al. 2002; Rymer et al. 2001).

There are two CaCC known to be expressed in the RPE. One is bestrophin-1, the product of the *BEST1* gene (Hartzell et al. 2008; Johnson et al. 2017). The role of bestrophin-1 in transepithelial transport is not fully clear. Patients with mutations in the *BEST1* gene show decreased light peaks in the EOG (Johnson et al. 2017; Guziewicz et al. 2017; Boon et al. 2009) that indicates a reduced basolateral Cl^- conductance. *Best1* knockout mice, however, show increased light rises in the DC-ERG (Marmorstein et al. 2006), which is an equivalent to EOG in humans. Indeed, some publications indicated an indirect contribution to the regulation of Cl^- transport. These publications discuss the possibility that a proportion of bestrophin-1 forms Cl^- channels in intracellular Ca^{2+} stores and thus that intracellular Ca^{2+} signaling responsible for regulation of transepithelial Cl^- transport (Neussert et al. 2010; Gomez et al. 2013; Barro Soria et al. 2009). A more recent paper indicated that bestrophin-1 ability to function as a volume-activated Cl^- channel might be a more important function for the RPE than its possible contribution to transepithelial Cl^- transport.

Another CaCC that is expressed in the RPE is anoctamin-2 (former TMEM16-B) (Keckeis et al. 2017a). Anoctamin-2 is localized in the RPE in the basolateral membrane and activated in response to ATP-dependent stimulation of P2Y receptors (Keckeis et al. 2017a). As the P2Y receptors are localized at the apical membrane of the RPE (Mitchell and Reigada 2008), it is discussed that ATP might be the aforementioned light-peak substance (Gallemore et al. 1988). It is likely that these Ca^{2+}-dependent Cl^- channels are further involved in adrenergic-stimulation of transepithelial Cl^- transport.

Together with intracellular Ca^+ increases, intracellular cAMP represents the second regulatory pathway for Cl^- transport (Miller and Farber 1984; Miller et al. 1982; Nao-i et al. 1990; Quinn et al. 2001). However, the data produced a contradictory picture of the cAMP effects. Whereas direct stimulation of intracellular cAMP levels seems to reduce transepithelial Cl^- transport (Miller and Farber 1984; Miller et al. 1982; Nao-i et al. 1990; Quinn et al. 2001; Kuntz et al. 1994), the stimulation of the RPE by adrenergic agonists increases Cl^- transport by cAMP (Quinn et al. 2001; Miller and Farber 1984). For the latter conclusion speaks also the observation of cAMP-activated Cl^- channels in the RPE (Hughes and Segawa 1993; Wills et al. 2000). In various species, the RPE expresses CFTR (cystic fibrosis transmembrane conductance regulator) (Blaug et al. 2003; Wills et al. 2000) that can function as a cAMP-activated Cl^- channel (Li et al. 1988). This raises the question whether patients with cystic fibrosis get a retinal degeneration. Patients with CFTR show changes in a component of the EOG, the so-called fast oscillation (Blaug et al. 2003; Constable et al. 2006). This observation points to the fact that

CFTR contributes to transepithelial Cl^- transport by the RPE. So far, however, retinal degeneration has not been described in these patients. It might be that another Cl^- channel compensates for the loss of CFTR activity, such as ClC2. Thus, it is discussed for other tissues that ClC2 might compensate for the loss of CFTR function in these patients. This hypothesis, however, could not be verified by investigation of a CFTR/ClC2 double knockout mouse (Zdebik et al. 2004). Thus, the cAMP-stimulated Cl^- transport follows a complicated and so far not understood pattern.

9.8 Indirect Regulation of Transepithelial Transport by Other Ion Channels

The RPE expresses a number of ion channels that might be indirectly involved in the regulation of transepithelial Cl^- transport across the RPE (Wimmers et al. 2007). This can occur by modulation of the intracellular Ca^{2+} signals that increase Cl^- transport. An ion channel that provides such a modulation is the voltage-dependent L-type Ca^{2+} channel. The L-type channel subtype $Ca_V1.3$ is expressed in the RPE (Mergler et al. 1998; Muller et al. 2014; Rosenthal et al. 2007; Rosenthal and Strauss 2002; Strauss et al. 1997, 2000; Strauss and Wienrich 1994; Ueda and Steinberg 1993; Wimmers et al. 2008a, b). It is localized to the basolateral membrane of the RPE. Its role in the regulation of Cl^- transport was found by analyzing the light peak of the EOG. $Ca_V1.3$ knockout mice show a reduced light peak in the DC-ERG (Wu et al. 2007). Systemic application of dihydropyridine (L-type channel blocker) also reduced the light-peak in the EOG of human subjects (Constable 2011). The effect of L-type channel activity on Cl^- transport occurs directly at the basolateral membrane in interaction with CaCCs. Activation of CaCCs by Ca^{2+} leads to depolarization of the basolateral membrane that in turn activates L-type channels. The subsequent Ca^{2+} influx into the RPE cell further increases Cl^- channel activity and thus the transepithelial Cl^- transport. In this way, L-type channels function as amplifiers of the Cl^- transport and their contribution to the EOG might be explained. Recently, anoctamin-4 that is expressed in the basolateral membrane of the RPE was identified as a Ca^{2+}-dependent ion channel that nonselectively conducts monovalent cations (Reichhart et al. 2019b). Furthermore, anoctamin-4 can also be activated by stimulation of P2Y receptors and the subsequent Ca^{2+} increase. Although not proven so far, it is likely that anoctamin-4 might also indirectly contribute to regulation of transepithelial transport. An increase in intracellular free Ca^{2+} by extracellular ATP activates anoctamin-4 that in turn might amplify the basolateral membrane depolarization together with activated Cl^- channels. Furthermore, activation of anoctamin-4 would provide a pathway for Na^+ to leave the RPE across the basolateral membrane. This would explain the long-debated additional transepithelial Na^+ transport in the RPE. Na^+ transport represents a considerably small proportion of the transported ions by the RPE and is directed from the retinal to the choroidal side of the RPE

(Tsuboi et al. 1986; Miller and Steinberg 1977a; Lasansky and De Fisch 1966; Hughes et al. 1998). The transport activities of the Na^+/K^+-ATPase and the Na^+/HCO_3^--cotransporter represent the entry mechanisms for Na^+ uptake into the RPE across the apical membrane and anoctamin-4 could provide the efflux pathway across the basolateral membrane.

9.9 Transport Through the RPE and Disease

There are not many diseases caused by reduced or disturbed transepithelial transport. As described, the knockout of ClC2 in mice evokes a retinal degeneration that resembles that of Retinitis pigmentosa (Bosl et al. 2001), a hereditary form of rod degeneration that represents the most common inherited cause of blindness (Wang et al. 2001; Chong and Bird 1999). A comparable pathology for human subjects, however, has not been reported so far. Better understood is significance of transepithelial transport in macular edema (Marmor 1999). Furthermore, given the fact that bestrophin-1 is a Ca^{2+}-dependent Cl^- channel in the basolateral membrane, the retinal dystrophy in Best's disease might be caused by a loss of transepithelial transport (Xiao et al. 2010; Johnson et al. 2017). Thus, the following section will deal with macular edema and Best's disease as examples for pathologies associated with changes in transepithelial transport.

9.10 Macular Edema

Macular edema is a condition defined by extracellular fluid accumulation in the macula (Marmor 1999). This can be caused, for example, by diabetic retinopathy (Wolfensberger 1999) or by mechanical insult to the eye (Wolfensberger et al. 2000; Tsuboi and Pederson 1985; Negi and Marmor 1983; Kirchhof and Sorgente 1989). In these scenarios, the fluid accumulation occurs by a decreased barrier function of the RPE and/or the endothelium of inner retinal vessels. Macular edema can efficiently be treated by administration of inhibitors for carbonic anhydrase (Wolfensberger et al. 1999, 2000; Wolfensberger 1999; Steinmetz et al. 1991; Korte and Smith 1993; Eichhorn et al. 1996; Marmor and Maack 1982; Marmor 1990, 1993). Blockers of carbonic anhydrase that reach the RPE via the blood stream lead to a reduction of intracellular pH and intracellular HCO_3^- concentration. This reduces the activity of the HCO_3^-/Cl^- exchanger in the basolateral membrane of the RPE because this transporter removes HCO_3^- from the RPE cytoplasm in exchange with Cl^-. By inhibition of carbonic anhydrase, less Cl^- is taken up to the RPE from the choroidal side across the basolateral membrane. Under physiological conditions, the activity of the $HCO3^-/Cl^-$ exchanger reduces the net Cl^- efflux across the basolateral membrane through Cl^- channels. By inhibition of the carbonic anhydrase, the reductive influence on the activity of the HCO_3^-/Cl^- exchanger

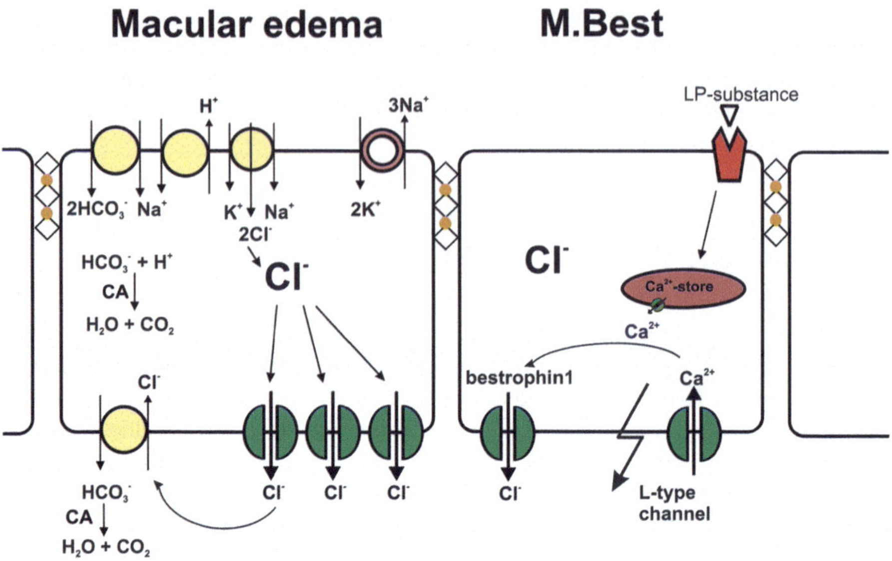

Fig. 9.4 Exemplary pathologic scenarios: Left cell: Macular Edema: Inhibition of carbonic anhydrase activity leads an increase in chloride efflux across the RPE basolateral membrane. Apical membrane: Na^+/HCO_3^- cotransporter, Na^+/H^+ exchanger, $Na^+/K^+/2Cl^-$ cotransporter, Na^+/K^+ ATPase, Basolateral membrane: Cl^-/HCO_3^- exchanger, ClC-2, voltage-dependent Cl^- channels of the ClC family, Bestrophin-1, Ca^{2+}-dependent Cl^- channel, CFTR, cystic fibrosis transmembrane conductance regulator, cAMP-dependent, CA, carbonic anhydrase. Right Cell: Morbus Best: A disturbed interaction between the mutant bestrophin-1 and the L-type channel might be responsible for the reduced light peak, which represents the Cl^- conductance of the basolateral side of the RPE. Apical membrane: LP, light peak. Basolateral membrane: bestrophin-1, Ca^{2+}-dependent Cl^- channel, L-Type channel: voltage-dependent Calcium channel $Ca_V1.3$

becomes smaller and the net Cl^- efflux increases across the RPE basolateral membrane (Marmor 1999; Wolfensberger et al. 1999, 2000). The resulting increase in Cl^- efflux out of the RPE increases the transepithelial transport of Cl^- across the RPE and therefore the elimination of water from the retina (Fig. 9.4, left cell).

9.11 Best's Disease

Best's vitelliforme macular dystrophy or Best's disease (M.Best) is a rare form of hereditary retinal degeneration. The autosomal dominantly inherited disease has an onset at juvenile ages. Characteristic for Best's disease is the bull's eye lesion resembling an egg yolk that has given the name vitelliform (Xiao et al. 2010; Johnson et al. 2017). Best's vitelliform macular dystrophy is caused by mutations in the *BEST1* gene (Kramer et al. 2000; Petrukhin et al. 1998). Besides Best's disease, mutations in the *BEST1* gene cause several other types of retinal degenerations (Boon et al. 2009). The diagnostic hallmark is the reduced light peak in the

EOG. Since the light peak is based on the basolateral membrane depolarization of the RPE resulting from activation of Cl^- channels (Arden and Constable 2006; Gallemore and Steinberg 1989b, 1990), it is likely that Best's disease is a retinal degeneration based on reduced transepithelial transport across the RPE. The *BEST1* gene product, bestrophin-1, was identified in heterologous expression as a Ca^{2+}-dependent and later also as a volume-activated Cl^- channel. Mutant bestrophin-1 showed a loss of Cl^- channel function (Sun et al. 2002) due to reduced protein trafficking from the Golgi to the plasma membrane (Milenkovic et al. 2011b, 2018; Johnson et al. 2013, 2015). Thus, these data led to the concept that a loss of a Ca^{2+}-dependent Cl^- channel, the mutant bestrophin-1, reduces both transepithelial transport of Cl^- and the reduction of the light-peak in the EOG. This concept, however, was challenged by a number of observations in animal models and by molecular analysis of the bestrophin-1 function. The *Best1* knockout mouse that should correspond with mutation-dependent loss of bestrophin-1 Cl^- channel function did not show a decrease in the light-peak amplitude of the DC-ERG but instead an increase (Marmorstein et al. 2006). Furthermore, freshly isolated RPE cells from the *Best1* knockout mouse showed no differences in the activity of CaCCs. A $Best^{W93C}$ knockin mouse model showed very moderate retinal degeneration and changes in the light-peak amplitude (Zhang et al. 2010). Here again, the Ca^{2+}-dependent Cl^- conductance remained unchanged compared to that of the wild-type cells, although the mutation W93C leads to loss of bestrophin-1 Cl^- channel activity in heterologoous expression systems (Sun et al. 2002). The analysis of the molecular function of bestrophin-1 indicated that bestrophin-1 is involved in the modulation of intracellular free Ca^{2+} as second messenger. Bestrophin-1 interacts with $Ca_V1.3$ L-type channels (Rosenthal et al. 2006; Reichhart et al. 2010; Milenkovic et al. 2011a), and it is likely that a proportion of bestrophin-1 is localized to cytosolic Ca^{2+} stores where it modulates the release of Ca^{2+} in response to activation of P2Y receptors (Neussert et al. 2010; Gomez et al. 2013; Barro-Soria et al. 2010). Thus, mutant bestrophin-1 might have indirect effects on transepithelial Cl^- transport. Since the $Ca_V1.3$ L-type channel knockout mouse shows a decreased light peak in the DC-ERG (Wu et al. 2007), a disturbed interaction between the mutant bestrophin-1 and the L-type channel might be responsible for the reduced EOG in patients (Fig. 9.4, right cell).

9.12 Summary

The evolution of light-sensitive organs developed structures composed of a pigmented cell and a light-sensitive cell. The pigmented cell, in the vertebrate eye the pigmented epithelium, has become an important partner for maintenance of photoreceptors that is involved visual function. Transepithelial transport of ions and water from the retina to the blood side across the RPE is one of the most important functions of the RPE. A transport of water driven by a transepithelial transport of Cl^- removes water from subretinal space and establishes the tight adhesion between the RPE and the retina. The transport of K^+ across the RPE from retina to blood side accompanies the Cl^- transport. Both transportation pathways are light-dependently regulated. The light dependence of Cl^- transport serves the removal of extracellular fluid that increases in the transition from dark to light. The light dependence of K^+ transport serves the K^+ homeostasis in the subretinal space. Two main properties of the RPE are essentially involved in disease: the barrier function and epithelial transport of water. A breakdown of water transport leads in animal models or possibly in patients with hereditary macular dystrophy to retinal degeneration. The barrier function of the RPE is essential for the immune privilege of the eye. However, the RPE barrier is an interface to the body system and exposed to various pathologic inputs such as diabetes, high blood pressure, or increased inflammatory activity. Future research must generate a clear picture how these impacts change the RPE barrier but also its epithelial function. It is essential to understand various vision-threatening diseases and to develop new therapies.

References

Ablonczy Z, Dahrouj M, Tang PH, Liu Y, Sambamurti K, Marmorstein AD, Crosson CE (2011) Human retinal pigment epithelium cells as functional models for the RPE in vivo. Invest Ophthalmol Vis Sci 52(12):8614–8620. https://doi.org/10.1167/iovs.11-8021

Adler AJ, Southwick RE (1992) Distribution of glucose and lactate in the interphotoreceptor matrix. Ophthalmic Res 24(4):243–252

Arden GB, Constable PA (2006) The electro-oculogram. Prog Retin Eye Res 25(2):207–248. https://doi.org/10.1016/j.preteyeres.2005.11.001

Arrindell EL, McKay BS, Jaffe GJ, Burke JM (1992) Modulation of potassium transport in cultured retinal pigment epithelium and retinal glial cells by serum and epidermal growth factor. Exp Cell Res 203(1):192–197

Ban Y, Rizzolo LJ (2000) Differential regulation of tight junction permeability during development of the retinal pigment epithelium. Am J Physiol Cell Physiol 279(3):C744–C750

Barro Soria R, Spitzner M, Schreiber R, Kunzelmann K (2009) Bestrophin-1 enables Ca2+-activated Cl- conductance in epithelia. J Biol Chem 284(43):29405–29412. https://doi.org/10.1074/jbc.M605716200

Barro-Soria R, Aldehni F, Almaca J, Witzgall R, Schreiber R, Kunzelmann K (2010) ER-localized bestrophin 1 activates Ca2+-dependent ion channels TMEM16A and SK4 possibly by acting as a counterion channel. Pflugers Arch 459(3):485–497. https://doi.org/10.1007/s00424-009-0745-0

Baylor D (1996) How photons start vision. Proc Natl Acad Sci U S A 93(2):560–565

Beatty S, Boulton M, Henson D, Koh HH, Murray IJ (1999) Macular pigment and age related macular degeneration. Br J Ophthalmol 83(7):867–877

Bell ET (1906) Experimental studies on the development of the eye and the nasal cavities in frog embryos. Anat Anz 29:185–194

Bialek S, Miller SS (1994) K+ and Cl- transport mechanisms in bovine pigment epithelium that could modulate subretinal space volume and composition. J Physiol 475(3):401–417

Bialek S, Joseph DP, Miller SS (1995) The delayed basolateral membrane hyperpolarization of the bovine retinal pigment epithelium: mechanism of generation. J Physiol 484(Pt 1):53–67

Blaug S, Quinn R, Quong J, Jalickee S, Miller SS (2003) Retinal pigment epithelial function: a role for CFTR? Doc Ophthalmol 106(1):43–50

Bok D (1993) The retinal pigment epithelium: a versatile partner in vision. J Cell Sci Suppl 17:189–195

Bok D, Hall MO (1971) The role of the pigment epithelium in the etiology of inherited retinal dystrophy in the rat. J Cell Biol 49(3):664–682

Boon CJ, Klevering BJ, Leroy BP, Hoyng CB, Keunen JE, den Hollander AI (2009) The spectrum of ocular phenotypes caused by mutations in the BEST1 gene. Prog Retin Eye Res 28 (3):187–205. https://doi.org/10.1016/j.preteyeres.2009.04.002

Bosl MR, Stein V, Hubner C, Zdebik AA, Jordt SE, Mukhopadhyay AK, Davidoff MS, Holstein AF, Jentsch TJ (2001) Male germ cells and photoreceptors, both dependent on close cell-cell interactions, degenerate upon ClC-2 Cl(-) channel disruption. EMBO J 20(6):1289–1299

Boulton M (1998) The role of melanin in the RPE. In: Marmor MF, Wolfensberger TJ (eds) The retinal pigment epithelium. Oxford University Press, Oxford, pp 68–65

Braekevelt CR, Hollenberg MJ (1970) The development of the retina of the albino rat. Am J Anat 127(3):281–301

Carter-Dawson L, Burroughs M (1992) Interphotoreceptor retinoid-binding protein in the cone matrix sheath. Electron microscopic immunocytochemical localization. Invest Ophthalmol Vis Sci 33(5):1584–1588

Chong NH, Bird AC (1999) Management of inherited outer retinal dystrophies: present and future. Br J Ophthalmol 83(1):120–122

Constable PA (2011) Nifedipine alters the light-rise of the electro-oculogram in man. Graefes Arch Clin Exp Ophthalmol 249(5):677–684. https://doi.org/10.1007/s00417-010-1604-6

Constable PA, Lawrenson JG, Arden GB (2006) Light and alcohol evoked electro-oculograms in cystic fibrosis. Doc Ophthalmol 113(2):133–143. https://doi.org/10.1007/s10633-006-9023-z

Daruwalla A, Choi EH, Palczewski K, Kiser PD (2018) Structural biology of 11-cis-retinaldehyde production in the classical visual cycle. Biochem J 475(20):3171–3188. https://doi.org/10.1042/BCJ20180193

Das A, Frank RN, Zhang NL, Turczyn TJ (1990) Ultrastructural localization of extracellular matrix components in human retinal vessels and Bruch's membrane. Arch Ophthalmol 108(3):421–429

Dawis S, Hofmann H, Niemeyer G (1985) The electroretinogram, standing potential, and light peak of the perfused cat eye during acid-base changes. Vision Res 25(9):1163–1177

Defoe DM, Ahmad A, Chen W, Hughes BA (1994) Membrane polarity of the Na(+)-K+ pump in primary cultures of Xenopus retinal pigment epithelium. Exp Eye Res 59(5):587–596

DiMattio J, Degnan KJ, Zadunaisky JA (1983) A model for transepithelial ion transport across the isolated retinal pigment epithelium of the frog. Exp Eye Res 37(5):409–420

Dmitriev AV, Govardovskii VI, Schwahn HN, Steinberg RH (1999) Light-induced changes of extracellular ions and volume in the isolated chick retina-pigment epithelium preparation. Vis Neurosci 16(6):1157–1167

Dornonville de la Cour M (1993) Ion transport in the retinal pigment epithelium. A study with double barrelled ion-selective microelectrodes. Acta Ophthalmol Suppl 209:1–32

Edelman JL, Miller SS (1991) Epinephrine stimulates fluid absorption across bovine retinal pigment epithelium. Invest Ophthalmol Vis Sci 32(12):3033–3040

Edelman JL, Lin H, Miller SS (1994a) Acidification stimulates chloride and fluid absorption across frog retinal pigment epithelium. Am J Physiol 266(4 Pt 1):C946–C956

Edelman JL, Lin H, Miller SS (1994b) Potassium-induced chloride secretion across the frog retinal pigment epithelium. Am J Physiol 266(4 Pt 1):C957–C966

Eichhorn M, Schreckenberger M, Tamm ER, Lutjen-Drecoll E (1996) Carbonic anhydrase activity is increased in retinal pigmented epithelium and choriocapillaris of RCS rats. Graefes Arch Clin Exp Ophthalmol 234(4):258–263

Frambach DA, Valentine JL, Weiter JJ (1988) Initial observations of rabbit retinal pigment epithelium-choroid-sclera preparations. Invest Ophthalmol Vis Sci 29(5):814–817

Frambach DA, Valentine JL, Weiter JJ (1989) Furosemide-sensitive Cl transport in bovine retinal pigment epithelium. Invest Ophthalmol Vis Sci 30(10):2271–2274

Fujii S, Gallemore RP, Hughes BA, Steinberg RH (1992) Direct evidence for a basolateral membrane Cl- conductance in toad retinal pigment epithelium. Am J Physiol 262(2 Pt 1): C374–C383

Gallemore RP, Steinberg RH (1989a) Effects of DIDS on the chick retinal pigment epithelium. I. Membrane potentials, apparent resistances, and mechanisms. J Neurosci 9 (6):1968–1976

Gallemore RP, Steinberg RH (1989b) Effects of DIDS on the chick retinal pigment epithelium. II. Mechanism of the light peak and other responses originating at the basal membrane. J Neurosci 9(6):1977–1984

Gallemore RP, Steinberg RH (1990) Effects of dopamine on the chick retinal pigment epithelium. Membrane potentials and light-evoked responses. Invest Ophthalmol Vis Sci 31(1):67–80

Gallemore RP, Steinberg RH (1993) Light-evoked modulation of basolateral membrane Cl-conductance in chick retinal pigment epithelium: the light peak and fast oscillation. J Neurophysiol 70(4):1669–1680

Gallemore RP, Griff ER, Steinberg RH (1988) Evidence in support of a photoreceptoral origin for the "light-peak substance". Invest Ophthalmol Vis Sci 29(4):566–571

Gallemore RP, Hernandez E, Tayyanipour R, Fujii S, Steinberg RH (1993) Basolateral membrane Cl- and K+ conductances of the dark-adapted chick retinal pigment epithelium. J Neurophysiol 70(4):1656–1668

Gallemore RP, Li JD, Govardovskii VI, Steinberg RH (1994) Calcium gradients and light-evoked calcium changes outside rods in the intact cat retina. Vis Neurosci 11(4):753–761

Gomez NM, Tamm ER, Straubeta O (2013) Role of bestrophin-1 in store-operated calcium entry in retinal pigment epithelium. Pflugers Arch 465(4):481–495. https://doi.org/10.1007/s00424-012-1181-0

Gonzalez-Fernandez F, Healy JI (1990) Early expression of the gene for interphotoreceptor retinol-binding protein during photoreceptor differentiation suggests a critical role for the interphotoreceptor matrix in retinal development. J Cell Biol 111(6 Pt 1):2775–2784

Gonzalez-Fernandez F, Van Niel E, Edmonds C, Beaver H, Nickerson JM, Garcia-Fernandez JM, Campohiaro PA, Foster RG (1993) Differential expression of interphotoreceptor retinoid-binding protein, opsin, cellular retinaldehyde-binding protein, and basic fibroblastic growth factor. Exp Eye Res 56(4):411–427

Griff ER (1990) Response properties of the toad retinal pigment epithelium. Invest Ophthalmol Vis Sci 31(11):2353–2360

Griff ER (1991) Potassium-evoked responses from the retinal pigment epithelium of the toad Bufo marinus. Exp Eye Res 53(2):219–228

Griff ER, Shirao Y, Steinberg RH (1985) Ba2+ unmasks K+ modulation of the Na+-K+ pump in the frog retinal pigment epithelium. J Gen Physiol 86(6):853–876

Guymer R, Luthert P, Bird A (1999) Changes in Bruch's membrane and related structures with age. Prog Retin Eye Res 18(1):59–90

Guziewicz KE, Sinha D, Gomez NM, Zorych K, Dutrow EV, Dhingra A, Mullins RF, Stone EM, Gamm DM, Boesze-Battaglia K, Aguirre GD (2017) Bestrophinopathy: an RPE-photoreceptor interface disease. Prog Retin Eye Res 58:70–88. https://doi.org/10.1016/j.preteyeres.2017.01.005

Hamann S (2002) Molecular mechanisms of water transport in the eye. Int Rev Cytol 215:395–431

Hamann S, Zeuthen T, la Cour M, Nagelhus EA, Ottersen OP, Agre P, Nielsen S (1998) Aquaporins in complex tissues: distribution of aquaporins 1-5 in human and rat eye. Am J Physiol 274(5 Pt 1):C1332–C1345

Hamann S, la Cour M, Lui GM, Bundgaard M, Zeuthen T (2000) Transport of protons and lactate in cultured human fetal retinal pigment epithelial cells. Pflugers Arch 440(1):84–92

Hartzell HC, Qu Z (2003) Chloride currents in acutely isolated Xenopus retinal pigment epithelial cells. J Physiol 549(Pt 2):453–469

Hartzell HC, Qu Z, Yu K, Xiao Q, Chien LT (2008) Molecular physiology of bestrophins: multifunctional membrane proteins linked to Best disease and other retinopathies. Physiol Rev 88(2):639–672. https://doi.org/10.1152/physrev.00022.2007
Hernandez EV, Hu JG, Frambach DA, Gallemore RP (1995) Potassium conductances in cultured bovine and human retinal pigment epithelium. Invest Ophthalmol Vis Sci 36(1):113–122
Hollyfield JG (1999) Hyaluronan and the functional organization of the interphotoreceptor matrix. Invest Ophthalmol Vis Sci 40(12):2767–2769
Hu JG, Gallemore RP, Bok D, Lee AY, Frambach DA (1994) Localization of NaK ATPase on cultured human retinal pigment epithelium. Invest Ophthalmol Vis Sci 35(10):3582–3588
Hu JG, Gallemore RP, Bok D, Frambach DA (1996) Chloride transport in cultured fetal human retinal pigment epithelium. Exp Eye Res 62(4):443–448. https://doi.org/10.1006/exer.1996.0049
Huang B, Karwoski CJ (1992) Light-evoked expansion of subretinal space volume in the retina of the frog. J Neurosci 12(11):4243–4252
Hughes BA, Segawa Y (1993) cAMP-activated chloride currents in amphibian retinal pigment epithelial cells. J Physiol 466:749–766
Hughes BA, Takahira M (1996) Inwardly rectifying K+ currents in isolated human retinal pigment epithelial cells. Invest Ophthalmol Vis Sci 37(6):1125–1139
Hughes BA, Shaikh A, Ahmad A (1995a) Effects of Ba2+ and Cs+ on apical membrane K+ conductance in toad retinal pigment epithelium. Am J Physiol 268(5 Pt 1):C1164–C1172
Hughes BA, Takahira M, Segawa Y (1995b) An outwardly rectifying K+ current active near resting potential in human retinal pigment epithelial cells. Am J Physiol 269(1 Pt 1):C179–C187
Hughes BA, Gallemore RP, Miller SS (1998) Transport mechanisms in the retinal pigment epithelium. In: Marmor MF, Wolfensberger TJ (eds) The retinal pigment epithelium. Oxford University Press, New York, pp 103–134
Ishida K, Panjwani N, Cao Z, Streilein JW (2003) Participation of pigment epithelium in ocular immune privilege. 3. Epithelia cultured from iris, ciliary body, and retina suppress T-cell activation by partially non-overlapping mechanisms. Ocul Immunol Inflamm 11(2):91–105
Jeffery G (1998) The retinal pigment epithelium as a developmental regulator of the neural retina. Eye 12:499–503
Johnson AA, Lee YS, Stanton JB, Yu K, Hartzell CH, Marmorstein LY, Marmorstein AD (2013) Differential effects of Best disease causing missense mutations on bestrophin-1 trafficking. Hum Mol Genet 22(23):4688–4697. https://doi.org/10.1093/hmg/ddt316
Johnson AA, Bachman LA, Gilles BJ, Cross SD, Stelzig KE, Resch ZT, Marmorstein LY, Pulido JS, Marmorstein AD (2015) Autosomal recessive bestrophinopathy is not associated with the loss of bestrophin-1 anion channel function in a patient with a novel BEST1 mutation. Invest Ophthalmol Vis Sci 56(8):4619–4630. https://doi.org/10.1167/iovs.15-16910
Johnson AA, Guziewicz KE, Lee CJ, Kalathur RC, Pulido JS, Marmorstein LY, Marmorstein AD (2017) Bestrophin 1 and retinal disease. Prog Retin Eye Res 58:45–69. https://doi.org/10.1016/j.preteyeres.2017.01.006
Joseph DP, Miller SS (1991) Apical and basal membrane ion transport mechanisms in bovine retinal pigment epithelium. J Physiol 435:439–463
Joseph DP, Miller SS (1992) Alpha-1-adrenergic modulation of K and Cl transport in bovine retinal pigment epithelium. J Gen Physiol 99(2):263–290
Keckeis S, Reichhart N, Roubeix C, Strauss O (2017a) Anoctamin2 (TMEM16B) forms the Ca2+-activated Cl- channel in the retinal pigment epithelium. Exp Eye Res 154:139–150. https://doi.org/10.1016/j.exer.2016.12.003
Keckeis S, Reichhart N, Roubeix C, Strauss O (2017b) Anoctamin2 (TMEM16B) forms the Ca(2+)-activated Cl(-) channel in the retinal pigment epithelium. Exp Eye Res 154:139–150. https://doi.org/10.1016/j.exer.2016.12.003
Keller SK, Jentsch TJ, Koch M, Wiederholt M (1986) Interactions of pH and K+ conductance in cultured bovine retinal pigment epithelial cells. Am J Physiol 250(1 Pt 1):C124–C137

Keller SK, Jentsch TJ, Janicke I, Wiederholt M (1988) Regulation of intracellular pH in cultured bovine retinal pigment epithelial cells. Pflugers Arch 411(1):47–52
Kennedy BG (1990) Na^+-K^+-Cl^- cotransport in cultured cells derived from human retinal pigment epithelium. Am J Physiol 259(1 Pt 1):C29–C34
Kenyon E, Yu K, la Cour M, Miller SS (1994) Lactate transport mechanisms at apical and basolateral membranes of bovine retinal pigment epithelium. Am J Physiol 267(6 Pt 1): C1561–C1573
Kenyon E, Maminishkis A, Joseph DP, Miller SS (1997) Apical and basolateral membrane mechanisms that regulate pHi in bovine retinal pigment epithelium. Am J Physiol 273(2 Pt 1):C456–C472
Kirchhof B, Sorgente N (1989) Pathogenesis of proliferative vitreoretinopathy. Modulation of retinal pigment epithelial cell functions by vitreous and macrophages. Dev Ophthalmol 16:1–53
Kita M, Marmor MF (1992) Effects on retinal adhesive force in vivo of metabolically active agents in the subretinal space. Invest Ophthalmol Vis Sci 33(6):1883–1887
Korte GE, Smith J (1993) Carbonic anhydrase type II in regenerating retinal pigment epithelium. A histochemical study in the rabbit. Experientia 49(9):789–791
Korte GE, Wanderman MC (1993) Distribution of Na+ K(+)-ATPase in regenerating retinal pigment epithelium in the rabbit. A study by electron microscopic cytochemistry. Exp Eye Res 56(2):219–229
Kramer F, White K, Pauleikhoff D, Gehrig A, Passmore L, Rivera A, Rudolph G, Kellner U, Andrassi M, Lorenz B, Rohrschneider K, Blankenagel A, Jurklies B, Schilling H, Schutt F, Holz FG, Weber BH (2000) Mutations in the VMD2 gene are associated with juvenile-onset vitelliform macular dystrophy (Best disease) and adult vitelliform macular dystrophy but not age-related macular degeneration. Eur J Hum Genet 8(4):286–292
Kuntz CA, Crook RB, Dmitriev A, Steinberg RH (1994) Modification by cyclic adenosine monophosphate of basolateral membrane chloride conductance in chick retinal pigment epithelium. Invest Ophthalmol Vis Sci 35(2):422–433
Kusaka S, Inanobe A, Fujita A, Makino Y, Tanemoto M, Matsushita K, Tano Y, Kurachi Y (2001) Functional Kir7.1 channels localized at the root of apical processes in rat retinal pigment epithelium. J Physiol 531(Pt 1):27–36
la Cour M (1985) The retinal pigment epithelium controls the potassium activity in the subretinal space. Acta Ophthalmol Suppl 173:9–10
la Cour M (1989) Rheogenic sodium-bicarbonate co-transport across the retinal membrane of the frog retinal pigment epithelium. J Physiol 419:539–553
la Cour M (1991a) Kinetic properties and Na+ dependence of rheogenic Na(+)-HCO3- co-transport in frog retinal pigment epithelium. J Physiol 439:59–72
la Cour M (1991b) pH homeostasis in the frog retina: the role of Na+:HCO3- co-transport in the retinal pigment epithelium. Acta Ophthalmol (Copenh) 69(4):496–504
la Cour M (1992) Cl- transport in frog retinal pigment epithelium. Exp Eye Res 54(6):921–931
la Cour M, Lund-Andersen H, Zeuthen T (1986) Potassium transport of the frog retinal pigment epithelium: autoregulation of potassium activity in the subretinal space. J Physiol 375:461–479
la Cour M, Lin H, Kenyon E, Miller SS (1994) Lactate transport in freshly isolated human fetal retinal pigment epithelium. Invest Ophthalmol Vis Sci 35(2):434–442
Lamb TD, Pugh EN Jr (2004) Dark adaptation and the retinoid cycle of vision. Prog Retin Eye Res 23(3):307–380. https://doi.org/10.1016/j.preteyeres.2004.03.001
Lasansky A, De Fisch FW (1966) Potential, current, and ionic fluxes across the isolated retinal pigment epithelium and choriod. J Gen Physiol 49(5):913–924
LaVail MM (1976) Rod outer segment disc shedding in relation to cyclic lighting. Exp Eye Res 23 (2):277–280
LaVail MM (1980) Circadian nature of rod outer segment disc shedding in the rat. Invest Ophthalmol Vis Sci 19(4):407–411

Li M, McCann JD, Liedtke CM, Nairn AC, Greengard P, Welsh MJ (1988) Cyclic AMP-dependent protein kinase opens chloride channels in normal but not cystic fibrosis airway epithelium. Nature 331(6154):358–360

Li JD, Gallemore RP, Dmitriev A, Steinberg RH (1994a) Light-dependent hydration of the space surrounding photoreceptors in chick retina. Invest Ophthalmol Vis Sci 35(6):2700–2711

Li JD, Govardovskii VI, Steinberg RH (1994b) Light-dependent hydration of the space surrounding photoreceptors in the cat retina. Vis Neurosci 11(4):743–752

Lin H, Miller SS (1991) pHi regulation in frog retinal pigment epithelium: two apical membrane mechanisms. Am J Physiol 261(1 Pt 1):C132–C142

Lin H, Miller SS (1994) pHi-dependent Cl-HCO3 exchange at the basolateral membrane of frog retinal pigment epithelium. Am J Physiol 266(4 Pt 1):C935–C945

Linsenmeier RA, Steinberg RH (1983) A light-evoked interaction of apical and basal membranes of retinal pigment epithelium: c-wave and light peak. J Neurophysiol 50(1):136–147

Linsenmeier RA, Steinberg RH (1984) Delayed basal hyperpolarization of cat retinal pigment epithelium and its relation to the fast oscillation of the DC electroretinogram. J Gen Physiol 83 (2):213–232

Linsenmeier RA, Steinberg RH (1986) Mechanisms of hypoxic effects on the cat DC electroretinogram. Invest Ophthalmol Vis Sci 27(9):1385–1394

Linsenmeier RA, Steinberg RH (1987) Mechanisms of azide induced increases in the c-wave and standing potential of the intact cat eye. Vision Res 27(1):1–8

Lu W, Reigada D, Sevigny J, Mitchell CH (2007) Stimulation of the P2Y1 receptor up-regulates nucleoside-triphosphate diphosphohydrolase-1 in human retinal pigment epithelial cells. J Pharmacol Exp Therap 323(1):157–164. https://doi.org/10.1124/jpet.107.124545

Maminishkis A, Jalickee S, Blaug SA, Rymer J, Yerxa BR, Peterson WM, Miller SS (2002) The P2Y(2) receptor agonist INS37217 stimulates RPE fluid transport in vitro and retinal reattachment in rat. Invest Ophthalmol Vis Sci 43(11):3555–3566

Marmor MF (1990) Control of subretinal fluid: experimental and clinical studies. Eye 4:340–344

Marmor MF (1993) Mechanisms of retinal adhesion. Prog Retin Eye Res 12:179–203

Marmor MF (1999) Mechanisms of fluid accumulation in retinal edema. Doc Ophthalmol 97 (3-4):239–249

Marmor MF, Maack T (1982) Enhancement of retinal adhesion and subretinal fluid resorption by acetazolamide. Invest Ophthalmol Vis Sci 23(1):121–124

Marmorstein AD (2001) The polarity of the retinal pigment epithelium. Traffic 2(12):867–872

Marmorstein LY, Wu J, McLaughlin P, Yocom J, Karl MO, Neussert R, Wimmers S, Stanton JB, Gregg RG, Strauss O, Peachey NS, Marmorstein AD (2006) The light peak of the electroretinogram is dependent on voltage-gated calcium channels and antagonized by bestrophin (best-1). J Gen Physiol 127(5):577–589. https://doi.org/10.1085/jgp.200509473

Martinez-Morales JR, Rodrigo I, Bovolenta P (2004) Eye development: a view from the retina pigmented epithelium. Bioessays 26(7):766–777. https://doi.org/10.1002/bies.20064

Maruiwa F, Naoi N, Nakazaki S, Sawada A (1999) Effects of bicarbonate ion on chick retinal pigment epithelium: membrane potentials and light-evoked responses. Vision Res 39 (1):159–167

Mazzoni F, Safa H, Finnemann SC (2014) Understanding photoreceptor outer segment phagocytosis: use and utility of RPE cells in culture. Exp Eye Res 126:51–60. https://doi.org/10.1016/j.exer.2014.01.010

McGrail KM, Sweadner KJ (1986) Immunofluorescent localization of two different Na,K-ATPases in the rat retina and in identified dissociated retinal cells. J Neurosci 6(5):1272–1283

Mergler S, Steinhausen K, Wiederholt M, Strauss O (1998) Altered regulation of L-type channels by protein kinase C and protein tyrosine kinases as a pathophysiologic effect in retinal degeneration. FASEB J 12(12):1125–1134

Milenkovic VM, Krejcova S, Reichhart N, Wagner A, Strauss O (2011a) Interaction of bestrophin-1 and Ca2+ channel beta-subunits: identification of new binding domains on the bestrophin-1 C-terminus. PLoS One 6(4):e19364. https://doi.org/10.1371/journal.pone.0019364

Milenkovic VM, Rohrl E, Weber BH, Strauss O (2011b) Disease-associated missense mutations in bestrophin-1 affect cellular trafficking and anion conductance. J Cell Sci 124(Pt 17):2988–2996. https://doi.org/10.1242/jcs.085878
Milenkovic A, Milenkovic VM, Wetzel CH, Weber BHF (2018) BEST1 protein stability and degradation pathways differ between autosomal dominant Best disease and autosomal recessive bestrophinopathy accounting for the distinct retinal phenotypes. Hum Mol Genet 27 (9):1630–1641. https://doi.org/10.1093/hmg/ddy070
Miller SS, Edelman JL (1990) Active ion transport pathways in the bovine retinal pigment epithelium. J Physiol 424:283–300
Miller S, Farber D (1984) Cyclic AMP modulation of ion transport across frog retinal pigment epithelium. Measurements in the short-circuit state. J Gen Physiol 83(6):853–874
Miller SS, Steinberg RH (1977a) Active transport of ions across frog retinal pigment epithelium. Exp Eye Res 25(3):235–248
Miller SS, Steinberg RH (1977b) Passive ionic properties of frog retinal pigment epithelium. J Membr Biol 36(4):337–372
Miller SS, Hughes BA, Machen TE (1982) Fluid transport across retinal pigment epithelium is inhibited by cyclic AMP. Proc Natl Acad Sci U S A 79(6):2111–2115
Mitchell CH, Reigada D (2008) Purinergic signalling in the subretinal space: a role in the communication between the retina and the RPE. Purinergic Signal 4(2):101–107. https://doi.org/10.1007/s11302-007-9054-2
Muller C, Mas Gomez N, Ruth P, Strauss O (2014) CaV1.3 L-type channels, maxiK Ca(2+)-dependent K(+) channels and bestrophin-1 regulate rhythmic photoreceptor outer segment phagocytosis by retinal pigment epithelial cells. Cell Signal 26(5):968–978. https://doi.org/10.1016/j.cellsig.2013.12.021
Nabi IR, Mathews AP, Cohen-Gould L, Gundersen D, Rodriguez-Boulan E (1993) Immortalization of polarized rat retinal pigment epithelium. J Cell Sci 104(Pt 1):37–49
Nao-i N, Gallemore RP, Steinberg RH (1990) Effects of cAMP and IBMX on the chick retinal pigment epithelium. Membrane potentials and light-evoked responses. Invest Ophthalmol Vis Sci 31(1):54–66
Negi A, Marmor MF (1983) The resorption of subretinal fluid after diffuse damage to the retinal pigment epithelium. Invest Ophthalmol Vis Sci 24(11):1475–1479
Neussert R, Muller C, Milenkovic VM, Strauss O (2010) The presence of bestrophin-1 modulates the Ca2+ recruitment from Ca2+ stores in the ER. Pflugers Arch 460(1):163–175. https://doi.org/10.1007/s00424-010-0840-2
Oakley B 2nd (1977) Potassium and the photoreceptor-dependent pigment epithelial hyperpolarization. J Gen Physiol 70(4):405–425
Padnick-Silver L, Linsenmeier RA (2002) Quantification of in vivo anaerobic metabolism in the normal cat retina through intraretinal pH measurements. Vis Neurosci 19(6):793–806
Pattnaik BR, Hughes BA (2012) Effects of KCNQ channel modulators on the M-type potassium current in primate retinal pigment epithelium. Am J Physiol Cell Physiol 302(5):C821–C833. https://doi.org/10.1152/ajpcell.00269.2011
Peterson WM, Meggyesy C, Yu K, Miller SS (1997) Extracellular ATP activates calcium signaling, ion, and fluid transport in retinal pigment epithelium. J Neurosci 17(7):2324–2337
Petit C (2001) Usher syndrome: from genetics to pathogenesis. Annu Rev Genomics Hum Genet 2:271–297
Petrukhin K, Koisti MJ, Bakall B, Li W, Xie G, Marknell T, Sandgren O, Forsman K, Holmgren G, Andreasson S, Vujic M, Bergen AA, McGarty-Dugan V, Figueroa D, Austin CP, Metzker ML, Caskey CT, Wadelius C (1998) Identification of the gene responsible for best macular dystrophy. Nat Genet 19(3):241–247
Philp NJ, Yoon H, Grollman EF (1998) Monocarboxylate transporter MCT1 is located in the apical membrane and MCT3 in the basal membrane of rat RPE. Am J Physiol 274(6 Pt 2):R1824–R1828

Polato F, Becerra SP (2016) Pigment epithelium-derived factor, a protective factor for photoreceptors in vivo. Adv Exp Med Biol 854:699–706. https://doi.org/10.1007/978-3-319-17121-0_93

Quinn RH, Miller SS (1992) Ion transport mechanisms in native human retinal pigment epithelium. Invest Ophthalmol Vis Sci 33(13):3513–3527

Quinn RH, Quong JN, Miller SS (2001) Adrenergic receptor activated ion transport in human fetal retinal pigment epithelium. Invest Ophthalmol Vis Sci 42(1):255–264

Reichenbach A, Henke A, Eberhardt W, Reichelt W, Dettmer D (1992) K+ ion regulation in retina. Can J Physiol Pharmacol 70(Suppl):S239–S247

Reichhart N, Milenkovic VM, Halsband CA, Cordeiro S, Strauss O (2010) Effect of bestrophin-1 on L-type Ca2+ channel activity depends on the Ca2+ channel beta-subunit. Exp Eye Res 91 (5):630–639. https://doi.org/10.1016/j.exer.2010.08.001

Reichhart N, Schoberl S, Keckeis S, Alfaar AS, Roubeix C, Cordes M, Crespo-Garcia S, Haeckel A, Kociok N, Fockler R, Fels G, Mataruga A, Rauh R, Milenkovic VM, Zuhlke K, Klussmann E, Schellenberger E, Strauss O (2019a) Anoctamin-4 is a bona fide Ca(2+)-dependent non-selective cation channel. Sci Rep 9(1):2257. https://doi.org/10.1038/s41598-018-37287-y

Reichhart N, Schöberl S, Keckeis S, Alfaar AS, Roubeix C, Cordes M, Crespo-Garcia S, Haeckel A, Kociok N, Föckler R, Fels G, Mataruga A, Rauh R, Milenkovic VM, Zühlke K, Klussmann E, Schellenberger E, Strauß O (2019b) Anoctamin-4 is a bona fide Ca2+-dependent non-selective cation channel. Sci Rep 9(1):2257. https://doi.org/10.1038/s41598-018-37287-y

Rizzolo LJ (1997) Polarity and the development of the outer blood-retinal barrier. Histol Histopathol 12(4):1057–1067

Rizzolo LJ (1999) Polarization of the Na+, K(+)-ATPase in epithelia derived from the neuroepithelium. Int Rev Cytol 185:195–235

Rosenthal R, Strauss O (2002) Ca2+-channels in the RPE. Adv Exp Med Biol 514:225–235

Rosenthal R, Bakall B, Kinnick T, Peachey N, Wimmers S, Wadelius C, Marmorstein A, Strauss O (2006) Expression of bestrophin-1, the product of the VMD2 gene, modulates voltage-dependent Ca2+ channels in retinal pigment epithelial cells. FASEB J 20(1):178–180. https://doi.org/10.1096/fj.05-4495fje

Rosenthal R, Heimann H, Agostini H, Martin G, Hansen LL, Strauss O (2007) Ca2+ channels in retinal pigment epithelial cells regulate vascular endothelial growth factor secretion rates in health and disease. Mol Vis 13:443–456

Rymer J, Miller SS, Edelman JL (2001) Epinephrine-induced increases in [Ca2+](in) and KCl-coupled fluid absorption in bovine RPE. Invest Ophthalmol Vis Sci 42(8):1921–1929

Segawa Y, Hughes BA (1994) Properties of the inwardly rectifying K+ conductance in the toad retinal pigment epithelium. J Physiol 476(1):41–53

Shimura M, Yuan Y, Chang JT, Zhang S, Campochiaro PA, Zack DJ, Hughes BA (2001) Expression and permeation properties of the K(+) channel Kir7.1 in the retinal pigment epithelium. J Physiol 531(Pt 2):329–346

Sparrow JR, Hicks D, Hamel CP (2010) The retinal pigment epithelium in health and disease. Curr Mol Med 10(9):802–823

Stamer WD, Bok D, Hu J, Jaffe GJ, McKay BS (2003) Aquaporin-1 channels in human retinal pigment epithelium: role in transepithelial water movement. Invest Ophthalmol Vis Sci 44 (6):2803–2808

Steinberg RH (1985) Interactions between the retinal pigment epithelium and the neural retina. Doc Ophthalmol 60(4):327–346

Steinberg RH, Linsenmeier RA, Griff ER (1983) Three light-evoked responses of the retinal pigment epithelium. Vis Res 23(11):1315–1323

Steinmetz RL, Fitzke FW, Bird AC (1991) Treatment of cystoid macular edema with acetazolamide in a patient with serpiginous choroidopathy. Retina 11(4):412–415

Strauss O (2005) The retinal pigment epithelium in visual function. Physiol Rev 85(3):845–881

Strauss O, Wienrich M (1994) Ca(2+)-conductances in cultured rat retinal pigment epithelial cells. J Cell Physiol 160(1):89–96

Strauss O, Mergler S, Wiederholt M (1997) Regulation of L-type calcium channels by protein tyrosine kinase and protein kinase C in cultured rat and human retinal pigment epithelial cells. FASEB J 11(11):859–867

Strauss O, Buss F, Rosenthal R, Fischer D, Mergler S, Stumpff F, Thieme H (2000) Activation of neuroendocrine L-type channels (alpha1D subunits) in retinal pigment epithelial cells and brain neurons by pp60(c-src). Biochem Biophys Res Commun 270(3):806–810

Streilein JW (2003) Ocular immune privilege: therapeutic opportunities from an experiment of nature. Nat Rev Immunol 3(11):879–889. https://doi.org/10.1038/nri1224

Sumita R (1961) The fine structure of Bruch's membrane in the choroid. Acta Soc Ophthalmol Jpn 28:1188

Sun H, Tsunenari T, Yau KW, Nathans J (2002) The vitelliform macular dystrophy protein defines a new family of chloride channels. Proc Natl Acad Sci U S A 99(6):4008–4013

Tsuboi S (1987) Measurement of the volume flow and hydraulic conductivity across the isolated dog retinal pigment epithelium. Invest Ophthalmol Vis Sci 28(11):1776–1782

Tsuboi S, Pederson JE (1985) Experimental retinal detachment. X. Effect of acetazolamide on vitreous fluorescein disappearance. Arch Ophthalmol 103(10):1557–1558

Tsuboi S, Pederson JE (1988) Volume flow across the isolated retinal pigment epithelium of cynomolgus monkey eyes. Invest Ophthalmol Vis Sci 29(11):1652–1655

Tsuboi S, Manabe R, Iizuka S (1986) Aspects of electrolyte transport across isolated dog retinal pigment epithelium. Am J Physiol 250(5 Pt 2):F781–F784

Ueda Y, Steinberg RH (1993) Voltage-operated calcium channels in fresh and cultured rat retinal pigment epithelial cells. Invest Ophthalmol Vis Sci 34(12):3408–3418

Wang Q, Chen Q, Zhao K, Wang L, Traboulsi EI (2001) Update on the molecular genetics of retinitis pigmentosa. Ophthalmic Genet 22(3):133–154

Wiederholt M, Zadunaisky JA (1984) Decrease of intracellular chloride activity by furosemide in frog retinal pigment epithelium. Curr Eye Res 3(4):673–675

Wills NK, Weng T, Mo L, Hellmich HL, Yu A, Wang T, Buchheit S, Godley BF (2000) Chloride channel expression in cultured human fetal RPE cells: response to oxidative stress. Invest Ophthalmol Vis Sci 41(13):4247–4255

Wimmers S, Karl MO, Strauss O (2007) Ion channels in the RPE. Prog Retin Eye Res 26 (3):263–301. https://doi.org/10.1016/j.preteyeres.2006.12.002

Wimmers S, Coeppicus L, Rosenthal R, Strauss O (2008a) Expression profile of voltage-dependent Ca2+ channel subunits in the human retinal pigment epithelium. Graefes Arch Clin Exp Ophthalmol 246(5):685–692. https://doi.org/10.1007/s00417-008-0778-7

Wimmers S, Halsband C, Seyler S, Milenkovic V, Strauss O (2008b) Voltage-dependent Ca2+ channels, not ryanodine receptors, activate Ca2+-dependent BK potassium channels in human retinal pigment epithelial cells. Mol Vis 14:2340–2348

Wolfensberger TJ (1999) The role of carbonic anhydrase inhibitors in the management of macular edema. Doc Ophthalmol 97(3-4):387–397

Wolfensberger TJ, Dmitriev AV, Govardovskii VI (1999) Inhibition of membrane-bound carbonic anhydrase decreases subretinal pH and volume. Doc Ophthalmol 97(3-4):261–271

Wolfensberger TJ, Chiang RK, Takeuchi A, Marmor MF (2000) Inhibition of membrane-bound carbonic anhydrase enhances subretinal fluid absorption and retinal adhesiveness. Graefes Arch Clin Exp Ophthalmol 238(1):76–80

Wu J, Marmorstein AD, Striessnig J, Peachey NS (2007) Voltage-dependent calcium channel CaV1.3 subunits regulate the light peak of the electroretinogram. J Neurophysiol 97 (5):3731–3735. https://doi.org/10.1152/jn.00146.2007

Xiao Q, Hartzell HC, Yu K (2010) Bestrophins and retinopathies. Pflugers Arch 460(2):559–569. https://doi.org/10.1007/s00424-010-0821-5

Xue L, Gollapalli DR, Maiti P, Jahng WJ, Rando RR (2004) A palmitoylation switch mechanism in the regulation of the visual cycle. Cell 117(6):761–771

Yamashita H, Yamamoto T (1991) Distribution of chloride ion in intercellular space of retinal pigment epithelium—effects of various agents. Jpn J Ophthalmol 35(1):42–50

Yang D, Pan A, Swaminathan A, Kumar G, Hughes BA (2003) Expression and localization of the inwardly rectifying potassium channel Kir7.1 in native bovine retinal pigment epithelium. Invest Ophthalmol Vis Sci 44(7):3178–3185

Yoon H, Fanelli A, Grollman EF, Philp NJ (1997) Identification of a unique monocarboxylate transporter (MCT3) in retinal pigment epithelium. Biochem Biophys Res Commun 234 (1):90–94

Yuan Y, Shimura M, Hughes BA (2003) Regulation of inwardly rectifying K+ channels in retinal pigment epithelial cells by intracellular pH. J Physiol 549(Pt 2):429–438

Zdebik AA, Cuffe JE, Bertog M, Korbmacher C, Jentsch TJ (2004) Additional disruption of the ClC-2 Cl(-) channel does not exacerbate the cystic fibrosis phenotype of cystic fibrosis transmembrane conductance regulator mouse models. J Biol Chem 279(21):22276–22283. https://doi.org/10.1074/jbc.M309899200

Zeuthen T, Hamann S, la Cour M (1996) Cotransport of H+, lactate and H2O by membrane proteins in retinal pigment epithelium of bullfrog. J Physiol 497(Pt 1):3–17

Zhang X, Hughes BA (2013) KCNQ and KCNE potassium channel subunit expression in bovine retinal pigment epithelium. Exp Eye Res 116:424–432

Zhang Y, Stanton JB, Wu J, Yu K, Hartzell HC, Peachey NS, Marmorstein LY, Marmorstein AD (2010) Suppression of Ca2+ signaling in a mouse model of Best disease. Hum Mol Genet 19 (6):1108–1118. https://doi.org/10.1093/hmg/ddp583

Zhang X, Yang D, Hughes BA (2011) KCNQ5/K(v)7.5 potassium channel expression and subcellular localization in primate retinal pigment epithelium and neural retina. Am J Physiol Cell Physiol 301(5):C1017–C1026. https://doi.org/10.1152/ajpcell.00185.2011

Chapter 10
Ion Transport in the Choroid Plexus Epithelium

Laura Øllegaard Johnsen, Helle Hasager Damkier, and Jeppe Praetorius

Abstract The epithelial cells of the choroid plexus secrete fluid at a very high rate. Therefore, the modest-sized tissue has been studied for decades as a model for epithelial secretion. It was soon observed that the choroid plexus epithelium differs from many other secretory epithelia in the overall orchestration of main ion transport mechanisms, most prominently the luminal membrane expression of the Na^+,K^+-ATPase. A renewed interest in the mechanisms of ion and fluid transport of the tissue has emerged in the recent years. This development is spurred by the aspiration for therapeutic control of cerebrospinal fluid secretion in diseases with disturbed fluid or ionic balance in the central nervous system. This chapter describes long-established features of choroid plexus ion transport, emerging areas of research in the choroid plexus physiology as well as issues of current controversy.

Keywords Cerebrospinal fluid secretion · Neuroepithelia · pH regulation · Transport proteins · Regulation of secretion

10.1 Introduction

The choroid plexus (CP) was suggested to secrete cerebrospinal fluid (CSF) as early as 1854 by Le Faivre (Faivre 1854) and later by Cushing (Cushing 1914), while the first direct experimental evidence of CSF secretion by the CP was presented by de Rougemont (de Rougemont et al. 1960). Their key observation is still among the firmest arguments for CSF secretion by the CP epithelium (CPE): that CSF is not a simple plasma ultrafiltrate.

As mentioned above, the CPE is believed to secrete the majority of the intraventricular CSF (Cserr 1971). The CSF provides buoyancy to decrease the effective weight of the brain from 1.5 kg to 50 g (Livingston 1949), which helps protect the

L. Ø. Johnsen · H. H. Damkier · J. Praetorius (✉)
Department of Biomedicine, Health, Aarhus University, Aarhus C, Denmark
e-mail: johnsen@biomed.au.dk; hd@biomed.au.dk; jp@biomed.au.dk

K. L. Hamilton, D. C. Devor (eds.), *Ion Transport Across Epithelial Tissues and Disease*, Physiology in Health and Disease,
https://doi.org/10.1007/978-3-030-55310-4_10

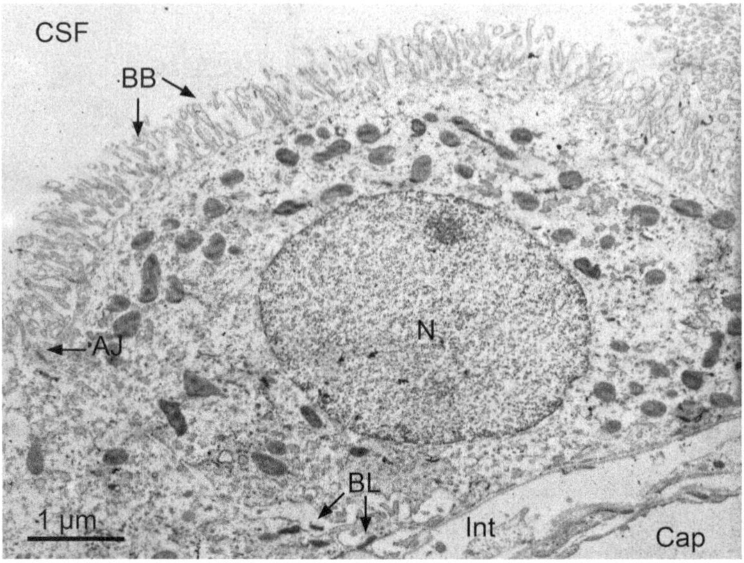

Fig. 10.1 Choroid plexus (CP) structure. An electron microscopy image of the mouse CP shows the position of the cuboidal epithelial cells with nuclei (N), and brush border (BB) between the ventricle lumen with cerebrospinal fluid (CSF) and the interstitium (Int) with capillaries (Cap). At the junction between the basal and lateral membrane a basolateral labyrinth is seen (BL)

brain parenchyma mechanically. The unproblematic access to the newly formed fluid and a very high transport rate are the main reasons the CPE has served as a model system for secretory epithelia. Also, the transport characteristics of CPE have been studied in relation to disease states and conditions, such as CPE papilloma, hydrocephalus, idiopathic intracranial hypertension, Alzheimer's disease, and a range of infectious and immunological disorders (Safaee et al. 2013; Supuran 2015; Marques et al. 2016; Watters et al. 1969; Silverberg et al. 2002).

The CPE morphology has been reviewed in depth elsewhere (Cserr 1971; Redzic and Segal 2004; Mortazavi et al. 2014). In brief, the blood equilibrates with the interstitial fluid through fenestrated capillaries. The cuboidal epithelial cells form a simple monolayer (Fig. 10.1) connected by typical contact complexes. The CSF in the ventricle lumen is separated from the lateral intercellular and basal spaces by tight junctions (TJs) close to the luminal membrane of the cell layer. CPE cells form adherence junctions below the TJs, and desmosomes appear scattered along the lateral surfaces. Focal adhesions and hemi-desmosomes adhere the CPE cells to the basement membrane. Numerous microvilli and tufts of motile cilia are present on the luminal surface of CPE cells. At the transition zone between the lateral and basal domains, a basolateral labyrinth is formed by the neighboring cells that produce numerous interdigitating protrusions thereby extending the basolateral surface area roughly tenfold.

10.2 Transport Proteins of Probable Relevance to Transepithelial Transport

10.2.1 Transcellular Ion Transport

The cellular transport of solutes as well as CSF secretion by the CPE has been studied in a variety of species and by a plethora of methods. Pioneering studies were carried out by use of radioactive ionic tracers, selective transport inhibitors, direct measurements of ionic compositions, and ionic requirement for transport. Later, patch clamping, molecular cloning, immunolocalization, and transgenic animals were exploited to uncover the mechanism in CSF secretion by the CPE.

In humans, a relatively high blood flow of 3 ml/g min nourishes the approximately 2 g of CPE tissue, which produces 500 ml CSF daily (Cserr 1971). These numbers bring the CPE among the most efficiently secreting mammalian epithelia. The fluid secretion by the CPE depends primarily on transepithelial movement of Na^+ (Pollay and Curl 1967; Welch 1963; Davson and Segal 1970), which is accompanied by Cl^-, HCO_3^- (Wright 1972) and water transport. In nascent CSF, the concentrations of Na^+ and HCO_3^- are higher, whereas Cl^- and K^+ levels are lower than what would be expected from a plasma ultrafiltrate (de Rougemont et al. 1960; Ames et al. 1964) (Fig. 10.2). The nascent CSF is approximately 5 mOsM hyperosmolar compared to the interstitial fluid beneath the epithelium (Davson et al. 1987), and there is a 5 mV lumen positive transepithelial potential difference across the monolayer (Held et al. 1964; Husted and Reed 1977; Welch and Sadler 1965). The membrane potential in CPE cells is reportedly –40 to –47 mV in mice (Roepke et al. 2011; Christensen et al. 2017).

The potent Na^+,K^+-ATPase inhibitor, ouabain, efficiently reduces both the Na^+ transport and the CSF secretion when administered luminally to the CPE via artificial CSF (Welch 1963; Davson and Segal 1970; Wright 1972) (Fig. 10.2). The general carbonic anhydrase inhibitor acetazolamide also inhibits CSF secretion efficiently, which apparently implicates HCO_3^- metabolism as a central aspect of CSF secretion (Tschirgi et al. 1954; Kister 1956; Davson and Luck 1957; Welch 1963; Ames et al. 1965; Segal and Burgess 1974; Pollay and Davson 1963; Vogh et al. 1987; Murphy and Johanson 1989a, b). Indeed, stilbene Cl^- and HCO_3^- transport inhibitors reduce Cl^- transport from the blood to the CPE and to the CSF (Deng and Johanson 1989).

A role for basolateral Na^+/H^+ exchange in CSF secretion by the CPE was indicated by the amiloride-sensitivity of the Na^+ transport from the blood side (Davson and Segal 1970). Intraperitonal amiloride decreased both the Na^+ influx into both the CPE and the CSF secretion even after ruling out competing renal effects of the drug by renal artery ligature (Johanson and Murphy 1990; Murphy and Johanson 1989a, b). Finally, the $Na^+,K^+,2Cl^-$ cotransport inhibitor bumetanide strongly reduced CSF secretion when applied from the luminal side (Bairamian et al. 1991; Javaheri and Wagner 1993; Keep et al. 1994). Furosemide only inhibits choroid plexus K^+ and Cl^- transport when applied from the CSF side, which most likely reflects luminal K^+,Cl^- cotransport activity (Zeuthen 1994; Zeuthen and

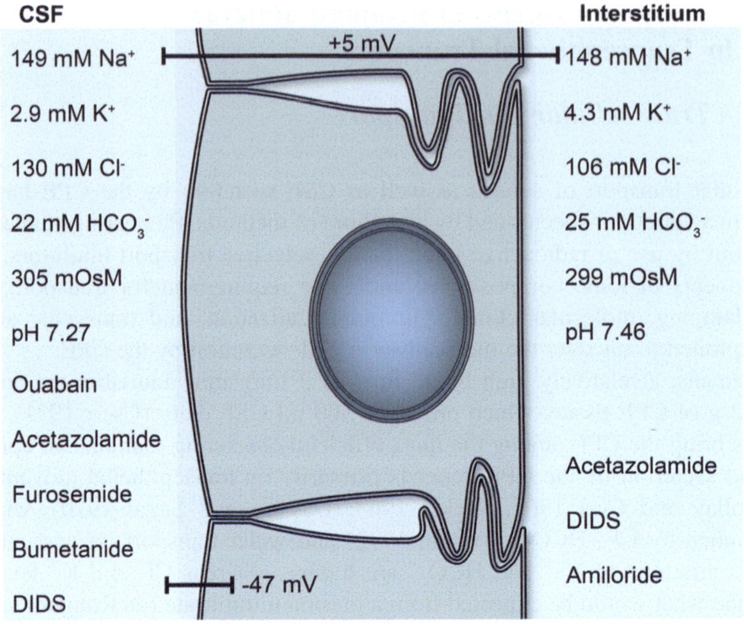

Fig. 10.2 Sites of action for inhibitors of CSF secretion by the CPE and the ionic composition of the interstitial fluid and the nascent cerebrospinal fluid. The inhibitors furosemide, ouabain, acetazolamide and DIDS are reported to inhibit CSF secretion by the CPE from the ventricle facing/luminal membrane, while amiloride, acetazolamide and DIDS are inhibitors acting from the blood facing/basolateral membrane (see text for details). The transepithelial potential difference amounts to 5 mV lumen positive (Held et al. 1964) and the membrane potential in moue CPECs is –47 mV (Christensen et al. 2017). The concentration of simple ions in the CSF and blood side are from Davson et al. (1987)

Wright 1981). Thus, early studies provided strong evidence that CSF secretion by the CPE depends on the concerted action of the Na^+,K^+-ATPase, a Na^+/H^+ exchanger, an anion exchanger, carbonic anhydrase, and cotransport through NKCC.

10.2.2 Na^+,K^+-ATPase Complexes

One striking feature of the CPE is the luminal membrane expression of the Na^+,K^+-ATPase. Ouabain applied from the ventricular side potently inhibits secretion by the CPE (Welch 1963; Davson and Segal 1970; Wright 1972), and the Na^+,K^+-ATPase was indeed demonstrated to be positioned in the luminal membrane in CPE cells (Masuzawa et al. 1984; Siegel et al. 1984). The CPE expresses multiple subunits of the Na^+,K^+-ATPase, such as the α1, β1, and β2 subunits, as well as the accessory FXYD protein phospholemman (Watts et al. 1991; Gonzalez-Martinez et al. 1994;

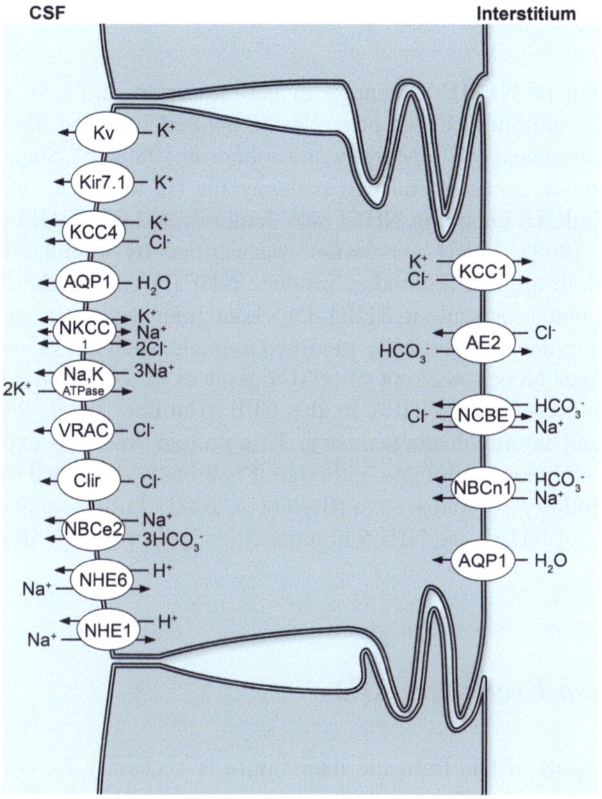

Fig. 10.3 Distribution of transport mechanisms believed to be directly or indirectly implicated in CSF secretion by the CPE. The luminal membrane is occupied by the Na^+,K^+-ATPase, the water channel AQP1, three K^+ channels, two Cl^- conductances, cotransporters NKCC1, and KCC4. Acid/base transporters NHE1 and NBCe2 are also localized to the luminal membrane. Fewer types of transport proteins seem present in the basolateral membrane domain. The KCC3 cotransporter has been suggested at the basolateral membrane along with modest numbers of AQP1 channels. Multiple acid base transporters are expressed in the basolateral membrane: the anion exchanger AE2, the cotransporter NBCn1 (sometimes a luminal membrane protein) and the NCBE, which is either a Na^+ dependent Cl/HCO_3^- exchanger or a Na^+:HCO_3^- cotransporter. Carbonic anhydrase 2 (CAII) is cytosolic, whereas the membrane attached CAXII is localized to the basolateral membrane

Feschenko et al. 2003). From the position in the luminal membrane, the Na^+,K^+-ATPase directly extrudes Na^+ into the CSF and is therefore often seen as the pivotal transport process in secretion by the CPE (Fig. 10.3). In contrast to most other transporting epithelia, the K^+ recycling occurs across the luminal membrane (see below). In addition to extruding Na^+ to the CSF, the Na^+,K^+-ATPase builds the electrochemical driving force for many other transport processes.

10.2.3 Na^+/H^+ Exchangers, NHE1 and NHE6

The involvement of Na^+/H^+ exchange in Na^+ transport and CSF secretion was indicated by the inhibitory effect of amiloride applied from the blood side of the CPE (Davson and Segal 1970; Murphy and Johanson 1989a, b). Such a mechanism would also provide a basolateral exit pathway for H^+ to sustain luminal HCO_3^- secretion. The mRNA encoding NHE1 was demonstrated first by RT-PCR analysis (Kalaria et al. 1998). NHE1 expression was verified by immunohistochemistry. Interestingly, both applied antibodies produce NHE1 signal in the luminal membrane domain that is absent in NHE1 knockout mice and only localizes to the basolateral membrane in genetically modified animals, such as Na^+-dependent Cl^-/HCO_3^- exchanger Ncbe knockout mice (Damkier et al. 2009). We have recently reported the expression of NHE6 in the CPE (Damkier et al. 2018) by mass spectrometry and immunohistochemistry. This protein was also expressed in the luminal plasma membrane domain, although it is broadly described as a membrane protein of the endo-lysosomal system (Brett et al. 2002). Future studies to define the respective roles of NHE1 and NHE6 in intracellular pH or even CSF pH regulation are warranted.

10.2.4 Anion Exchanger AE2

The cellular import of Cl^- from the interstitium is necessary for continuous CSF secretion. A role for classical anion exchangers was indicated by the sensitivity of Cl^- transport from the interstitial side to the CSF by basolateral application of the stilbene derivative DIDS (Deng and Johanson 1989; Frankel and Kazemi 1983). The DIDS-sensitive Cl^- import proved to be HCO_3^- dependent, suggesting the action of a Cl^-/HCO_3^- exchanger (Hughes et al. 2010). Interestingly, the anion exchanger AE2 was originally cloned from the mouse CPE and immunolocalized to the basolateral membrane domain of CPECs in several species (Lindsey et al. 1990; Alper et al. 1994; Praetorius and Nielsen 2006). The classical Cl^-/HCO_3^- exchangers AE1-AE3 belong to the *slc4* gene family of HCO_3^- transporters, and mediate electroneutral, DIDS-sensitive anion exchange (Alper 2009). At present, AE2 seems to be the only basolateral Cl^- entry mechanism in CPE cells.

10.2.5 Na^+-Dependent HCO_3^- Transporters NBCe2, Ncbe and NBCn1

The presence of CO_2/HCO_3^- is a requirement for efficient CSF secretion (Haselbach et al. 2001), indicating the central involvement of HCO_3^- transporters in CSF secretion. In addition to the effect of basolateral DIDS, injection of DIDS into the brain ventricles also affects the secretion by CPE cells, mainly affecting the CSF HCO_3^- concentration (Nattie and Adams 1988). Electrogenic transport of HCO_3^- across the luminal membrane has been demonstrated (Saito and Wright 1984), and the Na^+ dependence of transport indicated an outwardly directed Na^+:HCO_3^- cotransport mechanism (Johanson et al. 1992a, b). In support of this notion, DIDS inhibits Na^+ secretion from CPE cells in culture (Mayer and Sanders-Bush 1993). The mechanism responsible for this luminal transport seems to be the electrogenic Na^+:HCO_3^- cotransporter 2, NBCe2 (Bouzinova et al. 2005; Millar and Brown 2008). Indeed, we recently reported that NBCe2 is required for luminal alkaline extrusion and is a major mechanism in restoring CSF pH after acidification (Christensen et al. 2018).

As indicated above, a bulk of early evidence suggested that basolateral NaCl import was mediated by combined AE2 and NHE1 activity. Therefore, the immunolocalization of the Na^+-dependent Cl^-/HCO_3^- exchanger Ncbe and the Na^+-HCO_3^- cotransporter NBCn1 to the basolateral membrane domain of CPE cells was unexpected (Bouzinova et al. 2005; Damkier et al. 2009; Praetorius et al. 2004; Praetorius and Nielsen 2006). Ncbe mediates by far most of the Na^+-dependent HCO_3^- uptake into isolated CPE cells and, taken together with the small brain ventricle volume observed in Ncbe knockout mice, this suggests a strong dependence of CSF secretion on Ncbe activity (Jacobs et al. 2008). It should be mentioned that Ncbe knockout greatly affects the cellular expression of various membrane transporters such as the Na^+,K^+-ATPase and AQP1, as well as anchoring proteins (Christensen et al. 2013; Damkier et al. 2009; Damkier and Praetorius 2012). Thus, it has not been directly demonstrated that Ncbe is implicated in CSF secretion.

10.2.6 Carbonic Anhydrases

Carbonic anhydrases catalyze the conversion of H_2O and CO_2 to H_2CO_3, which then undergoes spontaneous conversion into H^+ and HCO_3^- (Boron and Boulpaep 2012). The activity of carbonic anhydrases seems to be crucial for secretion by the CPE, as the inhibitor acetazolamide strongly reduces CSF production (Ames et al. 1965; Davson and Segal 1970; Welch 1963). This led to the hypothesis that HCO_3^- secretion by the CPE depended on intracellular formation from CO_2 and H_2O concomitant with basolateral H^+ extrusion through NHE1. At the time these experiments were performed, the carbonic anhydrase activity was considered to be strictly cytosolic (i.e., mediated by CAII). However, membrane bound carbonic anhydrases CAXII and CAIX are expressed in the CPE (Kallio et al. 2006). With the demonstration of carbonic anhydrases with extracellular activity it is likely that the inhibitory effect of acetazolamide on CSF secretion is at least partly caused by inhibition of transcellular HCO_3^- transport rather than by blockage of intracellular hydration of CO_2.

In case the intracellular CAII dominates catalytic activity and forms intracellular H^+ and HCO_3^- as part of the CSF secretion, the continued luminal secretion of

HCO_3^- would require basolateral extrusion of H^+. In the absence of NHE expression in the basolateral membrane (Damkier et al. 2009), there is no known transport mechanism in the basolateral membrane to support this transport. Of all known proteins in CPE cells, only the basolateral Na^+:HCO_3^- importers Ncbe and NBCn1 would mediate a similar effect on intracellular pH. In case the basolateral extracellular carbonic anhydrase activity is the primary target of acetazolamide, it is likely that the transport rates of AE2, Ncbe or NBCn1 are reduced resulting in diminished CSF secretion. Thus, it would be interesting to know whether acetazolamide acts primarily on extracellular or intracellular carbonic anhydrases.

10.2.7 Na^+,K^+,2Cl^- Cotransporter 1, NKCC1

The implication of NKCC1 function in CSF secretion was suggested by its localization to the luminal membrane of CPE cells (Plotkin et al. 1997; Praetorius and Nielsen 2006), and the inhibition of CSF secretion by bumetanide applied to the CSF side of the epithelium (Bairamian et al. 1991; Javaheri and Wagner 1993; Keep et al. 1994). The explanation for the inhibitory effect of bumetanide remains controversial. The NKCC proteins transport Na^+,K^+, and Cl^- into cells, driven by the inward Na^+ and Cl^- gradients in most cell types. Inward transport has been experimentally supported (Bairamian et al. 1991), and implicated in regulatory volume increase (Wu et al. 1998). The luminal membrane of CPE cells, however, is exposed to a low extracellular K^+ concentration that could neutralize or reverse the net inward driving force (Keep et al. 1994). Recently, NKCC1 was again shown to work in both inward and outward transporting modes in CPE cells in vitro and in vivo (Steffensen et al. 2018; Gregoriades et al. 2018); thus, its exact role in CSF secretion remains much debated.

10.2.8 K^+,Cl^- Cotransporters

Furosemide-sensitive K^+ and Cl^- transport established a luminal K^+,Cl^- cotransport in the amphibian CPE (Zeuthen 1994; Zeuthen and Wright 1981; Zeuthen 1991), but it is unknown whether these proteins play a role in mammalian systems. The K^+,Cl^- cotransporters (KCCs) transport the ions outward in an electroneutral manner, driven by the chemical K^+ gradient. The mRNA encoding KCC1 has been detected in the CPE, but the protein expression has not yet been shown in the CPE (Kanaka et al. 2001). KCC4 is reportedly expressed in the CPE and was immunolocalized to the luminal membrane, where it might contribute to K^+ recycling (Karadsheh et al. 2004; Li et al. 2002). KCC3a seems to be expressed in the basolateral membrane domain of CPE cells, from where it would be the only suggested mechanism for K^+ extrusion in a transcellular K^+ absorption from the CSF (Pearson et al. 2001). Mass spectrometry identification of K^+,Cl^--cotransporters in the mouse CPE and qPCR indicates the expression of only KCC1 and KCC4 (Damkier et al. 2018; Steffensen et al. 2018).

10.2.9 K^+ Channels

The luminal membrane of CPE cells display K^+ conductances that are thought to sustain CSF secretion through recycling of K^+ (Zeuthen and Wright 1981). An inward-rectifying conductance (Kir) and an outward-rectifying conductance (Kv) have been reported (Kotera and Brown 1994). The inward-rectifying conductance might well be carried by the Kir7.1 channels expressed in the CPE (Döring et al. 1998), whereas the outward-rectifying conductance may implicate Kv1.1/ Kv1.3 channels (Speake et al. 2004) as well as KCNQ1 and KCNE2 channels (Roepke et al. 2011). All these K^+ channels were immunolocalized to the luminal plasma membrane domain (Nakamura et al. 1999; Roepke et al. 2011; Speake et al. 2004).

10.2.10 Cl^- Channels and Related Proteins

Cl^- and HCO_3^- are both secreted into CSF mainly by electrogenic mechanisms in the luminal membrane, and, thus, Cl^- channels seem to mediate the majority of the Cl^- transport. Two separate Cl^- conductances have been characterized in CPE cells. The first is an inward-rectifying protein kinase A-activated conductance, and the other is a volume-sensitive conductance (Kajita and Brown 1997; Kibble et al. 1997; Kibble et al. 1996). The main electrogenic anion efflux pathway in the luminal membrane of the amphibian CPE is a cAMP-activated anion conductance with high HCO_3^- permeability (Saito and Wright 1983, 1984).

In the mammalian CPE, the Clir channels also possess a high HCO_3^- conductance (Kibble et al. 1996). The volume-regulated anion conductance seems most active during cell volume regulation (Kibble et al. 1996, 1997) and might be insignificant in relation to CSF secretion (Millar et al. 2007). The molecular identities of the Cl^- channels have not been resolved, but studies have ruled out that CFTR and Clc-2 mediate the Clir activity (Kibble et al. 1996; Kibble et al. 1997; Speake et al. 2002). Mass spectrometry identification of Cl^- pathways in the mouse CPE indicates the expression of voltage-dependent anion channels (Vdac), more subunits of a volume regulated Lrrc channel and H^+/Cl^- exchangers (Clcn) in mouse CPECs (Damkier et al. 2018).

10.2.11 Na^+ Channels

The lack of apparent NHE polypeptides in the basolateral membrane domain of CPE cells opened the possibility of that other amiloride sensitive Na^+ transport mechanisms reside in the membrane in order to explain the effect of basolateral drug administration on CSF production. One such mechanism is the epithelial Na^+ channel (ENaC), which is normally restricted to quite tight epithelia. Several studies by Leenen and colleagues provide evidence for molecular and functional expression of ENaC in the CPE (Amin et al. 2005, 2009; Leenen 2010; Van Huysse et al. 2012; Wang et al. 2010; Leenen et al. 2015), mainly as a candidate luminal membrane channel. The inward electrochemical Na^+ gradient would make ENaC a candidate for mediating the basolateral amiloride-sensitive Na^+ entry. The published immunohistochemical analyses, however, yielded cytosolic anti-ENaC staining and did not support a primary basolateral expression of ENaC (Amin et al. 2009). Thus, the involvement of ENaC in basolateral Na^+ import seems unlikely. It should be noted that we and other researchers fail to detect the expression of all of the ENaC subunits or their function in CPE cells (own unpublished results and personal communications with Peter D. Brown).

10.2.12 Transcellular Water Transport

The water transport across the CPE seems to require transepithelial Na^+ transport (Ames et al. 1965; Pollay and Curl 1967; Welch 1963). Most widely, the water movement is believed to be driven by the 5 mOsM luminal hyperosmolarity (Davson and Segal 1970). The aquaporin 1 (AQP1) is highly H_2O-permeable and is robustly expressed in the luminal membrane domain of CPE cells and with much lower abundance in the basolateral membrane (Nielsen et al. 1993; Praetorius and Nielsen 2006). The basolateral AQP1 immunoreactivity is not negligible, as one might think from the published images. The basolateral AQP1 signal is actually similar to that of the nearby capillary endothelial cells. Thus, bilateral AQP1 might constitute a transcellular route for H_2O on its own to support CSF secretion.

The CPE H_2O permeability is reduced by 80% in AQP1 knockout mice (Oshio et al. 2005). The CSF secretion rate is, however, decreased by only 35% (Oshio et al. 2005), which could be explained in several ways. By comparison to the proximal tubule, AQP1 knockout may result in altered osmotic driving force in compensation for the reduced permeability (Vallon et al. 2000; Schnermann et al. 1998). Also, if the fact that CPE secretes only 70–80% of the CSF is taken into consideration, the effect of AQP1 knockout on secretion by CPE is underestimated. The paradigm of secondary active water transport by solute-water cotransporters, such as KCC cotransporters, was actually studied in the CPE early on in non-mammalian species (Zeuthen 1991; Zeuthen et al. 2016; Charron et al. 2006). Recently, the same features were ascribed to NKCC1 in the mouse choroid plexus (Steffensen et al.

2018). However, it is not demonstrated that lack of NKCC1 has an effect similar to or exceeding that of AQP1 knockout on water transport or CSF secretion.

10.2.13 Paracellular Transport

The serial arrangement of semi-leaky tight junctions (TJs) and the putatively transport-limiting morphology of the basolateral labyrinth can probably explain the intermediate range electrical resistance of $\sim$170 Ω cm^2 in the amphibian CPE and 50–200 Ω cm^2 in various mammalian cultured cells (Wright 1972, Damkier et al. 2013). The permeability of TJs is determined by the number of parallel strands as well as the expression of pore-forming claudins (Kriegs et al. 2007). Claudin-2 is the only TJ protein known to be pore-forming (Krug et al. 2012), among the array of claudins expressed by CPE cells (Wolburg et al. 2001; Kratzer et al. 2012). As the only claudin, claudin-2 has been described to permeate both cations and H_2O (Rosenthal et al. 2010, 2016). Thus, the paracellular route should be considered to contribute to the 20% of the transepithelial H_2O permeability that is not ascribed to AQP1. The transepithelial pathways of H_2O remain to be essential themes in the study of CSF secretion by the CPE.

10.3 The Choroid Plexus Epithelium in Cerebrospinal Fluid Secretion

A bulk of evidence supports a role for the CPE in the secretion of CSF (de Rougemont et al. 1960; Cserr 1971; Spector et al. 2015; Damkier et al. 2013; Hladky and Barrand 2016). It is estimated that the CPE contributes 70–80% of the CSF, with the remaining arising from the brain interstitial fluid and thereby indirectly from the blood-brain barrier (Redzic et al. 2005). This central paradigm is, surprisingly, still debated despite the following evidence in favor of CSF secretion by the CPE.

The CSF composition is incompatible with an origin as a plasma ultrafiltrate. The 5 mOsM hyperosmolarity of the CSF compared to blood plasma is evidence against a hydrostatic process (Davson and Purvis 1954). Also the higher-than-expected concentrations of Na^+ and HCO_3^- and lower K^+ and Cl^- concentrations in the CSF are arguments against filtration (Hughes et al. 2010), as is the 5 mV lumen positive transepithelial potential difference across the CPE (Welch and Sadler 1965).

The CP is shown to secrete fluid. Firstly, an increase in hematocrit from the arteries to the veins demonstrates a net loss of fluid from the blood during the perfusion of the CP (Davson et al. 1987). Secondly, when oil is applied on top of the CPE, fluid forms beneath the oil in vivo. Thirdly, the ionic contents of fluid produced by the luminal surface of the CPE closely resembled the composition of

CSF of the cisterna magna (de Rougemont et al. 1960). Finally, the basal rate by which the CPE produces fluid is unparalleled by any other mammalian epithelium (Damkier et al. 2013).

Ion transport inhibitors alter Na^+ transport and fluid secretion. Overall CSF secretion and CPE transport rates are equally affected by, e.g., ouabain, azetazolamide, amiloride and DIDS (Ames et al. 1965; Pollay et al. 1985; Wright 1978; Zeuthen and Wright 1978; Davson and Segal 1970; Davson and Luck 1957; Johanson et al. 1990, 1992a, b; Johanson and Murphy 1990; McCarthy and Reed 1974; Murphy and Johanson 1989a, b; Tschirgi et al. 1954; Vogh and Godman 1985; Welch 1963).

Ventricle volumes are affected by genetic disruption of CPE transporters. Knock-out of ion transporters Ncbe or NBCn2, and the water channel AQP1 reduce the ventricle volume and/or the CSF secretion rate (Jacobs et al. 2008; Kao et al. 2011; Oshio et al. 2005). Thus, there is ample evidence for fluid secretion by the CPE and a general dependence of CSF volume on secretion by the CPE across the many research laboratories and among a plethora of techniques.

10.4 Regulation of Cerebrospinal Fluid (CSF) Secretion

The rate of CSF secretion is classically believed to be relatively constant. Nevertheless, both the CPE cells and the vasculature supplying the CPE express several receptors capable of regulating the activity of transport proteins implicated in CSF secretion (Nilsson et al. 1992; Chodobski and Szmydynger-Chodobska 2001). Indeed, factors originating from the blood as well as nerves supplying the CPE are implicated in regulating the production of CSF (Lindvall and Owman 1981; Faraci et al. 1988; Salpietro et al. 2014). In addition, paracrine factors and mediators from the CSF itself seem to affect CSF formation through action on the CPE, as detailed in the following sections. Direct action on receptors in the CPECs will most likely change CSF secretion rate dependent on the transporters involved.

Action on receptors in the vasculature can certainly increase the supply of solutes to the CPE without hindrance through the leaky capillaries but the CPEC still must be the rate-limiting step in the secretion. Therefore, there is likely not a linear connection between increased blood supply by vasodilation of the choroidal arteries and increased CSF secretion. The opposite situation of vasoconstriction is probably more likely to affect secretion rate by limiting supply of solutes to the transporters in the basolateral membrane of the CPECs. It is difficult to directly study perfusion of the choroid plexus while studying CSF secretion. One study using MRI found an age-dependent decrease in CP permeability (Bouzerar et al. 2013) that is in line with the decreased CSF secretion described in animal models. More studies are, however, necessary to define the correlation between CP perfusion and CSF secretion under normal circumstances.

10.4.1 Arginine Vasopressin

The neuropeptide arginine vasopressin (AVP) is produced mainly in the hypothalamus; however, the neuropeptide is also an example of a factor which not only reaches the CP from the plasma and CSF but also is produced by the CP itself (Hallbeck et al. 1999; Szmydynger-Chodobska et al. 2006). The CP is also a site of action of AVP, as the AVP type 1a (V_{1a}) receptor is highly expressed in the luminal membrane of the CPE, in addition to the receptor expression in the choroidal arterioles (Szmydynger-Chodobska et al. 2004; Segal et al. 1992; Hernando et al. 2001). The production and release of AVP from the CPE cells are stimulated by centrally released angiotensin II (Ang II) (Szmydynger-Chodobska et al. 2004).

Systemic administration of AVP has been shown to increase the water content in the brain tissue and, hence, has the potential to exacerbate various forms of brain injury resulting in cerebral edema (Szczepanska-Sadowska et al. 1984; Trabold et al. 2008). Additionally, it has been shown that increased CSF osmolality and dehydration are followed by an increase in AVP concentration (Dogterom et al. 1978; Zerbe and Robertson 1983; Szmydynger-Chodobska et al. 2006). Contrary to the general effect of AVP on the brain, the overall mechanism of AVP on the CP is a decreased CSF formation rate, presumably due to the AVP-mediated reduced blood flow to the CP (through inhibition of the vascular V_1-receptor) (Faraci et al. 1990). This decrease in CSF production occurs despite an increased water permeability in the CP capillaries (Raichle and Grubb 1978). Other studies have suggested that AVP's inhibitory effect on the CP is mediated through regulation of membrane transporters including a reduction of Cl^- efflux (Johanson et al. 1999), but a more thorough investigation of the exact effect of AVP on the CP remains to be conducted.

10.4.2 Aldosterone

Besides the production in the adrenal glands, the mineralocorticoid aldosterone is also produced in the hypothalamus, and high-affinity mineralocorticoid receptors (MR) have also been described in the CP (de Kloet et al. 2000). In general, aldosterone plays an important role in the systemic Na^+ and H_2O balance, mainly through its effect on the kidneys. It is generally accepted that aldosterone is present within the CSF in concentrations correlating to plasma levels and removal of the adrenal glands diminishes brain aldosterone content, suggesting that the aldosterone present in the CSF primarily is derived from the adrenal glands (Weber 2003). Intracerebroventricular administration of aldosterone has been shown to cause elevated systemic blood pressure and CSF production through increased Na^+-K^+-ATPase activity (Kageyama and Bravo 1988).

Cortisol and aldosterone have similar affinity for MR; however, plasma levels of cortisol exceed aldosterone levels. The CP expresses both MR and glucocorticoid receptors (GR) to which cortisol binds equally well. Normally, specificity of aldosterone binding to MR is secured by enzymatic conversion of cortisol to the inactive cortisone by 11β-Hydroxysteroid dehydrogenase II (HDS-11βII). Both human and rabbit CP expresses HDS-11βI (converts cortisone to cortisol) but not HDS-11βII (Sinclair et al. 2007). The fact that the CP expresses both MR and GR but does not express HDS-11βII is puzzling, as HDS-11βII enzymes in general are expressed in aldosterone-selective tissue (Seckl 1997). Therefore, no well-defined mechanism for aldosterone in the choroid plexus has yet been established, and there is no specific mechanism of action that aldosterone applies upon binding to MR.

The effect of cortisol levels and HDS-11βI on CSF secretion by the choroid plexus has, however, been suggested to play a role in the puzzling disorder idiopathic intracranial hypertension (IIH). IIH is a disorder characterized by increased volume of CSF leading to increased intracranial pressure. The patients suffer from severe headaches and optic nerve compression resulting in visual disturbances. In severe cases the patients require cannulation of CSF or fenestration of the optic nerve sheath. The exact mechanism behind the disorder has not been found, but lowering CSF production in general by various diuretics, including furosemide and acetazolamide, is often the treatment of choice. Patients suffering from IIH are primarily female and almost always obese. Interestingly, a clinical study revealed that following weight loss the overall levels of HDS-11βI activity fell and did so in correlation with the fall in intracranial pressure (Sinclair et al. 2010). The change in CSF cortisone levels correlated with the weight loss. This suggested the involvement of increased CSF cortisone and HDS-11βI activity in IIH and thereby in CSF production. Understanding the molecular mechanism behind the increase in CSF production by activation of HDS-11βI needs to be further investigated.

10.4.3 Angiotensin II

The CPE cells are related to the renin-angiotensin-aldosterone axis. Renin, angiotensinogen, and angiotensin-converting enzyme (ACE) are all produced by the CPE cells (Chai et al. 1987; Imboden et al. 1987; Inagami et al. 1980), and in addition, the Ang II receptors AT_{1A} and AT_{1B} are expressed in the CP (Chen et al. 1997; Johren and Saavedra 1996). In regard to the peripheral blood circulation, Ang II is a potent vasoconstrictor and systemic administration of Ang II causes a decrease in CP blood flow (Maktabi et al. 1990). Intracerebroventricular administration of Ang I and II also causes a decreased blood flow to the CP but does not affect the cerebral blood flow. Ang II mediates the release of AVP from the CPE cell, by binding to the AT_1 receptor (Chodobski et al. 1998). After AVP has been released from the CPE cell, it binds to the V_1 receptors and causes a decrease in blood flow and CSF production rate (see Sect. 10.4.1). To further support this, it has been shown that intraventricular administration of ACE inhibitors causes increased CSF production in rat (Vogh and Godman 1989).

10.4.4 Growth Factors

The CP is known to secrete several different growth factors (Chodobski and Szmydynger-Chodobska 2001). This includes fibroblast growth factor 2 (FGF2), insulin-like growth factor I and II (IGF-I and IGF-II), vascular endothelial growth factor (VEGF), nerve growth factor (NGF), and transforming growth factor α and β (TGF-α and TGF-β) (reviewed in Kaur et al. 2016). The growth factors have several effects on the CPE; the IGF-II has, e.g., been proposed to have an autocrine regulatory mechanism on the growth of the CPE cells by binding to its receptors expressed on the CPE (Nilsson et al. 1996). Other growth factors such as VEGF and TGF-β have been shown to be involved in the maintenance of the CP, where VEGF upholds the vasculature of the CP in addition to having a role in the permeability of the blood vessels, and TGF-β has a role in repairing tissue secondary to injury (Maharaj et al. 2008). Furthermore, the growth factor FGF2, secreted by the CP, indirectly reduces the formation of CSF via promotion of AVP release from the CPE (Johanson et al. 2004; Szmydynger-Chodobska et al. 2002).

10.4.5 Cyclic Adenosine Monophosphate

It is generally accepted that specific hormone-receptor interactions stimulate the membrane bound adenylyl cyclase (AC) which facilitates the conversion of ATP to cyclic adenosine monophosphate (cAMP). cAMP has been shown to act as a second messenger to mediate hormonal actions in several secretory epithelia, e.g., the CP (Deng and Johanson 1992). As previously mentioned, the fluid secretion by the CP primarily depends on transepithelial movement of Na^+, Cl^-, and HCO_3^- followed by H_2O, and the transport across the CP of Cl^- and HCO_3^- has been found to be regulated by cAMP; hence, increased intracellular levels of cAMP have been associated with increased CSF production (Deng and Johanson 1992; Saito and Wright 1984). Several factors have been shown to regulate the AC conversion of ATP to cAMP. These factors include adrenaline, noradrenaline, histamine, serotonin, vasoactive intestinal peptide (VIP), and prostaglandins PGE_2 and PGI_2 (reviewed in (Serezani et al. 2008)). However, the full mechanism of cAMP-regulated CSF secretion is yet to be established (Banizs et al. 2007).

10.4.6 Nervous Regulation

Both the arteries supplying the CP and the epithelial cells are rich in nerve fibers from the autonomic nervous system. The choroidal arteries are innervated by vasodilatory parasympathetic fibers originating from facial and glossopharyngeal ganglia, vasocontractory fibers from the sympathetic nervous system originating in the cervical ganglia, and sensory fibers from the trigeminal nerve (Hara et al. 1993). The CP epithelial cells also receive autonomic innervation, both sympathetic and parasympathetic, through binding of different neurotransmitters to receptors located on the CPE cells (Lindvall and Owman 1981; Nilsson et al. 1991; Rotter et al. 1979). The sympathetic nervous system is thought to have a tonic inhibitory influence on CSF production. It acts by decreasing CSF production through adrenergic activation of β1 receptors located on the CPE cells, which again is believed to cause a decreased activity in carbonic anhydrase (Vogh and Godman 1985). To further support this, sympathetic denervation has been shown to increase CSF production and intracranial pressure (Lindvall et al. 1978).

In addition, the CPE cells receive innervation from parasympathetic cholinergic nerve fibers which some studies indicate reduces CSF production. The mechanism of the cholinergic agents is most likely mediated through nitric oxide which decreases the Na^+-K^+-ATPase activity (Ellis et al. 2000). Another study, however, indicated that irritation of the parasympathetic glossopharyngeal nerve following subarachnoid hemorrhage increases CSF secretion, but the following destruction of the appertaining ganglion decreased secretion by actions directly on the CPE (Aydin et al. 2014). The cellular pathway for the increased secretion was, however, not studied. The exact role of the autonomic nervous system on both vasculature and the CPE in terms of CSF secretion is complicated. Understanding the pathways in both health and disease could be a means for targeting treatments such as the hydrocephalus that occurs in many patients after subarachnoid hemorrhage (O'Kelly et al. 2009).

10.4.7 Other Mediators

Endothelin 1 (ET-1) is a potent vasoconstrictor which is both synthesized by and has autocrine actions on the CPE cells (Hemsen and Lundberg 1991). Systemic administration of ET-1 has been shown to decrease choroidal blood flow and slightly decrease CSF production (Granstam et al. 1993; Schalk et al. 1992). Proinflammatory cytokines are generated as a response to brain injury and infection by various cells, for example, macrophages, lymphocytes, fibroblasts, and endothelial cells. The CPE cells are also capable of producing proinflammatory cytokines in response to per se infection (Schwerk et al. 2011). By binding of the cytokines to specific receptors on the cell surface and through different signaling pathways, inflammation causes altered gene transcription and protein expression (Block et al. 2007; Bonow et al. 2009). Furthermore, infection and inflammation can cause an increased CSF production (Bauer et al. 2008); however, the underlying mechanism have not yet been established.

The natriuretic peptide system has antagonizing properties against the renin-angiotensin-aldosterone system, and increased levels of atrial natriuretic peptide (ANP) and brain natriuretic peptide have been found in patients with increased intracranial pressure (Yamasaki et al. 1997; Kirchhoff et al. 2006). Intraventricular administration of ANP has been shown to reduce intracranial pressure and CSF production by up to 35% by a cGMP-dependent mechanism (Zorad et al. 1998; Steardo and Nathanson 1987). As mentioned above, a lot of different factors and mediators (surely more mediators than covered here), which both originate from the CP itself or reaches the CP through blood or CSF, have a regulatory effect on the CP's CSF secretion. However, it is important to bear in mind that the CP has a constitutively active CSF secretion, contrary to other secretory tissues, e.g., the pancreas, which relies upon hormones and neurotransmitters to activate secretion. Thus, all of the above-mentioned factors are only modulatory on the CSF secretion and several underlying mechanisms still need to be established.

10.5 The Choroid Plexus Epithelium Regulates Cerebrospinal Fluid pH

The neuronal excitability is greatly affected by pH changes in the brain parenchyma (Baron et al. 1985). Firing of action potentials builds up the acid content in the interstitium, which are usually buffered and cleared and do not lead to functional damage. In some circumstances, acid production can protect brain function. The high neuronal activity during seizures builds up an acidic environment, which activates acid-sensing ion channels. This leads to inhibition of the neuronal activity and, thus, to breaking the seizure (Ziemann et al. 2008). In febrile seizures, hyperventilation during the development of fever leads to respiratory alkalosis (Schuchmann et al. 2006). The seizure is broken by the central pH decrease that follows the increased neuronal activity as well as the respiratory decline. Thus, one might expect the CSF to be a regulator of brain pH with a higher capacity for buffering alkalosis than acidosis. As will be described below, this is not the case.

Proteins play a central role in buffering the effects of sudden changes in pH of the blood plasma and other fluid spaces. In contrast, the CSF contains only small amounts of protein, but CSF pH is nevertheless maintained within a quite narrow range. Fluctuations in plasma pH and pCO_2 are followed over time by similar changes in CSF pH (Lee et al. 1969). While CO_2 crosses all brain barriers rapidly, H^+ and HCO_3^- mainly cross the CPE via membrane transporters. One would expect this to result in sizable fluctuations in CSF pH in, e.g., respiratory acidosis because of the lack of protein buffers. However, in respiratory acidosis the decrease in blood pH surpasses that of CSF pH, indicating the presence of an efficient buffer such as the open CO_2/HCO_3^- buffer system. In support of such buffering, Hasan and collaborators found that inhalation of 5% CO_2 for 4 h led to a threefold higher increase in CSF HCO_3^- compared to plasma HCO_3^- in dogs (Hasan and Kazemi 1976). The robust expression of a HCO_3^- export protein NBCe2 in the luminal plasma membrane of the CPE spurred the idea that this tissue could serve a CSF pH regulatory function (Bouzinova et al. 2005) (Fig. 10.4).

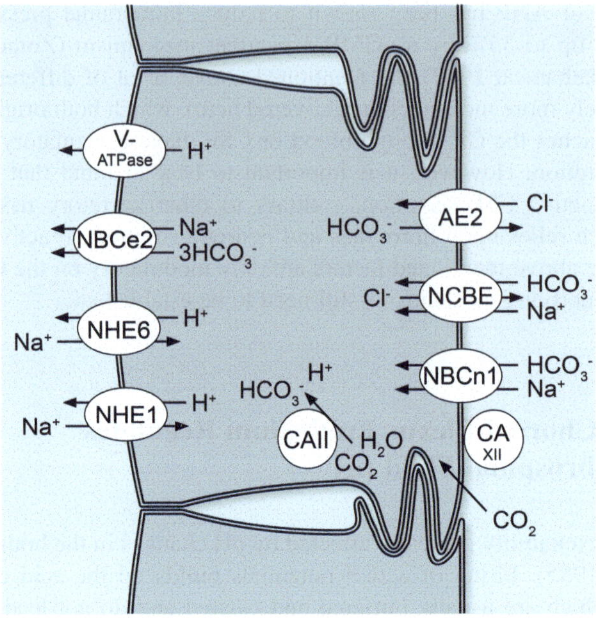

Fig. 10.4 Molecular machinery thought to be involved in CSF pH regulation by the CPE. While the basolateral AE2 and NCBE may sustain luminal transport of acid/base equivalents, they are more likely to load the CPECs with mainly Na^+ and Cl^-. At the luminal membrane, NHE1 might be a mechanism to acidify CSF, while NBCe2 may be a potent base extruder to the CSF. The *SLC4A11* gene product has also been localized to the luminal membrane in humans, but no functional evidence has been produced to implicate this transporter in CSF pH regulation. The V-ATPase has been demonstrated in low copy numbers in the CPECs, but its function did not seem to affect CSF pH even during cAMP stimulation (Christensen et al. 2017)

NBCe2 transports Na^+ and HCO_3^- out of the CPE cells, most likely with a 1:3 stoichiometry (Millar and Brown 2008), and with its membrane localization it constitutes an ideal candidate for the HCO_3^- secretion into the CSF. As discussed above, two sources of HCO_3^- for the secretion have been proposed. The classic view is an origin from the hydration of intracellular CO_2. The high levels of carbonic anhydrase II in the CPE would efficiently convert CO_2 to HCO_3^- to supply the luminal transporter with HCO_3^- (Johanson et al. 1992a, b). The problem with this explanation is the apparent lack of NHE proteins in the basolateral membrane. The alternative view is transcellular passage of HCO_3^-, where, e.g., NCBE mediates the import of HCO_3^- across the basolateral membrane (Praetorius et al. 2004).

CSF pH seems to respond more to respiratory acidosis than to alkalosis, indicating that the acid extrusion by the CPE is less efficient than the alkaline export (Kazemi et al. 1967). Nevertheless, the CPE cells express several proteins that could be involved in acid extrusion to the CSF. In the luminal membrane, Na^+/H^+ exchangers NHE1 (and NHE6) export H^+ in exchange for Na^+ (Damkier et al. 2009) and thereby are candidates for mechanisms to lower CSF pH. NHE1 is involved in the regulation of intracellular pH in the CPE cells (Damkier et al.

2009), but its putative role in CSF pH regulation has not been investigated. Recently, we demonstrated that a fraction of the vacuolar H^+-ATPase (V-ATPase) is expressed in the luminal membrane of CPE cells (Christensen et al. 2017). Although this pump is involved in extracellular acidification elsewhere, our study rejects the V-ATPase as a significant contributor to CSF pH regulation (Christensen et al. 2017). The $2Cl^-$/H^+ antiporter ClC-7 was also immunolocalized to the luminal membrane of CPE cells in the same study. However, the highly electrogenic nature of the transport by this protein makes ClC7 highly unlikely as an efficient acid extruder. It is unknown whether AE2 plays a role on CSF pH regulation, but the basolateral expression and the export of HCO_3^- in exchange for Cl^- (Alper et al. 1994) could be required to sustain luminal H^+ secretion.

10.6 Future Perspectives

The movement of water across the CPE from the interstitium to the brain ventricles remains a central area of research in CPE biology. In epithelia, water can be moved by both transcellular and paracellular pathways, and both processes may also be present in CSF secretion by the CPE. The luminal membrane of the CPE expresses AQP1 in abundance, and AQP1 is therefore regarded as a major water efflux route from the CPE cells to the CSF. Studies on AQP1 knockout mice have defined this aquaporin as a major determinant of the transepithelial water permeability. The low AQP1 expression in the basolateral plasma membrane spurred the assumption that the basolateral membrane has high intrinsic water permeability by other means to sustain luminal secretion. Claudin-2 is expressed in the CPE tight junctions and is the only known water-permeable claudin. This led to the speculation that a paracellular route mediates a significant part of the water transport to the CSF.

The bottleneck for transcellular Na^+ transport seems to be the basolateral membrane. Despite extensive research, the mechanism mediating basolateral Na^+-entry into the CPE cells remains elusive. For a long time, NHE1 was believed to constitute such a basolateral Na^+ loading mechanism by similarity to other transporting epithelia and because of the amiloride sensitivity of CSF secretion. However, we localized NHE1 and later NHE6 to the luminal membrane domain in CPE cells from normal animals by immunostaining and functional assays. The ENaC channel is more amiloride sensitive than NHEs and has been suggested as an alternative Na^+-loader. However, evidence for basolateral expression and channel function has not been presented. We have demonstrated the molecular and functional expression of NCBE in the basolateral membrane of CPE cells and suggested this transporter as a major Na^+ importer. Still, we have not demonstrated this directly and fail to explain the amiloride sensitivity of CSF secretion.

The origin of secreted HCO_3^- also remains an open question in CPE biology. Most researchers believe that the secreted HCO_3^- is made by the intracellular conversion from CO_2 and H_2O catalyzed by carbonic anhydrase II. However, the expression of two Na^+-dependent HCO_3^- loaders in the basolateral membrane indicates that secreted HCO_3^- might take a transepithelial route. A shift in paradigm toward transcellular HCO_3^- movement occurred first in gastrointestinal acid/base physiology, where, e.g., the duodenal epithelium achieves potent HCO_3^- secretion this way. Unfortunately, membrane impermeant carbonic anhydrase inhibitors have thus far not been applied consistently to distinguish between intracellular and extracellular carbonic anhydrase activity in epithelial transport. Exploiting genetic approaches may be more successful in identifying the major source of HCO_3^- for secretion in the CPE.

The contribution of the CPE to CSF pH regulation represents a promising subfield of the current CPE research. Further studies are warranted to identify which of the many luminal membrane acid-base transporters in the CPE cells that take part in CSF pH regulation to sustain normal brain function.

References

Alper SL (2009) Molecular physiology and genetics of Na^+-independent SLC4 anion exchangers. J Exp Biol 212(Pt 11):1672–1683

Alper SL, Stuart-Tilley A, Simmons CF, Brown D, Drenckhahn D (1994) The fodrin-ankyrin cytoskeleton of choroid plexus preferentially colocalizes with apical Na^+K^+-ATPase rather than with basolateral anion exchanger AE2. J Clin Invest 93(4):1430–1438

Ames A 3rd, Sakanoue M, Endo S (1964) Na, K, Ca, Mg, and Cl concentrations in choroid plexus fluid and cisternal fluid compared with plasma ultrafiltrate. J Neurophysiol 27:672–681

Ames A 3rd, Higashi K, Nesbett FB (1965) Effects of Pco2 acetazolamide and ouabain on volume and composition of choroid-plexus fluid. J Physiol 181(3):516–524

Amin MS, Wang H, Reza E, Whitman SC, Tuana BS, Leenen FHH (2005) Distribution of epithelial sodium channels and mineralocorticoid receptors in cardiovascular regulatory centers in rat brain. Am J Physiol 289:R1787–R1797

Amin MS, Reza E, Wang H, Leenen FH (2009) Sodium transport in the choroid plexus and salt-sensitive hypertension. Hypertension 54(4):860–867

Aydin MD, Kanat A, Turkmenoglu ON, Yolas C, Gundogdu C, Aydin N (2014) Changes in number of water-filled vesicles of choroid plexus in early and late phase of experimental rabbit subarachnoid hemorrhage model: the role of petrous ganglion of glossopharyngeal nerve. Acta Neurochirurgica 156(7):1311–1317

Bairamian D, Johanson CE, Parmelee JT, Epstein MH (1991) Potassium cotransport with sodium and chloride in the choroid plexus. J Neurochem 56(5):1623–1629

Banizs B, Komlosi P, Bevensee MO, Schwiebert EM, Bell PD, Yoder BK (2007) Altered pH_i regulation and Na^+/HCO_3^- transporter activity in choroid plexus of cilia-defective $Tg737^{orpk}$ mutant mouse. Am J Physiol Cell Physiol 292(4):C1409–C1416

Baron R, Neff L, Louvard D, Courtoy PJ (1985) Cell-mediated extracellular acidification and bone resorption: evidence for a low pH in resorbing lacunae and localization of a 100-kD lysosomal membrane protein at the osteoclast ruffled border. J Cell Biol 101(6):2210–2222

Bauer DF, Tubbs RS, Acakpo-Satchivi L (2008) Mycoplasma meningitis resulting in increased production of cerebrospinal fluid: case report and review of the literature. Childs Nerv Syst 24 (7):859–862

Block ML, Zecca L, Hong JS (2007) Microglia-mediated neurotoxicity: uncovering the molecular mechanisms. Nat Rev Neurosci 8(1):57–69

Bonow RH, Aid S, Zhang Y, Becker KG, Bosetti F (2009) The brain expression of genes involved in inflammatory response, the ribosome, and learning and memory is altered by centrally injected lipopolysaccharide in mice. Pharmacogenomics J 9(2):116–126

Boron WF, Boulpaep EL (eds) (2012) Medical physiology. Saunders, Philadelphia

Bouzerar R, Chaarani B, Gondry-Jouet C, Zmudka J, Baledent O (2013) Measurement of choroid plexus perfusion using dynamic susceptibility MR imaging: capillary permeability and age-related changes. Neuroradiology 55(12):1447–1454

Bouzinova EV, Praetorius J, Virkki LV, Nielsen S, Boron WF, Aalkjaer C (2005) Na^+-dependent HCO_3^- uptake into the rat choroid plexus epithelium is partially DIDS sensitive. Am J Physiol Cell Physiol 289(6):C1448–C1456

Brett CL, Wei Y, Donowitz M, Rao R (2002) Human Na^+/H^+ exchanger isoform 6 is found in recycling endosomes of cells, not in mitochondria. Am J Physiol Cell Physiol 282(5):C1031–C1041

Chai SY, McKinley MJ, Mendelsohn FA (1987) Distribution of angiotensin converting enzyme in sheep hypothalamus and medulla oblongata visualized by *in vitro* autoradiography. Clin Exp Hypertens A 9(2–3):449–460

Charron FM, Blanchard MG, Lapointe JY (2006) Intracellular hypertonicity is responsible for water flux associated with Na^+/glucose cotransport. Biophys J 90(10):3546–3554

Chen X, Li W, Yoshida H, Tsuchida S, Nishimura H, Takemoto F, Okubo S, Fogo A, Matsusaka T, Ichikawa I (1997) Targeting deletion of angiotensin type 1B receptor gene in the mouse. Am J Physiol 272(3 Pt 2):F299–F304

Chodobski A, Szmydynger-Chodobska J (2001) Choroid plexus: target for polypeptides and site of their synthesis. Microsc Res Tech 52(1):65–82

Chodobski A, Szmydynger-Chodobska J, Johanson CE (1998) Vasopressin mediates the inhibitory effect of central angiotensin II on cerebrospinal fluid formation. Eur J Pharmacol 347 (2–3):205–209

Christensen IB, Gyldenholm T, Damkier HH, Praetorius J (2013) Polarization of membrane associated proteins in the choroid plexus epithelium from normal and slc4a10 knockout mice. Front Physiol 4:344. https://doi.org/10.3389/fphys.2013.00344

Christensen HL, Paunescu TG, Matchkov V, Barbuskaite D, Brown D, Damkier HH, Praetorius J (2017) The V-ATPase is expressed in the choroid plexus and mediates cAMP-induced intracellular pH alterations. Physiol Rep 5(1). https://doi.org/10.14814/phy2.13072

Christensen HL, Barbuskaite D, Rojek A, Malte H, Christensen IB, Fuchtbauer AC, Fuchtbauer EM, Wang T, Praetorius J, Damkier HH (2018) The choroid plexus sodium-bicarbonate cotransporter NBCe2 regulates mouse cerebrospinal fluid pH. J Physiol 596(19):4709–4728

Cserr HF (1971) Physiology of the choroid plexus. Physiol Rev 51(2):273–311

Cushing H (1914) Studies on the cerebro-spinal fluid: I. Introduction. J Med Res 31(1):1–19

Damkier HH, Praetorius J (2012) Genetic ablation of Slc4a10 alters the expression pattern of transporters involved in solute movement in the mouse choroid plexus. Am J P hysiol Cell Physiol 302(10):C1452–C1459

Damkier HH, Prasad V, Hubner CA, Praetorius J (2009) Nhe1 is a luminal Na^+/H^+ exchanger in mouse choroid plexus and is targeted to the basolateral membrane in Ncbe/Nbcn2-null mice. Am J Physiol Cell Physiol 296(6):C1291–C1300

Damkier HH, Brown PD, Praetorius J (2013) Cerebrospinal fluid secretion by the choroid plexus. Physiol Rev 93(4):1847–1892

Damkier HH, Christensen HL, Christensen IB, Wu Q, Fenton RA, Praetorius J (2018) The murine choroid plexus epithelium expresses the $2Cl^-/H^+$ exchanger ClC-7 and Na^+/H^+ exchanger NHE6 in the luminal membrane domain. Am J Physiol Cell Physiol 314(4):C439–C448

Davson H, Luck CP (1957) The effect of acetazoleamide on the chemical composition of the aqueous humour and cerebrospinal fluid of some mammalian species and on the rate of turnover of ^{24}Na in these fluids. J Physiol 137(2):279–293

Davson H, Purvis C (1954) Cryoscopic apparatus suitable for studies on aqueous humour and cerebro-spinal fluid. J Physiol 124(2):12–13P

Davson H, Segal MB (1970) The effects of some inhibitors and accelerators of sodium transport on the turnover of 22Na in the cerebrospinal fluid and the brain. J Physiol 209(1):131–153

Davson H, Welch K, Segal MB (eds) (1987) Physiology and pathophysiology of the CSF. Churchill Livingston, Edinburgh

de Kloet ER, Van Acker SA, Sibug RM, Oitzl MS, Meijer OC, Rahmouni K, de Jong W (2000) Brain mineralocorticoid receptors and centrally regulated functions. Kidney Int 57 (4):1329–1336

Deng QS, Johanson CE (1989) Stilbenes inhibit exchange of chloride between blood, choroid plexus and the cerebrospinal fluid. Brain Res 510:183–187

Deng QS, Johanson CE (1992) Cyclic AMP alteration of chloride transport into the choroid plexus-cerebrospinal fluid system. Neurosci Lett 143(1–2):146–150

de Rougemont J, Ames A 3rd, Nesbett FB, Hofmann HF (1960) Fluid formed by choroid plexus; a technique for its collection and a comparison of its electrolyte composition with serum and cisternal fluids. J Neurophysiol 23:485–495

Dogterom J, van Wimersma Greidanus TB, De Wied D (1978) Vasopressin in cerebrospinal fluid and plasma of man, dog, and rat. Am J Physiol 234(5):E463–E467

Döring F, Derst C, Wischmeyer E, Karschin C, Schneggenburger R, Daut J, Karschin A (1998) The epithelial inward rectifier channel Kir 7.1 displays unusual K^+ permeation properties. J Neurosci 18:8625–8636

Ellis DZ, Nathanson JA, Sweadner KJ (2000) Carbachol inhibits Na^+-K^+-ATPase activity in choroid plexus via stimulation of the NO/cGMP pathway. Am J Physiol Cell Physiol 279(6): C1685–C1693

Faivre J (1854) Structure du conarium et des plexus choroïde chez l'hommes et des animaux. Gaz Med Paris 9:555–556

Faraci FM, Mayhan WG, Farrell WJ, Heistad DD (1988) Humoral regulation of blood flow to choroid plexus: role of arginine vasopressin. Circ Res 63(2):373–379

Faraci FM, Mayhan WG, Heistad DD (1990) Effect of vasopressin on production of cerebrospinal fluid: possible role of vasopressin (V1)-receptors. Am J Physiol 258(1 Pt 2):R94–R98

Feschenko MS, Donnet C, Wetzel RK, Asinovski NK, Jones LR, Sweadner KJ (2003) Phospholemman, a single-span membrane protein, is an accessory protein of Na,K-ATPase in cerebellum and choroid plexus. J Neurosci 23 (6):2161–2169

Frankel H, Kazemi H (1983) Regulation of CSF composition—blocking chloride-bicarbonate exchange. J Appl Physiol Respir Environ Exerc Physiol 55(1 Pt 1):177–182

Gonzalez-Martinez LM, Avila J, Marti E, Lecuona E, Martin-Vasallo P (1994) Expression of the beta-subunit isoforms of the Na,K-ATPase in rat embryo tissues, inner ear and choroid plexus. Biol Cell 81 (3):215–222

Granstam E, Wang L, Bill A (1993) Vascular effects of endothelin-1 in the cat; modification by indomethacin and L-NAME. Acta Physiologica Scand 148(2):165–176

Gregoriades JMC, Madaris A, Alvarez FJ, Alvarez-Leefmans FJ (2018) Genetic and pharmacologic inactivation of apical NKCC1 in choroid plexus epithelial cells reveals the physiological function of the cotransporter. Am J Physiol Cell Physiol. https://doi.org/10.1152/ajpcell.00026.2018

Hallbeck M, Hermanson O, Blomqvist A (1999) Distribution of preprovasopressin mRNA in the rat central nervous system. J Comp Neurol 411(2):181–200

Hara H, Zhang QJ, Kuroyanagi T, Kobayashi S (1993) Parasympathetic cerebrovascular innervation: an anterograde tracing from the sphenopalatine ganglion in the rat. Neurosurgery 32 (5):822–827

Hasan FM, Kazemi H (1976) Dual contribution theory of regulation of CSF HCO3 in respiratory acidosis. J Appl Physiol Respir Environ Exerc Physiol 40(4):559–567

Haselbach M, Wegener J, Decker S, Engelbertz C, Galla HJ (2001) Porcine Choroid plexus epithelial cells in culture: regulation of barrier properties and transport processes. Microsc Res Tech 52(1):137–152

Held D, Fencl V, Pappenheimer JR (1964) Electrical potential of cerebrospinal fluid. J Neurophysiol 27:942–959

Hemsen A, Lundberg JM (1991) Presence of endothelin-1 and endothelin-3 in peripheral tissues and central nervous system of the pig. Regul Pept 36(1):71–83

Hernando F, Schoots O, Lolait SJ, Burbach JP (2001) Immunohistochemical localization of the vasopressin V1b receptor in the rat brain and pituitary gland: anatomical support for its involvement in the central effects of vasopressin. Endocrinology 142(4):1659–1668

Hladky SB, Barrand MA (2016) Fluid and ion transfer across the blood-brain and blood-cerebrospinal fluid barriers; a comparative account of mechanisms and roles. Fluids Barriers CNS 13(1):19. https://doi.org/10.1186/s12987-016-0040-3

Hughes AL, Pakhomova A, Brown PD (2010) Regulatory volume increase in epithelial cells isolated from the mouse fourth ventricle choroid plexus involves Na^+-H^+ exchange but not Na^+-K^+-$2Cl^-$ cotransport. Brain Res 1323:1–10

Husted RF, Reed DJ (1977) Regulation of cerebrospinal fluid bicarbonate by the cat choroid plexus. J Physiol 267:411–428

Imboden H, Harding JW, Hilgenfeldt U, Celio MR, Felix D (1987) Localization of angiotensinogen in multiple cell types of rat brain. Brain Res 410(1):74–77

Inagami T, Celio MR, Clemens DL, Lau D, Takii Y, Kasselberg AG, Hirose S (1980) Renin in rat and mouse brain: immunohistochemical identification and localization. Clin Sci 59 Suppl 6:49s–51s

Jacobs S, Ruusuvuori E, Sipila ST, Haapanen A, Damkier HH, Kurth I, Hentschke M, Schweizer M, Rudhard Y, Laatikainen LM, Tyynela J, Praetorius J, Voipio J, Hubner CA (2008) Mice with targeted Slc4a10 gene disruption have small brain ventricles and show reduced neuronal excitability. Proc Natl Acad Sci U S A 105(1):311–316

Javaheri S, Wagner KR (1993) Bumetanide decreases canine cerebrospinal fluid production. In vivo evidence for NaCl cotransport in the central nervous system. J Clin Invest 92(5):2257–2261

Johanson CE, Murphy VA (1990) Acetazolamide and insulin alter choroid plexus epithelial cell [Na^+], pH, and volume. Am J Physiol 258(6 Pt 2):F1538–F1546

Johanson CE, Sweeney SM, Parmelee JT, Epstein MH (1990) Cotransport of sodium and chloride by the adult mammalian choroid plexus. Am J Physiol 258(2 Pt 1):C211–C216

Johanson CE, Murphy VA, Dyas M (1992a) Ethacrynic acid and furosemide alter Cl, K, and Na distribution between blood, choroid plexus, CSF, and brain. Neurocheml Res 17 (11):1079–1085

Johanson CE, Parandoosh Z, Dyas ML (1992b) Maturational differences in acetazolamide-altered pH and HCO3 of choroid plexus, cerebrospinal fluid, and brain. Am J Physiol 262(5 Pt 2): R909–R914

Johanson CE, Preston JE, Chodobski A, Stopa EG, Szmydynger-Chodobska J, McMillan PN (1999) AVP V1 receptor-mediated decrease in Cl^- efflux and increase in dark cell number in choroid plexus epithelium. Am J Physiol 276(1 Pt 1):C82–C90

Johanson C, McMillan P, Tavares R, Spangenberger A, Duncan J, Silverberg G, Stopa E (2004) Homeostatic capabilities of the choroid plexus epithelium in Alzheimer's disease. Cerebrospinal Fluid Res 1(1):3. https://doi.org/10.1186/1743-8454-1-3

Johren O, Saavedra JM (1996) Expression of AT1A and AT1B angiotensin II receptor messenger RNA in forebrain of 2-wk-old rats. Am J Physiol 271(1 Pt 1):E104–E112

Kageyama Y, Bravo EL (1988) Hypertensive mechanisms associated with centrally administered aldosterone in dogs. Hypertension 11(6 Pt 2):750–753

Kajita H, Brown PD (1997) Inhibition of the inward-rectifying Cl^- channel in rat choroid plexus by a decrease in extracellular pH. J Physiol 498(Pt 3):703–707

Kalaria RN, Premkumar DR, Lin CW, Kroon SN, Bae JY, Sayre LM, LaManna JC (1998) Identification and expression of the Na^+/H^+ exchanger in mammalian cerebrovascular and choroidal tissues: characterisation by amiloride-sensitive [^{3}H]MIA binding and RT-PCR analysis. Brain Res Mol Brain Res 58:178–187

Kallio H, Pastorekova S, Pastorek J, Waheed A, Sly WS, Mannisto S, Heikinheimo M, Parkkila S (2006) Expression of carbonic anhydrases IX and XII during mouse embryonic development. BMC Dev Biol 6:22. https://doi.org/10.1186/1471-213X-6-22

Kanaka C, Ohno K, Okabe A, Kuriyama K, Itoh T, Fukuda A, Sato K (2001) The differential expression patterns of messenger RNAs encoding K-Cl cotransporters (KCC1,2) and Na-K-2Cl cotransporter (NKCC1) in the rat nervous system. Neuroscience 104(4):933–946

Kao L, Kurtz LM, Shao X, Papadopoulos MC, Liu L, Bok D, Nusinowitz S, Chen B, Stella SL, Andre M, Weinreb J, Luong SS, Piri N, Kwong JMK, Newman D, Kurtz I (2011) Severe neurologic impairment in mice with targeted disruption of the electrogenic sodium bicarbonate cotransporter NBCe2 (Slc4a5 gene). J Biol Chem 286 (37):32563–32574

Karadsheh MF, Byun N, Mount DB, Delpire E (2004) Localization of the KCC4 potassium-chloride cotransporter in the nervous system. Neuroscience 123(2):381–391

Kaur C, Rathnasamy G, Ling EA (2016) The choroid plexus in healthy and diseased brain. J Neuropathol Exp Neurol 75(3):198–213

Kazemi H, Shannon DC, Carvallo-Gil E (1967) Brain CO2 buffering capacity in respiratory acidosis and alkalosis. J Appl Physiol Respir Environ Exerc Physiol 22(2):241–246

Keep RF, Xiang J, Betz AL (1994) Potassium cotransport at the rat choroid plexus. Am J Physiol 267(6 Pt 1):C1616–C1622

Kibble JD, Tresize AO, Brown PD (1996) Properties of the cAMP-activated Cl^- conductance in choroid plexus epithelial cells isolated from the rat. J Physiol 496:69–80

Kibble JD, Garner C, Kajita H, Colledge WH, Evans MJ, Radcliff R, Brown PD (1997) Whole-cell Cl^- conductances in mouse choroid plexus epithelial cells do not require CFTR expression. Am J Physiol 272:C1899–C1907

Kirchhoff C, Stegmaier J, Bogner V, Buhmann S, Mussack T, Kreimeier U, Mutschler W, Biberthaler P (2006) Intrathecal and systemic concentration of NT-proBNP in patients with severe traumatic brain injury. J Neurotrauma 23(6):943–949

Kister SJ (1956) Carbonic anhydrase inhibition. VI. The effect of acetazolamide on cerebrospinal fluid flow. J Pharmacol Exp Therap 117(4):402–405

Kotera T, Brown PD (1994) Evidence for two types of potassium current in rat choroid plexus epithelial cells. Pflügers Archiv 427:317–324

Kratzer I, Vasiljevic A, Rey C, Fevre-Montange M, Saunders N, Strazielle N, Ghersi-Egea JF (2012) Complexity and developmental changes in the expression pattern of claudins at the blood-CSF barrier. Histochem Cell Biol 138(6):861–879

Kriegs JO, Homann V, Kinne-Saffran E, Kinne RK (2007) Identification and subcellular localization of paracellin-1 (claudin-16) in human salivary glands. Histochem Cell Biol 128(1):45–53

Krug SM, Gunzel D, Conrad MP, Lee IF, Amasheh S, Fromm M, Yu AS (2012) Charge-selective claudin channels. Ann N Y Acad Sci 1257:20–28

Lee JE, Chu F, Posner JB, Plum F (1969) Buffering capacity of cerebrospinal fluid in acute respiratory acidosis in dogs. Am J Physiol 217(4):1035–1038

Leenen FH (2010) The central role of the brain aldosterone-"ouabain" pathway in salt-sensitive hypertension. Biochimica Biophysica Acta 1802(12):1132–1139

Leenen FH, Hou X, Wang HW, Ahmad M (2015) Enhanced expression of epithelial sodium channels causes salt-induced hypertension in mice through inhibition of the alpha2-isoform of Na^+, K^+-ATPase. Physiol Rep 3(5). https://doi.org/10.14814/phy2.12383

Li H, Tornberg J, Kaila K, Airaksinen MS, Rivera C (2002) Patterns of cation-chloride cotransporter expression during embryonic rodent CNS development. Eur J Neurosci 16 (12):2358–2370

Lindsey AE, Schneider K, Simmons DM, Baron R, Lee BS, Kopito RR (1990) Functional expression and subcellular localization of an anion exchanger cloned from choroid plexus. Proc Natl Acad Sci U S A 87(14):5278–5282

Lindvall M, Owman C (1981) Autonomic nerves in the mammalian choroid plexus and their influence on the formation of cerebrospinal fluid. J Cereb Blood Flow Metab 1(3):245–266

Lindvall M, Edvinsson L, Owman C (1978) Sympathetic nervous control of cerebrospinal fluid production from the choroid plexus. Science 201(4351):176–178

Livingston RB (1949) Cerebrospinal fluid. In: Fulton JF (ed) A textbook of physiology. Saunders, Philadelphia, pp 916–980

Maharaj AS, Walshe TE, Saint-Geniez M, Venkatesha S, Maldonado AE, Himes NC, Matharu KS, Karumanchi SA, D'Amore PA (2008) VEGF and TGF-beta are required for the maintenance of the choroid plexus and ependyma. J Exp Med 205(2):491–501

Maktabi MA, Heistad DD, Faraci FM (1990) Effects of angiotensin II on blood flow to choroid plexus. Am J Physiol 258(2 Pt 2):H414–H418

Marques F, Sousa JC, Brito MA, Pahnke J, Santos C, Correia-Neves M, Palha JA (2016) The choroid plexus in health and in disease: dialogues into and out of the brain. Neurobiol Dis. https://doi.org/10.1016/j.nbd.2016.08.011

Masuzawa T, Ohta T, Kawamura M, Nakahara N, Sato F (1984) Immunohistochemical localization of Na^+, K^+-ATPase in the choroid plexus. Brain Res 302(2):357–362

Mayer SE, Sanders-Bush E (1993) Sodium-dependent antiporters in choroid plexus epithelial cultures from rabbit. J Neurochem 60:1308–1316

McCarthy KD, Reed DJ (1974) The effect of acetazolamide and furosemide on cerebrospinal fluid production and choroid plexus carbonic anhydrase activity. J Pharmacol Exp Ther 189 (1):194–201

Millar ID, Brown PD (2008) NBCe2 exhibits a 3 HCO_3^- :1 Na^+ stoichiometry in mouse choroid plexus epithelial cells. Biochem Biophys Res Comm 373:550–554

Millar ID, Bruce JI, Brown PD (2007) Ion channel diversity, channel expression and function in the choroid plexuses. Cerebrospinal Fluid Res 4:8

Mortazavi MM, Griessenauer CJ, Adeeb N, Deep A, Bavarsad Shahripour R, Loukas M, Tubbs RI, Tubbs RS (2014) The choroid plexus: a comprehensive review of its history, anatomy, function, histology, embryology, and surgical considerations. Childs Nerv Syst 30(2):205–214

Murphy VA, Johanson CE (1989a) Alteration of sodium transport by the choroid plexus with amiloride. Biochim Biophys Acta 979(2):187–192

Murphy VA, Johanson CE (1989b) Acidosis, acetazolamide, and amiloride: effects on ^{22}Na transfer across the blood-brain and blood-CSF barriers. J Neurochem 52(4):1058–1063

Nakamura N, Suzuki Y, Sakuta H, Ookata K, Kawahara K, Hirose S (1999) Inwardly rectifying K^+ channel Kir7.1 is highly expressed in thyroid follicular cells, intestinal epithelial cells and choroid plexus epithelial cells: implication for a functional coupling with Na^+,K^+-ATPase. Biochem J 342:329–336

Nattie EE, Adams JM (1988) DIDS decreases CSF HCO_3^- and increases breathing in response to CO2 in awake rabbits. J Appl Physiol (1985) 64(1):397–403

Nielsen S, Smith BL, Christensen EI, Agre P (1993) Distribution of the aquaporin CHIP in secretory and resorptive epithelia and capillary endothelia. Proc Natl Acad Sci USA 90 (15):7275–7279

Nilsson C, Fahrenkrug J, Lindvall-Axelsson M, Owman C (1991) Epithelial cells purified from choroid plexus have receptors for vasoactive intestinal polypeptide. Brain Res 542(2):241–247

Nilsson C, Lindvall-Axelsson M, Owman C (1992) Neuroendocrine regulatory mechanisms in the choroid plexus-cerebrospinal fluid system. Brain Res Brain Res Rev 17(2):109–138

Nilsson C, Hultberg BM, Gammeltoft S (1996) Autocrine role of insulin-like growth factor II secretion by the rat choroid plexus. Eur J Neurosci 8(3):629–635

O'Kelly CJ, Kulkarni AV, Austin PC, Urbach D, Wallace MC (2009) Shunt-dependent hydrocephalus after aneurysmal subarachnoid hemorrhage: incidence, predictors, and revision rates. Clinical article. J Neurosurg 111(5):1029–1035

Oshio K, Watanabe H, Song Y, Verkman AS, Manley GT (2005) Reduced cerebrospinal fluid production and intracranial pressure in mice lacking choroid plexus water channel Aquaporin-1. FASEB J 19(1):76–78

Pearson MM, Lu J, Mount DB, Delpire E (2001) Localization of the K^+-Cl^- cotransporter, KCC3, in the central and peripheral nervous systems: expression in the choroid plexus, large neurons and white matter tracts. Neurosci 103(2):481–491

Plotkin MD, Kaplan MR, Peterson LN, Gullans SR, Hebert SC, Delpire E (1997) Expression of the Na^+-K^+-$2Cl^-$ cotransporter BSC2 in the nervous system. Am J Physiol 272(1 Pt 1):C173–C183

Pollay M, Curl F (1967) Secretion of cerebrospinal fluid by the ventricular ependyma of the rabbit. Am J Physiol 213(4):1031–1038

Pollay M, Davson H (1963) The passage of certain substances out of the cerebrospinal fluid. Brain 86:137–150

Pollay M, Hisey B, Reynolds E, Tomkins P, Stevens FA, Smith R (1985) Choroid plexus Na^+/K^+-activated adenosine triphosphatase and cerebrospinal fluid formation. Neurosurgery 17 (5):768–772

Praetorius J, Nielsen S (2006) Distribution of sodium transporters and aquaporin-1 in the human choroid plexus. Am J Physiol Cell Physiol 291(1):C59–C67

Praetorius J, Nejsum LN, Nielsen S (2004) A SLC4A10 gene product maps selectively to the basolateral membrane of choroid plexus epithelial cells. Am J Physiol 286:C601–C610

Raichle ME, Grubb RL Jr (1978) Regulation of brain water permeability by centrally-released vasopressin. Brain Res 143(1):191–194

Redzic ZB, Segal MB (2004) The structure of the choroid plexus and the physiology of the choroid plexus epithelium. Adv Drug Deliv Rev 56(12):1695–1716

Redzic ZB, Preston JE, Duncan JA, Chodobski A, Szmydynger-Chodobska J (2005) The choroid plexus-cerebrospinal fluid system: from development to aging. Curr Top Dev Biol 71:1–52

Roepke TK, Kanda VA, Purtell K, King EC, Lerner DJ, Abbott GW (2011) KCNE2 forms potassium channels with KCNA3 and KCNQ1 in the choroid plexus epithelium. FASEB J 25:4264–4273

Rosenthal R, Milatz S, Krug SM, Oelrich B, Schulzke JD, Amasheh S, Gunzel D, Fromm M (2010) Claudin-2, a component of the tight junction, forms a paracellular water channel. J Cell Sci 123 (Pt 11):1913–1921. https://doi.org/10.1242/jcs.060665

Rosenthal R, Gunzel D, Krug SM, Schulzke JD, Fromm M, Yu AS (2016) Claudin-2-mediated cation and water transport share a common pore. Acta Physiol (Oxf). https://doi.org/10.1111/apha.12742

Rotter A, Birdsall NJ, Burgen AS, Field PM, Hulme EC, Raisman G (1979) Muscarinic receptors in the central nervous system of the rat. I. Technique for autoradiographic localization of the binding of ^{3H}propylbenzilylcholine mustard and its distribution in the forebrain. Brain Res 180 (2):141–165

Safaee M, Oh MC, Bloch O, Sun MZ, Kaur G, Auguste KI, Tihan T, Parsa AT (2013) Choroid plexus papillomas: advances in molecular biology and understanding of tumorigenesis. Neuro Oncol 15(3):255–267

Saito Y, Wright EM (1983) Bicarbonate transport across the frog choroid plexus and its control by cyclic nucleotides. J Physiol 336:635–648

Saito Y, Wright E (1984) Regulation of bicarbonate transport across the brush border membrane of the bull-frog choroid plexus. J Physiol 350:327–342

Salpietro V, Mankad K, Kinali M, Adams A, Valenzise M, Tortorella G, Gitto E, Polizzi A, Chirico V, Nicita F, David E, Romeo AC, Squeri CA, Savasta S, Marseglia GL, Arrigo T, Johanson CE, Ruggieri M (2014) Pediatric idiopathic intracranial hypertension and the underlying endocrine-metabolic dysfunction: a pilot study. J Pediatr Endocrinol Metab 27 (1–2):107–115

Schalk KA, Faraci FM, Heistad DD (1992) Effect of endothelin on production of cerebrospinal fluid in rabbits. Stroke 23(4):560–563

Schnermann J, Chou CL, Ma T, Traynor T, Knepper MA, Verkman AS (1998) Defective proximal tubular fluid reabsorption in transgenic aquaporin-1 null mice. Proc Natl Acad Sci UA 95 (16):9660–9664

Schuchmann S, Schmitz D, Rivera C, Vanhatalo S, Salmen B, Mackie K, Sipila ST, Voipio J, Kaila K (2006) Experimental febrile seizures are precipitated by a hyperthermia-induced respiratory alkalosis. Nat Med 12(7):817–823

Schwerk C, Adam R, Borkowski J, Schneider H, Klenk M, Zink S, Quednau N, Schmidt N, Stump C, Sagar A, Spellerberg B, Tenenbaum T, Koczan D, Klein-Hitpass L, Schroten H (2011) *In vitro* transcriptome analysis of porcine choroid plexus epithelial cells in response to Streptococcus suis: release of pro-inflammatory cytokines and chemokines. Microbes Infect 13 (11):953–962

Seckl JR (1997) 11beta-Hydroxysteroid dehydrogenase in the brain: a novel regulator of glucocorticoid action? Front Neuroendocrinol 18(1):49–99

Segal MB, Burgess AM (1974) A combined physiological and morphological study of the secretory process in the rabbit choroid plexus. J Cell Sci 14(2):339–350

Segal MB, Chodobski A, Szmydynger-Chodobska J, Cammish H (1992) Effect of arginine vasopressin on blood vessels of the perfused choroid plexus of the sheep. Prog Brain Res 91:451–453

Serezani CH, Ballinger MN, Aronoff DM, Peters-Golden M (2008) Cyclic AMP: master regulator of innate immune cell function. Am J Respir Cell Mol Biol 39(2):127–132

Siegel GJ, Holm C, Schreiber JH, Desmond T, Ernst SA (1984) Purification of mouse brain (Na^+ + K^+)-ATPase catalytic unit, characterization of antiserum, and immunocytochemical localization in cerebellum, choroid plexus, and kidney. J Histochem Cytochem 32(12):1309–1318

Silverberg GD, Huhn S, Jaffe RA, Chang SD, Saul T, Heit G, Von Essen A, Rubenstein E (2002) Downregulation of cerebrospinal fluid production in patients with chronic hydrocephalus. J Neurosurg 97(6):1271–1275

Sinclair AJ, Onyimba CU, Khosla P, Vijapurapu N, Tomlinson JW, Burdon MA, Stewart PM, Murray PI, Walker EA, Rauz S (2007) Corticosteroids, 11beta-hydroxysteroid dehydrogenase isozymes and the rabbit choroid plexus. J Neuroendocrinol 19(8):614–620

Sinclair AJ, Walker EA, Burdon MA, van Beek AP, Kema IP, Hughes BA, Murray PI, Nightingale PG, Stewart PM, Rauz S, Tomlinson JW (2010) Cerebrospinal fluid corticosteroid levels and cortisol metabolism in patients with idiopathic intracranial hypertension: a link between 11beta-HSD1 and intracranial pressure regulation? J Clin Endocrinol Metab 95(12):5348–5356

Speake T, Kajita H, Smith CP, Brown PD (2002) Inward-rectifying anion channels are expressed in the epithelial cells of choroid plexus isolated from ClC-2 'knock-out' mice. J Physiol 539:385–390

Speake T, Kibble JD, Brown PD (2004) Kv1.1 and Kv1.3 channels contribute to the delayed-rectifying K^+ conductance in rat choroid plexus epithelial cells. Am J Physiol 286:C611–C620

Spector R, Keep RF, Robert Snodgrass S, Smith QR, Johanson CE (2015) A balanced view of choroid plexus structure and function: focus on adult humans. Exp Neurol 267:78–86

Steardo L, Nathanson JA (1987) Brain barrier tissues: end organs for atriopeptins. Science 235 (4787):470–473

Steffensen AB, Oernbo EK, Stoica A, Gerkau NJ, Barbuskaite D, Tritsaris K, Rose CR, MacAulay N (2018) Cotransporter-mediated water transport underlying cerebrospinal fluid formation. Nat Commun 9(1):2167. https://doi.org/10.1038/s41467-018-04677-9

Supuran CT (2015) Acetazolamide for the treatment of idiopathic intracranial hypertension. Expert Rev Neurother 15(8):851–856

Szczepanska-Sadowska E, Simon-Oppermann C, Gray DA, Simon E (1984) Plasma and cerebrospinal fluid vasopressin and osmolality in relation to thirst. Pflugers Archiv 400(3):294–299

Szmydynger-Chodobska J, Chun ZG, Johanson CE, Chodobski A (2002) Distribution of fibroblast growth factor receptors and their co-localization with vasopressin in the choroid plexus epithelium. Neuroreport 13(2):257–259

Szmydynger-Chodobska J, Chung I, Kozniewska E, Tran B, Harrington FJ, Duncan JA, Chodobski A (2004) Increased expression of vasopressin v1a receptors after traumatic brain injury. J Neurotrauma 21(8):1090–1102

Szmydynger-Chodobska J, Chung I, Chodobski A (2006) Chronic hypernatremia increases the expression of vasopressin and voltage-gated Na channels in the rat choroid plexus. Neuroendocrinol 84(5):339–345

Trabold R, Krieg S, Scholler K, Plesnila N (2008) Role of vasopressin V(1a) and V2 receptors for the development of secondary brain damage after traumatic brain injury in mice. J Neurotrauma 25(12):1459–1465

Tschirgi RD, Frost RW, Taylor JL (1954) Inhibition of cerebrospinal fluid formation by a carbonic anhydrase inhibitor, 2-acetylamino-1,3,4-thiadiazole-5-sulfonamide (diamox). Proc Soc Exp Biol Med 87(2):373–376

Vallon V, Verkman AS, Schnermann J (2000) Luminal hypotonicity in proximal tubules of aquaporin-1-knockout mice. Am J Physiol Renal Physiol 278(6):F1030–F1033

Van Huysse JW, Amin MS, Yang B, Leenen FH (2012) Salt-induced hypertension in a mouse model of Liddle syndrome is mediated by epithelial sodium channels in the brain. Hypertension 60(3):691–696

Vogh BP, Godman DR (1985) Timolol plus acetazolamide: effect on formation of cerebrospinal fluid in cats and rats. Can J Physiol Pharmacol 63(4):340–343

Vogh BP, Godman DR (1989) Effects of inhibition of angiotensin converting enzyme and carbonic anhydrase on fluid production by ciliary process, choroid plexus, and pancreas. J Ocular Pharmacol 5(4):303–311

Vogh BP, Godman DR, Maren TH (1987) Effect of $AlCl_3$ and other acids on cerebrospinal fluid production: a correction. J Pharmacol Exp Ther 243(1):35–39

Wang HW, Amin MS, El-Shahat E, Huang BS, Tuana BS, Leenen FH (2010) Effects of central sodium on epithelial sodium channels in rat brain. Am J Physiol Regul Integr Comp Physiol 299 (1):R222–R233

Watters GV, Page L, Lorenzo AV, Cutler RW, Barlow CF (1969) Relationship between cerebrospinal fluid (CSF) formation, absorption and pressure in human hydrocephalus. Trans Am Neurol Assoc 94:153–156

Watts AG, Sanchez-Watts G, Emanuel JR, Levenson R (1991) Cell-specific expression of mRNAs encoding Na^+,K^+-ATPase alpha- and beta-subunit isoforms within the rat central nervous system. Proc Natl Acad Sci U S A 88 (16):7425–7429

Weber KT (2003) Aldosteronism revisited: perspectives on less well-recognized actions of aldosterone. J Lab Clin Med 142(2):71–82

Welch K (1963) Secretion of cerebrospinal fluid by choroid plexus of the rabbit. Am J Physiol 205:617–624

Welch K, Sadler K (1965) Electrical potentials of choroid plexus of the rabbit. J Neurosurg 22:344–351

Wolburg H, Wolburg-Buchholza K, Liebnera S, Engelhardt H (2001) Claudin-1, claudin-2 and claudin-11 are present in tight junctions of choroid plexus epithelium of the mouse. Neurosci Lett 307:77–80

Wright EM (1972) Mechanisms of ion transport across the choroid plexus. J Physiol 226 (2):545–571

Wright EM (1978) Transport processes in the formation of the cerebrospinal fluid. Rev Physiol Biochem Pharmacol 83:3–34

Wu Q, Delpire E, Hebert SC, Strange K (1998) Functional demonstration of Na^+-K^+-$2Cl^-$ cotransporter activity in isolated, polarized choroid plexus cells. Am J Physiol 275(6 Pt 1): C1565–C1572

Yamasaki H, Sugino M, Ohsawa N (1997) Possible regulation of intracranial pressure by human atrial natriuretic peptide in cerebrospinal fluid. Eur Neurol 38(2):88–93

Zerbe RL, Robertson GL (1983) Osmoregulation of thirst and vasopressin secretion in human subjects: effect of various solutes. Am J Physiol 244(6):E607–E614

Zeuthen T (1991) Secondary active transport of water across ventricular cell membrane of choroid plexus epithelium of Necturus maculosus. J Physiol 444:153–173

Zeuthen T (1994) Cotransport of K^+, Cl^- and H_2O by membrane proteins from choroid plexus epithelium of Necturus maculosus. J Physiol 478(Pt 2):203–219
Zeuthen T, Wright EM (1978) An electrogenic Na^+/K^+ pump in the choroid plexus. Biochimica et Biophys Acta 511(3):517–522
Zeuthen T, Wright EM (1981) Epithelial potassium transport: tracer and electrophysiological studies in choroid plexus. J Membrane Biol 60:105–128
Zeuthen T, Gorraitz E, Her K, Wright EM, Loo DD (2016) Structural and functional significance of water permeation through cotransporters. Proc Natl Acad Sci U S A 113(44):E6887–E6894
Ziemann AE, Schnizler MK, Albert GW, Severson MA, Howard MA 3rd, Welsh MJ, Wemmie JA (2008) Seizure termination by acidosis depends on ASIC1a. Nat Neurosci 11(7):816–822
Zorad S, Alsasua A, Saavedra JM (1998) Decreased expression of natriuretic peptide A receptors and decreased cGMP production in the choroid plexus of spontaneously hypertensive rats. Mol Chem Neuropathol 33(3):209–222

Chapter 11
Transport Functions of Ectoderm Epithelial Cells Forming Dental Enamel

Michael L. Paine, Alan Boyde, and Rodrigo S. Lacruz

Abstract The development of enamel encapsulates fundamental cellular processes associated with the intricate life stages of the ectodermal epithelial cells known as ameloblasts that regulate its growth. Ameloblasts are tall and narrow cells with reversed polarity and advance as a front producing the enamel matrix with unique properties. Following the development of the full thickness of the enamel, ameloblasts then switch their elongated morphology becoming a cell with the appearance of transporting epithelium. Enamel is then mineralized to reach a mineral content of ~96% by weight. The last 10 years have seen a surge in studies aiming to identify the transport machinery used by ameloblasts to generate such mineralization levels. While the majority of the mineral content of enamel is dominated by Ca^{2+} (36%) and Pi (18%), there are a number of other important contributors to the development of the carbonated hydroxyapatite enamel crystals, including HCO_3^-, Cl^-, Mg^{2+}, and others. Yet there is a clear interplay among the channels, pumps, exchangers, and transporters involved in ion transport in ameloblasts. Part of these connections arise from the need to ensure an intensification of mineral growth alongside careful monitoring of extracellular pH. Although the molecular components of transporting ameloblasts are emerging, much remains to be done. Here, we describe the current model for ion transport by ameloblasts.

M. L. Paine
Center for Craniofacial Molecular Biology, University of Southern California School of Dentistry, Los Angeles, CA, USA
e-mail: paine@usc.edu

A. Boyde
Dental Physical Sciences, Institute of Dentistry, Bart's and The London School of Medicine and Dentistry, Queen Mary University of London, London, UK
e-mail: a.boyde@qmul.ac.uk

R. S. Lacruz (✉)
Department of Molecular Pathobiology, New York University College of Dentistry, New York, NY, USA
e-mail: rodrigo.lacruz@nyu.edu

K. L. Hamilton, D. C. Devor (eds.), *Ion Transport Across Epithelial Tissues and Disease*, Physiology in Health and Disease,
https://doi.org/10.1007/978-3-030-55310-4_11

Keywords Ameloblasts · Transepithelial ion · Transport · Enamel crystals · pH regulation

11.1 Introduction

Enamel is formed by specialized epithelial cells of ectodermal origin known as ameloblasts. Cohorts of ameloblasts move outward away from the dentine leaving behind a space rich in a protein matrix they secrete (Boyde 1989). Long and thin hydroxyapatite crystals are forming in this organic mix. At the opposite (basal) side of the secreted matrix, a number of cell layers support the advancing ameloblasts. This basal region is in proximity to capillaries, providing ions and nutrients to the cells (Sasaki et al. 1990). At the apical (distal) end, ameloblasts are in near-direct contact with the enamel layer (Fig. 11.1), and lack capillary vessels. After the matrix thickness has been completed and ameloblasts stop advancing, they transform into cells with characteristics of transporting epithelia, also undergoing cyclic modulations in morphology (see below), a necessary process for mineralization of the crystals (Smith and Nanci 1995). Therefore, discussions on enamel physiology

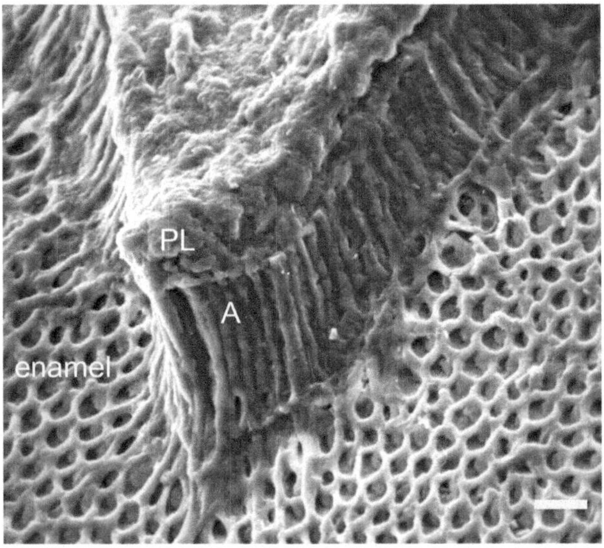

Fig. 11.1 Ameloblasts on developing enamel: Electron micrograph of the developing enamel of a dog deciduous molar with the enamel organ partly dissected away to reveal the pitted developing enamel surface. Shown here are the tall columnar ameloblasts with the papillary later of the enamel organ at top center. The preparation was air-dried from ethanol, resulting in more shrinkage of the cells than the enamel. Samples were coated with aluminum by evaporation and imaged in 10 kV in SE SEM image, Stereoscan 1 SEM. PL, papillary layer; A, ameloblasts. Scale bar = ~ 7 μm

and the transport characteristics of the ameloblasts necessitate an understanding of the temporal requirements to build crystals, as these go hand in hand.

11.2 Key Features of Enamel Formation

Fully matured enamel is the hardest vertebrate tissue. During evolution, ameloblast cells have evolved a three-stage process to form enamel *de novo*, best characterized in rodent enamel as the most studied model. These stages are: (1) *secretory*, (2) *transition*, (3) *maturation* (see Box 11.1) (Smith 1998). These three biological stages can be recognized by the morphology of the ameloblasts and supporting cells of the enamel organ with important biological functions (Boyde 1989; Liu et al. 2016). Secretory stage ameloblasts are encompassed by stratum intermedium cells, and are also supported by the stellate reticulum cells, whereas maturation ameloblasts are associated with the papillary layer (ex-stratum intermedium) that is heavily vascularized (Fig. 11.2). The secretory and maturation stages in turn achieve the following functional steps in a smooth transition (Simmer and Fincham, 1995): (a) secretion of organic matrix, (b) crystal nucleation, (c) crystal elongation,

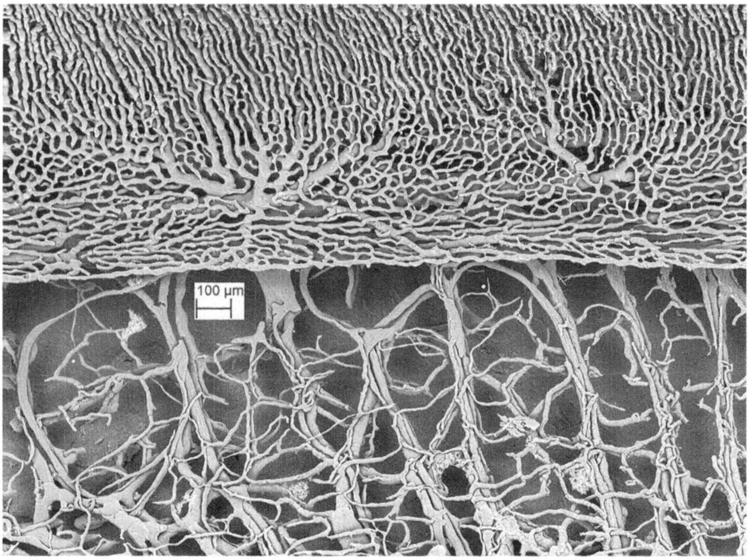

Fig. 11.2 Capillary blood supply network to the rat incisor enamel organ during the maturation stage: Plastic corrosion cast made by injecting Mercox (methacrylate) resin into the vasculature, polymerised *in situ*, then all tissues dissolved away with 15% NaOH solution as described in Hodde et al. (1983). Image acquired in 20 kV BSE SEM image of gold sputter-coated sample. View is from the enamel/tooth side. Note the very rich capillary network of the enamel organ at top. The lower part of the image shows feeder vessels from the sides of the forming tooth organ. There are many other feeder vessels running the length of the tooth "outside" the enamel organ

(d) removal of matrix, and (e) enamel maturation (i.e., increased enamel mineralization) (Simmer and Fincham 1995).

Box 11.1 Key Characteristics of Ameloblasts in Secretory and Maturation Stages

Secretory	Maturation
• Ameloblasts are long (~60–70 μM)	• Shortening of ameloblasts (~40 μM)
• Synthesis and secretion of matrix proteins	• Marked decrease in secretion of matrix proteins
• Large ER networks	• Loss of Tomes' process
• Ameloblasts have a distal (apical) extension (Tomes' process) a secretory site that influences the organization of crystals	• Removal of matrix • Cyclic modulation between RA and SA
• Enamel crystals are thin and surrounded by organic matrix	• Increased transport function
• Bulk of secreted product is amelogenin (AMELX)	• Enamel crystals increase in thickness
• Gene expression markers: AMELX, ABMN, ENAM, MMP20	• Gene expression markers: ODAM, AMTN, KLK4

Secretory stage ameloblasts are tall columnar cells enriched in endoplasmic reticulum tubular structures consistent with a key role in protein synthesis (Sasaki et al. 1990). The rich proteinaceous matrix they secrete is dominated by a serine rich species known as amelogenin (AMEL) (Smith and Nanci 1996). This and other enamel specific proteins (reviewed in Lacruz et al. 2017) serve as a scaffold for the organized growth of the crystals. The short transition stage is marked by a reduction in height as well as the loss of a number of ameloblasts to apoptosis (Kallenbach 1974; Smith and Warshawsky 1977). Maturation stage ameloblasts are complex cells not only because of a number of morphological changes (e.g., decrease in height) and changes in ultrastructural appearance, but also because they undergo modulations in their shape unusual in biological systems (Smith 1998). The most abundant type of maturation stage ameloblast cell, the ruffled-ended ameloblasts (RA), displays distal membrane infoldings (aka ruffled border) (Reith and Boyde 1981; Smith 1998). In a period of just a few hours, RA lose these complex infoldings and acquire a smooth distal appearance—smooth ameloblasts (SA)—temporarily, before recreating the infoldings, an operation that is repeated a few cycles (Smith et al. 1987; Smith 1998) (Fig. 11.3). The peculiarities of these RA are starting to come to light. In particular, there has been an appreciation for their role in transcellular/vectorial ion transport (Reith and Boyde 1979; Smith 1998; Josephsen et al. 2010; Damkier et al. 2014; Bronckers et al. 2015a; Bronckers 2017).

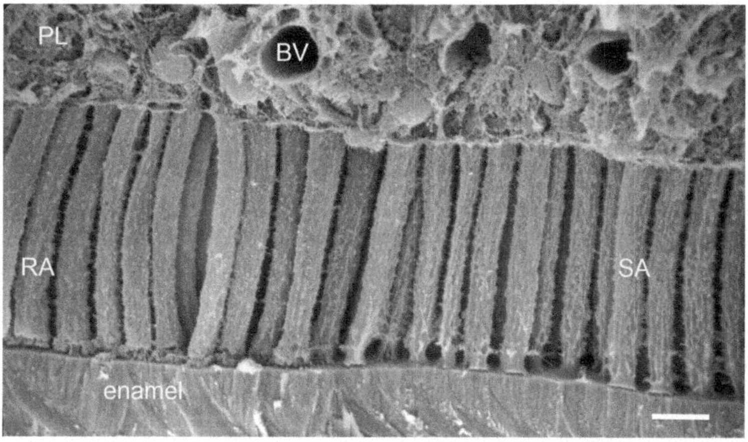

Fig. 11.3 Maturation stage ameloblasts: Rat lower incisor maturation ameloblasts exposed in a longitudinal dry fracture through critical point dried, glutaraldehyde perfusion fixed tissue, gold sputter coated, imaged in 10 kV in SE SEM image. Note the empty capillary blood vessels in the papillary layer at top. Ruffle-ended ameloblast are seen at left with well-defined terminal bar apparatus distally (near the enamel) contrasting with the spaces between smooth ameloblasts at right lacking distal terminal webs. Numerous microvilli can be appreciated covering parts of the outer membrane. The ameloblasts are ~ 40 μm tall. BV, blood vessel; PL, papillary layer; RA, ruffled ended ameloblasts; SA, smooth ended ameloblasts. Scale bar = ~ 9.5 μm

11.3 Growth of Enamel Crystals

The only known function of ameloblasts is to control and promote the growth of enamel crystals (Boyde 1989). Once matured, enamel is acellular and cannot undergo remodeling. The unit cell of enamel crystals, considered a calcium hydroxyapatite (Hap) with the chemical formula Ca_{10} $(PO_4)_6$ $(OH)_2$, contains impurities in the form of F^-, HCO_3^-, Mg^{2+}, Cl^- and other elements (Simmer and Fincham 1995).

Nucleation events are the initial steps in forming enamel Hap crystals from a solution, although details of the earliest events of this process remain a subject of study (Smith et al. 2016). Nucleation involves the arrangement of a small number of ions, atoms, or molecules, into a pattern typical of a solid crystal creating a site that is the base for additional particles to be deposited allowing crystal growth. This incipient crystal growth develops in the "enamel fluid" that is immediately adjacent to the distal pole of ameloblasts in the secretory stage (Aoba and Moreno 1987; Aoba 1996). Ion transport is low during the secretory stage, relative to maturation, yet the composition of the enamel fluid near the secretory ameloblasts already differs from that of serum, highlighting the specialized nature of this compartment (Aoba and Moreno 1987). Elongation (expansion along the *c* axis) of the growing crystals is facilitated by the enamel matrix proteins that surround them (Smith and Nanci 1996). In the maturation stage, as transepithelial transport raises and crystals then expand in width, crystal growth continues at a higher rate (Smith et al. 2005) as newly transported ions into the enamel fluid precipitate and become incorporated, in an organized fashion, into the already present crystal structures. The downside of these chemical reactions is the unwelcome release of H^+ into the enamel fluid (Simmer and Fincham 1995; Lacruz et al. 2010b) (but see ref. Josephsen et al. 2010). If these H^+ are no cleared, the pH of the fluid can become acidic hampering further crystal growth (Lacruz et al. 2010b). Thus a pH buffering system is essential in enamel (Lacruz et al. 2010b; Jalali et al. 2014), as will be discussed later.

Over the last ~10 years, enamel research has experienced a surge in studies dedicated to understanding the transport capacity of ameloblasts and adjacent cells of the enamel organ. Below, an overview of the current model for ion transport by ameloblasts is described. In most cases, ion transport is considered a transcellular activity (Smith 1998; Lacruz 2017). This scenario is a logical one as the careful stoichiometric assembly of specific amounts of ions into the crystals must be controlled. However, some activities may also benefit from paracellular movement given the morphological changes in maturation between RA to SA (Smith 1998). During the RA phase, distal tight junctions prevent the interstitial fluid to reach the enamel. In the SA phase these junctions become leaky or disappear, enabling the diffusion of ions and molecules from the interstitial space not only toward the enamel but also in the opposite direction (Fig. 11.3).

11.4 The Abundant Ions and Their Transport

11.4.1 Calcium

It has been estimated that Ca^{2+} represents about 36% of enamel by weight, followed by phosphate (Pi) at ~18% (Moreno and Aoba 1987) with ~86% of the Ca^{2+} entering the enamel during the maturation stage (Smith 1998). There have been a number of important reviews detailing the mechanisms associated with the transport of Ca^{2+} by the enamel organ (Bawden 1989; Takano 1995; Hubbard 2000; Nurbaeva et al. 2016). A generalized view proposed in those studies is that Ca^{2+} transport follows transcellular route rather than a paracellular or passive route (Hubbard 2000; Nurbaeva et al. 2016). A landmark approach, because of the steps taken to define the transcellular pathway, was reported by Hubbard in a series of papers (Hubbard 1996, 1998; Hubbard et al. 2011). The main consideration of this model was to describe Ca^{2+} transport as a "transcytosis" pathway requiring entry, ferry of Ca^{2+} across the cell, and extrusion steps (Hubbard 2000). This model implied, with limited available data, that store-operated Ca^{2+} entry channels were likely involved in Ca^{2+} influx into ameloblasts (Hubbard 2000). We have since confirmed Hubbard's predictions in several studies (Nurbaeva et al. 2015a; Eckstein et al. 2017, 2019; Nurbaeva et al. 2018). Ca^{2+} transport in enamel is also one of the most studied in recent times and hence requires a more in-depth discussion.

11.4.1.1 SOCE in Enamel Cells

Store-operated Ca^{2+} entry (SOCE) pathway is mediated by the Ca^{2+} release activated Ca^{2+} (CRAC) channels, a highly specialized Ca^{2+} influx channel (Prakriya and Lewis 2015). CRAC channels are formed by two endoplasmic reticulum (ER) single pass membrane proteins known as stromal interaction molecules (STIM) with two known homologues (STIM1, STIM2) (Roos et al. 2005; Zhang et al. 2005). These proteins act as luminal Ca^{2+} sensors that activate and gate the opening of the channel pore formed by the ORAI proteins (ORAI1-3) (Feske et al. 2006; Hoth and Niemeyer 2013; Amcheslavsky et al. 2015). Activation of the ER sensors is initiated following a decline of luminal [Ca^{2+}], resulting in a number of conformational changes in STIM leading to the binding and activation of ORAI (Smyth et al. 2010; Putney et al. 2016). This loss of ER [Ca^{2+}] is stimulated by agonist-receptor mediated interaction at the plasma membrane, resulting in the formation of phospholipase C and inositol-1,4,5-triphosphate (IP_3) that in turn bind to the IP_3 receptors in the ER membrane releasing stored Ca^{2+}. All homologues of ORAI as well as both STIM genes are expressed in the enamel organ cells with higher expression in maturation (Nurbaeva et al. 2015a). ORAI1 and ORAI2 have been localized to the basolateral membrane of maturation stage ameloblasts with STIM1 and STIM2 showing cytosolic signals (Nurbaeva et al. 2015a; Eckstein et al. 2019). As discussed, stimulation of SOCE requires ER Ca^{2+} release. In enamel cells, the IP_3 receptors show higher expression than ryanodine receptors (Nurbaeva et al. 2015a), another family of Ca^{2+} release channels, as the latter are found in low abundance in enamel cells, suggesting that the IP_3 receptors are the likely release channels in these cells, but this remains to be tested.

SOCE induction can be stimulated *in vitro* by the SERCA (sarco-endoplasmic reticulum Ca^{2+}-ATPase) inhibitor thapsigargin, which passively depletes ER Ca^{2+}. This has revealed that maturation enamel organ cells have higher SOCE than secretory cells (Nurbaeva et al. 2015a, 2018). These results are similar to those obtained when enamel cells of each stage are stimulated with physiological agonists such as ATP, acetylcholine, and cholecystokinin (Nurbaeva et al. 2018). Importantly, blocking CRAC channels with the specific inhibitors synta-66, GSK 7975A in primary enamel cells (Nurbaeva et al. 2015a, 2018), or with BTP2 and synta-66 in the ameloblast cell line LS8 cells (Nurbaeva et al. 2015b) either via thapsigargin or agonist stimulation, markedly reduces or nearly abolishes Ca^{2+} influx. This provides a solid pharmacological support for the key role of SOCE in enamel cells. However, physiological activation of SOCE remains unclear, but it is noteworthy that enamel cells synthesize cholecystokinin, which may be acting as an autocrine system, but this remains to be tested (Nurbaeva et al. 2018).

Additional support for the key role of SOCE in enamel derives from the fact that patients with mutations in the *STIM1* or *ORAI1* genes show abnormal and hypomineralized enamel (McCarl et al. 2009; Picard et al. 2009; Wang et al. 2014; Eckstein et al. 2019) highlighting an important association between CRAC channels and *amelogenesis imperfecta,* a clinical term that encompasses a number of dental disease phenotypes. More recently, animal models with a conditional deletion of *Stim1/2* added further support showing severely hypomineralized enamel and lacked SOCE (Eckstein et al. 2017; Furukawa et al. 2017), whereas conditional deletion of *Orai1* in mice showed only moderate effects in enamel (Eckstein et al. 2019). This latter study, however, reported that other Orai2 and Orai3 homologues can compensate for lack of Orai1 maintaining reduced Ca^{2+} influx (Eckstein et al. 2019). Interestingly, lack of STIM1/2 or ORAI1 in enamel cells resulted in changes in mitochondrial morphology and function, highlighting the connection between SOCE and mitochondria (Eckstein et al. 2017; Eckstein et al. 2019). This association has also been described in other cell types with the extent of SOCE mediated Ca^{2+} signals being modified by mitochondria in their role as Ca^{2+} sinks (Watson and Parekh 2012). The rapid Ca^{2+} uptake by mitochondria following ATP stimulation in enamel cells (Eckstein et al. 2018) also suggests that the mitochondrial Ca^{2+} uniporter (MCU) and its associated complex might be expressed in enamel cells. The link between mitochondria and Ca^{2+} and phosphate transport will be discussed below. Combined, pharmacological and molecular evidence suggest that Ca^{2+} influx in enamel cells is likely dominated by ORAI channels and their activators. However, this does not prevent the possibility that SOCE modulators may also be involved (see below regarding TRPM7).

11.4.1.2 Ca^{2+} Pumps, Exchangers and Buffers

Cytosolic Ca^{2+} buffers in enamel cells include SERCA2, the main ATP-dependent pump translocating cytosolic Ca^{2+} into the ER lumen (Hovnanian 2007). Other buffer proteins include classical buffers such as parvalbumin, and calretinin, as well as the buffer/sensor proteins such as calbindin (9 and 28 kDa), calmodulin, and calcineurin (reviewed Nurbaeva et al. 2016).

Modulation of cytosolic Ca^{2+} by plasma membrane Ca^{2+}-ATPases (PMCA) has been a subject of recent interest given discrepancies in previous reports on the localization of these low-affinity Ca^{2+} pumps. Paine's group indicated that three of four PMCA coding genes (*ATP2B*1, 3 and 4) were upregulated during the secretory stage with PMCA1 and PMCA4 localized along the basolateral membrane in secretory and maturation (Robertson et al. 2017). Regardless of the dominant PMCA isoform expressed, these pumps are likely to play a housekeeping role in clearing cytosolic Ca^{2+} during secretory and maturation given their response to modest increase in cytosolic $[Ca^{2+}]$ (Nurbaeva et al. 2016; Robertson et al. 2017). However, no functional data measuring PMCA activity in enamel cells has been reported, as is the case for the Ca^{2+} exchangers.

Two families of Ca^{2+} exchangers have been reported in enamel cells: the Na^{+}/ Ca^{2+} exchanger (NCX1-3) (Okumura et al. 2010) and the Na^{+}/ Ca^{2+}/K^{+} exchanger (NCKX1-6) (Hu et al. 2012; Lacruz et al. 2012a, b). Of these, molecular and clinical data support an essential role for NCKX4 in modulating Ca^{2+} extrusion as evidenced by the severe dental phenotype in patients with mutations in the coding gene *SLC24A4* and in *Slc4a4*$^{-/-}$ mice (Parry et al. 2013; Wang et al. 2014). NCKX4 localization in apical membranes also support this role (Hu et al. 2012; Wang et al.

2014). For NCX, it is more likely that they play a housekeeping Ca^{2+} clearance function (Nurbaeva et al. 2016; Lacruz et al. 2017) as no human mutations in the coding genes are known to affect enamel.

11.4.2 Phosphate

An important difference between ideal hydroxyapatite and the apatite found in enamel is that the latter includes the substitution of HPO_4^{2-} for PO_4^{-3} (Simmer and Fincham 1995). Despite the role of phosphate in the stabilization of the incipient enamel crystals (Simmer and Fincham 1995), available data regarding phosphate transport by enamel cells is limited. RT-PCR analysis of tooth organs identified the mRNA expression of the electrogenic cotransporter *Npt2b* (Onishi et al. 2007) and two genome-wide studies reported additional members of not only the NaPi family (NaPi-IIa-c) but also PiT-1 and PiT-2 (Lacruz et al. 2012a; Yin et al. 2014). The NaPi family of cotransporters are involved in the inward translocation of divalent inorganic phosphate (HPO_4^{2-}) in exchange for two or three Na^+ to generate the required electrochemical gradient (Lederer and Wagner 2019). NaPi-IIb (coded by *SLC34A2*), was found to be significantly upregulated during maturation in rat enamel cells (Lacruz et al. 2012b). Immunostaining of anti-NaPi-2b showed strong signals in the distal membranes of maturation stage ameloblasts (Bronckers et al. 2015a). This is intriguing, and Bronckers and colleagues (Bronckers et al. 2015a) suggested that NaPi-2b could translocate Pi outward, arguing against the known physiological role of NaPi proteins as inward Pi movers (Lederer and Wagner 2019). PiT-1 and PiT-2 proteins are also Na^+-coupled Pi transporters, and their mRNA was expressed in both secretory and maturation stage enamel organ cells (Lacruz et al. 2012a; Yin et al. 2014). Against these molecular data, radiolabeled studies described greater ^{32}Pi incorporation into secretory ameloblasts than in maturation cells (Robinson et al. 1974). As is the case for the Ca^{2+} extrusion system, functional studies on Pi transporters in enamel are lacking.

11.4.3 Transcellular Movement of Ca^{2+} and Pi

Historically, Ca^{2+} transport across the columnar polarized ameloblasts was considered to be mediated a by gradient facilitated by PMCA extrusion at the apical pole (Bawden 1989). Hubbard also discussed the possibility, by referencing other epithelial systems, that calbindins may act as a ferry of bound Ca^{2+} ions, vectorially, but Hubbard also discounted this notion (Hubbard et al. 2011). Instead, Hubbard proposed a model in which ER tubules may act as tunnels that facilitate the safe transport of Ca^{2+} across the long ameloblasts (Hubbard 2000). This may be a possibility but it remains untested, although the basolateral localization of ORAI proteins and extended positioning of ER tubules in maturation ameloblasts make this a feasible model. An alternative scenario that has not been clearly defined in enamel cells is the possible contribution of secretory vesicles. These vesicles are known to move across the cytosol of a number of cells primed by elevations in intracellular [Ca^{2+}] (Scheenen et al. 1998; Thevenod 2002; Messenger et al. 2014), similar to those resulting from activation of SOCE. Once these vesicles fuse to the distal membranes, they can exocytose/release Ca^{2+} in the low millimolar range (Scheenen et al. 1998; Thevenod 2002; Messenger et al. 2014). Moreover, a recent and intriguing possibility that would partly solve the issue of both Pi and Ca^{2+} transport in ameloblasts is the untested notion that ameloblasts package Ca-Pi in mitochondria delivering these complexes via mitochondrial-derived vesicles (MDV) to the extra-cellular space. MDVs have come to light recently as important contributors to calcium phosphate storage/transport in osteoblasts associated with the mineralization of the extracellular matrix (Boonrungsiman et al. 2012). Immunofluorescence of the mitochondrial outer membrane TOMM20 protein that we have published recently in maturation stage ameloblasts (Eckstein et al. 2017) showed discretely localized TOMM20 labelled bodies near the apical cell pole, posing the question of whether MDV can be found in enamel cells. We cannot confirm that these structures contain Ca/Pi as the tissues had been decalcified, but this remains an important area of study in our laboratory.

11.5 The Lesser (but Important) Ions in Enamel and Their Transport

Mg^{2+}, Cl^-, Fe^{3+}, F^-, and others are found in less than 1% in fully formed enamel. Mg^{2+} is particularly important as it a well-known growth inhibitor of Hap (Simmer and Fincham 1995). In enamel, there are 52 Ca^{2+} for each Mg^{2+} (Simmer and Fincham 1995). Recently, the chanzyme TRPM7 (transient receptor melastanin member 7) has been implicated in enamel development (Nakano et al. 2016; Ogata et al. 2017). TRPM7 was first described as a Mg^{2+} channel with inward characteristics (Nadler et al. 2001; Duan et al. 2018), but it is now known to also conduct Ca^{2+}, Mn^+, and other ions (Faouzi et al. 2017). It contains a kinase domain that appears to modulate SOCE (Faouzi et al. 2017), making it an interesting candidate to affect Ca^{2+} dynamics in enamel cells. The subcellular localization of TRPM7 in ameloblasts is not well defined. Other Mg^{2+} transporters described in enamel cells include cyclin and CBS (cystathionine beta synthase) domain divalent metal cation transport mediator 4 (CNNM4), a protein mediating Mg^{2+} extrusion (Funato et al. 2014). Mutations in *CNNM4* result in Jalili syndrome, and patients show enamel defects (Parry et al. 2009). The subcellular localization of CNNM4 in ameloblasts revealed strong basolateral signals (Yamazaki et al. 2013). Of note, defects in the tight junction protein CLDN16 (claudin 16), result in hypercalciuria and hypomagnesaemia with nephrocalcinosis, and patients also present with dental phenotype (Yamaguti et al. 2017).

CFTR (cystic fibrosis conductance transmembrane regulator) has emerged as an important mediator of Cl^- transport in enamel cells (Sui et al. 2003). CFTR localizes largely to the apical pole of maturation stage ameloblasts and mutations in the *CFTR* gene cause enamel defects (Bronckers et al. 2015b; Eckstein et al. 2018) despite the low amount of Cl^- in enamel (Young 1974). It may be the case that the outward transport of Cl^- by CFTR is required to maintain a gradient outside the cells to facilitate HCO_3^- movement (see below). Other Cl^- channels expressed by ameloblasts include, among a number of CLC members, CLCN7 (chloride voltage-gated channel 7), which was found expressed in lysosomes of maturation stage ameloblasts likely (Lacruz et al. 2013a). Together with the vacuolar-type ATPase (V-ATPase) pumps which generate H^+ influx into the lumen of lysosomes, CLCN7 might be required for luminal acidification (Lacruz et al. 2013a). Interestingly, a study by Josephsen and colleagues found that V-ATPase was also expressed in the distal membranes of RA cells, and suggested that this mechanism would help clear H^+ from the cell by moving them into the enamel zone (Josephsen et al. 2010).

Fe^{3+}-induced staining is visible in the surface enamel of most (but not all) rodent incisors and throughout the dentition in many shrews (Insectivora). The ferritin heavy chain 1 (FTH1) has been shown to be expressed in enamel organ cells of rodent incisors being upregulated in maturation (Lacruz et al. 2011, 2012a). Data on Fe^{3+} or its transport on human enamel is poor. Regarding F^-, there is a wealth of information on the effects of F^- in enamel crystal formation and ameloblasts, which is dependent on the F^- concentration, duration of exposure of cells to F^-, and the developmental stage affected (Fejerskov et al. 1990; DenBesten 1999; Robinson et al. 2006; Bronckers et al. 2009a; DenBesten and Li 2011). It is also known that incorporation of F^- at low levels in enamel, forming fluohydroyapatite, increases its resistance to acid-induced solubility (Fejerskov et al. 1990). However, no F^- transporters are known in mammals and so some of us have suggested that F^- transport may occur via Cl^- channels (Lacruz et al. 2013b, 2017), although data to support this are lacking.

11.5.1 Bicarbonate and pH

Despite the low solubility of Hap crystals and their stability, forming enamel crystals comes at a cost. The precipitation of one mole of Hap releases ~ 8 moles of H^+ in the fluid (Simmer and Fincham 1995). This mild acidification could prevent additional crystal growth so a buffering system of the pH is required (but see ref. Josephsen et al. 2010). It is now well documented that the pH of the enamel changes from near neutral in the secretory stage to mildly acidic during maturation, with the most acidic value being ~6.3 (Sasaki et al. 1991; Smith et al. 1996; Takagi et al. 1998). Additional changes occur during the modulations from RA to SA (Lyman and Waddell 1977). The pH buffering system in ameloblasts is therefore best described during maturation given the increased requirements for neutralizing pH at this stage (Smith et al. 1996).

The interplay between HCO_3^- (a base) and that of the conjugate acid CO_2 (commonly as waste product of mitochondrial respiration) is key to understanding the role of HCO_3^- as a pH buffer. Acid/base control by ameloblasts, as is the case in other epithelia, has to be stringently monitored. This is likely the reason for multiple acid/base associated proteins being expressed by ameloblasts, as discussed below (Paine et al. 2008; Lacruz et al. 2017). Bicarbonate (HCO_3^-) is not able to diffuse across cell membranes. Hence, the bulk of pH buffering by ameloblasts uses membrane proteins to translocate bicarbonate (HCO_3^-) into and out of the cell (Lacruz et al. 2017), a process dominated by proteins encoded by the SCL gene family (solute carriers).

Fourteen SLC genes and their associated proteins are known to be involved widely in HCO_3^- transport in epithelial cells (Alka and Casey 2014), although not all of these have been identified in enamel cells at this point. At the basal pole of ameloblasts, the electrogenic Na^+-dependent HCO_3^- co-transporter NBCe1 (encoded by *SLC4A4*) is considered a key modulator of the inward transport of bicarbonate (Lacruz et al. 2010c). The Na^+-independent Cl^-/HCO_3^- exchanger AE2 (encoded by *SLC4A2*), largely confined to the basolateral pole, contributes to bicarbonate transfer to the enamel zone in exchange for Cl^- (Lyaruu et al. 2008; Paine et al. 2008; Bronckers et al. 2009b; Josephsen et al. 2010). Mutations in the coding genes for AE2 and NBCe1 result in enamel deficiencies (Bronckers et al. 2009b; Jalali et al. 2014).

Additional SLC genes expressed by the ameloblasts with known functions in HCO_3^- transport comprise several members of the *SLC26A* family, including (protein name in brackets): *SLC26A1* (SAT1); *SLC26A3* (DRA), *SLC26A4* (Pendrin), *SLC26A6* (PAT1), and *SLC26A7* (SUT1) (Bronckers et al. 2011; Jalali et al. 2015; Yin et al. 2015). These proteins can function as Cl^-/HCO_3^- exchangers, although they are also known to transport other anions (Alper and Sharma 2013). A feature of some of these proteins is their direct interactions with CFTR and carbonic anhydrases (CA) (Alka and Casey 2014), as also shown by co-IP of SLC26A and CFTR in enamel cells (Yin et al., 2015). The close association with CFTR, as discussed above, may then facilitate the exchange of Cl^- for HCO_3^- by the various proteins encoded by SLC26A. The role of HCO_3^- transporting ion channels other than CFTR (i.e., bestrophin, anoctamin 1, GABA) by ameloblasts is unknown.

Bicarbonate may also be generated intracellularly by carbonic anhydrase 2 (CA2) as well as in the enamel compartment by the secreted CA6, among a number of CA isozymes expressed by ameloblasts (Lin et al. 1994; Toyosawa et al. 1996; Smith et al. 2006; Josephsen et al. 2010; Lacruz et al. 2010a). The enzymatic reactions mediated by CA, while decreasing the CO_2 load, also result in the generation of H^+. Clearance of H^+ out of ameloblasts is likely mediated by the exchanger NHE1, and may also be cleared by pumping into acidic vesicles (e.g., lysosomes) via V-type ATPase (Josephsen et al. 2010). This pump has also been identified in the distal membranes of maturation ameloblasts (Josephsen et al. 2010).

11.6 Summary

The last decade has witnessed a rise in studies aiming at decoding the transport machinery used by ameloblast cells to form and mineralize enamel. Perhaps the most noticeable feature of ameloblasts is the critical need to balance the spatial-temporal transport of ions, while maintaining pH, so that they assemble in a coordinated fashion in the extracellular space with a reasonable stoichiometric repetitiveness. Many research questions remain unclear. For example, the subcellular localization and associated function of a number of components are yet to be clarified. But based on our own and published models (Josephsen et al. 2010; Bronckers et al. 2015a; Bronckers 2017; Lacruz 2017; Eckstein and Lacruz 2018), a summary model of the transport system of ameloblasts can be proposed (Fig. 11.4). A caveat is that, except for studies on SOCE and a few others, this and other models rely on protein or mRNA expression patterns, lacking functional/experimental data.

A strong candidate for Ca^{2+} uptake in ameloblasts is SOCE mediated by STIM1 and ORAI1 (Nurbaeva et al. 2015a, 2018; Eckstein et al. 2017, 2019). ORAI appears to dominate the bulk of Ca^{2+} influx, although whether there are modulators of ORAI function remains to be elucidated. With it, it would seem reasonable that IP_3R mediated ER Ca^{2+} release is also operating in these cells based on expression data and that SERCA clearly handles the ER refilling (Franklin et al. 2001; Nurbaeva et al., 2015a). Although some advances have been made regarding physiological activators of SOCE (Nurbaeva et al. 2018), this remains open-ended. Calcium extrusion is handled by NCX and PMCA (Takano et al. 1986; Bawden 1989; Takano 1995; Okumura et al. 2010), in a housekeeping fashion, but neither can compensate for the lack of NCKX4, which has a critical role in Ca^{2+} extrusion resulting in severe enamel defects (Hu et al. 2012; Parry et al. 2013; Wang et al. 2014).

Pi transport remains understudied, but NaPi proteins appear to be the primary inward Pi transporters. Here, the subcellular localization of NaPi2b at the distal membranes (Bronckers et al. 2015a) seems at odds with the role of NaPi as a mediator of Pi uptake. Pi is needed for enamel crystal formation, so removing this Pi out of the enamel zone when NCKX4, NCX and PMCA are delivering Ca^{2+} is a challenging notion. Clearly more research on Pi transport is needed to elucidate a clear uptake/extrusion system in ameloblasts.

Besides the two major ions Ca^{2+} and Pi, ameloblasts work hard to maintain pH balance (Lacruz et al. 2010b) using a battery of proteins that generate bicarbonate (CA) (Lin et al. 1994; Smith et al. 2006; Lacruz et al. 2010a), that can uptake bicarbonate (NBCe1) (Lacruz et al. 2010c; Jalali et al. 2014), or that can transport it out of the cell, in general at the expense of Cl^- (Lyaruu et al. 2008; Paine et al. 2008; Jalali et al. 2015; Yin et al. 2015). CFTR, expressed at the distal membranes of maturation stage ameloblasts (Bronckers et al. 2015b; Eckstein et al. 2018), in all likelihood moves Cl^- out of the cell and into the enamel space as a sort of currency for it then to re-enter the cell in lieu of bicarbonate that is extruded. The proton load can be cleared via NHE1 at the lateral cell membrane (Josephsen et al. 2010) as these protons can diffuse to the basal region. The presence of V-ATPase in distal membrane is also problematic in principle (Josephsen et al. 2010), as this adds to the

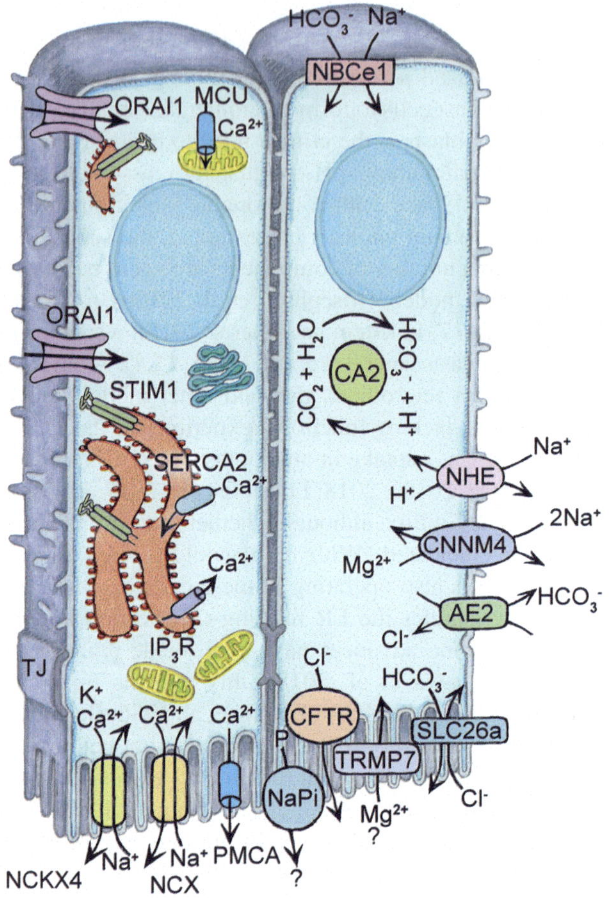

Fig. 11.4 Model for transepithelial ion transport in maturation stage ameloblasts: Ameloblasts are represented here during the ruffled-border stage in maturation, when cells increase their transport functions. The cell at left depicts the main components of Ca^{2+} handling, whereas the cells at right represents proteins involved in the maintenance of pH balance in the extracellular space. The Ca^{2+} handling toolkit in ameloblasts includes ORAI1 as the main Ca^{2+} influx channel, activated by STIM1 upon depletion of ER Ca^{2+} stores. Activation of STIM1 likely involves the IP_3 receptors and ER refilling is provided by SERCA. Our unpublished data also suggest that the mitochondrial Ca^{2+} uniporter (MCU) might be involved in buffering SOCE. Ca^{2+} extrusion is handled largely by NCKX4, with NCX and PMCA being associated with a housekeeping role in the outward movement of Ca^{2+}. Bicarbonate transport and pH regulation (cell at right) are important functions in enamel cells. HCO_3^- uptake is mediated by NBCe1 at the basal pole and is also generated within the cells by CA2. HCO_3^- transfer outside the cell is provided by a number of exchangers, including AE2 and several members of the SLC26a gene family (SAT1, DRA, Pendrin, PAT1, SUT1). Potentially, the CFTR outward transport of Cl^- may work in tandem with the SLC26a proteins which would extrude HCO_3^- in exchange for Cl^-. H^+ generated by mitochondrial respiration and the activity of CA2 are transferred outside the cell by NHE. Mg^{2+} extrusion is likely mediated by CNNM4, with TRPM7 possibly mediating its uptake distally, although the localization of TRPM7 is yet to be elucidated in detail. Likewise, the proposal that NaPi plays a role in the outward transfer of Pi requires additional testing

ongoing acidification of the enamel zone by nucleation and the functioning of the secreted CA6. Thus, the suggestion that V-ATPase might be required to slow down crystal growth to enable the removal of organics (Josephsen et al. 2010) should be tested.

Mg^{2+} is important in many cellular functions and the ubiquitous Mg^{2+} chanzyme TRPM7 has been reported to be expressed by ameloblasts although its localization is yet unclear with rather strong cytosolic signals (Nakano et al., 2016). At any rate, this is a likely candidate for Mg^{2+} uptake with a putative role of CNNM4 extruding Mg^{2+} at the basolateral region (Yamazaki et al. 2013).

In summary, a model for the main known transport functions of ameloblasts and the proteins involved in this process has been presented. A number of gaps in knowledge were discussed. For example, it would be useful to test more broadly how the cyclic modulations from RA to SA modulate ion transport as well as the role and composition of the ruffled-border. Also, it would be interesting to learn how the known components, and others yet to be described, of the emerging transport system of ameloblasts, can function in tandem with other proteins. This seems to be the case as often deficiencies in one protein affect the functioning of others, at least insofar as the readout of mineral composition may be reflecting such alterations (Lacruz et al. 2010c; Bronckers et al. 2017). Redundancy also appears to be a component of several functions, including HCO_3^- transport, given the abundant battery of proteins with similar function. Yet an important facet of enamel mineralization research is the need for increased studies with the aim of functionally testing the roles of the molecular components of ameloblasts, as highlighted recently (Lacruz 2017; Varga et al. 2018).

Acknowledgments The authors are grateful to Kirk Hamilton and Daniel Devor for their kind invitation to contribute to this volume. The work presented here was funded by the National Institute of Dental and Craniofacial Research (NIH/NIDCR) awards to R.S.L. (R01DE025639 and R01DE027679) and DE019629 and DE024704 to M.P.

References

Alka K, Casey JR (2014) Bicarbonate transport in health and disease. IUBMB Life 66:596–615

Alper SL, Sharma AK (2013) The SLC26 gene family of anion transporters and channels. Mol Aspects Med 34:494–515

Amcheslavsky A, Wood ML, Yeromin AV et al (2015) Molecular biophysics of Orai store-operated Ca^{2+} channels. Biophys J 108:237–246

Aoba T (1996) Recent observations on enamel crystal formation during mammalian amelogenesis. Anat Rec 245:208–218

Aoba T, Moreno EC (1987) The enamel fluid in the early secretory stage of porcine amelogenesis: chemical composition and saturation with respect to enamel mineral. Calcif Tissue Int 41:86–94

Bawden JW (1989) Calcium transport during mineralization. Anat Rec 224:226–233

Boonrungsiman S, Gentleman E, Carzaniga R et al (2012) The role of intracellular calcium phosphate in osteoblast-mediated bone apatite formation. Proc Natl Acad Sci U S A 109:14170–14175

Boyde A (1989) Enamel. In: Berkowitz BKB, Boyde A, Frank RM, Hohling HJ, Moxham BJ, Nalbandian J, Tonge CH (eds) Teeth, vol 6. Springer, Berlin, pp 309–473

Bronckers AL (2017) Ion transport by ameloblasts during amelogenesis. J Dent Res 96:243–253
Bronckers AL, Lyaruu DM, DenBesten PK (2009a) The impact of fluoride on ameloblasts and the mechanisms of enamel fluorosis. J Dent Res 88:877–893
Bronckers AL, Lyaruu DM, Jansen ID et al (2009b) Localization and function of the anion exchanger Ae2 in developing teeth and orofacial bone in rodents. J Exp Zool B Mol Dev Evol 312B:375–387
Bronckers AL, Guo J, Zandieh-Doulabi B et al (2011) Developmental expression of solute carrier family 26A member 4 (SLC26A4/pendrin) during amelogenesis in developing rodent teeth. Eur J Oral Sci 119(Suppl 1):185–192
Bronckers AL, Lyaruu D, Jalali R et al (2015a) Ameloblast modulation and transport of Cl^-, Na^+, and K^+ during amelogenesis. J Dent Res 94:1740–1747
Bronckers AL, Lyaruu DM, Guo J et al (2015b) Composition of mineralizing incisor enamel in cystic fibrosis transmembrane conductance regulator-deficient mice. Eur J Oral Sci 123:9–16
Bronckers AL, Jalali R, Lytton J (2017) Reduced protein expression of the $Na^+/Ca^{2+}+K^+$-exchanger (SLC24A4) in apical plasma membranes of maturation ameloblasts of fluorotic mice. Calcif Tissue Int 100:80–86
Damkier HH, Josephsen K, Takano Y et al (2014) Fluctuations in surface pH of maturing rat incisor enamel are a result of cycles of H^+-secretion by ameloblasts and variations in enamel buffer characteristics. Bone 60:227–234
DenBesten PK (1999) Biological mechanisms of dental fluorosis relevant to the use of fluoride supplements. Community Dent Oral Epidemiol 27:41–47
DenBesten P, Li W (2011) Chronic fluoride toxicity: dental fluorosis. Monogr Oral Sci 22:81–96
Duan J, Li Z, Li J et al (2018) Structure of the mammalian TRPM7, a magnesium channel required during embryonic development. Proc Natl Acad Sci U S A 115:E8201–E8210
Eckstein M, Lacruz RS (2018) CRAC channels in dental enamel cells. Cell Calcium 75:14–20
Eckstein M, Vaeth M, Fornai C et al (2017) Store-operated Ca^{2+} entry controls ameloblast cell function and enamel development. JCI Insight 2:e91166
Eckstein M, Aulestia FJ, Nurbaeva MK et al (2018) Altered Ca^{2+} signaling in enamelopathies. Biochim Biophys Acta 1865:1778–1785
Eckstein M, Vaeth M, Aulestia FJ, et al (2019) Differential regulation of Ca^{2+} influx by ORAI channels mediates enamel mineralization. Sci Signal 12:eaav4663
Faouzi M, Kilch T, Horgen FD et al (2017) The TRPM7 channel kinase regulates store-operated calcium entry. J Physiol 595:3165–3180
Fejerskov O, Manji F, Baelum V. (1990) The nature and mechanisms of dental fluorosis in man. J Dent Res 69(Spec No):692–700; discussion 721
Feske S, Gwack Y, Prakriya M et al (2006) A mutation in Orai1 causes immune deficiency by abrogating CRAC channel function. Nature 441:179–185
Franklin IK, Winz RA, Hubbard MJ (2001) Endoplasmic reticulum Ca^{2+}-ATPase pump is up-regulated in calcium-transporting dental enamel cells: a non-housekeeping role for SERCA2b. Biochem J 358:217–224
Funato Y, Yamazaki D, Mizukami S et al (2014) Membrane protein CNNM4-dependent Mg^{2+} efflux suppresses tumor progression. J Clin Invest 124:5398–5410
Furukawa Y, Haruyama N, Nikaido M et al (2017) Stim1 regulates enamel mineralization and ameloblast modulation. J Dent Res 96:1422–1429
Hodde KC, Boyde A, Reid SA et al (1983) The vascular architecture of the rat incisor enamel organ. Beitr elektronenmikroskop Direktabb Oberfl (Munster) 16:431–442
Hoth M, Niemeyer BA (2013) The neglected CRAC proteins: Orai2, Orai3, and STIM2. Curr Top Membr 71:237–271
Hovnanian A (2007) SERCA pumps and human diseases. Subcell Biochem 45:337–363
Hu P, Lacruz RS, Smith CE et al (2012) Expression of the sodium/calcium/potassium exchanger, NCKX4, in ameloblasts. Cells Tissues Organs 196:501–509

Hubbard MJ (1996) Abundant calcium homeostasis machinery in rat dental enamel cells. Up-regulation of calcium store proteins during enamel mineralization implicates the endoplasmic reticulum in calcium transcytosis. Eur J Biochem 239:611–623

Hubbard MJ (1998) Enamel cell biology. Towards a comprehensive biochemical understanding. Connect Tissue Res 38:17–32. Discussion 35–41

Hubbard MJ (2000) Calcium transport across the dental enamel epithelium. Crit Rev Oral Biol Med 11:437–466

Hubbard MJ, McHugh NJ, Mangum JE (2011) Exclusion of all three calbindins from a calcium-ferry role in rat enamel cells. Eur J Oral Sci 119(Suppl 1):112–119

Jalali R, Guo J, Zandieh-Doulabi B et al (2014) NBCe1 (SLC4A4) a potential pH regulator in enamel organ cells during enamel development in the mouse. Cell Tissue Res 358:433–442

Jalali R, Zandieh-Doulabi B, DenBesten PK et al (2015) Slc26a3/Dra and Slc26a6 in Murine Ameloblasts. J Dent Res 94:1732–1739

Josephsen K, Takano Y, Frische S et al (2010) Ion transporters in secretory and cyclically modulating ameloblasts: a new hypothesis for cellular control of preeruptive enamel maturation. Am J Physiol Cell Physiol 299:C1299–C1307

Kallenbach E (1974) Fine structure of rat incisor ameloblasts in transition between enamel secretion and maturation stages. Tissue Cell 6:173–190

Lacruz RS (2017) Enamel: molecular identity of its transepithelial ion transport system. Cell Calcium 65:1–7

Lacruz RS, Hilvo M, Kurtz I et al (2010a) A survey of carbonic anhydrase mRNA expression in enamel cells. Biochem Biophys Res Commun 393:883–887

Lacruz RS, Nanci A, Kurtz I et al (2010b) Regulation of pH during amelogenesis. Calcif Tissue Int 86:91–103

Lacruz RS, Nanci A, White SN et al (2010c) The sodium bicarbonate cotransporter (NBCe1) is essential for normal development of mouse dentition. J Biol Chem 285:24432–24438

Lacruz RS, Smith CE, Chen YB et al (2011) Gene-expression analysis of early- and late-maturation-stage rat enamel organ. Eur J Oral Sci 119(Suppl 1):149–157

Lacruz RS, Smith CE, Bringas P Jr et al (2012a) Identification of novel candidate genes involved in mineralization of dental enamel by genome-wide transcript profiling. J Cell Physiol 227:2264–2275

Lacruz RS, Smith CE, Moffatt P et al (2012b) Requirements for ion and solute transport, and pH regulation during enamel maturation. J Cell Physiol 227:1776–1785

Lacruz RS, Brookes SJ, Wen X et al (2013a) Adaptor protein complex 2-mediated, clathrin-dependent endocytosis, and related gene activities, are a prominent feature during maturation stage amelogenesis. J Bone Miner Res 28:672–687

Lacruz RS, Smith CE, Kurtz I et al (2013b) New paradigms on the transport functions of maturation-stage ameloblasts. J Dent Res 92:122–129

Lacruz RS, Habelitz S, Wright JT et al (2017) Dental enamel formation and implications for oral health and disease. Physiol Rev 97:939–993

Lederer E, Wagner CA (2019) Clinical aspects of the phosphate transporters NaPi-IIa and NaPi-IIb: mutations and disease associations. Pflugers Arch 471:137–148

Lin HM, Nakamura H, Noda T et al (1994) Localization of H^+-ATPase and carbonic anhydrase II in ameloblasts at maturation. Calcif Tissue Int 55:38–45

Liu H, Yan X, Pandya M et al (2016) Daughters of the enamel organ: development, fate, and function of the stratum intermedium, stellate reticulum, and outer enamel epithelium. Stem Cells Dev 25:1580–1590

Lyaruu DM, Bronckers AL, Mulder L et al (2008) The anion exchanger Ae2 is required for enamel maturation in mouse teeth. Matrix Biol 27:119–127

Lyman GE, Waddell WJ (1977) pH gradients in the developing teeth of young mice from autoradiography of [14C]DMO. Am J Physiol 232:F364–F367

McCarl CA, Picard C, Khalil S et al (2009) ORAI1 deficiency and lack of store-operated Ca^{2+} entry cause immunodeficiency, myopathy, and ectodermal dysplasia. J Allergy Clin Immunol 124:1311–1318, e1317
Messenger SW, Falkowski MA, Groblewski GE (2014) Ca^{2+}-regulated secretory granule exocytosis in pancreatic and parotid acinar cells. Cell Calcium 55:369–375
Moreno EC, Aoba T (1987) Calcium bonding in enamel fluid and driving force for enamel mineralization in the secretory stage of amelogenesis. Adv Dent Res 1:245–251
Nadler MJ, Hermosura MC, Inabe K et al (2001) LTRPC7 is a Mg.ATP-regulated divalent cation channel required for cell viability. Nature 411:590–595
Nakano Y, Le MH, Abduweli D et al (2016) A critical role of TRPM7 as an ion channel protein in mediating the mineralization of the craniofacial hard tissues. Front Physiol 7:258
Nurbaeva MK, Eckstein M, Concepcion AR et al (2015a) Dental enamel cells express functional SOCE channels. Sci Rep 5:15803
Nurbaeva MK, Eckstein M, Snead ML et al (2015b) Store-operated Ca^{2+} entry modulates the expression ofenamel genes. J Dent Res 94:1471–1477
Nurbaeva MK, Eckstein M, Feske S et al (2016) Ca^{2+} transport and signalling in enamel cells. J Physiol 595:3015–3039
Nurbaeva MK, Eckstein M, Devotta A et al (2018) Evidence that calcium entry into calcium-transporting dental enamel cells is regulated by cholecystokinin, acetylcholine and aTP. Front Physiol 9:801
Ogata K, Tsumuraya T, Oka K et al (2017) The crucial role of the TRPM7 kinase domain in the early stage of amelogenesis. Sci Rep 7:18099
Okumura R, Shibukawa Y, Muramatsu T et al (2010) Sodium-calcium exchangers in rat ameloblasts. J Pharmacol Sci 112:223–230
Onishi T, Okawa R, Ogawa T et al (2007) Phex mutation causes the reduction of npt2b mRNA in teeth. J Dent Res 86:158–162
Paine ML, Snead ML, Wang HJ et al (2008) Role of NBCe1 and AE2 in secretory ameloblasts. J Dent Res 87:391–395
Parry DA, Mighell AJ, El-Sayed W et al (2009) Mutations in CNNM4 cause Jalili syndrome, consisting of autosomal-recessive cone-rod dystrophy and amelogenesis imperfecta. Am J Hum Genet 84:266–273
Parry DA, Poulter JA, Logan CV et al (2013) Identification of mutations in SLC24A4, encoding a potassium-dependent sodium/calcium exchanger, as a cause of amelogenesis imperfecta. Am J Hum Genet 92:307–312
Picard C, McCarl CA, Papolos A et al (2009) STIM1 mutation associated with a syndrome of immunodeficiency and autoimmunity. N Engl J Med 360:1971–1980
Prakriya M, Lewis RS (2015) Store-operated calcium channels. Physiol Rev 95:1383–1436
Putney JW, Steinckwich-Besancon N, Numaga-Tomita T et al (2016) The functions of store-operated calcium channels. Biochim Biophys Acta 1864:900–906
Reith EJ, Boyde A (1979) The enamel organ, a control gate for calcium influx into the enamel. J Dent Res 58:980
Reith EJ, Boyde A (1981) The arrangement of ameloblasts on the surface of maturing enamel of the rat incisor tooth. J Anat 133:381–388
Robertson SYT, Wen X, Yin K et al (2017) Multiple calcium export exchangers and pumps are a prominent feature of enamel organ cells. Front Physiol 8:336
Robinson C, Hiller CR, Weatherell JA (1974) Uptake of 32P-labelled phosphate into developing rat incisor enamel. Calcif Tissue Res 15:143–152
Robinson C, Yamamoto K, Connell SD et al (2006) The effects of fluoride on the nanostructure and surface pK of enamel crystals: an atomic force microscopy study of human and rat enamel. Eur J Oral Sci 114 Suppl 1:99–104; discussion 127–109, 380
Roos J, DiGregorio PJ, Yeromin AV et al (2005) STIM1, an essential and conserved component of store-operated Ca^{2+} channel function. J Cell Biol 169:435–445

Sasaki T, Goldberg M, Takuma S et al (1990) Cell biology of tooth enamel formation. Functional electron microscopic monographs. Monogr Oral Sci 14:1–199
Sasaki S, Takagi T, Suzuki M (1991) Cyclical changes in pH in bovine developing enamel as sequential bands. Arch Oral Biol 36:227–231
Scheenen WJ, Wollheim CB, Pozzan T et al (1998) Ca^{2+} depletion from granules inhibits exocytosis. A study with insulin-secreting cells. J Biol Chem 273:19002–19008
Simmer JP, Fincham AG (1995) Molecular mechanisms of dental enamel formation. Crit Rev Oral Biol Med 6:84–108
Smith CE (1998) Cellular and chemical events during enamel maturation. Crit Rev Oral Biol Med 9:128–161
Smith CE, Nanci A (1995) Overview of morphological changes in enamel organ cells associated with major events in amelogenesis. Int J Dev Biol 39:153–161
Smith CE, Nanci A (1996) Protein dynamics of amelogenesis. Anat Rec 245:186–207
Smith CE, Warshawsky H (1977) Quantitative analysis of cell turnover in the enamel organ of the rat incisor. Evidence for ameloblast death immediately after enamel matrix secretion. Anat Rec 187:63–98
Smith CE, McKee MD, Nanci A (1987) Cyclic induction and rapid movement of sequential waves of new smooth-ended ameloblast modulation bands in rat incisors as visualized by polychrome fluorescent labeling and GBHA-staining of maturing enamel. Adv Dent Res 1:162–175
Smith CE, Issid M, Margolis HC et al (1996) Developmental changes in the pH of enamel fluid and its effects on matrix-resident proteinases. Adv Dent Res 10:159–169
Smith CE, Chong DL, Bartlett JD et al (2005) Mineral acquisition rates in developing enamel on maxillary and mandibular incisors of rats and mice: implications to extracellular acid loading as apatite crystals mature. J Bone Miner Res 20:240–249
Smith CE, Nanci A, Moffatt P (2006) Evidence by signal peptide trap technology for the expression of carbonic anhydrase 6 in rat incisor enamel organs. Eur J Oral Sci 114 Suppl 1:147–153; discussion 164–145, 380–141
Smith CE, Hu Y, Hu JC et al (2016) Ultrastructure of early amelogenesis in wild-type, Amelx$^{-/-}$, and Enam$^{-/-}$ mice: enamel ribbon initiation on dentin mineral and ribbon orientation by ameloblasts. Mol Genet Genomic Med 4:662–683
Smyth JT, Hwang SY, Tomita T et al (2010) Activation and regulation of store-operated calcium entry. J Cell Mol Med 14:2337–2349
Sui W, Boyd C, Wright JT (2003) Altered pH regulation during enamel development in the cystic fibrosis mouse incisor. J Dent Res 82:388–392
Takagi T, Ogasawara T, Tagami J et al (1998) pH and carbonate levels in developing enamel. Connective Tiss Res 38:181–187
Takano Y (1995) Enamel mineralization and the role of ameloblasts in calcium transport. Connect Tissue Res 33:127–137
Takano Y, Ozawa H, Crenshaw MA (1986) Ca-ATPase and ALPase activities at the initial calcification sites of dentin and enamel in the rat incisor. Cell Tissue Res 243:91–99
Thevenod F (2002) Ion channels in secretory granules of the pancreas and their role in exocytosis and release of secretory proteins. Am J Physiol Cell Physiol 283:C651–C672
Toyosawa S, Ogawa Y, Inagaki T et al (1996) Immunohistochemical localization of carbonic anhydrase isozyme II in rat incisor epithelial cells at various stages of amelogenesis. Cell Tissue Res 285:217–225
Varga G, DenBesten P, Racz R et al (2018) Importance of bicarbonate transport in pH control during amelogenesis—need for functional studies. Oral Dis 24:879–890
Wang S, Choi M, Richardson AS et al (2014) STIM1 and SLC24A4 are critical for enamel maturation. J Dent Res 93:94S–100S
Watson R, Parekh AB (2012) Mitochondrial regulation of CRAC channel-driven cellular responses. Cell Calcium 52:52–56

Yamaguti PM, Neves FA, Hotton D et al (2017) Amelogenesis imperfecta in familial hypomagnesaemia and hypercalciuria with nephrocalcinosis caused by CLDN19 gene mutations. J Med Genet 54:26–37

Yamazaki D, Funato Y, Miura J et al (2013) Basolateral Mg^{2+} extrusion via CNNM4 mediates transcellular Mg^{2+} transport across epithelia: a mouse model. PLoS Genet 9:e1003983

Yin K, Hacia JG, Zhong Z et al (2014) Genome-wide analysis of miRNA and mRNA transcriptomes during amelogenesis. BMC Genomics 15:998

Yin K, Lei Y, Wen X et al (2015) SLC26A gene family participate in pH regulation during enamel maturation. PLoS One 10:e0144703

Young RA (1974) Implications of atomic substitutions and other structural details in apatites. J Dent Res 53:193–203

Zhang SL, Yu Y, Roos J et al (2005) STIM1 is a Ca^{2+} sensor that activates CRAC channels and migrates from the Ca^{2+} store to the plasma membrane. Nature 437:902–905

The manufacturer's authorised representative in the EU is Springer Nature Customer Service Centre GmbH, Europaplatz 3, 69115 Heidelberg, Germany. If you have any concerns regarding our products, please contact ProductSafety@springernature.com

Printed and bound by CPI Group (UK) Ltd, Croydon, CR0 4YY
30/09/2026
02225452-0015